In Case of a Transportation Emergency Involving a Compressed Gas

In the United States call

CHEMTREC
(Chemical Transportation Emergency Center)

800-424-9300
Day or Night—Toll-Free

In Canada call

TEAP
(Transportation Emergency Assistance Plan)

Valleyfield, Quebec	514-373-8330
Maitland, Ontario	613-348-3616
Niagara Falls, Ontario	416-356-8310
Sarnia, Ontario	519-339-3711
Copper Cliff, Ontario	705-682-2881
Edmonton, Alberta	403-477-8339
Vancouver, British Columbia	604-929-3441

For further information see page 9

HANDBOOK OF
Compressed Gases

HANDBOOK OF

Compressed Gases

Second Edition

COMPRESSED GAS ASSOCIATION, INC.
New York, New York

VNR VAN NOSTRAND REINHOLD COMPANY
NEW YORK   CINCINNATI   ATLANTA   DALLAS   SAN FRANCISCO
LONDON   TORONTO   MELBOURNE

6440 - 6283

CHEMISTRY

Van Nostrand Reinhold Company Regional Offices:
New York Cincinnati Atlanta Dallas San Francisco

Van Nostrand Reinhold Company International Offices:
London Toronto Melbourne

Library of Congress Catalog Card Number: 80-19969
ISBN: 0-442-25419-9

Manufactured in the United States of America

Published by Van Nostrand Reinhold Company
135 West 50th Street, New York, N.Y. 10020

Published simultaneously in Canada by Van Nostrand Reinhold Ltd.

15 14 13 12 11 10 9 8 7 6 5 4 3 2 1

Library of Congress Cataloging in Publication Data

Compressed Gas Association.
  Handbook of compressed gases.

  Includes bibliographies and index.
  1. Gases, Compressed. I. Title.
TP761.C65C6   1980   665.8   80-19969
ISBN 0-442-25419-9

Preface

The first edition of the *Handbook of Compressed Gases* was published in 1966. It has served well as a source of basic information about compressed gases, their transportation, uses, safety considerations and rules and regulations pertaining to them, as is evidenced by sales of nearly 9500 copies. Because of changes in technology and the development of new gases, the Compressed Gas Association, Inc. (CGA) has prepared a second edition of the *Handbook* which provides some new information and updates the information found in the first edition.

The content and authority for this second edition stems primarily from the many CGA committees made up of technically qualified people from member companies, as was also the case with the first edition. Just as the Association is concerned with the broadest of interests relating to gases, so is the book designed to serve the needs of the broadest group of users—including businessmen, members of the medical profession, public officials, students as well as engineers and scientists seeking basic guideposts in an unfamiliar terrain. Just as the Association is a technical organization interested in both adequate data and safe and sound utilization for gases, so do the two parts of the book, I and II, deal primarily with general information about their safe handling, containment and transportation on the one hand, and information about individual gases on the other.

CGA is a nonprofit service organization incorporated in the state of New York. Founded in 1913 as the Compressed Gas Manufacturers' Association, the CGA includes among its membership some 300 companies, including many of North America's largest firms devoted primarily to producing and distributing compressed, liquefied and cryogenic gases, as well as leading chemical industry companies which have compressed and liquefied gas divisions. Manufacturers and distributors of portable containers, cargo tanks, tank cars, medical equipment in the fields of anesthesia and respiration and accessories are also included. The Association's interests are increasingly international in scope since many overseas companies are included in a special membership classification. The primary objective of the CGA can best be understood by reference to the organization's first constitutional purpose: "To promote, develop, represent and coordinate technical and standardization activities in the compressed gas industries, including end uses of products, in the interest of safety and efficiency, and to the end that they may serve to the fullest extent the best interest of the public."

Any interested businesses, corporations, associations and individuals may join the CGA and participate in its activities.

More than 40 technical committees made up of technically qualified representatives from member companies conduct the most fundamental work of the CGA. Some, like the Ammonia Committee or the Atmospheric Gases Committee, deal with problems connected with individual gases; others, like the committees on medical equipment and gases and the halogenated hydrocarbons, deal with gas families and their uses; still other groups of committees are concerned with all types and sizes of containers and the various modes of transportation needed to convey products from the producer to the user. Committees are formally organized and take on projects flexibly, as befits an overall field that cuts across many professional and business specialties.

Primarily, the CGA's technical committees work to develop standards for the guidance of the industry, regulatory agencies and the public. In this work they frequently collaborate with other organizations concerned with safety standards involving compressed gases, such as the American National Standards Institute (ANSI), the International Standards Organization (ISO), the National Fire Protection Association (NFPA), the American Society of Anesthesiologists (ASA), the National L-P Gas Association (NLPGA) and many others. For a company to have its technically qualified management personnel serve on CGA committees represents both a substantial and vital contribution to the public and to the welfare of the industry, as well as a prized honor.

In a particularly significant part of its work, the CGA acts as an advisor to any interested regulatory agencies that include the Department of Transportation (DOT), the Interstate Commerce Commission (ICC), the Food and Drug Administration (FDA), the Canadian Transport Commission (CTC) and state, provincial, municipal and local agencies concerned with the safe handling of compressed gases. The Association has also long maintained a close working relationship with the Bureau of Explosives (B of E) of the Association of American Railroads (AAR), which serves as the official representative of all United States and Canadian railroads to the DOT and CTC, respectively. CGA recommendations have consequently played an essential and constructive role, down through the years, in developing the whole system of practices and regulations for transporting compressed gases throughout North America—the continent which benefits from the world's most highly advanced compressed gas technology. The CGA has thus served industry and the public through more than 60 years, since the time of its founding as the Compressed Gas Manufacturers' Association in 1913.

Through the decades of its past activities and accomplishments, the CGA has seen profound changes develop increasingly within the industry as the march of technological progress has steadily accelerated. Fluorocarbon gases, for example, were unknown in their common air-conditioning applications of today when the Association began early in the century. Another even more portentous development has been the rise of the entire field of cryogenics, based unavoidably on gases, and crucial to progress in such divergent fields as surgery, food preservation, low-temperature refrigeration and the exploration and utilization of space. It is indicative of today's widening horizons that the CGA Constitution currently interprets "compressed gas" to mean gases in their gaseous, liquid or solid state.

To report in full on this field, continuously expanding and changing, would require nothing less than loose-leaf binding with daily bulletins and lots of shelf space. However, in the field, there is an expanding core of essential knowledge that people need to refer to most frequently and most widely, and that underlies advances unfolding today and tomorrow. The second edition of this *Handbook* is an attempt to put that updated and expanded essential core in the hands of the many and diverse people to whom it can be most helpful.

Robert L. Swope and Frank J. Heller gave valuable service as the two successive chairmen of CGA's Handbook Committee through the preparation of this second edition. A number of new features and sections of the second edition resulted from their ideas and efforts, with which they were assisted by the individual members of their committee.

Robert E. Lenhard devoted many months of notable accomplishment to helping complete the second edition manuscript and guide it through production, in the capacity of consultant to CGA after having retired as CGA's managing director. Throughout all phases of work on the second edition, principal editorial reponsibility has been carried by The Hudson Group, Inc., under the project direction of Gene R. Hawes, president of The Hudson Group.

We are indebted to the many individual members of CGA committees who have contributed and reviewed data in our combined effort to make this volume an authentic and correct source of information. We will be surprised however if this book contains no errors and will be glad to be advised of any detected by readers.

C. H. Glasier, President
Compressed Gas Association, Inc.

Contents

HANDBOOK OF
Compressed Gases

Handling, Shipping and Storing Compressed Gases in Compliance with Standards for Safety

Introduction

Part I of the *Handbook* presents two types of material. Basic introductory information concerning compressed gases is given in Chapters 1 through 3. Major standards for safety, primarily in the use and shipping of compressed gases, are set forth in Chapters 4 through 10. Chapter titles indicate the areas covered by each chapter.

The standards in Chapters 4 through 10 have been developed by the Compressed Gas Association, Inc. (CGA). The Association issues standards and recommendations on a number of more highly specialized aspects of compressed gases as well. Some of these standards and recommendations are referred to in notes in the chapters and in the sections of Part II. A full list of the current CGA publications, which contain all of the Association's standards, appears in Appendix C.

A detailed explanation of the International System of metric units (SI) and the conversion factors used in this book can be found at the end of Chapter 1. A list of abbreviations is given at the end of the introduction to Part II.

1

The Compressed and Liquefied Gases Today

Gases in compressed or liquefied form play countless indispensable roles in modern technology. Oxygen and acetylene often are used to cut the steel for structures and machinery, after oxygen has first helped to produce stronger and cheaper steels. Acetylene welding and brazing of familiar metals have long been very common; other flammable gases, like hydrogen, are equally essential for welding certain metals, and some of the newer metals and alloys (like stainless steels, titanium and zirconium) can be welded only in an inert gas atmosphere. Propane and other fuel gases are also widely used in metal cutting.

Anhydrous ammonia is used to fertilize croplands pressed for greater productivity with lower costs and higher yields in the face of a growing world food shortage. Carbon dioxide hangs in special containers on thousands of plant and laboratory walls as the most widely used dry extinguisher for chemical and electrical fires.

Nitrous oxide, cyclopropane and ethylene are among some anesthetic gases of present-day surgery, while giving oxygen is both standard first aid in many emergencies and standard treatment for maladies including heart ailments. Oxygen given in the new "hyperbaric" chambers (ones in which pressures exceed atmospheric pressure) has begun bringing patients literally back from the dead.

Propane and butane provide precisely controllable heat for innumerable processes in industry; similarly, low temperatures used in in-dustrial processing and scientific research are produced by refrigerant gases and gases liquefied into cryogenic fluids below temperatures of -238 degrees Fahrenheit or -150 degrees Celsius (abbreviated -238 F (-150 C) in this book).

Many chemicals essential to the making of plastics, synthetic rubber and modern drugs are compressed or liquefied gases—among them, butadiene, chlorine, vinyl chloride, acetylene, ammonia and the methylamines.

Compressed gases make it possible to explore vast unknown realms underseas, within and without submarines. And not only do gases keep space-voyaging astronauts alive; liquefied gases produce the power that drives rocket vehicles out into space.

Even everyday life in industrialized nations today depends on gases for technological advantages so common that they are taken for granted. All refrigerating and air-conditioning machines, in homes, stores, offices, plants, schools, hospitals, warehouses, autos, trucks and trains, use refrigerant gases—often, the fluorocarbon gases—while the "dry ice" that solid carbon dioxide represents has largely replaced its natural namesake in the bulk handling of perishable foods. It is, of course, bubbles of carbon dioxide dissolved under pressure that make up the "pop" in all kinds of soda pop or "carbonated" beverages.

In a like manner, nitrous oxide and some of the LP-gases provide the push behind all kinds

3

of pressure-packaged products which foam or spray out of hand-size containers—foams of whipped dessert toppings and shaving lathers, and sprays of perfumes, shampoos, suntan lotions, paints, insecticides and many other useful liquids. Millions of children and adults instantly recognize the traces of chlorine gas in the water of swimming pools that guard them against infection. Propane and butane— the chief "liquefied petroleum" or "LP" gases— again are the compact portable fuels in millions of homes that lie beyond gas mains. LP-gas powers whole fleets of trucks and buses instead of gasoline or diesel fuels, especially in the midwest and western United States, while LP-gas is used heavily on farms for powering vehicles, burning weeds and many other purposes. And in the heart of cities, signs lighting the way to refreshment or entertainment contain in their scripts of tubular glass the bright, glowing colors of neon, argon, helium or krypton.

In sum, to a steadily increasing degree, many of the powers of modern technology, as well as many necessities and conveniences of modern life, stem from mastery of the production and use of compressed and liquefied gases.

TRENDS TOWARD LARGER USAGE

Current trends point to an accelerating development of the uses of compressed gas. Many such growing uses apply to human physiology. Use of gases to sustain life in hostile environments outside the earth's atmosphere has barely begun to be developed, and will surely mushroom with widening exploration. In medicine, cryosurgery with liquefied nitrogen is becoming a standard operating technique and oxygen therapies keep multiplying, while hyperbaric oxygenation treatments have been introduced only by more advanced hospitals.

Gases are handled more and more as cold liquids and cryogenic fluids with steadily improved equipment and materials for low-temperature work, largely because many gases are most compact and exert little or no pressure when cooled down to liquids. Methane or nat-ural gas liquefied at about -260 F (-162 C), already referred to today as "LNG" (liquefied natural gas), may be expected to become as commonplace as LP-gas. Compatible shipping, handling and receiving facilities for low-temperature liquid transportation of gases appear with growing frequency. A host of new materials and methods for containing and piping cryogenic fluids has been originated in recent years, and will certainly be augmented in years ahead by many new materials now completely unknown.

PURPOSE OF THIS HANDBOOK

Today's very large and growing use of compressed gas in almost every area of industry, in medicine and even in many facets of daily life, gives rise to widespread needs for authoritative information which the *Handbook of Compressed Gases* attempts to meet. The *Handbook* is designed primarily for use by the many persons in business, industry, government, education, transportation, the health professions and the armed services who deal with gases. In order to be of the greatest utility, it is written to be understood by the person without professional training in the sciences or engineering—to serve for reading and reference beyond just the engineering office, the research laboratory and the college library. The book gives basic information about each gas having present commercial importance, and general information about all gases that is essential for persons who are working in any way with gases.

The first two chapters of Part I introduce persons who deal with gases, as well as students, to the whole compressed gas field and to the American and Canadian regulatory bodies active within it. The remaining chapters of Part I give the safe handling and container standards that have been developed by the compressed gas industry itself. Part II presents detailed information about individual gases.

Reference use of Parts I and II is facilitated by their organization into many sections and subsections. In Part I there is a chapter on each specific area of standardization, such as cylinder valve connections, safety relief devices and safe handling of cylinders. In Part II there is a section on each gas. Because of important notes and qualifications, it is essential to read the Part II introduction before using any of the information on individual gases.

WHAT THE COMPRESSED AND LIQUEFIED GASES ARE

Whether or not we consider any compound or element to be a gas depends, of course, on the cosmic coincidence of temperatures and pressures stable within fortunately narrow limits that are found on the surface of the planet Earth. Accordingly, to us a gas is any substance that boils at atmospheric pressure and any temperature between the absolute zero of outer space through some 40 to perhaps 80 F (4.4-26.7 C). Eleven of the 92 elements (not including transuranium elements) happen to have such boiling points, as do apparently unlimited numbers of compounds and mixtures like air. (The 11: hydrogen, nitrogen, oxygen, fluorine, chlorine, and the 6 inert gases—helium, neon, argon, krypton, xenon and radon.)

Gases are defined as "compressed" for practical reasons of transportation, storage and use. The definition carrying greatest weight in the United States is the one that has been adopted by the Department of Transportation, the federal regulatory body empowered by act of Congress to provide regulations for their safe transportation in interstate commerce. The DOT definition currently reads: (with parenthetic information added by *Handbook* editors)

> . . . any material or mixture having in the container an absolute pressure exceeding 40 psi (pounds per square inch) at 70 F (275.8 kPa at 21.1 C) or, regardless of the pressure at 70 F (21.1 C), having an absolute pressure exceeding 104 psi at 130 F (717 kPa at 54.4 C), or any liquid flammable material having a vapor pressure exceeding 40 psi absolute at 100 F (275.8 kPa at 37.8 C) as determined by ASTM (American Society for Testing and Materials) Test D-323.

Absolute pressure, incidentally, is the pressure in a container that would appear on an ordinary gage plus the local atmospheric pres-

sure of some 15 psi (14.696 psi at sea level and 32 F (101.32 kPa at 0 C) is the generally accepted standard value).

In America, then, a compressed gas is generally taken to be any substance which, when enclosed in a container, gives a pressure reading of at least:

(1) either 25 psig (pounds per square inch, gage pressure) at 70 F (172.4 kPa at 21.1 C); or over 89 psig at 130 F (613.6 kPa at 54.4 C); or

(2) if the contained substance is flammable, 25 psig at 100 F (172.4 kPa at 37.8 C).

Two major groups of gases that differ in physical state when contained result from this definition, and from the range of boiling points among gases. They are:

(1) Gases which do not liquefy in containers at ordinary terrestrial temperatures and under pressures attained in commercially used containers, which range up to 2000 to 2500 psi (13,789 to 17,237 kPa).

(2) Gases which do become liquids to a very large extent in containers at ordinary temperatures and at pressures from 25 to 2500 psi (172.4 to 17,237 kPa).

The first group, commonly called the *nonliquefied gases*, are elements or compounds that have relatively low boiling points, say, from -150 F (-101.1 C) on down. However, the nonliquefied gases, of course, become liquids if cooled to temperatures below their boiling points. Those that liquefy at "cryogenic" temperatures (temperatures from absolute zero, or -459.7 F, up to around -238 F (-273.16 to -150 C)—through the upper limit set for the cryogenic range varies widely from one authority to another), are therefore also known as *cryogenic fluids*.

The second group, called for some years the *liquefied gases*, are elements or compounds that have boiling points relatively near atmospheric temperatures, from about -130 F to 25 F or 30 F (-90 C to -3.9 C or -1.1 C). The liquefied gases solidify at cryogenic temperatures, and only one of them—carbon dioxide—has as yet come into wide use in solid form.

Oxygen, helium and nitrogen are examples

of gases in wide use both as nonliquefied gases and cryogenic fluids. With respective boiling points of -297 F, -425 F and -320 F (-182.8 C, -253.9 C, -195.5 C), they are charged into high-pressure steel cylinders at more than 2000 psig at 70 F (13,789 kPa at 21.1 C) for shipment and use as nonliquefied gases. However, when shipped as cryogenic fluids, they are cooled down to liquid form and charged into special insulated containers that keep them below their boiling points and operate at gage pressures of less than 75 psig or even less than 1 psig (517.1 or 6.9 kPa).

Examples of widely used liquefied gases and their boiling points are anhydrous ammonia (-28 F), chlorine (-29 F), propane (-44 F) (-33.4 C, -33.9 C, -42.4 C), and carbon dioxide (which sublimes directly from a solid to a gas at -109 F (-78.3 C) and one atm, and liquefies only under pressure). When charged into cylinders at typical pressures ranging from 85 psig for chlorine to 860 psig for carbon dioxide (586 to 5929 kPa), these are all largely liquids. In cylinders filled with them to the maximum amounts permitted for shipment, only a small vapor space is left inside to allow for expansion in case of heating.

A third physical state in the container is represented by only one widely used gas—acetylene, which is sometimes referred to as a "dissolved" gas (in contrast to a "nonliquefied" or "liquefied" gas). The industry recommends that free acetylene should not ordinarily be handled at pressures greater than 15 psig (103 kPa) because, if handled at higher pressures without special equipment, it can decompose with explosive violence. In consequence, acetylene cylinders are packed with an inert porous material which is then saturated with acetone. Acetylene charged into the cylinder dissolves in the acetone and in solution will not decompose at or below the maximum authorized shipping pressure of 250 psig at 70 F (1724 kPa at 21.1 C). Some consumers of acetylene in bulk for chemical processing make acetylene at their plants by reacting calcium carbide and water rather than having it supplied in the special acetylene cylinders.

THE MAJOR FAMILIES OF COMPRESSED GASES

Compressed and liquefied gases are also often described according to loosely knit families to which they belong through common origins, properties or uses.

Atmospheric gases comprise one of the leading families. Its bulkiest member is nitrogen, constituting 78 percent air by volume, and oxygen (21 percent of the air) is its second most abundant member. All animal life, of course, utterly depends on these two elements, needing the first for nutrition and the other for respiration. Most of the remaining 1 percent of the atmosphere consists of a subfamily of gases sharing the property of chemical inertness, the inert gases: chiefly argon, with minute amounts of helium, neon, krypton, xenon and radon. The last four are frequently called the "rare gases" of the atmosphere, due to their scarcity. Hydrogen also occurs minutely in the atmosphere, as do a large variety of trace constituents, small amounts of carbon dioxide and large amounts of water vapor.

Nitrogen, oxygen, argon and the rare gases are commercially produced by cooling the air down to liquid form and then distilling off "fractions" of it having different boiling points, much as petroleum fractions are distilled at higher temperatures. The process is called fractionation. Some high-purity helium is obtained in fractionation, but most helium used today comes from wells of natural gas in which it occurs in concentrations of a few percent.

Fuel gases burned in air or with oxygen to produce heat make up a large family of gases related through the major use. Its members are notably the hydrocarbon gases—especially the leading LP-gases, propane and butane. Methane, the largest component of natural gas, is another leading representative of the family, and such welding gases as acetylene and hydrogen are somewhat special representatives. New members now joining the fuel gas family in a current development in the field are the inhibited methylacetylene-propadiene mixtures, which resemble LP-gas in a number of ways.

An opposite application relates members of another extensive family, that of refrigerant gases. A refrigerant gas should be one that liquefies easily under pressure, for it works by being compressed to a liquid mechanically and then by absorbing large amounts of heat as it circulates in cooling coils and vaporizes back into gas. Dry or anhydrous ammonia does liquefy under low pressure, and was the earliest widely used refrigerant. Any liquefied gas, though, is a strong candidate for membership in the refrigerant family; even some gases bordering on cryogenic fluids, like methane, have been classified as refrigerant gases by one of the main organizations in the field, ASHRAE (the American Society of Heating Refrigerating and Air Conditioning Engineers).

Among the most popular refrigerant gases today are the fluorocarbons, a family of almost indefinitely large size since they are any of the endless series of hydrocarbons which have been fluorinated. Fluorocarbons serve well as refrigerants because most of them are chemically inert to a large extent and they can be selected, mixed or compounded to provide almost any physical properties desired in particular refrigerant applications.

Nitrous oxide illustrates how confused family ties among the gases can be, for it is a prominent member of the family of gases used in medicine, as well as a respected propellant gas and a reliable refrigerant gas. Cyclopropane and ethylene are also gases that, like nitrous oxide, serve widely in medicine as anesthetics. Oxygen enjoys large and growing application in medicine, where it is employed alone as well as mixed with carbon dioxide or helium for many kinds of inhalation therapy.

Gases considered in the United States to be members of the poison gas family are generally those that the DOT has classified as poison gases to insure public safety in interstate shipment. Two of these gases that have commercial importance are included in Part II of this volume, hydrogen cyanide and phosgene. Both are shipped as liquefied gases, and their inhalation hazards are incidental to their use as intermediates in the chemical industry. The two gases serve as intermediates in the production of wide varieties of compounds, some of the most familiar being "Plexiglas" and other acrylic plastics (in the case of hydrogen cyanide) and barbiturate drugs (in the case of phosgene). As is generally the case with gases designated class A or "extremely dangerous" by the DOT, the potential danger posed by hydrogen cyanide and phosgene is chiefly through inhalation. Hydrogen cyanide concentrations of from 100 to 200 parts per million (ppm) in air, if inhaled for 30 to 60 minutes, can be fatal. Inhaling phosgene concentrations of some 700 ppm can be fatal within several minutes. Extensive safety requirements and practices have been developed for shipping and handling these and other poison gases, however, and accidents do not occur with them when all recommended protective measures are taken.

Still other gases that are important commercially have no marked family ties; many of these are used in chemical industry processing. The methylamines, for example, serve as chemical intermediate sources of reactive organic nitrogen, and methyl mercaptan helps synthesize insecticides. They are liquefied gases. Another liquefied gas, sulfur dioxide, is widely employed as a preservative and bleach in processing foods, as a bleach in manufacturing sulfite papers and artificial silk, and as an additive to irrigating water to improve the yield of alkaline soils in the American Southwest. Carbon monoxide, a nonliquefied gas, serves chiefly in the making of other chemicals, such as ethylene, and also in refining high-purity nickel metal. Fluorine, another nonliquefied gas, is used to make certain fluorides—among them sulfur hexafluoride, which has high dielectric strength and is widely employed as an insulating gas in electrical equipment.

BASIC SAFETY PRECAUTIONS TAKEN AGAINST COMPRESSED GAS HAZARDS

The very powers that make the compressed and liquefied gases widely useful in modern life—high heat output in combustion of some gases, high reactivity in chemical processing of others, extremely low temperatures available from some, and the economy of handling them all in compact form at high pressure or low temperature—these powers often can also represent hazards if the gases are not handled with full knowledge and care.

Part I of this volume presents the specific and detailed standards that have been developed by the compressed gas industry to insure safety, and readers are urged to consult the full recommendations there for any aspect of the field with which they may be concerned. At this point, however, it may be helpful to outline briefly the chief hazards posed by compressed gases and the main precautions taken against them.

Practically all gases can act as simple asphyxiants by displacing the natural oxygen in the air. The chief precaution taken against this potential hazard is adequate ventilation of all enclosed areas in which unsafe concentrations may build up. A second precaution is to avoid entering unventilated areas that may contain high concentrations of gas without first putting on breathing apparatus with a self-contained or hose-line air supply. A number of gases do have characteristic odors which can warn of their presence in air; others, however, like the atmospheric gases, have no odor or color whatever. Warning labels are required for compressed and liquefied gas shipping containers; similar warning signs are placed at the approaches to areas in which the gases are regularly stored and used. Unauthorized persons should be kept away from such areas.

Some gases can also have a toxic effect on the human system, either through being inhaled, or through having high vapor concentrations or having liquefied gas come in contact with the skin or the eyes. Adequate ventilation of enclosed areas similarly serves as the chief precaution against high concentrations of gases which can exert toxic effects. In addition, for unusually toxic gases, automatic devices can be purchased or built to monitor the gas concentration constantly and set off alarms if the concentration should approach a danger point. Precautions against skin or eye contact with liquefied gases that are toxic or very cold, or both, include thorough knowledge and training for all personnel handling such gases, the development of foolproof procedures and equipment for handling them and special protective clothing and equipment (such as protective garments, gloves and face shields).

With the flammable gases, it is necessary to guard against the possibility of fire or explosion. Ventilation again represents a prime precaution against these hazards, together with safe procedures and equipment to detect possible leaks. Should fire break out, suitable fire extinguishing apparatus and preparation can help limit damage. Care should also be taken to keep any flammable gas from reaching any source of ignition or heat, such as sparking electrical equipment or sparks struck by ordinary tools, boiler rooms or open flames.

Oxygen poses a combustion hazard of a special kind; though it does not itself ignite, it lowers the ignition point of flammable substances and greatly accelerates combustion. It should not be allowed closer than 10 ft (3.048 m) to any flammable substance, including grease and oil, and should be stored no closer than that to cylinders or tanks containing flammable gases.

Hazards resulting from the possible rupture of a cylinder or other vessel containing gas at high pressure can be avoided by careful handling of containers at all times. For example, cylinders should never be struck or allowed to fall. Should the cylinder valve be broken off when the cylinder is charged at high pressure, it could become a projectile. Under most circumstances, the relatively small opening through the remaining portion of the valve is not large enough to cause "rocketing." Cylinders should not be dragged or rolled across the floor; they should be moved by a hand truck. Also, when they are upright on a hand truck or floor or vehicle, they should be chained securely to keep them from falling over. Moreover, cylinders should not be heated to the point at which any part of their outside surface exceeds a temperature of 130 F (54.4 C), and never with a torch or other open flame. Similar precautions should be taken with larger shipping and storage containers. Initial protection against the possibility of vessel rupture is provided by the demanding requirements and recommendations for compressed gas container construction, testing and retesting.

The compressed and liquefied gases are handled today in the United States and Canada in enormous volumes and in a great variety of applications with a very high record of industrial and public safety. The compressed gas industry itself has led in the development of safe equipment and practices, acting both out of public interest and enlightened self-interest.

In addition, the public and the industry are protected by many government authorities which make regulations about shipping, storage, labeling and other matters in the interest of public safety. The industry cooperates fully with these bodies, and has often served them by helping develop regulatory codes that are sound and thoroughly safe for all concerned. The main regulatory authorities with responsibilities in the compressed gas field are described in the next chapter.

For assistance in any transportation emergency involving compressed gases (as well as other chemicals) the Chemical Manufacturers Association has set up CHEMTREC (Chemical Transportation Emergency Center) with a toll-free number listed on the front and back flyleaves of this book. CHEMTREC provides immediate advice for those at the scene of emergencies, then promptly contacts the shipper of the chemicals involved for more detailed assistance and appropriate follow-up.

CHEMTREC operates 24 hours a day, 7 days a week. It is not a general information source for chemicals but is designed to deal with chemical transportation emergencies. For further information write to: Manager, Chemical Transportation Emergency Center, 1825 Connecticut Avenue, N.W., Washington, DC 20009; or phone: 202-328-4218.

In Canada a similar plan called TEAP (Transportation Emergency Assistance Plan) is operated as a public service by the Canadian Chemical Producers' Association through the cooperation of the member companies who operate the Regional Control Centers (RCCs). There are a number of RCCs throughout Canada. Their numbers are listed on the front and back flyleaves of this book, and they are open 24 hours a day, 7 days a week. For further information write to: The Canadian Chemical Producers' Association, 350 Sparks Street, Suite 505, Ottawa KIR 758.

METRIC UNITS OF MEASUREMENT

The specific types of metric units used in this book to designate metric equivalents of quantitative values given in U.S. units are "SI units" (International System of Units). The SI units have been adopted as an American National Standard by the American National Standards Institute (ANSI) in a booklet designated as "ANSI Z210.1" which was originally approved on March 15, 1973. The use of SI units for metric applications has also been approved by the National Bureau of Standards (NBS), the American Society for Testing and Materials (ASTM), the American Society of Mechanical Engineers (ASME) and other organizations.

The SI system of metric units is an attempt to standardize systems of measurement used by different countries throughout the world.

The SI units used for different kinds of properties important in the compressed gas field are as follows:

Pressure	kilopascal	kPa
Temperature	degree Celsius	C
Density	kilogram per cubic meter	kg/m^3
Volume	cubic meter	m^3
Specific volume	cubic meter per kilogram	m^3/kg

The conversion factors given on the following page have been used to obtain SI metric units from U.S. units.

Category	U.S. Unit	Multiplied by	SI Metric Unit
Pressure	lb/in.^{2} (psi)	6.894757	kPa
Pressure	kg/cm^2	98.06650	kPa
Pressure	atm	101.325	kPa
Temperature	F	(F − 32)/1.8	C*
Density	lb/cu ft	16.01846	kg/m^3
Volume	cu ft	0.02831685	m^3
Specific volume	cu ft/lb	0.06242796	m^3/kg
Heat	Btu/lb	2.326	kJ/kg
Heat	Btu/cu ft	37.25895	kJ/m^3
Heat	Btu/gal	278.7163	kJ/m^3
Specific heat	Btu/(lb)(F)	4.1868	kJ/(kg)(C)
Mass	lb	0.4535924	kg
Length	inch	0.0254	m
Length	foot	0.3048	m
Length	mile	1.609344	km

*The recommended SI unit of temperature is the degree Kelvin (K), but degree Celsius (C) values are acceptable for commonly used temperature measurements. A degree difference on the Celsius scale is the same as a degree difference on the Kelvin scale. 0 K equals −273.15 C.

CHAPTER 2

Regulatory Authorities for Compressed Gases in the United States and Canada

Persons who produce, supply or use compressed gases, as well as those involved in compressed gas transportation, must comply with a variety of governmental safety regulations in the United States and Canada. These regulations are issued and enforced by regulatory bodies on the federal, state or provincial, and local levels of government in the two countries.

Federal regulation applies chiefly to land shipment of compressed gas between states or provinces and to shipment by water and by air, and sets detailed specifications for shipping containers and various shipping practices. Federal regulation also sets standards affecting the safety and health of workers involved with the manufacture, shipment and use of compressed gases as well as other industries. State and provincial and local regulation applies mainly to the storage and use of compressed gases though it extends in some cases to certain transportation matters within the locality, province or state.

The major regulatory bodies or kinds of bodies on each level of government in the two nations, and important aspects of their powers, are as follows. Regulation of shipping rates or charges by various governmental bodies is not treated here, although compressed gas shipment is subject to rate regulations of certain federal and state or provincial authorities.

FEDERAL REGULATORY AUTHORITIES

DOT in United States; CTC in Canada

The two most influential agencies regulating compressed gas shipments in North America are the Department of Transportation (DOT) of the United States Government and the Canadian Transport Commission (CTC) of the Canadian Government. The DOT transportation regulations prior to 1967 were administered by the Interstate Commerce Commission (ICC) and the CTC was formerly known as the Board of Transport Commissioners (BTC). Both agencies issue requirements for shipping compressed gases by rail in the case of gases which they classify as "dangerous" articles or "hazardous materials" in interstate or interprovincial commerce. Their regulations require that the designated gases be shipped in containers which comply with certain specifications, and that the containers be equipped with pressure (safety) relief devices as stipulated, tested by methods identified, filled within prescribed maximum amounts, and in some instances be boxed or transported in particular ways.

It was at the turn of the century that the United States Congress charged the Interstate

11

Commerce Commission with the promulgation of regulations, including container specifications, for the safe transportation of hazardous commodities by rail and authorized the ICC to utilize the services of the organization now known as the Bureau of Explosives of the Association of American Railroads. The Bureau had already been functioning for many years as a source of safety requirements for the railroads; it began working in close cooperation with the ICC, and has continued to do so down through the present time with the DOT. It has long worked in cooperation with the CTC as well, for the Association of American Railroads (AAR) includes Canadian as well as U.S. rail carriers.

Today, the Bureau acts as a central point of coordination and communication for the railroads on the one hand, and the DOT and the CTC on the other. It also serves all three groups as a central office for registration, licensing and inspection. (As this book goes to press the DOT is considering changes in the delegations to the Bureau of Explosives in the regulations. The latest edition of the regulations should be checked for current information on this subject.) Its director and chief inspector also holds the title of "Agent," and he serves in that capacity as the official agent and attorney for all AAR member rail carriers with the DOT and the CTC (and similarly acts on behalf of some steamship lines and freight forwarders as well).

In part because the Bureau of Explosives represents both the Canadian and U.S. railroads, substantially identical requirements have been adopted by the CTC for Canada and by the DOT for the United States concerning compressed gas shipment by rail across state or province lines. Canadian needs and interests are thus taken into account in the setting of CTC regulations, as are the U.S. needs and interests in the setting of DOT regulations. In consequence, compressed gas cylinders, tank cars and portable tanks can move freely between the two countries, and the shipping, charging, labeling and other practices authorized for the one country largely meet the requirements of the other

country. The result represents a remarkable achievement in international cooperation that realizes substantial economies and increased economic acitvity for both nations.

One major difference between the DOT and the CTC is that the DOT regulates interstate motor vehicle transport of gases as well as rail transport in the United States, while the CTC regulates only rail shipment of gases in Canada. It is the Canadian Provincial Governments rather than the Canadian Federal Government that set regulations on shipping gases by motor vehicles in Canada.

The complete regulations of the DOT including those applying to compressed gases, are published in their current form under the title: "Code of Federal Regulations, 49, Transportation, Parts 100 to 199," published by the Office of the Federal Register, National Archives and Records Service, Central Services Administration and can be obtained from Superintendent of Documents, United States Government Printing Office, Washington, DC 20402.

The DOT regulations are also published in part under the title, "Tariff BOE-6000, Publishing Hazardous Materials Regulations of the Department of Transportation, Including Specifications for Shipping Containers," Parts 170-179. This document and supplements are available from the Bureau of Explosives, 1920 L St. N.W., Washington, D.C. 20036.

DOT regulations pertaining to motor vehicle shipment of dangerous articles are also published by the American Trucking Associations, Inc. under the title, "Dangerous Articles Tariff 111-B," and can be obtained from the American Trucking Associations, Inc., Attn: Traffic Department, 1616 P Street, N.W., Washington, DC 20036.

The complete regulations of the CTC including those applying to compressed gases are published in their current form under the title:

"Regulations for the Transportation of Dangerous Commodities by Rail," prescribed by General Order No. 1974-1-Rail of the Canadian Transport Commission, July 31, 1974; published by Information Canada, Ottawa, Ontario.

United States Coast Guard; Transport Canada

Water shipment of compressed gases is regulated in the United States by the Coast Guard which is part of the DOT, and in Canada by the Transport Canada, both of which are federal government organizations in their respective countries. The regulations of each of these bodies generally provide for compressed gas shipment by water in the same kinds of containers that meet DOT and CTC requirements for rail shipment. As the Department of Transport notes, however, the packing requirements for shipment by sea are in some cases more stringent than those set for shipment by rail.

Additional regulations of the two bodies give requirements and specifications for shipping certain compressed gases by tankships and tank barges. Ammonia, chlorine and liquefied petroleum gas are the main gases that have been shipped in bulk aboard such tank vessels (though others for which tank-barge shipment in particular is growing include inhibited butadiene, anhydrous dimethylamine, liquefied hydrogen, liquefied oxygen, methyl chloride and vinyl chloride).

Current regulations of the United States Coast Guard are published in full as:

"Tariff BOE-6000 Publishing United States Coast Guard Regulations Governing the Transportation or Storage of Explosives or Other Dangerous Articles or Substances, and Combustible Liquids on Board Vessels."

International Air Transport Association

Regulations for the air transport of compressed gases made by federal agencies in the United States and Canada, with still further joint requirements of almost all major airlines, are incorporated in an annual publication of the International Air Transport Association, to which airlines belong. The volume reflects in addition the national regulations of many other countries, and is published in several different languages. Its twenty-first annual edition was published and effective December 1, 1978. The publication is entitled, "IATA Restricted Articles Regulations," and may be obtained from the IATA, P.O. Box 160, 1216 Cointrim-Geneva, Switzerland.

OTHER AREAS OF FEDERAL REGULATION

A few special areas of the compressed gas field are subject to further regulation by other federal government bodies in the United States and Canada, as in the labeling and purity requirements prescribed for gases used in medicine under national laws concerning drugs. Another special area of labeling regulation in the United States arose with the enactment of the Hazardous Substances Labelling Act; the provisions of this act apply only to those gases under pressure which go into the home, as with aerosol propellants. The act does not apply generally to compressed gases, as sometimes has been assumed mistakenly. Information about regulation in these special areas is available from the various appropriate authorities in the two countries, such as the Federal Food and Drug Administration in the United States and the Department of Health and Welfare in Canada.

The United States Federal Government regulates matters affecting the safety and health of employees in all industry, including compressed gases, through the Department of Labor, Occupational Safety and Health Act (OSHA) of Congress dated December 29, 1970, Public Law 91-596, known as the Williams–Steiger Act.

REGULATION BY STATES AND PROVINCES

States of the United States and the Canadian Provinces vary widely in their regulatory provisions for compressed gases. These governments generally have developed regulations applying to compressed gases through their chief fire safety officer, pressure vessel authorities or industry and labor commission. However, in some instances, it may prove to be the Railway Commission or the Industrial Accident Commission which has developed the main body of applicable regulations. Certain states have adopted the DOT regulations in full or in part. These are the DOT regulations concerning the shipment of dangerous articles, as well as the Federal Motor Carrier Safety Regulations of the DOT. Also, many of the states have based their fire codes for precautions with liquefied petroleum gas on the recommendations of the National Fire Protection Association given in NFPA Pamphlet No. 58, "Storage and Handling of Liquefied Petroleum Gases." Also widely reflected in state, as well as local regulations, are several other NFPA codes as listed in detail in Appendix A.

State or provincial regulation of compressed gas concerns primarily its storage and use, though some state regulation in the United States governs its transportation. In Canada, as noted previously, the provinces (rather than the national government) regulate shipment by motor vehicle.

Producers, distributors and users of compressed gases should obtain current and full information on state or province regulations directly from the appropriate governmental offices. A summary of certain aspects of current compressed gas regulation by states of the United States appears in Appendix A to this volume.

LOCAL REGULATION

Municipalities, towns and other local governments in the United States and Canada have also adopted regulations applying to compressed gas storage, use and transportation. These will frequently be issued under the authority of officials like the fire commissioner, the head of the department of building or the zoning commissioner. Inquiries about local regulations should again be made with the governments concerned.

CHAPTER 3

Compressed Gas
Containers

Intended primarily for persons unfamiliar with the compressed gas field, this chapter describes and illustrates the major kinds of containers used today for compressed and liquefied gases (including those liquefied at very low or cryogenic temperatures). It does not represent standards published by Compressed Gas Association, Inc., as do Chapters 4 through 7 and Chapter 10 of Part I. The CGA, however, through its technical committees, continually reviews and recommends changes in existing DOT and CTC specifications, or proposes new specifications to keep pace with new developments in the industry. This chapter treats the various kinds of containers in sections as follows:

Section A. Cylinders and Small Containers

Section B. Regulators and Control Valves for Cylinders and other Containers

Section C. Containers for Shipping by Rail in Bulk

Section D. Containers for Shipping by Highway in Bulk

Section E. Containers for Shipping by Water in Bulk

Section F. Liquefied Cold Compressed Gas Containers

Section G. Storage Containers

Safety relief devices for these different types of containers are not dealt with in this chapter because they are discussed in detail in Chapter 5

of Part I. Similarly, readers should see Chapter 4 of Part I for information on the safe handling of cylinders and other containers; Chapter 6, for the labeling, requalifying, repair and disposition of cylinders; Chapter 7, for cylinder valve connection standards; and Part I, Chapter 8 for methods of unloading bulk shipments of liquefied compressed gas.

DOT, CTC AND ASME REGULATIONS FOR COMPRESSED GAS CONTAINERS

Most of the shipping and storage containers for compressed gases made in North America are built to comply with one major source of detailed specifications. These are the identical regulations adopted by the United States Department of Transportation and the Canadian Transport Commission for Canada identified in Chapter 2 of Part I, and the "Boiler and Pressure Vessel Code" (particularly Section VIII, on unfired pressure vessels) of the American Society of Mechanical Engineers. (The 1980 edition of the code is available from the ASME at 345 E. 47th St., New York, NY, 10017.) The DOT and CTC regulations are obligatory for containers shipped in interstate or interprovincial commerce, and apply mainly to shipping containers; the ASME code applies mainly to stationary storage vessels (though truck cargo tanks and portable tanks must conform to the ASME code under DOT and CTC specifications).

Inspectors examine compressed gas containers built to the respective regulations as they are made, and approve them for container markings (usually stamped) which indicate that the marked container fulfills the regulatory specifications identified in the marks. Inspectors authorized by the Department of Transportation are required for high-pressure cylinders. Inspectors are required to file a report on each individual container that is made under the specifications of these regulations before the container goes into service, and each container made is identified and thus registered by serial number and manufacturer's symbol. Compliance with one or more of these regulations is often stipulated in state and insurance company requirements as well as in federal regulations.

SECTION A

Cylinders and Small Containers

Cylinders for compressed gases are generally defined in the DOT and CTC specifications as containers having a maximum water capacity of 1000 lb (453.6 kg) or less. This is approximately the equivalent of 120 gal (454.2 liters).

The newer 3AX, 3AAX and 3T specifications permit the use of larger cylinders, a popular size having a water capacity of approximately 5000 lb (2268.0 kg). They are made in a wide variety of lengths, diameters and proportions, and range in capacities from their authorized maximums down to a cubic foot or less. The 3AX, 3AAX and 3T cylinders have a minimum water capacity of 1000 lb (453.6 kg) with no specified upper volumetric limit.

Cylinders broad and squat in proportions are generally made for low-pressure service, while cylinders tall and thin in proportions are in-

In left foreground, cryogenic tank car of the AAR-240W type for transporting nonflammable gases in the form of cryogenic fluids; the car carries the gases at pressures below 25 psig, and is not subject to DOT regulations. Cryogenic storage tanks of large and small capacity appear at the rear, and a truck cargo tank shows partly at the right, in this view of the shipping area of an air separation plant producing atmospheric gases.

Compressed gas cylinders of a great many of the different sizes and styles in which they are made are shown here, ranging from some of the smallest at bottom to some of the largest at top. The narrow, thin, seamless cylinders are generally those built for high-pressure service; the broader, thick welded cylinders are in general ones made for service at moderate or low pressures. Valve protection caps are in place on some of the cylinders shown, while others appear without valves screwed into the threaded openings for them. Some compressed gas cylinders are made with concave or dished heads, instead of the convex heads used for all the cylinders illustrated here.

tended primarily for high-pressure service. Cylinders are most often made with flat bottoms, or are fitted with foot-rings so that they may stand on end and be securely fastened in an upright position while their contents are being withdrawn. Some cylinders, though, have hemispherically rounded sealed ends and are designed for withdrawal of contents while securely fastened in a horizontal or slanted position. One end of the cylinder (or both ends, in some special cases) is tapered into a neck which is tapped with screw threads for attachment of the cylinder valve.

Cylinders are the type of compressed gas containers most widely authorized for different means of shipment. They are the most generally accepted type of container for air shipment in the case of most gases.

CYLINDER MANUFACTURE

Cylinders are made from seamless tubing, brazed or welded tubing, billets in the billet process, or flat sheets drawn to cylindrical shapes in large punch-press dies. Sealed ends are made either by spinning in a lathe under a flame at red or white heat, by forging or by die-drawing. The sealed ends of some cylinders are closed by spinning or forging. In some instances the sealed end of such cylinders is drilled and then "plugged" with an additional metal piece.

Provision for Interchangeable Use of DOT and CTC Specification Cylinders

Under the DOT and CTC regulations, a cylinder meeting any one specification may be authorized for shipping a number of different gases, or any one gas may be authorized for shipment in a number of different specification cylinders. This differs from the usual European practice, in which a cylinder meeting a single set of cylinder specifications is authorized for shipping only a single designated gas. The one exception in North American practice concerns acetylene, which is authorized for shipment as a "dissolved" gas only in acetylene cylinders— those meeting specification DOT-8 or DOT-8AL, CTC-8 or CTC-8AL. In turn, 8 or 8AL cylinders must not be charged with any gas but acetylene.

Required Cylinder Markings

Under the DOT and CTC regulations, cylinders must be marked to indicate the specification under which they were made, the service pressure for which they were designed, a serial number provided by the maker or user, a symbol indicating the inspector (if required) and a symbol indicating the maker or manufacturer. The markings are usually stamped into the shoulder of the cylinder (the part sloping up to the neck) or into the top surface of the neck itself. The complete detailed marking requirements for any given cylinder will be found under the appropriate specification in the Code of Federal Regulations (CFR), Title 49, Part 178.

The markings of a typical cylinder might be arranged as follows on one side of the shoulder:

DOT-3A 2015

462

XY

CGA

In this case the DOT specification is 3A, the service pressure is 2015 psig at 70 F (13,893 kPa at 21.1 C), the manufacturer is XY, the user is CGA and the serial number is 462. These same markings could be arranged in a horizontal line around the shoulder and might appear as follows:

DOT-3A 2015   462   XY   CGA

Still other markings are required. Cylinders are tested as part of the final inspection after being manufactured, and the month and year of the initial qualifying test are required as stamped markings (appearing as 5-79 for May 1979) on the shoulder area. The date stamp should be placed so that the dates of subsequent retestings for required requalification can be added later. Moreover, the DOT and CTC specifications require that the word "SPUN" or "PLUG" be stamped near the specification mark where an end closure has been made by spinning, or by spinning, drilling and plugging. The symbol of the inspector (if required) is commonly placed between the month and year of the test date. Because of space limitations, the test date and other markings are sometimes on the opposite side of the shoulder from the markings previously described. The complete markings on a cylinder might be as follows:

DOT-3A 2015—SPUN

462   5-AB-79

XY

CGA

AB represents the symbol of the inspector.

If a plus (+) sign appears immediately after

Permanent markings stamped into the shoulder or head of a new cylinder in accordance with DOT requirements, as shown by the cylinder above, include: the DOT specification to which the cylinder was made ("DOT-3AA"), followed by the service pressure in psig ("2015"); on the next line, the serial number of the individual cylinder, assigned by the manufacturer ("Z45015"); next, the identifying mark of the cylinder's manufacturer ("PST"); and last, the month and date of the qualification test required before the cylinder can go into service ("7, 60" for July 1960, with the "7" and "60" separated in this case by the identifying mark of the inspector who made the first test). Months and years of subsequent hydrostatic tests required periodically for requalification follow the date of the initial test, being stamped below or beside the first date on the shoulder.

the test date marking of a specification 3A, 3AX, 3AA, 3AAX or 3T cylinder, it means that the cylinder is authorized for charging up to 10 percent in excess of the marked service pressure. The requirements for this authorization will be found in the CFR, Title 49, Part 173, Subpart G, paragraph 173.302 (4) (c).

If a five-pointed star is stamped in the cylinder shoulder after the most recent test date

(DOT-3A or 3AA cylinders), it indicates that the cylinder may be retested every ten years instead of every five years. The details of the requirements qualifying a cylinder for the ten-year retest interval will be found in the CFR, Title 49, Part 173, Subpart B, paragraph 173.34 (e) (15).

For cylinders stamped DOT-3 or DOT-3E without any following pressure marking the service pressure is 1800 psig (12,411 kPa). Such cylinders manufactured in recent years should be stamped DOT-3E 1800.

Stamped markings on cylinders should not be changed except in conformance with the requirements of the CFR, Title 49, Part 173, Subpart B, paragraph 173.34 (c).

Common Cylinder Types

Among types of cylinders manufactured in the largest quantities are the following:

DOT-3A. These are seamless carbon-steel cylinders and are produced usually for high-pressure service in excess of 150 psig (1034 kPa) and ranging as high as 15,000 psig (103,400 kPa).

DOT-3AA. These are seamless alloy-steel cylinders used generally for the same purposes as 3A cylinders. The alloy steels are heat treated as specified so that the metal withstands higher working stresses than 3A cylinders, permitting a lighter weight for a given size and service pressure.

DOT-3B. These are seamless carbon-steel cylinders like the 3A cylinders, but made for a lower service pressure range of 150 to 500 psig (1034 to 3447 kPa).

DOT-3E. These are seamless carbon-steel cylinders made for a service pressure of 1800 psig (12,410 kPa) and limited in size to 2 inches (5.08 cm) in diameter and a length of 2 ft (60.96 cm).

DOT-3HT. These are seamless alloy-steel cylinders made for inside service in the aircraft industry. They are limited to a water capacity of 150 lb (68.04 kg) and are made for a service pressure of 900 psig (6205 kPa) or higher.

DOT-8 or 8AL. These are seamless or welded steel cylinders only for acetylene service, and made to the single authorized service pressure of 250 psig (1724 kPa); heads are attached by welding or brazing by the dipping process; DOT-8 specification cylinders may have a longitudinal seam if it is forge lap welded; welded circumferential seam authorized if body has no longitudinal seam. These cylinders also contain a permanent porous inner filling material which is saturated with an acetylene solvent, usually acetone.

DOT-3AX. These are carbon-steel cylinders, similar to DOT-3A cylinders, except that the minimum volumetric capacity is 1000 lb (453.6 kg) of water, and the minimum service pressure is 500 psig (3447 kPa).

DOT-3AAX. These are alloy-steel cylinders, similar to DOT-3AA cylinders, except that the minimum volumetric capacity is 1000 lb (453.6 kg) of water, and the minimum service pressure is 500 psig (3447 kPa).

DOT-3T. These are alloy-steel cylinders with a minimum volumetric capacity of 1000 lb (453.6 kg) of water, and a minimum service pressure of 1800 psig (12,411 kPa).

DOT-4A. These are forge-welded steel cylinders with a volumetric capacity of not over 1000 lb (453.6 kg) of water, and made for a service pressure of 150 to 500 psig (1034 to 3447 kPa).

DOT-4B. These are welded and brazed steel cylinders with a volumetric capacity of not over 1000 lb (453.6 kg) of water, and made for a service pressure of 150 to 500 psig (1034 to 3447 kPa).

DOT-39. These are seamless welded or brazed steel or aluminum cylinders with a maximum volumetric capacity of 55 lb (24.9 kg) of water, and a maximum service pressure of 500 psig (3447 kPa); or a maximum volumetric capacity of 10 lb (4.54 kg) of water for service pressures over 500 psig (3447 kPa). Spherical containers are covered by this specification. DOT-39 cylinders are nonreusable and nonrefillable.

Small Containers Exempt from Cylinder Requirements

Under the DOT and CTC regulations limited quantities of compressed gases are excepted from the requirements under certain conditions. The details of these exceptions will be found in CFR, Title 49, Subpart B, Paragraph 172.101; and Subpart G, paragraph 173.306.

SECTION B

Regulators and Control Valves for Cylinders and Other Containers

To insure proper discharge of compressed gases from cylinders and other containers, the correct regulators and control valves must be used in the case of any nonliquefied or liquefied compressed gas. This section primarily treats the regulators and control valves used for cylinders.

Nonliquefied Gases

To insure safe removal of a nonliquefied gas from a cylinder, the pressure must be reduced to a safe value. This is most commonly done with an automatic pressure regulator like that shown.

Cylinders are delivered to users with cylinder

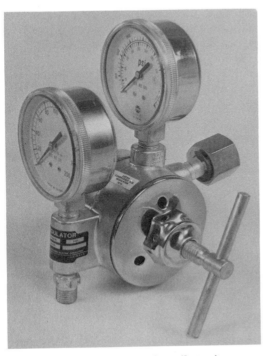

Automatic pressure regulator (2-stage).

valves complying with national standards (see Chapter 7, Section A, "American–Canadian Standard Inlet and Outlet Connections"). The regulator is attached to the cylinder valve outlet with the regulator connection. Basically, the regulator consists of a spring-loaded (or gas-loaded) diaphragm that controls the opening or closing of the delivery orifice. The diaphragm control can be set by hand to maintain a constant delivery pressure at any value over the range for which it is designed. Once the delivery pressure is set, the diaphragm acts to open or close the delivery orifice to keep the delivery pressure constant. Gages attached to the regulator show the delivery pressure chosen and the cylinder pressure. A flow-control valve that is also part of the regulator controls the volume of gas delivered at the chosen delivery pressure.

Essential Factors in Choosing Regulators

The choice of a regulator for use in a specific gas service depends on four factors: (1) design and materials for safe and trouble-free operations with the specific gas and pressures involved; (2) the range of delivery pressures required; (3) the degree of accuracy of delivery pressure to be maintained; and (4) the flow rate required. There are two basic types of automatic pressure regulators: the single-stage, and the double or two-stage. Generally, a two-stage regulator will deliver a more constant pressure under more widely varying operating conditions than will a single-stage regulator. A single-stage regulator will show a slight variation in delivery pressure as the cylinder pressure drops, as well as a drop in delivery pressure as the flow rate is increased. The two-stage regulator is less subject to such delivery-pressure variations. The single-stage regulator will also have a higher "lockup" pressure (pressure increase above delivery set point necessary to stop flow) than the two-stage regulator at relatively large flow rates.

Liquefied Gases

Liquefied gases are present in a cylinder in liquid form at higher pressure levels and as a gas at lower pressure levels; usually, the gas phase of the liquefied gas is drawn for use. The cylinder pressure will remain constant at the vapor pressure of the material as long as there is any liquid remaining in the cylinder. Therefore, a single-stage automatic pressure regulator may be used for a constant delivery pressure while there is still liquid in the cylinder. For example, 80 percent of the contents of a carbon dioxide cylinder can be drawn off at room temperature before the cylinder pressure drops below the vapor pressure and there is no liquid phase of the gas left. For withdrawing at a constant delivery pressure all of the liquefied gas in a cylinder, a two-stage regulator should be used.

Certain problems can develop when removing the gas phase of liquefied gas. Rapid removal of the gas may cause the liquid to cool too rapidly, causing the pressure and flow to drop below the required level. To prevent this, cylinders may be heated in a water bath with the temperature no higher than 130 F (54.4 C).

For the controlled removal of the liquid phase of a liquefied gas, a manual flow-control valve is used. Special liquid flow regulators are also available. Removal of the liquid *must* be done at the vapor pressure of the material. Care must be taken to prevent blockage of the gas line downstream from a user's heat exchanger, which would cause excessive pressure buildup in the heat exchanger and the cylinder. Safety relief devices should be installed in all transfer lines to relieve sudden and dangerous hydrostatic or vapor-pressure buildups.

Handling and Use of Automatic Regulators

When an automatic pressure regulator is attached to a cylinder, the threads must not be forced. If the regulator does not fit, it must in no way be forced. The poor fit of a regulator may indicate that it is the wrong regulator for that specific type of gas. However, users should also make sure in advance that the correct regulator for the gas service intended is chosen by acting on the advice of a responsible source, such as the supplier or producer of the gas or regulators.

Use the following procedure to obtain the proper delivery pressure from an automatic regulator:

(1) After the regulator has been attached to the cylinder valve outlet, rotate the delivery-pressure adjusting screw counterclockwise until it turns freely.

(2) Open the cylinder valve slowly until the tank gage on the regulator registers the cylinder pressure. At this point the cylinder pressure should be checked to see if it is at the expected value. A large error may indicate leakage in the cylinder valve.

(3) With the flow-control valve at the regulator outlet closed, turn the delivery-pressure adjusting screw clockwise until the required delivery pressure is reached. Control of flow can be regulated by means of the flow-control valve installed in the regulator outlet or by a supplementary valve placed by the user in a pipeline downstream from the regulator. The regulator itself should not be used as a flow control by adjusting the pressure to obtain different flow rates. This defeats the purpose of the pressure regulator. In some cases where higher flows are obtained in this manner, the pressure setting may be in excess of the design pressure of the user's system of piping and devices for employing the gas.

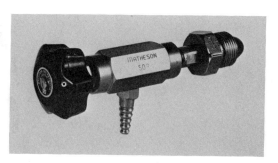

Manual needle valve.

Manual Flow Controls

Manual flow controls may be used when an intermittent flow is needed and an operator will be present at all times. A manual flow control, illustrated in the photograph, is simply a valve operated manually to deliver the proper amount of gas. A very fine control of the flow of gas

Cylinder valves.

can be obtained, but remember that dangerous pressures can build up in a closed system or in one that becomes plugged, since no means are provided for the automatic prevention or release of excessive pressures.

Cylinder Valves

There are a number of different types of cylinder valves affixed to the cylinders that are delivered to users. Regulators are attached directly to these cylinder vlaves by the user. Four commonly used types of cylinder valves are illustrated. Many of these valves or outlets have been standardized by the Compressed Gas Association for the different families of gases to prevent the interchange of regulator equipment between gases which are not compatible. These standards have been adopted by the American National Standards Institute (ANSI) (see Part I, Chapter 7, Section A).

Use of adapters to change the outlet size of a cylinder valve defeats the whole purpose of standardizing the valve outlets. Adapters should be used with care only on gases definitely known to be compatible. Equipment for certain gases, such as oxygen, should never be used with other gases. Gases which are oil-pumped can cause an oil film to coat internal parts of gas delivery systems. Introduction of oxygen into such an oil-coated system can cause a fire or explosion.

SECTION C

Containers for Shipping by Rail in Bulk

Bulk rail shipment of compressed gases is authorized under the DOT and CTC regulations in containers of three major kinds: single-unit tank cars; multi-unit [or "ton multi-unit" (TMU)] tank cars; and specification 107A tank cars that consists of clustered and fixed cylindrical tubes that extend the length of the car and are manifolded in a common header. Adaptations of existing tank car specifications, and proposed specifications for new types of cars, must be approved by the Tank Car Committee of the Association of American Railroads.

Single-Unit Tank Cars

Single-unit tank cars authorized for compressed gas shipment by the DOT or the CTC carry the gas in a single large pressure tank which is permanently mounted to the car frame. These cars resemble, generally, the familiar railroad tank cars used for oil and other nonpressurized liquids. Single-unit pressure tank cars for use with some of the cooler liquefied gases (such as sulfur dioxide or carbon dioxide) are often made with insulation (lagging) between inner and outer metal shells. Tank car types frequently required for compressed gases are those complying with DOT or CTC specifications 105A, 112A and 114A. [The number following a designation like 105A, as in "105A 500-W," is the marked test pressure of the tank—500 psig (3447 kPa), in this example. The highest test pressure authorized for a single-unit tank car in current regulations is 600 psig (4137 kPa)].

Single-unit tank cars that had been built under DOT or CTC specifications until a few years ago had been made with capacities ranging up to about 85,000 lb (38,536 ka) of water or some 10,000 gal (37.85 m^3).

Oversize Single-Unit Tank Cars

A number of oversize single-unit tank cars, popularly called "jumbo" cars, have been built and put into compressed gas service under DOT specifications or exemptions in the last few years. These cars have been introduced primarily for liquefied gases that are shipped in very large quantities, like propane and ammonia, and have capacities in exceptional cases as large as 50,000 or even 60,000 gal (189.3 or even 227.1

A group of tank cars leaving a plant in Texas.

m³) per car. However, a fairly common capacity for jumbo cars today is 30,000 or 32,000 gal (113.6 to 121.1 m³). Tank cars built after November 30, 1970 must not exceed 34,500 gal (130.6 m³) capacity or 263,000 lb (119,294 kg) gross weight on rail. Existing tank cars may not be converted to exceed 34,500 gal (130.6 m³) capacity or 263,000 lb (119, 294 kg) gross weight on rail.

Extremely large tank car capacities have been obtained within the maximum height and width allowances for railroad cars by making the big-

A 33,000 gallon L. P. gas tank car complete with thermal coating head shield and interlocking top and bottom couplers.

gest oversize cars far longer than usual and swelling out the lower tank contours in long "belly" bulges. One design enlarges capacity through a so-called "figure-eight" cross section that consists of two intersecting circles, one above the other.

The largest oversize tank cars are usually restricted to main lines in order to avoid possible weight and other problems they might en-counter in traversing spur lines not built for fast, heavy-duty service.

Multi-Unit or "TMU" Tank Cars

Multi-unit tank cars consist of a kind of flat-bed railroad car that carries 15 large cylindrical pressure tanks crosswise on the car. The filled tanks are lifted onto or off the car by crane or

A multi-unit tank car for rail transport of compressed gas, as above, carries 15 tanks of at least 1000 lb (453.6 kg) water capacity each; the tanks are popularly called "ton multi-unit" or "TMU" tanks because they were first introduced to carry one ton of liquefied chlorine per tank. TMU tanks are usually loaded and unloaded with a hoist, and the cars are almost always shipped with the full 15 tanks in place because the rail shipping charge per car is the same regardless of the number of tanks carried. TMU tanks are also authorized for the highway shipment of some compressed gases.

The trailer-mounted truck cargo tank at left and the tube trailer at right aboard a railroad flatcar illustrate how railroad "piggyback" shipment can be used for combined rail and highway transport of compressed gases.

hoist. Specifications 106A and 110A in the DOT and CTC codes are the principal ones for multi-unit tanks, and the codes limit their water capacity to a minimum of 1500 lb (680.4 kg) and a maximum of 2600 lb (1179.3 kg) which is about 180 to 310 gal (0.68 to 1.17 m³). 106A tanks have forged-welded heads formed convex to pressure. Present authorized test pressures are 500 to 800 psig (3447 and 5516 kPa) for 110A type.

Their popular name, "ton containers" or "ton multi-unit tanks," (which is used throughout the *Handbook*) stems from the fact that these tanks were first introduced to transport a ton of liquefied chlorine apiece.

TMU tanks are uninsulated.

107A Tank Cars Made Up of Clustered Tubular Tanks

Tank car specification 107A in the DOT and CTC regulations provides for cars carrying clustered sets of long tubular tanks advantageous for transporting bulk quantities of nonliquefied gases (like helium, argon or air) at high pressure. The multiple tubular tanks are either hollow forged, drawn or seamless tubing. They are permanently mounted on the car, and are connected or manifolded to a common header at one end of the car. A common diameter for the tubes is 30 in. (76.2 cm), and as many as 30 or more tubes may be installed on a single car frame. Test pressures for the 107A tank cars range up to 3500 psig (24,132 kPa) and capacities up to a quarter-million standard cubic feet (7079.1 m³) of gas (in the case of helium).

"Piggyback" Rail Transport

Cargo tanks on truck semitrailers and tube trailers are also often transported "piggyback" on railroad flatcars in bulk shipment of compressed gases by rail.

SECTION D

Containers for Shipping by Highway in Bulk

Three major types of containers are authorized for the bulk shipment of compressed gases by highway under the DOT regulations: truck or truck semitrailer cargo tanks; portable tanks; and tube trailers.

Cargo Tanks

Cargo tanks are large-capacity tanks permanently mounted on truck bodies or semitrailer bodies. The specifications most often authorized for compressed gas cargo tanks in the DOT regulations are MC-330 and MC-331. Tanks of each type must comply with the ASME pressure vessel code, and must have a design pressure of not less than 100 psig nor more than 500 psig. Either type may be insulated for use with such liquefied gases as carbon dioxide or nitrous oxide and either may be fitted with refrigerating and heating coils. Cargo

A relatively small compressed gas cargo tank mounted directly on a truck body (instead of on a semitrailer, the common mounting for larger tanks) for local deliveries of LP-gas. The tank is uninsulated.

tank capacities range up to as large as 10,000 gal or more per tank; no minimum capacity is given for them in the specifications. In some instances, truck tractors draw double tank trailers of compressed gas.

This moderately large, insulated cargo tank illustrates semitrailer mounting widely used for highway shipment of compressed gas; the tank shown is typical of those built primarily for liquefied carbon dioxide.

Portable Tanks

Portable tanks complying with specification DOT-51 are also authorized for shipping many compressed gases. The DOT-51 specification provides for steel tanks of at least 1000 lb water capacity (about 120 gal) with service pressures of not less than 100 psig nor more than 500 psig. The cylindrical DOT-51 tanks are often made with flat skid mountings attached, and are commonly called "skid tanks." Portable tanks are shipped primarily by truck, but they also are used to some extent in rail shipment. The U.S. water shipment regulations limit the maximum gross weight of full portable tanks to 20,000 lb.

TMU Tanks on Trucks

For some gases, DOT regulations authorize shipment of TMU or multi-unit tanks by motor vehicle as well as by rail. TMU tanks shipped on trucks must be securely chocked or clamped while in transit, and adequate facilities must be present for handling tanks where transfer in transit is necessary.

Tube Trailers

High-pressure nonliquefied gases like oxygen and nitrogen at atmospheric temperatures are often shipped by highway in tube trailers. These are truck semitrailers on which a number of very long gas cylinders have been mounted and

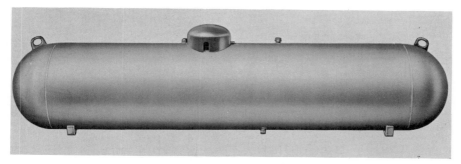

Portable tanks used for highway and rail shipment of compressed gas are made in many different sizes above the specified minimum of 1000 lb (453.6 kg) water capacity (about 120 gal) (0.454 m^3) and in a wide variety of shapes, as this unusually long portable tank suggests. Their outfitting with flat skid mountings, as illustrated above, has led to the common practice of calling them also, "skid tanks."

manifolded in a common header. Tube trailer service pressures are as high as 2000 psig (13,790 kPa) or more. The tubular cylinders of the trailers are often made according to 3A or 3AA cylinder specifications, or to cylinder specification 3AX, 3AAX or 3T (3AX, 3AAX and 3T specifications provide for containers some 22 in. (55.88 cm) in diameter instead of the customary $9\frac{5}{8}$ in. (24.46 cm) diameter of 3A and 3AA cylinders; also, 3AX, 3AAX and 3T cylinders have a minimum size of 1000 lb (455.6 kg) water capacity under the DOT regulations. Tube trailers have been built to carry as much as 45,000 standard cu ft (1274.3 m³) of oxygen or 128,000 standard cu ft (3624.6 m³) of helium.

SECTION E

Containers for Shipping by Water in Bulk

Practically all types of containers authorized for shipping compressed gases on land are also authorized under some conditions for water shipment; for example, even single-unit tank cars are approved for cargo vessels or railroad car ferry vessels, and tank trailers are approved for cargo vessels and trailerships, in the cases of some gases.

Tankships and Tank Barges

Some tankships and many tank barges are built with fixed pressure tanks primarily for bulk water transport of compressed and liquefied gases. Regulations of the United States Coast Guard and of the Canadian Transport Commission (see Chapter 2, Part I) set forth detailed requirements for fixed tanks or barges

Fixed pressure tanks built into tank barges like the one above are used for inland waterway shipment of increasing numbers of gases. Among gases shipped bulk by barge are anhydrous ammonia, inhibited butadiene, chlorine, anhydrous dimethylamine, liquefied hydrogen, LP-gas methyl chloride and vinyl chloride.

Tankship built in 1978 in France has a capacity of 125,000 cubic meters of liquefied natural gas (LNG). It is currently in service transporting LNG from Algeria to the United States.

used for shipping anhydrous ammonia and chlorine. Special permission of these regulatory agencies has authorized tank vessels for other compressed gases; the first of any importance was a ship equipped to transport LP-gases in intercoastal services; and more recently ocean-going tankers that carry liquefied methane in insulated and refrigerated tanks from North America to Europe, and from North Africa to North America. Among other gases that are shipped in substantial quantities by tank barge are inhibited butadiene, anhydrous dimethyl-amine, liquefied hydrogen, methyl chloride and vinyl chloride.

SECTION F

Liquefied Cold Compressed Gas Containers

A wide variety of containers has been developed for shipping gases liquefied at very low or cryogenic temperatures—ranging from about −250 F (−156.7 C) down to the neighborhood

The row of individual small-volume containers for gases liquefied at cryogenic temperatures is made up of insulated spherical flasks with bright-metal outside casings and, fourth from left in the row, a cylinder of the DOT-4L type. Larger cryogenic containers include the insulated, dolly-mounted portable tank at right and larger truck cargo tanks in background.

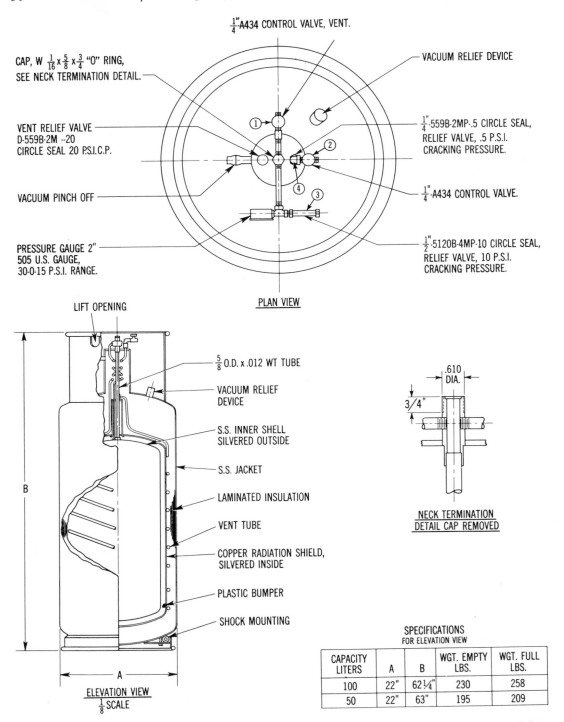

$\frac{1}{4}$"-A434 CONTROL VALVE, VENT.

CAP, W $\frac{1}{16}$ x $\frac{5}{8}$ x $\frac{3}{4}$ "O" RING,
SEE NECK TERMINATION DETAIL.

VACUUM RELIEF DEVICE

VENT RELIEF VALVE
D-559B-2M --20
CIRCLE SEAL 20 P.S.I.C.P.

$\frac{1}{4}$"-559B-2MP-.5 CIRCLE SEAL,
RELIEF VALVE, .5 P.S.I.
CRACKING PRESSURE.

VACUUM PINCH OFF

$\frac{1}{4}$"-A434 CONTROL VALVE.

PRESSURE GAUGE 2"
505 U.S. GAUGE,
30-0-15 P.S.I. RANGE.

$\frac{1}{2}$"-5120B-4MP-10 CIRCLE SEAL,
RELIEF VALVE, 10 P.S.I.
CRACKING PRESSURE.

PLAN VIEW

LIFT OPENING

$\frac{5}{8}$ O.D. x .012 WT TUBE

VACUUM RELIEF
DEVICE

S.S. INNER SHELL
SILVERED OUTSIDE

S.S. JACKET

LAMINATED INSULATION

VENT TUBE

COPPER RADIATION SHIELD,
SILVERED INSIDE

PLASTIC BUMPER

SHOCK MOUNTING

.610
DIA.

3/4"

NECK TERMINATION
DETAIL CAP REMOVED

ELEVATION VIEW
$\frac{1}{8}$ SCALE

SPECIFICATIONS
FOR ELEVATION VIEW

CAPACITY LITERS	A	B	WGT. EMPTY LBS.	WGT. FULL LBS.
100	22"	62¼"	230	258
50	22"	63"	195	209

Cross-sectional view of a cryogenic cylinder of the DOT-4L type designed for use with liquid helium at −452 F the lowest boiling point of any known substance. The cylinder is insulated both by a vacuum and a laminated, high-efficiency insulating material.

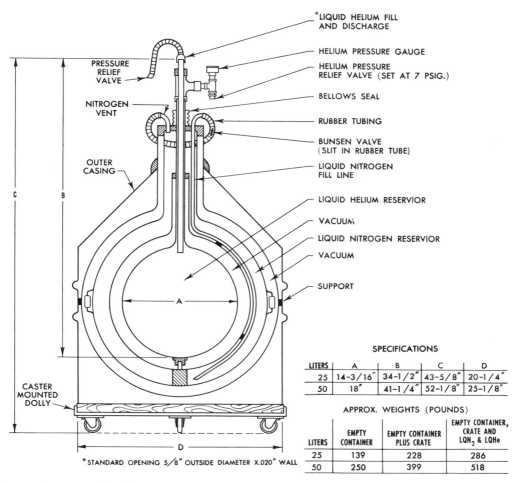

	*LIQUID HELIUM FILL AND DISCHARGE
	HELIUM PRESSURE GAUGE
	HELIUM PRESSURE RELIEF VALVE (SET AT 7 PSIG.)
PRESSURE RELIEF VALVE	BELLOWS SEAL
NITROGEN VENT	RUBBER TUBING
	BUNSEN VALVE (SLIT IN RUBBER TUBE)
	LIQUID NITROGEN FILL LINE
OUTER CASING	LIQUID HELIUM RESERVIOR
	VACUUM
	LIQUID NITROGEN RESERVIOR
	VACUUM
	SUPPORT
CASTER MOUNTED DOLLY	

SPECIFICATIONS

LITERS	A	B	C	D
25	14-3/16"	34-1/2"	43-5/8"	20-1/4"
50	18"	41-1/4"	52-1/8"	25-1/8"

APPROX. WEIGHTS (POUNDS)

LITERS	EMPTY CONTAINER	EMPTY CONTAINER PLUS CRATE	EMPTY CONTAINER, CRATE AND LQN$_2$ & LQHe
25	139	228	286
50	250	399	518

*STANDARD OPENING 5/8" OUTSIDE DIAMETER X.020" WALL

Cross-sectional view of a spherical, small-volume flask for liquid helium which is insulated by liquid nitrogen at −320 F (−195.6 C) and by double vacuum layers.

Cryogenic tank car of the DOT-113A type for transporting liquefied hydrogen at −423 F (−252.8 C).

Heavily insulated 11,000 gallon cryogenic cargo tank for highway shipment of liquid helium.

of absolute zero, or −459.69 F (−273.16 C). The containers usually have high-efficiency insulation, and most of them dissipate heat absorbed in the contained cryogenic fluid by venting small amounts of vapor.

Containers for cryogenic fluids include specifications 4L cylinders of the DOT and CTC regulations; small nonregulated shipping containers of various kinds; the DOT and CTC specification 113A tank car, a single-unit car designed primarily for liquefied hydrogen; a nonregulated tank car complying with Association of American Railroads specification AAR-240W, intended mainly for liquefied oxygen, nitrogen or argon; various kinds of nonregulated cargo tank trucks and tank semitrailers; and the liquefied methane tankers and liquefied hydrogen tank barges mentioned in the preceding section, which are authorized by DOT exemption. (Nonregulated in the sense used here means not subject to DOT or CTC regulations, because the containers indicated as nonregulated are ones for nonflammable and very cold liquefied gases carried at pressures below 25 psig (172.4 kPa). A number of the regulated and nonregulated shipping containers are described in more detail in the Part II sections on gases which are often shipped as cryogenic fluids, such as helium, hydrogen, oxygen, nitrogen and argon.

Among proposed specifications for additional cryogenic containers that are being developed by Compressed Gas Association, Inc., as this volume goes to press is CGA-341, [which in some modified form may become known as MC-338 when it is approved by the DOT] for truck cargo tanks; and 51L, for portable tanks.

SECTION G

Storage Containers

Compressed and liquefied gases are often stored by users of smaller quantities in the shipping containers—in banks or storerooms of cylinders, in TMU tanks and portable tanks, in high-pressure tube trailers or 107A tank cars.

Stationary storage tanks into which gases are transferred from shipping containers are most often steel pressure vessels conforming to the ASME code. Storage tanks for relatively low-pressure liquefied or nonliquefied gases are

made in a great variety of sizes, designs and materials.

Gases formerly shipped only in nonliquefied, high-pressure form are today increasingly shipped and stored as compact cryogenic fluids. Cryogenic storage systems often include vaporizing units for converting the very cold fluid to gas at atmospheric temperatures and pressures.

Among the largest and most striking types of storage tanks now used for liquefied gases are huge underground caverns or refrigerated, lined and capped pits dug into the ground. These are developed most often for such fuel gases as propane-butane. One such pit storage facility being developed in New Jersey for liquefied natural gas will hold, when full, one billion standard cubic feet of gas (283,000 m^3).

Compact liquid oxygen storage system typical of those at hospitals is located outdoors and has standby supply of higher-pressure gaseous oxygen in manifolded cylinders.

Aerial view of tanks for typical storage installation for propane or ammonia in open farm country; siding and unloading gear for receiving large shipments by rail appear at left bottom, while truck-loading pumps to provide for local deliveries show at right center.

Low, earth-covered dome caps 160,000-barrel frozen pit for storing LP-gas liquefied at moderately low temperature and pressure. The large storage pit, for which the surrounding earth serves as insulation, is located in the western United States. A similar storage pit for liquefied natural gas in the Northeast has a capacity of one billion cubic feet (283,000 m^3) of gas at standard conditions.

CHAPTER 4

Safe Handling of
Compressed Gases

The very properties that make compressed gases useful in almost every area of modern life can also make them dangerous when mishandled. Years of experience with compressed gases have led to practices and equipment which, if employed, result in complete safety. This chapter presents the basic standards developed by Compressed Gas Association, Inc., for the safe handling of compressed gases. The standards as phrased by the CGA reflect many years of very broad experience in dealing with the gases. The chapter's four sections are as follows:

Section A. General Requirements for Safe Handling

Section B. Oxygen-Deficient Atmospheres

Section C. Safe Handling Requirements for Gases Used Medicinally

Section D. Safe Handling of Cylinders for Underwater Breathing

Essential knowledge for anyone involved with compressed gases in quantities of conventional cylinder size and larger is given in Section A. Section B gives the precautions that should be taken in oxygen-deficient atmospheres. Section C covers essential information for all persons working with compressed gases in hospitals, clinics and emergency treatment facilities for medical purposes. Section D gives essential information for persons involved in underwater diving.

SECTION A

General Requirements for Safe Handling

1. INTRODUCTION

1.1. Compressed gas containers, when constructed according to the applicable Department of Transportation (DOT) Specifications and maintained in accordance with DOT Hazardous Materials Regulations may be considered safe for the purposes for which they are intended. Accidents occurring during the transportation, handling, use and storage of these containers can almost invariably be traced to

failure to follow the DOT Regulations or to abuse or mishandling of the containers.

1.2. The following rules compiled by the Compressed Gas Association, Inc., are primarily for the guidance of users of compressed gases in containers and are based upon accident prevention experience. Some precautions are also applicable to gas suppliers. It should not be assumed that every acceptable safety precaution is contained herein, or that unusual circumstances may not require further or additional procedures. The best safety precaution is to ensure that the container handler or product user is adequately trained in the safe handling of the container as well as in the proper usage of the product.

2. REGULATIONS APPLICABLE TO COMPRESSED GASES IN CONTAINERS

2.1. The interstate transportation of compressed gases in containers is regulated by the Federal Department of Transportation through the modal Administrators: highway transportation by the Federal Highway Administration; rail transportation by the Federal Railroad Administration; water transportation by the Coast Guard; and air transportation by the Federal Aviation Administration. Some states and municipalities also regulate the intrastate and intracity transportation of compressed gases through state and local transportation authorities.

2.2. Federal regulations covering the interstate highway and rail transportation of compressed gases in containers are published in the United States Code of Federal Regulations, Title 49, Part 170 to 179. (1) The water transportation of compressed gases is published in the Code of Federal Regulations, Title 46, Part 146. Air transportation of compressed gases is covered by the Code of Federal Regulations, Title 14, Part 103. (2) In addition, these transportation regulations for compressed gases are published for carrier and shipper use by carrier associations. Highway and rail regulations are published by the Bureau of Explosives, Association of American Railroads, under the title. "Hazardous Materials Regulations of the Department of Transportation Including Specifications for Shipping Containers." Water transportation regulations for compressed gases are also published by the Bureau of Explosives entitled, "Coast Guard Department of Transportation Regulations Governing the Transportation or Storage of Explosives or Other Dangerous Articles or Substances, and Combustible Liquids on Board Vessels." Air transportation of com-

pressed gases is covered by the carrier tariffs of the Air Transport Association for domestic shipments and of the International Air Transport Association for foreign shipments. For state and municipal regulations applying to the transportation of compressed gases, the appropriate local transportation authority must be consulted.

2.3. DOT Regulations require that compressed gases be shipped in containers manufactured to DOT specifications and maintained in accordance with DOT Regulations. In Canada, containers in which compressed gases are shipped by rail must comply with the specifications and regulations of the Canadian Transport Commission.

2.4. Persons filling and maintaining compressed gas containers shall be adequately trained so that filled compressed gas containers shall comply with DOT Regulations governing the following subjects:

(a) The type of containers in which each gas may be shipped.

(b) The charging of containers as to amount of gas and conditions of filling.

(c) The requirements for marking and labeling of containers in transportation.

(d) The requirements for qualifying, maintaining and requalifying containers.

(e) The conditions under which a container may be transported.

2.5. Matters affecting the safety and health of employees in all industry, including those engaged in the manufacture, storage, handling and usage of compressed gases are regulated by the United States Government, Department of Labor, Occupational Safety and Health Act of the Congress dated December 29, 1970, Public Law 91-596, known as the Williams–Steiger Act. Many of the states and municipalities have, or will have, their own regulations to cover these same fields.

2.6. American National Standards Institute Standards:

(a) The storage, handling and use of compressed gases in containers for welding and cutting shall comply with ANSI Z 49.1. (13)

(b) Valve outlet connections shall be in accordance with ANSI B57.1. (5)

3. SAFE HANDLING RULES FOR COMPRESSED GAS CONTAINERS
(in general usage)

3.1. Filling

3.1.1. Containers must not be filled except by the owner or with the owner's consent.

3.1.2. Containers to be filled for transportation of compressed gas shall comply with DOT Regulations. Filling shall be in accordance with DOT Regulations.

3.1.3. Each cylinder must bear the proper DOT label or alternative marking (4) required for the compressed gas contained. Content identification must be applied before filling, or before removal from the filling manifold or, when mixtures are analyzed. Content identification may be applied after filling and analyzing, and must be present during transportation and delivery to the user.

3.1.4. Compressed gases should not be transferred from one container to another container except by a gas manufacturer or a gas distributor who compresses gases into containers by compressors or pumps.

3.1.5. Compressed gas containers must not contain gases capable of combining chemically with each other or with the container material so as to endanger its integrity.

3.1.6. The gas service shall not be changed without first removing the original content, and if residues are present, cleaning and purging. Changing gas service shall be in accordance with CGA Pamphlet C-10 (18).

3.1.7. Rules pertaining to the storage and handling of containers apply with equal force to the storage and handling of spheres and drums where the alternate use of these containers is authorized by DOT Regulations.

3.1.8. A checklist or operating procedure should be posted or made available for the guidance of container fillers.

3.2. Maintenance

3.2.1. Containers and their appurtenances shall be maintained only by the container owner or his authorized representative.

3.2.2. The prescribed markings stamped into containers shall not be removed or changed without authority from the Bureau of Explosives.

3.2.3. The user shall not deface or remove any markings, labels, decals, tags or stencil marks applied by the supplier and used for identification of content.

3.2.4. The user shall not change, modify, tamper with, obstruct or repair the pressure-relief devices in container valves or in containers.

3.2.5. The user shall not repair or alter containers or container valves.

3.2.6. *Painting.*

3.2.6.1. The user shall not paint containers unless authorized by the owner.

3.2.6.2. Containers may be painted by the gas suppliers to permit the suppliers to help recognize their containers and to segregate them more readily in their handling operations. Color shall not be used to identify container content. Medical gas containers shall be color marked in accordance with Section C (CGA Pamphlet C-9). (18)

3.2.7. The distributor or user shall notify the owner or supplier of the container if any condition has occurred which might permit any harmful foreign substance to enter the container or valve giving details and container serial number.

3.2.8. A leaking or defective container shall not be offered for shipment by common or contract carrier. Consult the supplier for advice under these circumstances.

3.2.9. Compressed gas containers that have been exposed to fires shall not be shipped if they still contain compressed gas. Consult the gas supplier under these circumstances. Fire-damaged cylinders must be inspected in accordance with CGA Pamphlet C-6 (16) and reconditioned when practical by the supplier in accordance with DOT Section 173.34 (f).

3.2.10. When containers or valves are severely corroded, the supplier shall be notified and his instructions followed.

3.2.11. Any other damage noted that might impair the safety of the container shall be called to the attention of the gas supplier before the return of the container.

3.3 Container Usage

3.3.1. Where the user is responsible for the handling of the container and connecting it for use, such containers shall carry a legible label or marking identifying the content. Refer to American National Standard Method of Marking Portable Compressed Gas Containers to Identify the Material Contained, Z48.1 (3), and CGA Pamphlet C-7 "Guide to the Preparation of Precautionary Labeling and Marking of Compressed Gas Containers." (4) Containers not bearing a legible label identifying the contents shall not be used and shall be returned to the gas supplier.

3.3.2. Where removable caps are provided by the gas supplier for valve protection, the user shall keep such caps on containers at all times except when containers are connected to dispensing equipment.

3.3.3. Containers shall not be used as rollers, supports or for any purpose other than to contain the content as received.

3.3.4. The user shall keep container valves closed at all times (charged or empty) except when the container is in use. By "in use" it is meant when gas is flowing from the container, when the container gas is maintaining pressure in a supply line or when the container is standing by during and between operations utilizing the gas.

3.3.5. Containers shall not be placed where they might become part of an electrical circuit. When the containers are used in conjunction with electric welding, compressed gas containers shall not be grounded. These precautions will prevent burning by electric welding arc.

3.3.6. Compressed gas containers shall not be subjected to an atmospheric temperature above 130 F (54.4 C). A flame shall never be permitted to come in contact with any part of a compressed gas container. If ice or snow accumulate on a container, thaw at room temperature, or thaw with water at temperature not exceeding 130 F (54.4 C).

3.3.7. Containers shall not be subjected to artificially created low temperatures without approval of the supplier. Many steels undergo significantly decreased impact resistance and ductility at low temperatures. The use of non-cryogenic liquefied gas containers at extremely low ambient temperatures shall require the use of check valves to prevent back flow due to the low vapor pressure of the liquefied gas.

3.3.8. If a container leaks, and the leak cannot be remedied by tightening a valve gland or packing nut, the valve shall be closed and a tag attached stating that the container is unserviceable. If the gas is toxic, utilize appropriate gas mask or self-contained breathing apparatus. If it is flammable, keep away from ignition sources. Remove the leaking container outdoors to a well-ventilated location or place under an exhaust ventilating system suitable for the product. If the gas is flammable or toxic, place an appropriate sign at the container warning against these hazards. The gas supplier shall be notified and his instructions followed as to the return or disposition of the container.

3.3.9. When in doubt about the proper handling of a compressed gas container or its content, the manufacturer or supplier of the gas shall be consulted.

3.3.10. Before returning empty containers, the valve shall be closed and container valve protection caps, if used, shall be replaced. Cylinders equipped with valve-outlet caps or plugs, or sealing valve caps shall be returned with these caps or plugs in a gas-tight condition. Empty containers shall be labeled or marked according to DOT Regulations.

3.3.11. Nonrefillable containers such as Specification DOT-2P, DOT-2Q and DOT-39 containers, shall not be refilled with any material after use of the original contents. After usage such containers shall be disposed of in accordance with the container manufacturers' or fillers' recommendations.

3.3.12. The use of oxygen and fuel-gas containers, including storage and handling by the user, when used for welding, cutting, brazing and allied processes, shall comply with the requirements of ANSI Z 49.1. (13)

3.4. Moving Containers

3.4.1. Containers shall not be dragged or slid. Where practical the user should use a suitable hand truck, fork truck, roll platform or similar device with container secured for transporting.

3.4.2. Containers shall not be dropped or permitted to strike against each other or other surfaces violently.

3.4.3. Lifting.

3.4.3.1. Caps shall not be used for lifting containers except for the use of hand trucks which grip the cylinder cap for lifting onto the hand truck. (This shall not be interpreted to prohibit cylinders with caps from being suspended during cylinder manufacturing operations, or tilting cylinders to an upright position.)

3.4.3.2. Magnets shall not be used for lifting containers.

3.4.3.3. Ropes, Chains or Slings shall not be used to suspend containers unless provisions at time of manufacture have been made on the container for appropriate lifting attachments, such as lugs.

3.4.3.4. Where appropriate lifting attachments have not been provided on the container, suitable cradles or platforms to hold the containers shall be used for lifting.

3.5. Storing Containers

3.5.1. Containers shall be stored in accordance with all state and local regulations and in accordance with appropriate standards of the Occupational Safety and Health Administration, the Compressed Gas Association and the National Fire Protection Association (see Part 7 References).

3.5.2. Container storage areas in users' facilities shall be prominently posted with the name of the gases to be stored.

3.5.3. Where gases of different types are stored at the same location, containers should be grouped by types of gas, and the groups arranged to take into account the gases contained. Full and empty containers should be stored separately with the storage layout so planned that containers comprising old stock can be removed first with a minimum handling of other containers.

3.5.4. Storage rooms shall be well ventilated and dry. Where practicable, storage rooms should be of fire-resistive construction. Storage room temperatures shall not exceed 130 F (54.4 C). Storage in subsurface locations should be avoided.

3.5.5. Containers shall not be stored near readily ignitable substances such as gasoline or waste, or near combustibles in bulk, including oil.

3.5.6. Containers should not be exposed to continuous dampness and should not be stored near salt or other corrosive chemicals or fumes. Corrosion may damage the containers and may cause the valve protection caps to stick.

3.5.7. Containers shall be protected from any object that will produce a harmful cut or other abrasion in the surface of the metal. Containers shall not be stored near elevators, gangways, unprotected platform edges or in locations where heavy moving objects may strike or fall on them.

3.5.8. The user shall store containers standing upright where they are not likely to be knocked over, or containers shall be secured in an upright or in a horizontal position.

3.5.8.1. Liquefied gas containers used in welding and cutting shall be stored and used valve end up. Acetylene cylinders should be stored and used valve end up. Storage of acetylene cylinders valve end up will minimize possibility of solvent being discharged. ("Valve end up" includes conditions where the container axis may be inclined as much as 45° from the vertical.)

3.5.9. Containers may be stored in the open but should be protected from the ground beneath to prevent bottom corrosion. Containers may be stored in the sun except in localities where extreme temperatures prevail. If the supplier recommends storage in the shade for a particular gas, such recommendation shall be observed.

3.5.10. Containers in public areas should be protected against tampering.

3.5.11. Containers when stored inside shall not be located near exits, stairways or in areas normally used or intended for the safe exit of people.

3.6. Connecting Container and Withdrawing Content

3.6.1. Compressed gases shall be handled only by properly trained persons.

3.6.2. The user responsible for the handling of the container and connecting it for use shall check the identity of the gas by reading the label or other markings on the container before using. If container content is not identified by marking, the container shall be returned to the supplier without using. Container color shall not be relied upon for content identification.

3.6.3. Users shall keep removable caps in place until connecting container to equipment.

3.6.4. The user shall secure containers while connected to a portable welding, cutting, brazing or heating appliance or other portable utilization equipment to prevent them from being knocked over.

3.6.5. A suitable pressure-regulating device shall be used where gas is admitted to a system of lower-pressure rating than the supply pressure, and where, due to the gas capacity of the supply source, the system pressure rating may be exceeded. (*Note:* This is a requirement regardless of the possible presence of a pressure-relief device protecting the system.)

3.6.6. A suitable pressure-relief device shall be used to protect a system utilizing a compressed gas where the system has a pressure rating less than the compressed gas supply source and where, due to the gas capacity of the supply source, the system pressure rating may be exceeded.

3.6.7. Connections that do not fit shall not be forced. Threads on regulator connections or other auxiliary equipment shall match those on container valve outlet. Detailed, dimensioned drawings of standard container valve outlet and inlet connections are published in the "American National and Canadian Standard Compressed Gas Cylinder Valve Outlet and Inlet Connections" (ANSI-B57.1 and CSA-B96). (5)

3.6.8. Where compressed gas containers are connected to a manifold, such a manifold and its related equipment, such as regulators, shall be of proper design.

3.6.9. Regulators, gages, hoses and other appliances provided for use with a particular gas or group of gases shall not be used on containers containing gases having different chemical properties unless information obtained from the supplier indicates that this can be done safely. As an example, only pressure-regulating devices approved for use with oxygen shall be used in oxygen service.

3.6.10. Container valve shall be opened slowly. Valve outlets shall be pointed away from yourself and other persons. On valves without handwheels, the wrenches provided by, or recommended by, the gas supplier shall be used. On valves with handwheels, wrenches shall not be used, except when designed specifically for that purpose. Valve wheels shall not be hammered in attempting to open or close the valve. For valves that are hard to open, or frozen because of corrosion, the supplier shall be contacted for instructions.

3.6.11. Compressed gas shall not be used to dust off clothing. This may cause serious injury to the eyes or body, or create a fire hazard.

3.6.12. Compressed gases shall not be used where the container may be contaminated by the feedback of process materials unless protected by suitable traps or check valves.

3.6.13. Connections to piping, regulators, and other appliances shall be kept tight to prevent leakage. Where hose is used, it shall be kept in good condition.

3.6.14. Before a regulator is removed from a container, the container valve shall be closed and the regulator drained of gas pressure.

4. SAFE HANDLING RULES FOR COMPRESSED GAS CONTAINERS (by service classification)

4.1. Flammable Gases

4.1.1. Storage Outdoors or in Separate Building (without other occupancy):

4.1.1.1. Provisions should be made to protect flammable gases from hazardous exposure to and against hazardous exposure from adjoining buildings, equipment, property and concentrations of people.

4.1.1.2. Ventilating, Heating and Electrical Equipment in Separate Buildings:

4.1.1.2.1. Heating shall be by steam, hot water or other indirect means. Heating by flames or fire shall be prohibited.

4.1.1.2.2. Electrical equipment shall conform to the provisions of National Electric Code (NFPA No. 70), article 501 for Class 1 Division 2 locations.

4.1.1.2.3. Sources of ignition shall be prohibited.

4.1.1.2.4. Separate buildings shall be well ventilated.

4.1.2. Storage in Separate Rooms (without other occupancy):

4.1.2.1. The walls, partitions and ceilings of separate rooms shall be of noncombustible construction having a fire-resistance rating of at least one hour, and in accordance with the rating requirements of the NFPA standard applicable to the flammable gas involved. The walls or partitions shall be continuous from floor to ceiling and shall be securely anchored. At least one wall of the room shall be an exterior wall. Openings to other parts of the building shall be protected by a self-closing fire door for a Class B opening and having a fire-resistance rating of at least one hour. Windows in partitions shall be wired glass in metal frames with fixed sash. Installation shall be in accordance with the Standard for the Installation of Fire Doors and Windows (NFPA No. 80).

4.1.2.2. Separate rooms shall be well ventilated. Heating systems, electrical equipment and control of sources of ignition shall comply with the following:

4.1.2.2.1. Heating shall be by steam, hot water or other indirect means. Heating by flames or fire shall be prohibited.

4.1.2.2.2. Electrical equipment shall conform to the provisions of National Electrical Code (NFPA No. 70), Article 501 for Class 1 Division 2 locations.

4.1.3. Storage in Building and Rooms (with other occupancy):

4.1.3.1. Flammable gas containers stored inside of buildings with other occupancies shall be kept at least 20 ft (6.1 m) from flammable liquids and from highly combustible materials and similar substances, and not near arcing electrical equipment, open flames or other sources of ignition.

4.1.3.2. Flammable gas containers stored inside of industrial buildings at consumer sites, except those in use or those attached for use, shall be limited to a total gas capacity of 2000 cu ft (56.6 m³) of acetylene or nonliquefied flammable gas, or a total water capacity of 735 lb (333 kg) for liquefied petroleum gas or methylacetylene-propadiene stabilized. [*Note:* 735 lb (333 kg) water capacity is equivalent to about 309 lb (140 kg) of propane, 368 lb (167 kg) of methylacetylene-propadiene, stabilized, or 375 lb (170 kg) of butane.]

4.1.4. Fire Protection.

4.1.4.1. Adequate portable fire extinguishers of carbon dioxide or dry chemical types shall be available for fire emergencies at storage installations.

4.1.4.2. No smoking signs shall be posted around the storage area of buildings, or at entrance to special storage rooms.

4.1.4.3. A flame shall not be used for detection of flammable gas leaks. Flammable gas leak detector, soapy water, or other suitable solution shall be used.

4.1.5. Gaseous hydrogen systems at consumer sites shall comply with NFPA No. 50A. (9)

4.1.6. The usage of welding and cutting processes shall comply with NFPA No. 51B. (11) and ANSI Z49.1 (13).

4.1.7. Liquefied petroleum gases shall be stored and handled in compliance with NFPA No. 58. (12)

4.1.8. Liquefied natural gas shall be stored and handled in compliance with NFPA No. 59A. (19)

4.2. Oxygen (Including Oxidizing Gases)

4.2.1. Oxygen containers, valves, regulators, hose and other oxygen apparatus shall be kept free from oil or grease and shall not be handled with oily hands, oily gloves or with greasy equipment.

4.2.2. Oxygen containers in storage shall be separated from flammable gas containers or combustible materials (especially oil or grease) by a minimum distance of 20 ft (6.1 m) or by a noncombustible barrier at least 5 ft (1.5 m) high having a fire-resistance rating of at least one-half hour.

4.2.3. An oxygen manifold or oxygen bulk storage system at consumer sites and which has storage capacity of more than 20,000 cu ft (600 m³) of oxygen (measured at 14.7 psia at 70 F or 101 kPa at 21.1 C) including unconnected reserves at the site, shall comply with the provisions of the Standard for Bulk Oxygen Systems at Consumer Sites, NFPA No. 50. (8)

4.2.4. The oxygen concentration in work areas, other than in hyperbaric chambers, shall not exceed 23 percent by volume.

4.3. Acid and Alkaline Gases

4.3.1. Precautions shall be taken to avoid contacting the skin or eyes with acid or alkaline gases. Goggles or face shields, rubber (or other suitable chemically resistant material) gloves and aprons shall be worn. Long sleeves and trousers shall be worn. Open shoes or sneakers shall be prohibited.

4.3.2. Personnel handling and using acid and alkaline gases shall have available for immediate use in emergencies, gas masks or self-contained breathing apparatus of a design approved by U.S. Bureau of Mines, National Institute of Occupational Safety and Health or other approving authority for the particular gas service desired. Gas masks may be used only under conditions where the concentration of the acid or alkaline gas in excess of the gas mask rating will not be encountered and where the oxygen content of the atmosphere is not less than 19 percent by volume. Such equipment shall be located conveniently to the place of work, but kept out of the area most likely to be contaminated.

4.3.3. Areas in which acid or alkaline gases are filled or utilized shall be equipped with an emergency shower and eyewash fountain. Drenching with copious amounts of water is the accepted first-aid procedure in event of exposure of corrosive gases to the skin or eyes.

4.3.4. Total quantity of acid and alkaline gases on the users' sites should be limited to their foreseeable requirements.

4.3.5. Acid and alkaline gases should be utilized in a well-ventilated area.

4.3.6. The following is a partial listing of the acid and alkaline gases:

ammonia, boron trifluoride, chlorine, dimethylamine, ethylamine, fluorine, hydrogen bromide, hydrogen chloride, hydrogen sulfide, methylamine, nitrosyl chloride, sulfur dioxide and trimethylamine.

(Note: The following is a partial list of acid or alkaline liquids that are shipped in compressed gas containers: boron trichloride, hydrogen fluoride and hydrogen iodide.)

4.4. Highly Toxic Gases

4.4.1. Personnel handling and using highly toxic gases shall have available for immediate use in emergencies, gas masks or self-contained breathing apparatus of a design approved by U.S. Bureau of Mines, National Institute of Occupational Safety and Health, or other approving authority for the particular gas service desired. Gas masks may be used only under conditions where the concentrations of the toxic gas in excess of the gas mask rating will not be encountered, and where the oxygen content of the atmosphere is not less than 19 percent by volume. Such equipment shall be located conveniently to the place of work, but kept out of the area most likely to be contaminated.

4.4.2. Storage of highly toxic gases shall be outdoors, or in a separate noncombustible building without other occupancy, or in a separate room without other occupancy and of noncombustible construction with a fire-resistance rating of at least one hour. Storage locations shall be protected against tampering.

4.4.3. Highly toxic gases shall be filled and utilized only in forced ventilated areas or, preferably, in hoods with forced ventilation, or outdoors. Highly toxic gases emitted from equipment in high concentration shall be discharged into appropriate scrubbing equipment which will remove such toxic gases from effluent gas streams.

4.4.4. Before using a highly toxic gas, read all the label and data sheet information associated with the use of the particular highly toxic gas. All personnel working in the immediate area where these gases are handled shall be instructed as to the toxicity of the gases and methods of protection against harmful exposure and first-aid treatment in case of exposure. Personnel shall not be exposed to concentrations of highly toxic gases in excess of the time-weighted threshold limit values (TLV) as established by the Occupational Safety and Health Administration.

4.4.5. The total quantity of highly toxic gases on the users' sites should be limited to their foreseeable requirements.

4.4.6. The following are commonly handled highly toxic gases which have established threshold limit values of one part per million or less or require a DOT poison gas label:

carbonyl fluoride, chlorine, diborane, fluorine, germane, hydrogen cyanide, hydrogen selenide, nickel carbonyl (liquid), nitric oxide, nitrogen dioxide, ozone, phosgene, phosphine and stibine.

4.4.7. Because of the hazardous nature of highly toxic gases, persons handling such gases are advised to contact the supplier for more complete information with regard to usage and first-aid.

4.5. Liquefied Cold Compressed Gases

4.5.1. Liquefied cold compressed gases are gases which are handled in liquid form at relatively low pressures and extremely low temperatures, usually below -150 F (-101.1 C). Because of their low temperatures, liquefied cold compressed gases are handled in multiwall, vacuum-insulated cylinders, tank trucks, tank cars and storage tanks to minimize evaporation and venting of the gas. Some liquefied cold compressed gases in small quantities are also handled in open and in low-pressure thermostype containers in laboratory work. These containers are not pressurized sufficiently to bring them under the jurisdiction of the DOT Regulations covering compressed gases but they are regulated by the ATA (domestic) and the IATA (foreign) tariffs when transported by air.

4.5.2. Liquefied cold compressed gases and cold gases can cause frostbite injury upon contact with the body. When handling liquefied cold compressed gases, suitable eye protection, such as a face shield, safety glasses or safety goggles shall be worn to protect against the extremely cold liquid and gas. Hand protection, such as asbestos or leather gloves, shall be worn to prevent contact with cold liquid, cold gas and cold equipment or piping.

4.5.3. Liquefied cold compressed gas containers shall be stored and handled in well-ventilated areas to prevent excessive concentrations of the gas. Containers are equipped with pressure-relief devices which permit venting of gas intentionally.

4.5.4. Liquefied cold compressed gas containers shall be handled and stored in an upright position. The containers must not be dropped, tipped over or rolled on their sides.

4.5.5. Containers and equipment designed for a specific liquefied cold compressed gas service shall not be used for the storage of another liquefied cold compressed gas unless such service is approved by the supplier or by the container or equipment manufacturer.

4.5.6. Liquefied cold compressed gas containers shall be provided with pressure-relief devices adequate to prevent excessive pressures within the containers. Piping shall be equipped with pressure-relief devices where liquefied cold compressed gases or cold gas may be trapped between valves.

4.5.7. Only transfer lines designed for liquefied cold compressed gases shall be used. Transfer of liquefied cold compressed gases shall be performed slowly enough to minimize rapid cooling and contraction of warm containers and equipment.

4.5.8. Liquid oxygen containers, piping and equipment shall be kept clean and free of grease, oil and organic materials. Smoking and open flames shall not be permitted in areas where liquid oxygen is stored or transferred. Liquid oxygen systems at consumer sites shall comply with NFPA No. 50. (8)

4.5.9. Smoking, open flames and general-purpose electrical equipment shall be prohibited where liquid hydrogen is stored or handled. Liquid hydrogen shall be stored and transferred under positive pressure to prevent the infiltration and solidification of air or other gases.

4.5.9.1. Liquid hydrogen systems at consumer sites shall comply with NFPA No. 50B. (10)

4.5.10. Liquid helium and liquid neon shall be stored and transferred under positive pressure to prevent the infiltration and solidification of air and other gases.

4.5.11. Smoking, open flames and general-purpose electrical equipment shall be prohibited where liquefied natural gas is stored or handled. Liquefied natural gas shall be stored and transferred under positive pressure to prevent the infiltration of air or other gases.

4.5.11.1. Liquefied natural gas systems at utility plants and consumer sites shall comply with NFPA No. 59A. (19)

4.6. Inert Gases

Inert gases, such as argon, carbon dioxide, helium, krypton, neon, nitrogen and xenon, are simple asphyxiants which can displace the oxygen in air necessary to sustain life and can cause rapid suffocation due to oxygen deficiency. Self-contained breathing apparatus or air-line masks shall be worn in areas containing an oxygen-deficient atmosphere where the oxygen concentration is less than 19 percent by volume.

5. GENERAL PRECAUTIONS FOR TANK CARS

5.1. Shipment of compressed gases in tank cars shall comply with Part 173.314 of the DOT Regulations on Requirements for Compressed Gases in Tank Cars.

5.2. Qualification, maintenance, and use of tank cars with compressed gases shall comply with Part 173.31 of the DOT Regulations on Qualification, Maintenance and Use of Tank Cars.

5.3. Placarding and handling of tank cars with compressed gases shall comply with Part 174 of the DOT Regulations on Carriers by Rail Freight.

5.4. Tank cars for use with compressed gas shall comply with Part 179 of the DOT Regulations on Specifications for Tank Cars.

5.5. Care must be taken that compressed gases are loaded only in tank cars designed and suitable for the particular gas to be charged. Before a tank car can be charged with a compressed gas other than that for which commodity use has been approved by the Tank Car Committee of the Association of American Railroads, approval must be obtained from that committee by the car owner or party authorized by the owner.

5.6. Shippers' detailed instructions and diagrams for unloading should always be followed and all caution markings on the tank or dome must be read and observed. Angle valves should be opened slowly to avoid closing of excess-flow valves in the eduction pipe. The use of a hammer on valve or cover plate to release an excess-flow valve shall be avoided.

5.7. Unloading operations should be carefully supervised and should be performed only by reliable persons properly instructed and made responsible for careful compliance with all safety regulations. Operators should be provided with proper personal protective equipment.

5.8. Safe handling of DOT-106A and 110A tanks ("Ton containers") should follow the recommendations as specified for other containers where applicable.

5.9. Never tamper with the pressure-relief devices or the valves on tank cars. In the event of a leak in the tank car or fittings which cannot be stopped by simple adjustments, isolate the car if practicable, and telephone the supplier for instructions.

5.10. When possible, railway sidings on which compressed gas tank cars are placed for unloading should be level and be devoted solely to this purpose.

5.11. Derails should be placed at one or both ends of the unloading track approximately one car length from the car being unloaded, unless the car is protected by a closed and locked switch or gate.

5.12. Cars should be electrically grounded before unloading if content is flammable.

5.13. For further information on tank car

unloading, refer to "How to Receive and Unload Liquefied Compressed Gases." (7)

6. CARGO TANKS AND PORTABLE TANKS

6.1. Shipment of compressed gases in cargo tanks and portable tanks shall comply with Part 173.315 of the DOT Regulations on Compressed Gases in Cargo Tanks and Portable Tank Containers.

6.2. Qualification, maintenance and use of cargo tanks with compressed gas shall comply with Part 173.33 of the DOT Regulations on Cargo Tank Use Authorization.

6.3. Qualification, testing, maintenance and use of portable tanks with compressed gas shall comply with Part 173.32 of the DOT Regulations on Qualification, Testing, Maintenance and Use of Portable Tanks.

6.4. Placarding and handling of cargo tanks and portable tanks with compressed gas shall comply with Part 177 of the DOT Regulations on Shipments Made by Way of Common, Contract, or Private Carriers by Public Highway.

6.5. Cargo tanks for use with compressed gas shall comply with Subpart J of Part 178 of the DOT Regulations on Specifications for Containers for Motor Vehicle Transportation.

6.6. Portable tanks for use with compressed gas shall comply with Subpart H of Part 178 of the DOT Regulations on Specifications for Portable Tanks.

6.7. Cargo tanks mounted on motor vehicles are normally not handled by the gas user. However, in cases where the cargo tank is handled by the user, he shall consult the gas supplier for instructions on safe handling procedures.

7. REFERENCES

(1) "Title 49, Code of Federal Regulations Parts 100 to 199, U.S. Department of Transportation." Available from Superintendent of Documents, U.S. Government Printing Office, Washington, D.C. 20402 or republished by the Bureau of Explosives as "Hazardous Materials Regulations of the Department of Transportation including Specifications for Shipping Containers," available from the Bureau of Explosives, 1920 L Street NW, Washington, DC 20036.

(2) "Transportation of Dangerous Articles and Magnetized Materials," the Federal Aviation Administration, Washington, DC 20553.

(3) "American National Standard Method of Marking Portable Compressed Gas Containers to Identify the Material Contained." ANSI Z48.1 (Pamphlet C-4), Compressed Gas Association, Inc., and American National Standards Institute, 1430 Broadway, New York, NY 10018.

(4) "A Guide to the Preparation of Precautionary Labeling and Marking of Compressed Gas Containers," (Pamphlet C-7), Compressed Gas Association, Inc.

(5) "American National-Canadian Standard Compressed Gas Cylinder Valve Outlet and Inlet Connections," (Pamphlet V-1), Compressed Gas Association, Inc., and ANSI B57.1, American National Standards Institute, 1430 Broadway, New York, NY 10018.

(6) "Oxygen-Fuel Gas Systems for Welding and Cutting," NFPA Standard No. 51, National Fire Protection Association, 470 Atlantic Avenue, Boston, Mass. 02210.

(7) "How to Receive and Unload Liquefied Compressed Gases," Chapter 8 of this book.

(8) "Standard for Bulk Oxygen Systems at Consumer Sites," NFPA No. 50, National Fire Protection Association, 470 Atlantic Avenue, Boston, Mass. 02210.

(9) "Standard for Gaseous Hydrogen Systems at Consumer Sites," NFPA No. 50A, National Fire Protection Association, 470 Atlantic Avenue, Boston, Mass. 02210.

(10) "Standard for Liquefied Hydrogen Systems at Consumer Sites," NFPA No. 50B, National Fire Protection Association, 470 Atlantic Avenue, Boston, Mass. 02210.

(11) "Standard for Fire Prevention in Use of Cutting and Welding Processes," NFPA No. 51B, National Fire Protection Association, 470 Atlantic Avenue, Boston, Mass. 02210.

(12) "Standard for the Storage and Handling of Liquefied Petroleum Gases," NFPA No.

58, National Fire Protection Association, 470 Atlantic Avenue, Boston, Mass. 02210; and ANSI Z106.1, American National Standards Institute, 1430 Broadway, NY 10018.

(13) "Standard on Safety in Welding and Cutting," ANSI Z49.1, American Welding Society, 2501 Northwest 7th Street, Miami, Fla. 33125 and American National Standards Institute, 1430 Broadway, New York, NY 10018.

(14) "Matheson Gas Data Book," Matheson Gas Products, Division of Will Ross, Inc., East Rutherford, NJ 07073.

(15) "Effects of Exposure to Toxic Gases— First Aid and Medical Treatment," Matheson Gas Products, Division of Will Ross, Inc., East Rutherford, NJ 07073.

(16) "Standards for Visual Inspection of Compressed Gas Cylinders," (Pamphlet C-6), Compressed Gas Association, Inc.

(17) "Recommendations for Change of Service for Compressed Gas Cylinders Including Procedures for Inspection and Contaminant Removal," (Pamphlet C-10), Compressed Gas Association, Inc.

(18) "Standard Color-Marking of Compressed Gas Cylinders Intended for Medical Use in the United States," (Pamphlet C-9), Compressed Gas Association, Inc.

(19) "Standard for the Production, Storing and Handling of Liquefied Natural Gas (LNG)," NFPA No. 59A, National Fire Protection Association, 470 Atlantic Ave., Boston, Mass. 02210.

SECTION B

Oxygen-Deficient Atmospheres

The normal oxygen content of air is approximately 21 percent. Depletion of oxygen content in air, either by combustion or by displacement with inert gas, is a potential hazard to personnel throughout industry. A poorly recognized aspect of this hazard is the response of humans when exposed to an atmosphere as low in oxygen as 8 to 12 percent. In this case, unconsciousness can be immediate and without warning.

When the oxygen content of air is reduced to about 15 or 16 percent, the flame of ordinary combustible materials, including those commonly used as fuel for heat or light, will be extinguished. Somewhat below this concentration an individual breathing the air is mentally incapable of diagnosing the situation, as the symptoms of sleepiness, fatigue, lassitude, loss of coordination, errors in judgement and confusion will be masked by a state of "euphoria," giving the victim a false sense of security and well being.

Human exposure to atmospheres containing 12 percent or less oxygen, will bring about unconsciousness without warning and so quickly that the individual cannot help or protect himself. This is true if the condition is reached by immediate change of environment or by gradual depletion of oxygen.

Most personnel working in or around oxygen-deficient atmospheres rely on the buddy system for protection but the buddy is equally liable to asphyxiation if he enters the area to rescue his unconscious partner unless he is equipped with a portable air supply. The best protection is obtained by providing both the worker and his buddy with a portable supply of respirable air. Life lines are acceptable only if the area is free of obstructions and the buddy is capable of lifting his partner's weight rapidly and without straining himself. In practice this has seldom been possible.

If an oxygen-deficient atmosphere is known to exist (as can be determined by analysis) or is suspected take the following steps:

Use the buddy system. Use more than one buddy if required to remove a worker in an emergency.

Provide both the worker and his buddy with self-contained or air-line breathing equipment.

SECTION C

Safe Handling Requirements for Gases Used Medicinally

1. INTRODUCTION

1.1. This section relates to compressed medical gases and their containers. Its purpose is to promote safe practices, not only for the protection of those persons handling medical gases, but also for the general public as well, in order that these gases may be made available under proper and safe conditions.

1.2. This section has been prepared for the convenience and guidance of the medical and dental professions, their co-workers and all other personnel engaged in, or responsible for, the handling of compressed medical gases. In a facility which uses these products, the handling of compressed medical gases must be the responsibility of well-trained and reliable persons who are fully informed as to the potential hazards of these commodities to themselves and to others. In order to familiarize such personnel with compressed medical gases and their handling such personnel should study and follow the practices and principles herein outlined.

1.3. In order to present adequately these safe practice recommendations, particularly for the guidance of those whose knowledge of this subject may be limited, there are included herein certain related descriptive data regarding properties of the medical gases as well as information on compressed gas containers and regulations governing these commodities.

1.4. This section does not purport to give any guidance for the administration of medical gases which is quite properly the responsibility of the medical and dental professions.

2. MEDICAL GASES AND THEIR PROPERTIES

2.1. General

2.1.1. Medical gases are prepared under carefully controlled conditions to meet the purity specifications prescribed in the United States Pharmacopeia (USP). They are normally shipped under pressure in metal containers in accordance with the Regulations of the Department of Transportation (DOT).

2.1.2. In cylinders containing liquefied compressed gas and vapor in equilibrium, the pressure in the container is determined solely by the vapor pressure of the contained liquid at the equilibrium temperature. In common usage, cylinder pressure will drop if gas is used continuously, due to cooling of the liquid as gas is vaporized. However, if liquid remains when withdrawal stops, cylinder pressure will slowly return to the pressure corresponding to the temperature of the cylinder contents. The pressure in the cylinder charged with a liquefied compressed gas is therefore not an indication of the amount of gas remaining in the cylinder. The cylinder contents can be determined by weight or other suitable contents indicators.

2.1.3. In cylinders charged with a nonliquefied gas, the pressure in the container is related both to temperature and the amount of gas in the container. For nonliquefied compressed gases such as oxygen, helium, carbon dioxide-oxygen mixtures or helium-oxygen mixtures, cylinder content may be determined by pressure; i.e., at a given temperature when the pressure is reduced to half the original pressure the cylinder will be approximately half full. Cylinder contents may also be determined by weight using appropriate weight-volume conversion factors.

2.1.4. Some of the properties and pertinent facts about currently used medical gases and their mixtures are given in 2.2 through 2.10.

2.2. Air

Chemical formula	Mixture of nitrogen and oxygen with oxygen content between 19.5 and 23.5 percent.
Molecular weight	29 (approximately)
Color	Colorless
Odor	Odorless
Life support capability	Will support life. Grade D, Type I, air is the minimum grade to be used for general respiratory use. See CGA Pamphlet G-7 "Compressed Air for Human Respiration."
Physical state in full cylinder	Nonliquefied gas in conventional compressed gas cylinders such as DOT-3A or 3AA.
Number of liters in 1 kg at normal atmospheric pressure and 70 F	833 (approximately)
Number of liters in 1 oz at normal atmospheric pressure and 70 F	24 (approximately)
Density (kg/m^3; 70 F; 1 atm)	1.20 (approximately)
Density (lb/cu ft; 70 F; 1 atm)	0.0749 (approximately)
Combustion characteristics	Nonflammable; supports combustion.
Usual methods of manufacture	Compressed from surrounding air after drying and purifying or by mixing USP oxygen and USP nitrogen. *Note:* Where oxygen and nitrogen are mixed to make "air," the filling agency should verify the oxygen content of each cylinder.
Normal cylinder filling limit	Nonliquefied gas; 1900–2200 psig at 70 F (13,100–15,170 kPa at 21 C), depending upon the type of cylinder.
Pressure in normally charged cylinders	Nonliquefied gas; pressure in cylinders of air will vary as described in 2.1.3. At any given temperature the pressure will decrease proportionately as the cylinder contents are withdrawn.

2.3. Carbon Dioxide

Chemical formula	CO_2
Molecular weight	44.010
Color	Colorless
Odor	Pungent
Life support capability	Will not support life. Forms acid with body fluids.
Physical state in full cylinder	Liquefied gas below 88 F (31 C); nonliquefied gas at 88 F (31 C) and higher.
Number of liters in 1 kg at normal atmospheric pressure and 70 F	545
Number of liters in 1 oz at normal atmospheric pressure and 70 F	15.45

Specific gravity of gas compared to air	1.522
Density (kg/m^3; 70 F; 1 atm)	1.834
Density (lb/cu ft; 70 F; 1 atm)	0.1144
Combustion characteristics	Nonflammable; does not support combustion.
Usual methods of manufacture	1. Absorption and recovery from product of combustion of various carbonaceous materials.
	2. From natural sources such as springs or water containing the gas in solution.
	3. By thermal decomposition of carbonate.
	4. By action of acid on carbonates.
	5. As a by-product in fermentation processes.
Normal cylinder filling limit	68 percent of the weight of water that the cylinder will hold. (Normal industry practice has been to fill G carbon dioxide cylinders to 50 lb (22.7 kg).)
Pressure in normally charged cylinders	Pressure in cylinders of carbon dioxide charged to a filling density of 68 percent will vary with temperature approximately as shown in Table 1. At temperatures below 88 F (31 C) the pressure will drop with continuous withdrawal of gas. However, if liquid remains when withdrawal stops, cylinder pressure will slowly return to the pressure corresponding to the temperature of the cylinder contents. At temperatures of 88 F (31 C) or higher, the pressure at a given temperature will decrease proportionately as the cylinder contents are withdrawn.

TABLE 1

Temperature (F)	Approximate Gage Pressure in psig in a Cylinder of Carbon Dioxide When Both Liquid and Vapor Are Present
50	638
60	733
70	838
80	955
88 critical point	1057

2.4. Cyclopropane

Chemical formula	C_3H_6
Molecular weight	42.08
Color	Colorless
Odor	Characteristic
Life support capability	Strong anesthetic. Will not support life.

Physical state in full cylinder — Liquefied gas

Number of liters in 1 kg at normal atmospheric
pressure and 70 F — 574

Number of liters in 1 oz at normal atmospheric
pressure and 70 F — 16.28

Specific gravity of gas compared to air — 1.48

Density (kg/m³; 70 F; 1 atm) — 1.75

Density (lb/cu ft; 70 F; 1 atm) — 0.109

Combustion characteristics at atmospheric
pressure — Flammable

Ignition limits in air — 2.40–10.3 percent by volume

Ignition limits in oxygen — 2.48–60.0 percent by volume

Usual method of manufacture — Reduction of trimethylene chlorobromide in presence of zinc.

Normal cylinder filling limit — 55 percent of the weight of water that the cylinder will hold.

Pressure in normally charged cylinders — Pressure in cylinders of cyclopropane will vary with temperature approximately as shown in Table 2 as long as liquid and vapor are both present. The pressure will drop with continuous withdrawal of gas. However, if liquid remains when withdrawal stops, cylinder pressure will slowly return to the pressure corresponding to the temperature of the cylinder contents.

TABLE 2

Temperature (F)	Approximate Gage Pressure in psig in a Cylinder of Cyclopropane When Both Liquid and Vapor Are Present
50	53
60	65
70	79
80	86
90	116
100	134
110	154
120	174
130	199

2.5. Helium

Chemical formula — He

Atomic weight — 4.003

Color — Colorless

Odor — Odorless

Life support capability — Will not support life.

Physical state in cylinder — Nonliquefied gas in conventional compressed gas cylinder such as DOT-3A or 3AA.

Number of liters in 1 kg at normal atmospheric pressure and 70 F	6037
Number of liters in 1 oz at normal atmospheric pressure and 70 F	171.15
Specific gravity of gas compared to air	0.138
Density (kg/m^3; 70 F; 1 atm)	0.1656
Density (lb/cu ft; 70 F; 1 atm)	0.01034
Combustion characteristics	Nonflammable; does not support combustion.
Usual method of manufacture	From natural gas through a process of liquefaction and purification.
Normal cylinder filling limit	Nonliquefied gas; 1600 psig at 70 F (11,030 kPa at 21 C), depending upon the type of cylinder.
Pressure in normally charged cylinders	Pressure in cylinders of helium will vary as described in 2.1.3. At any given temperature, the pressure will decrease proportionately as the cylinder contents are withdrawn.

2.6. Nitrogen

Chemical formula	N_2
Molecular weight	28.013
Color	Colorless
Odor	Odorless
Life support capability	Will not support life.
Physical state in full cylinder	Nonliquefied gas in conventional compressed gas cylinders such as DOT-3A or 3AA. Pressurized liquid gas in DOT-4L cylinders.
Number of liters in 1 kg at normal atmospheric pressure and 70 F	861
Number of liters in 1 oz at normal atmospheric pressure and 70 F	24.42
Specific gravity of gas compared to air	0.967
Density (kg/m^3; 70 F; 1 atm)	1.1605
Density (lb/cu ft; 70 F; 1 atm)	0.07245
Combustion characteristics	Nonflammable; does not support combustion.
Usual method of manufacture	Separation from air.
Normal cylinder filling limit	Nonliquefied gas; 1900–2200 psig at 70 F (13,100–15,170 kPa at 21 C), depending upon the type of cylinder.
Pressure in normally charged cylinders	Nonliquefied gas; pressure in cylinders of nitrogen will vary as described in 2.1.3. At any given temperature the pressure will decrease proportionately as the cylinder contents are withdrawn. Pressurized liquid gas; up to 235 psig/(1620 kPa) in DOT-4L cylinders. Pressure in cylinder during normal operation is 75 psig/(520 kPa). During nonusage, pressure will increase very slowly over a period of 3 to 5 days to 235

psig/(1620 kPa), after which the cylinder will vent gas at a rate of 2 to 5 cu ft per hour (cfh) (0.06–0.14 m^3 per hour).

2.7. Nitrous Oxide

Chemical formula	N_2O
Molecular weight	44.013
Color	Colorless
Odor	Odorless
Life support capability	Anesthetic. Will not support life.
Physical state in full cylinder	Liquefied gas below 98 F/(37 C). Nonliquefied gas at 98 F/(37 C) and higher.
Number of liters in 1 kg at normal atmospheric pressure and 70 F	545
Number of liters in 1 oz at normal atmospheric pressure and 70 F	15.44
Specific gravity of gas compared to air	1.529
Density (kg/m^3; 70 F; 1 atm)	1.8357
Density (lb/cu ft; 70 F; 1 atm)	0.1146
Combustion characteristics	Nonflammable; supports combustion.
Usual method of manufacture	Thermal decomposition of ammonium nitrate.
Normal cylinder filling limit	68 percent of the weight of water that the cylinder will hold.
Pressure in normally charged cylinders	Pressure in cylinders of nitrous oxide charged to a filling density of 68 percent will vary with temperature approximately as shown in Table 3. In any cylinder of nitrous oxide at temperatures below 98 F (37 C), the pressure will drop with continuous withdrawal of gas. However, if liquid remains when withdrawal stops, the cylinder pressure will slowly return to the pressure corresponding to the temperature of the cylinder. At temperatures of 98 F (37 C), or higher, the pressure at a given temperature will decrease proportionately as the cylinder contents are withdrawn.

TABLE 3

Temperature (F)	Approximate Gage Pressure in psig in a Cylinder of Nitrous Oxide When Both Liquid and Vapor Are Present
50	575
60	660
70	745
80	850
90	960
97.50 critical point	1054

2.8. Oxygen

Chemical formula	O_2
Molecular weight	31.999
Color	Colorless
Odor	Odorless
Life support capability	Life supporting
Physical state in the cylinder	Nonliquefied gas in conventional compressed gas cylinders such as DOT-3A or 3AA. Pressurized liquid gas in DOT 4L cylinders.
Number of liters in 1 kg at normal atmospheric pressure and 70 F	754
Number of liters in 1 oz at normal atmospheric pressure and 70 F	21.4
Specific gravity of gas compared to air	1.1049
Density (kg/m^3; 70 F; 1 atm)	1.326
Density (lb/cu ft; 70 F; 1 atm)	0.08279
Combustion characteristics	Nonflammable; supports combustion.
Usual method of manufacture	Separation from air.
Normal cylinder filling limit	Nonliquefied gas; 1900–2200 psig at 70 F (13,100–15,170 kPa at 21 C), depending upon the type of cylinder.
Pressure in normally charged cylinders	Nonliquefied gas; pressures in cylinders of oxygen will vary as described in 2.1.3. At any given temperature the pressure will decrease proportionately as the cylinder contents are withdrawn.
	Pressurized liquid gas; up to 235 psig/(1620 kPa) in DOT-4L cylinders. Pressure in cylinder during normal operation is 75 psig/(520 kPa). During nonusage pressure will increase very slowly over a period of 3 to 5 days to 235 psig, after which the cylinder will vent gas at a rate of 2 to 5 cu ft per hour (cfh) (0.06–0.14 m^3 per hour).

2.9. Helium-Oxygen Mixture

2.9.1. Mixtures of helium and oxygen are supplied in a number of different compositions, the most usual mixture containing 80 percent helium and 20 percent oxygen. The mixtures exist in the cylinder as nonliquefied gas of homogeneous composition. They are ordinarily charged in cylinders to a pressure of 1600 psig at 70 F (11,030 kPa at 21 C). At any given temperature the pressure will decrease proportionately as the cylinder contents are withdrawn. Mixture in excess of 80 percent helium may not support life.

Note: The oxygen content of each cylinder should be verified by the filling agency.

2.9.2. One ounce of a mixture of 80 percent helium and 20 percent oxygen by volume is equivalent to about 71 liters at normal atmospheric pressure and temperature.

2.10. Carbon Dioxide-Oxygen Mixtures

2.10.1. Mixtures of carbon dioxide and oxygen are generally supplied in a number of different compositions ranging from 1 percent carbon dioxide-99 percent oxygen to 30 percent carbon dioxide-70 percent oxygen. These mixtures exist in the cylinder as nonliquefied gas of homogeneous composition. They are ordinarily charged in cylinders to a pressure of 1800–1900 psig at 70 F (12,410–13,100 kPa at 21 C), depending upon the percentage of carbon dioxide. At any given temperature the pressure will decrease proportionately as the cylinder contents are withdrawn. These mixtures may be injurious or nonlife supporting, depending upon concentration and duration of inhalation.

Note: The component content of each mixture should be verified by the filling agency in either of two ways: (1) the carbon dioxide content of each cylinder should be verified and the oxygen content confirmed in such a manner that verifies the integrity of each lot filled, or (2) the oxygen content of each cylinder should be verified and the carbon dioxide content confirmed in such a manner that verifies the integrity of each lot filled.

2.10.2. One ounce of the mixtures within the range of compositions given above is equivalent to about 21 liters at normal atmospheric pressure and temperature. The factor varies with the composition of the mixture.

3. RULES AND REGULATIONS PERTAINING TO MEDICAL GASES

3.1. DOT Regulations and Specifications

3.1.1. General.

3.1.1.1. The gases described in the foregoing pages are classified by the Department of Transportation as compressed gases, and as such their transportation is regulated by the United States Government under the provisions of an Act of Congress, dated September 6, 1960. Under this Act, cylinders in which medical gases are shipped in commerce subject to the jurisdiction of the Department of Transportation must comply with DOT specifications (formerly identified as ICC specifications). These specifications require, among other things, that the steel used in cylinders must meet certain chemical and physical requirements and that cylinders must pass a hydrostatic pressure test. In Canada, cylinders in which medical gases are shipped

must comply with the specifications of the Canadian Transport Commission.

3.1.1.2. DOT cylinder specifications and regulations applying to the shipment of compressed gases are published as Title 49 Code of Federal Regulations (CFR), Parts 170–179 and may be obtained from the Superintendent of Documents, U.S. Government Printing Office, Washington, DC 20402.*

3.1.2. Cylinder Filling Limits.

3.1.2.1. Because of the characteristic of any gas in a closed container to increase in pressure with rising temperature, the possibility always exists that a cylinder charged with gas at a safe pressure at normal temperatures would reach a dangerously high pressure at elevated temperatures. This is equally true whether the contents of the cylinder are in the gaseous or liquid state. In the latter case, the liquid may expand to such a degree that excess hydrostatic pressure develops within the cylinder. To prevent these conditions from occurring with normal usage, the DOT has drawn up regulations which limit the amount of gas that may be charged into a cylinder.

3.1.2.2. In addition to certain other restrictions the charging of cylinders is in general limited so that service pressure in the cylinder at 70 F (21 C) does not exceed the service pressure for which the cylinder is designed. However, certain types of cylinders are permitted to be charged with some nonliquefied, nonflammable gases, among which oxygen, helium, helium-oxygen mixtures and carbon dioxide-oxygen mixtures are included, to a pressure at 70 F (21 C) which is 10 percent in excess of their marked service pressure.

3.1.2.3. The charging of cylinders with medical gases other than carbon dioxide and nitrous oxide is also limited by the requirement that the pressure in the cylinder at 130 F (54.4 C) must not exceed one and one-fourth times the maximum permitted filling pressure at 70 F

---

*Also reprinted and available for a nominal charge from the Bureau of Explosives of The Association of American Railroads, 1920 L Street, N.W., Washington, DC 20036.

(21 C). The pressure in pressurized liquid oxygen cylinders is limited to one and one-fourth times the service pressure.

3.1.2.4. The charging of cylinders with liquefied gases is further limited by setting a maximum permitted filling density for each gas. The term "filling density" is defined as "the percent ratio of the weight of gas in a container to the weight of water that the container will hold at 60 F (16 C)." (49 CFR, Par. 173.300 (g).) Therefore, to determine the maximum amount of liquefied gas in pounds that may be charged into a container, multiply the water capacity in pounds by the maximum permitted filling density prescribed for the gas (1 lb of water at 60 F = 27.737 cu in.; 1 kg of water at 15 C = 1000.9 cm^3, water in air).

3.1.3. Retesting Cylinders.

3.1.3.1. DOT Regulations require that cylinders be subjected periodically to visual examinations and, with a few exceptions, must also be subjected to internal hydrostatic pressure and examined in the manner described in Chapter 6, Section D. The hydrostatic retest must be performed at five-year intervals unless the procedures for filling, handling and maintaining the cylinders meet other requirements of the regulations, in which case, the hydrostatic retest period may be extended to ten-year intervals. However, all cylinders charged with carbon dioxide or mixtures of carbon dioxide in excess of 30 percent must be retested every five years. These retests must be made at a minimum pressure which is specified in DOT Regulations for each type of cylinder; for example, DOT-3AA and 3A cylinders must be retested at a minimum pressure of 5/3 times the marked service pressure. Cylinders which have been in a fire must be removed from service until they have been subjected to the prescribed procedures which, in most cases, include reheat-treating and retesting.

3.1.3.2. If a cylinder leaks, shows evidence of damage that may weaken it appreciably or shows a permanent expansion which exceeds 10 percent of the total expansion in the retest, it must be condemned. Cylinders condemned because of excessive permanent expansion may be reheat-treated and retested.

3.1.3.3. Records must be kept giving data showing the results of the tests made on all cylinders. Each cylinder passing the test must be plainly and permanently stamped with the month and year of the test, for example, 4-62 for April, 1962. Dates of previous tests must not be obliterated. Additional marks may be required to indicate type of test and location.*

3.1.4. Marking and Labeling Cylinders.

3.1.4.1. Definite markings are prescribed by the DOT Regulations. For example, on DOT-3AA and 3A cylinders the following markings are required to be stamped plainly and permanently on the shoulder, top head or neck in letters or figures at least $\frac{1}{4}''$ high, if space permits:

(a) The DOT specification number followed by the service pressure, for example, DOT-3AA 2015.

(b) A serial number (except in the case of certain small sizes of cylinder which may be identified by lot numbers) and an identifying symbol (letters) to be located below the DOT mark. The symbol and numbers must be those of the manufacturer, and the symbol must be registered with the Bureau of Explosives. Additional symbols may be shown indicating purchaser and user.

(c) The inspector's official mark near the serial number and the date of test so placed that the dates of subsequent tests can be easily added.

(d) Spun and plugged cylinders must be specifically so marked.

3.1.4.2. The markings on cylinders must not be changed except as specifically provided in DOT Regulations. Attention is called to the fact that changing of serial numbers or ownership marks is prohibited unless a detailed report is filed with the Bureau of Explosives. Marking on cylinders must be kept in readable condition.

3.1.4.3. In addition to the above specified markings on cylinders, DOT Regulations require, with certain exceptions, that each pack-

---

*See Chapter 6, Section D for additional information about retest apparatus and its operation.

age containing compressed gas must bear an appropriate identifying label, as follows:

Oxygen	DOT Oxidizer Label (Yellow)
Other nonflammable gases	DOT Nonflammable Gas Label (Green)
Flammable gases	DOT Flammable Gas Label (Red)

When empty cylinders bearing these labels are to be shipped as such, the labels must be removed, obliterated, or completely covered by a DOT Empty Label.

3.1.4.4. DOT Regulations permit the use of a combination label-tag, one side of which contains the prescribed wording of the DOT label, while the other side is used as a shipping tag with space for the names and addresses of both the shipper and consignee. Medical gas manufacturers in general use on large cylinders such a combination label-tag attached to the cylinder cap and this tag is perforated so that when the cylinder is empty part of the tag may be torn off at the perforation, thus effectively obliterating the label wording. At the same time, that part of the tag remaining attached to the cylinder contains the return address of the supplier of the gas.

3.1.4.5. Small cylinders of medical gases are usually packed in outside containers such as cartons or crates. In this case the individual cylinders do not need to bear the DOT label, but such appropriate label must be attached to the outside package.

3.1.4.6. Outside packages through which the specification markings on inside containers are not visible must also be labeled or marked to the effect that the inside packages comply with prescribed specifications.

3.1.4.7. Appendix A of Section B, Chapter 6, outlines an alternative marking system which is acceptable on medical gas cylinders which are carried by private and contract motor carriers and are not overpacked. The DOT marking which is included in that system does not need to be covered, removed or obliterated when shipped as empty cylinders.

3.1.5. Cylinder Recharging.

3.1.5.1. DOT Regulations prohibit the shipment of cylinders unless they were charged by or with the consent of the owner. If this consent should be granted, the recharging of cylinders must comply in every respect with regulations that govern the charging of cylinders at a manufacturer's plant.

3.1.5.2. For information on the hazards involved in recharging cylinders by inexperienced operators see 4.5, entitled "Transfilling Cylinders."

3.2. Federal Food, Drug and Cosmetic Act

3.2.1. The shipment of medical gases must comply with the Federal Food, Drug and Cosmetic Act and its implementing regulations. The registration requirements of these regulations will be found in Code of Federal Regulations (CFR), Title 21, Part 207. Under the provisions of this Act the medical gases must conform to the standards of the United States Pharmacopeia and must be appropriately labeled.

3.2.2. The USP contains a section or monograph on each of the following gases—carbon dioxide, cyclopropane, helium, nitrogen, nitrous oxide and oxygen. Definite standards and methods of testing are prescribed in each monograph to assure a product of appropriate quality and purity.

3.2.3. Although mixtures of medical gases, such as carbon dioxide-oxygen or helium-oxygen are not specifically covered in the USP, the regulations provide that the individual active components must be clearly identified on the label and must meet their respective requirements.

3.2.4. The regulations further provide that the gases must be packaged and labeled as prescribed in the USP.

3.2.5. The regulations further provide that each cylinder bear a label containing the name and address of the manufacturer, packer, or distributor as well as an accurate statement of the quantity of the contents and warning and precautionary statements.

3.2.6. For the purposes of enforcing the Act

and its regulations, inspectors of the Federal
Food and Drug Administration are empowered
to make factory or warehouse inspections and
to enter and inspect any vehicles used for transportation of these commodities in interstate
commerce.

3.3. Color Code

3.3.1. A color code to aid in the identification of medical gas cylinders has been adopted
by the medical gas industry. CGA Pamphlet C-9
sets forth the color code used and the proper
application of these colors to the various gas
cylinders.

Kind of Gas	Color USA	Canada
Nitrogen	Black	Black
Oxygen	Green	White
Carbon dioxide	Gray	Gray
Nitrous oxide	Blue	Blue
Cyclopropane	Orange	Orange
Helium	Brown	Brown
Carbon dioxide-oxygen	Gray and green	Gray and white
Helium-oxygen	Brown and green	Brown and white
Air	Yellow	White and black
Oxygen-nitrogen (other than air)	Green and black	White and black

4. RECOMMENDED SAFE PRACTICES FOR HANDLING MEDICAL GASES*

4.1. General Rules

4.1.1. Never permit oil, grease or other readily
combustible substances to come in contact with
cylinders, valves, regulators, gages, hoses and
fittings. Oil and certain gases such as oxygen
or nitrous oxide may combine with explosive
violence.

*See Section A of this chapter for general precau-
tions for handling compressed gases.

4.1.2. Never lubricate valves, regulators, gages
or fittings with oil or any other combustible
substances.

4.1.3. Do not handle cylinders or apparatus
with oily hands or gloves.

4.1.4. Connections to piping, regulators and
other appliances should always be kept tight to
prevent leakage. Where hose is used, it should
be kept in good condition.

4.1.5. Never use an open flame to detect gas
leaks. Leak detection instruments or commercial leak detector solutions should be used.

4.1.6. Prevent sparks or flame from any
source from coming in contact with cylinders
and equipment.

4.1.7. Never interchange regulators or other
appliances used with one gas with similar equipment intended for use with other gases.

4.1.8. Fully open the cylinder valve when the
cylinder is in use.

4.1.9. Never attempt to mix gases in cylinders. (Mixtures should be obtained already prepared from recognized suppliers.)

4.1.10. Before placing cylinders in service
any paper wrapping should be removed so that
the cylinder label is clearly visible.

4.1.11. Identify the gas content by the label
on the cylinder before using. If the cylinder is
not identified to show the gas contained, return
the cylinder to the supplier without using.

4.1.12. Do not deface or remove any markings which are used for identification of content
of cylinder.

4.1.13. When returning empty cylinders,
close valve before shipment and see that cylinder valve protective caps and outlet caps or
plugs, if used, are replaced before shipping.
Cover DOT Green, Yellow or Red Gas Label
with an Empty Label, or if cylinder is provided with combination shipping and DOT tag,
remove lower portion (see Par. 3.1.4.3. and
3.1.4.7).

4.1.14. No part of any cylinder containing a
compressed gas should be subjected to a temperature above 130 F (54.4 C). A flame should
never be permitted to come in contact with
any part of a compressed gas cylinder (see Par.
3.1.2.3.).

4.1.15. The user shall not change, modify, tamper with, obstruct or repair the pressure relief devices in container valves or in containers.

4.1.16. Never attempt to repair or to alter cylinders.

4.1.17. Never use cylinders for any purpose other than to supply the contained gas as received from the supplier.

4.1.18. Cylinder valves should be closed at all times except when gas is actually being used.

4.1.19. Notify gas supplier if any condition has occurred which might have permitted any foreign substance to enter the cylinder or valve, giving details and cylinder number.

4.1.20. Do not place cylinders where they might become part of an electric circuit.

4.1.21. Cylinders should be repainted only by the supplier.

4.1.22. Compressed gases should be handled only by experienced and properly instructed persons.

4.2. Moving Cylinders

4.2.1. Where caps are provided for valve protection, such caps should be kept on cylinders when cylinders are moved.

4.2.2. Never drop cylinders nor permit them to strike each other violently.

4.2.3. Avoid dragging or sliding cylinders. It is safer to move large cylinders even short distances by using a suitable truck, making sure that the cylinder retaining chain or strap is fastened in place.

4.3. Storing Cylinders

Whenever cylinders of medical gases are stocked by hospitals, doctors or distributors, the question of storage is of great importance. Many cities and other governmental agencies have regulations covering the storage of medical gases. Persons storing medical gases should be familiar with these regulations and fully comply with them. The following are general recommendations for the storage of medical gas cylinders.

4.3.1. Cylinders should be stored in a definitely assigned location.

4.3.2. Full and empty cylinders should be stored separately with the storage layout so planned that cylinders comprising old stock can be removed first with a minimum of handling of other cylinders.

4.3.3. Storage rooms should be dry, cool and well ventilated. Where practical, storage rooms should be fire resistant.

4.3.4. Cylinders should be protected against excessive rise of temperature. Do not store cylinders near radiators or other sources of heat. Do not store cylinders near highly flammable substances such as oil, gasoline, waste etc. Keep sparks and flame away from cylinders.

4.3.5. Do not store reserve stocks of cylinders containing flammable gases in the same room with those containing oxygen or nitrous oxide, unless the following conditions can be met: (1) state and local regulations allow such storage, and (2) there is adequate separation by distance or flame-resistant partitioning walls.

4.3.6. Cylinders should never be stored in the operating room.

4.3.7. Small cylinders may best be stored in bins, grouped as to the various gases or mixtures of gases.

4.3.8. Large cylinders should be stored in such a manner that they are restrained from being knocked over.

4.3.9. DOT-4L cylinders must be stored in an upright position.

4.3.10. Be careful to protect cylinders from any object that will produce a cut or other abrasion in the surface of the metal. Do not store cylinders in locations where heavy moving objects may strike or fall on them. Where caps are provided for valve protection, such caps should be kept on cylinders in storage.

4.3.11. Cylinders should not be exposed to continuous dampness and should not be stored near corrosive chemicals or fumes. Rusting will damage the cylinders and may cause the valve protection caps to stick.

4.3.12. Never store cylinders where oil, grease or other readily combustible substance may come in contact with them. Oil and certain gases such as oxygen or nitrous oxide may combine with explosive violence.

4.3.13. Cylinders should be protected against tampering by unauthorized individuals.

4.3.14. Valves should be kept closed on empty cylinders to avoid entry of atmospheric contaminants.

4.4. Withdrawing Cylinder Content

4.4.1. CAUTION! The release of high-pressure gas from cylinders can be hazardous unless adequate means are provided for reducing the gas pressure to usable levels and for controlling the gas flow. Accordingly, pressure-reducing regulators should always be used when withdrawing the contents of gas cylinders, as such devices deliver a constant safe working pressure. Needle valves or similar devices without regulating mechanisms should not be used in place of pressure-reducing regulators, because excessive pressures may develop downstream of such devices and result in possible damage to equipment or injury to personnel.

4.4.2. Do not remove valve protection cap until ready to withdraw contents or to connect to a manifold.

4.4.3. Where compressed gas cylinders are connected to a manifold, such a manifold must be of proper design and equipped with one or more pressure regulators.

4.4.4. After removing the valve protection cap, slightly open the valve for an instant to clear the opening of possible dust and dirt. This should not be done with a cylinder containing flammable gas.

4.4.5. When opening valve, point the outlet away from you. Never use wrenches or tools except those provided or approved by the gas supplier. Never hammer the valve wheel in attempting to open or to close the valve.

4.4.6. Regulators, pressure gages and manifolds provided for use with a particular gas or group of gases must not be used with cylinders containing other gases.

4.4.7. Never use medical gases where the cylinder is liable to become contaminated by the feedback of other gases or foreign material unless protected by suitable traps or check valves.

4.4.8. It is important to make sure that the threads on regulator-to-cylinder valve connections or the pin indexing devices on yoke-to-cylinder valve connections are properly mated. Never force connections that do not fit.

4.4.9. Never permit gas to enter the regulating device suddenly. Always open the cylinder valve slowly.

4.4.10. Before the regulating device and the cylinder are disconnected, close the cylinder valve.

4.4.11. Cylinder valves should be closed at all times except when the gas is actually being used.

4.5. Transfilling Cylinders

4.5.1. There are serious hazards involved in transferring compressed gas from one cylinder to another. The Compressed Gas Association, Inc. recognizes these hazards and in past communications with hospital executives and professional anesthetists has urged that the practice of transferring medical gases from one cylinder to another be prohibited. This recommendation has also been made by the National Fire Protection Association which in its Pamphlet No. 56A entitled "Standard for the Use of Inhalation Anesthetics" states, "transfer of gas from one cylinder to another on the hospital site or by hospital personnel shall be prohibited."

4.5.2. In the interest of public safety, the Compressed Gas Association, Inc. urges that cylinders be returned to charging plants for refilling under recognized safe practices, and calls your attention to the following:

4.5.2.1. The hazard of overfilling small cylinders is always present when the charging is done by inexperienced operators who lack adequate knowledge of proper filling procedure and properties of the gas being handled. Filling capacities may vary for cylinders even though their sizes appear to be the same. Overfilling may result in cylinder rupture and damage. All operations involved in the transfer of gases from one container to another require experienced supervision and equipment maintenance of a high degree in order to avoid personal injury and property damage.

4.5.2.2. Unless proper precautions are

taken, a dangerous mixture of gases may occur when charging one cylinder from another. Manufacturers report that each year there are returned to them supposedly empty cylinders which actually contain ether or a gas other than that originally shipped. Some of these contaminating gases are flammable. Intermixture of flammable and oxidizing gases may cause a serious explosion. To avoid this, manufacturers have established definite procedures for detecting contaminating gases, and have developed special equipment for the thorough cleaning, when necessary, and preparing of all medical gas cylinders before they are recharged.

4.5.2.3. Cylinders which have been used for one type of gas may inadvertently or with intent be recharged by inexperienced or improperly trained operators with a gas other than that originally or last contained in the cylinder. Such practice will definitely cause contamination and may introduce a serious explosion or health hazard as well.

4.5.2.4. The importance of purity of medical gases cannot be overemphasized. This is recognized by the fact that the sale and distribution of gases which are adulterated or may be injurious to health are prohibited by the Federal Food, Drug and Cosmetic Act. Medical gas manufacturers are required to supply compressed gases labeled in accordance with the requirements of this Federal Act, and to furnish such gases in full compliance with standards of purity prescribed by the United States Pharmacopeia. Many states have similar laws affecting local distribution. The transfer of medical gases from one cylinder to another by inexperienced persons may adversely affect purity.

4.5.2.5. Safety relief devices, valves and parts must be inspected at frequent intervals to insure safe operation, and repairs or replacements made when defects are found. Manufacturers regularly engaged in the production of gases are best equipped to perform this essential maintenance work.

4.6. Use of Ether with Medical Gases

4.6.1. While ether is not a product of the medical gas industry, it is used so much in common with medical gases that recognition of its hazards is important. Ether is a highly volatile liquid, giving off, even at comparatively low temperatures, vapors which form, with air or oxygen, flammable and explosive mixtures. The vapors are heavier than air and may travel a considerable distance to a source of ignition and flash back. Spontaneously explosive peroxides sometimes form on long standing or exposure in bottles to sunlight. In order to prevent these peroxides from forming in an anesthesia machine, the ether vaporizer wicks should be rinsed and dried after each day's use and the residual ether in the vaporizer jar should be discarded and the jar thoroughly cleaned before it is replaced on the machine. In storing containers of ether, they should be safeguarded against mechanical injury and kept in unheated compartments away from sunlight and any source of ignition. When used in connection with nitrous oxide or oxygen in anesthesia, care must be taken to avoid exhausting the gas cylinders completely to prevent the possibility of ether being drawn back into the cylinder.

5. MEDICAL GAS EQUIPMENT

5.1. Standard Sizes and Capacities of Medical Gas Cylinders

5.1.1. Medical gases are supplied in standard cylinder styles having the dimensions and capacities as shown in Table 4.

5.2. Valve Outlet Connections

5.2.1. Style E and smaller cylinders are equipped with flush-type valves which are used with yoke connections. For details refer to references in 5.2.2. Style M and larger cylinders are equipped with valves having threaded outlet connections.

5.2.2. Detailed, dimensioned drawings of standard and alternate standard compressed gas cylinder valve outlets have been prepared by the Compressed Gas Association, Inc. and recognized by the American National Standards Institute and Canadian Standards Association as American National and Canadian Standards,

TABLE 4. TYPICAL MEDICAL GAS CYLINDERS VOLUME AND WEIGHT OF AVAILABLE CONTENTS*
All Volumes at 70 F (21.1 C)

Cylinder Style and Dimensions	Nominal Volume cu in./liter	Contents	Air	Carbon Dioxide	Cyclopropane	Helium	Nitrogen	Nitrous Oxide	Oxygen	Mixtures of Oxygen — Helium	Mixtures of Oxygen — CO₂
B 3½" o.d. × 13" 8.89 × 33cm	87/1.43	psig liters lb–oz kilograms		838 370 1–8 0.68	75 375 1–7¼ 0.66				1900 200 — -.		
D 4½" o.d. × 17" 10.8 × 43 cm	176/2.88	psig liters lb–oz kilograms	1900 375 — —	838 940 3–13 1.73	75 870 3–5½ 1.51	1600 300 — —	1900 370 — —	745 940 3–13 1.73	1900 400 — —	** 300 ** **	** 400 ** **
E 4¼" o.d. × 26" 10.8 × 66 cm	293/4.80	psig liters lb–oz kilograms	1900 625 — —	838 1590 6–7 2.92		1600 500 — —	1900 610 — —	745 1590 6–7 2.92	1900 660 — —	** 500 ** **	** 660 ** **
M 7" o.d. × 43" 17.8 × 109 cm	1337/21.9	psig liters lb–oz kilograms	1900 2850 — —	838 7570 30–10 13.9		1600 2260 — —	2200 3200 — —	7.45 7570 30–10 13.9	2200 3450 122 cu ft —	** 2260 ** **	** 3000 ** **
G 8½" o.d. × 51" 21.6 × 130 cm	2370/38.8	psig liters lb–oz kilograms	1900 5050 — —	838 12,300 50–0 22.7		1600 4000 — —		745 13,800 56–0 25.4		** 4000 ** **	** 5330 ** **
H or K 9¼" o.d. × 51" 23.5 × 130 cm	2660/43.6	psig liters lb–oz kilograms	2200 6550 — —			2200 6000 — —	2200 6400 — —	745 15,800 64 29.1	2200† 6900 244 cu ft —		

*These are computed contents based on nominal cylinder volumes and rounded to no greater variance than ±1 percent.

**The pressure and weight of mixed gases will vary according to the composition of the mixture.

†275 cu ft/7800 liter cylinders at 2490 psig are available upon request.

respectively. These drawings cover the Pin-Index Safety System for flush outlet valves and the valve threaded outlet connections published in CGA Pamphlet V-1, American National Standard–Canadian Standard Compressed Gas Cylinder Valve Outlet and Inlet Connections.*

5.2.2.1. CGA Pamphlet V-1 gives detailed information on the Pin-Index Safety System for Flush-Type Cylinder Valves. This system has been devised to prevent the interchangeability of medical gas cylinders equipped with flush-type valves as employed in Style E and smaller cylinders. The system consists of a combination of two pins projecting from the yoke assembly of the apparatus and so positioned as to fit into matching holes in the cylinder valve. All medical gas cylinders having flush-type valves and shipped by member companies of the Compressed Gas Association, Inc. are now drilled in conformance to this standard. The Association has strongly recommended that all new gas apparatus now being manufactured incorporate the system and that older, existing equipment be modified in accordance with it. All users of medical gas administering apparatus are urged to take full advantage of the safety features gained by adoption of the Pin-Index Safety System.

5.2.3. The use of standard valve outlet connections should be encouraged to the exclusion of adapters.

5.3. Regulators

5.3.1. The purpose of a regulator is to reduce the pressure of the gas as it is discharged from the cylinder to a usable pressure, and to control

the flow of the gas. Other terms are often used erroneously for the regulator including such names as "gage" and "valve." A "valve" is defined as a movable mechanism which opens and closes a passage. This term usually refers to the part on top of the cylinder to which the regulator is connected.

5.3.2. A "gage" is an instrument of measure, and most regulators are equipped with two gages. One gage measures the pressure of the gas in the cylinder in pounds per square inch. The other gage may register the reduced or working pressure in pounds per square inch or, when used with a flowmeter, it may measure the rate of discharge or flow of gas from the regulator in liters or gallons per minute.

5.3.3. There are several methods of adjusting the flow of gas from the regulators, the most common being by means of an adjusting screw on the bonnet of the regulator.

5.4. Piping and Manifold Systems

5.4.1. Piping and manifold systems for medical gases should be constructed only under the supervision of a competent engineer who is thoroughly familiar with the problems incident to piping compressed gases. Consultation with your gas or equipment supplier before installation of piping and manifold systems may often be helpful. Standards for piping and manifold systems are published in NFPA Pamphlet 56F "Standard for Nonflammable Medical Gas Systems"* and in the Compressed Gas Association Pamphlet P-2.1 "Standard for Medical-Surgical Vacuum Systems in Hospitals."**

*Obtainable for a nominal charge from the American National Standard Institute, Inc., 1430 Broadway, New York, NY 10018, and the Compressed Gas Association, Inc., 500 Fifth Avenue, New York, NY 10110.

*Obtainable for a nominal charge from the National Fire Protection Association, 470 Atlantic Avenue, Boston, MA 02210.

**Obtainable for a nominal charge from the Compressed Gas Association, Inc., 500 Fifth Avenue, New York, NY 10110.

SECTION D

Safe Handling of Cylinders for Underwater Breathing

1. INTRODUCTION

Both steel and aluminum compressed gas cylinders used for underwater diving service (SCUBA) are subjected to extremely harsh conditions. Due to the corrosive action of the seawater environment, the structural integrity of the cylinders may become weakened, or residual corrosion products could block the valve and regulator, and restrict the diver's air supply. Proper maintenance of these cylinders is therefore of the utmost importance to ensure their long life and satisfactory service combining safety with pleasure.

2. DESCRIPTION OF CYLINDERS

2.1. Cylinder Types

A variety of cylinders are in SCUBA diving use in the United States, all of which should have the following in common:

(a) They feature a mark, DOT (or ICC), which confirms that they are manufactured in compliance with Department of Transportation (or formerly Interstate Commerce Commission) specifications.

(b) They feature a manufacturing or retest date which indicates compliance with periodic retest regulations set forth by the Department of Transportation Hazardous Materials Regulations (49 CFR 173.34 (e)).

Only cylinders bearing such identification should be used. Cylinders should never be charged to a pressure higher than that designated on the cylinder, except for those cylinders with a plus sign that meet all requirements of 49 CFR 173.302 (c) which may be filled to a pressure 10 percent in excess of the marked service pressure. At the present time, the most common cylinders used for underwater diving are DOT 3AA1800, DOT 3AA2250, DOT 3AA-3000, DOT SP6498-2475, DOT SP6498-3000,

DOT SP6688-3000, and DOT SP6576-2750. In each case, the first three letters indicate that the cylinder is manufactured according to Department of Transportation requirements. The middle series of letters and digits indicates the applicable section of the DOT Specification or Special Permit to which the cylinders are manufactured. DOT Special Permits (SP) are being changed to Exemption (E) and both designations, DOT SP and DOT E are acceptable. The last four digits indicate the service pressure of the cylinder which is the maximum pressure permitted in the cylinder at 70 F. Cylinders marked with plus sign (+) following the test date may be charged to a pressure 10 percent in excess of the marked service pressure, provided all the requirements of 49 CFR 173.302 (c) are met. For example, "DOT 3AA 2250 . . . 6-75+" permits actual filling pressure to be as high as 2475 psi, at 70 F.

2.2. Coatings and Linings

The SCUBA cylinders used in the United States have a variety of external coatings to enhance either aesthetic and/or corrosion characteristics. Steel cylinders are usually galvanized (i.e., plated with zinc), coated with vinyl or epoxy or painted. Aluminum cylinders are either painted or anodized. In addition to the external coatings, some cylinders are internally lined with organic coatings. Cylinders with internal linings should be marked "LND" in the area of the service pressure for positive identification. These coatings and linings, if properly applied and maintained, provide a high degree of corrosion protection. It is of utmost importance that all coatings or linings be examined at least annually to ensure their continuity. If the coating or lining is damaged in some way, moisture can penetrate beneath it, creating potential corrosion sites. This applies to both the inside and the outside of the cylinder.

3. CHARGING OF CYLINDERS

SCUBA cylinders should be charged with "PURE COMPRESSED AIR OR RECONSTITUTED AIR ONLY." A more detailed discussion of air will be found in CGA Pamphlets G-7 and G-7.1. Charging should be done by a competent dealer or other reliable source having proper facilities, to ensure that compressed air suitable for breathing is free from moisture, oil and other impurities, in accordance with CGA Pamphlet G-7.1, "American National Standard Commodity Specification for Air," Z86.1. *Never put oxygen or other gases in any SCUBA cylinder normally used for air.* Explosion or physiological hazards may be significantly increased by such incorrect charging.

4. PRECAUTIONS IN USE OF CYLINDERS

4.1. Water

Cylinder valves should remain tightly closed when cylinders are empty to prevent water or atmospheric moisture, which promote corrosion, from entering the cylinder. Preferably, the valve should be closed while there is still positive air pressure in the cylinder.

4.2. Handling

A SCUBA cylinder contains between 50 and 100 cu ft of compressed air at a pressure in excess of 2000 psi when fully charged. Should a charged cylinder fail while under this pressure, severe damage and/or injury can result. Cylinders should not be dropped or roughly handled during transportation. Nor should they be dragged, as such action can cause substantial wear on the cylinder base or sidewall, significantly increasing the possible explosion hazard. When transporting cylinders, great care should also be taken to ensure that the valve mechanism is protected. Be sure the cylinders are firmly secured (with the valve protected) so they cannot shift about—the rocketing effect resulting from a sheared-off valve is potentially devastating.

4.3. Storage

Compressed air cylinders for SCUBA use should be stored in a cool place. Storage should be in the vertical position, well-secured in order to reduce the possibility of tipover. The cylinder should have a slightly positive pressure and the valve should be closed. Do not store a charged cylinder near a source of heat.

4.4. Lubrication

Cylinders, valves, and regulators *should not be lubricated or serviced other than by a qualified SCUBA equipment repair facility.* Certain oils and greases can have toxic effects on the human body and therefore their indiscriminate use by improperly trained individuals can be highly dangerous.

5. INSPECTION AND RETESTING

5.1. Annual Visual Inspection

At least once each year, the SCUBA cylinder should be visually inspected internally and externally in accordance with CGA Pamphlet C-6 "Standards for Visual Inspection of Compressed Gas Cylinders," by a competent inspection agency or station.

5.1.1. Interior. Visual inspection of the cylinder interior should include:

(1) Depressurization of the cylinder.

(2) Removal of the valve.

(3) Inspection of the interior with a light source affording adequate illumination.

(4) If corrosion has occurred, it should be removed by tumbling, shot blasting or other suitable means. Attention should be given to interior linings noted in 2.2 of this pamphlet followed by hydrostatic testing of the cylinder to assure its strength under pressure.

(5) If traces of oil or other noxious material are found, the charge station should be notified, and the material removed through suitable cleaning means before the cylinder is put back into service.

(6) Once the inspection has been completed and it is determined that all corrosion products,

oil and moisture have been removed, the cylinder should be charged to a positive pressure and stored vertically in a cool location.

5.1.2. Exterior. Visual inspection of the cylinder exterior should include:

(1) Removal of the boot to check for accumulation of salt, sand or possible corrosion products.

(2) A probe of any apparent blisters in the coating with a sharp object, to determine if unseen corrosion is taking place underneath the coating. Where the coating is broken, the surface should be cleaned down to base metal and recoated with a compatible coating material.

5.2. Periodic Hydrostatic Retest

A SCUBA cylinder must be hydrostatically pressure tested every five years by a qualified inspection station, following the rules and procedures of the Department of Transportation Hazardous Materials Regulations, Section 173.34 (e), and CGA Pamphlet C-1 "Methods for Hydrostatic Testing of Compressed Gas Cylinders." When the cylinder has been hydrostatically tested, a new test date must be stamped near the original DOT or ICC identification. A plus (+) sign authorizing filling to a pressure 10 percent in excess of the marked service pressure may only be added when elastic expansion and wall thickness determinations, visual inspection and other requirements of 173.302 (c) of DOT Regulations are followed. Refer to CGA Pamphlet C-5, "Cylinder Service Life-Seamless, High-Pressure Cylinders." Following hydrostatic retest procedures, the interior of the cylinder should be properly dried to remove all traces of moisture.

CHAPTER 5

Pressure-Relief Device Standards

Almost all compressed gas containers are fitted with devices that let gas escape should surrounding conditions (such as heat) cause the enclosed gas to increase in pressure to a dangerous degree. The standards that have been developed by the Compressed Gas Association, Inc. for these pressure-relief devices are given in this chapter. Its sections present the CGA pressure-relief device standards as follows:

Section A. Pressure-Relief Device Standards—Part 1: Cylinders for Compressed Gases
Section B. Pressure-Relief Device Standards—Part 2: Cargo and Portable Tanks for Compressed Gases
Section C. Pressure-Relief Device Standards—Part 3: Compressed Gas Storage Containers.

SECTION A

Pressure-Relief Device Standards—Part 1: Cylinders for Compressed Gases

1. INTRODUCTION

This Standard represents the minimum requirements for pressure-relief devices considered to be appropriate and adequate for use on cylinders having capacities of 1000 lb (453.6 kg) of water, or less and DOT-3AX, 3AAX and 3T cylinders having capacities over 1000 lb (453.6 kg) of water and which comply with the specifications and charging and maintenance regulations of the Department of Transportation (DOT) or the corresponding specifications and regulations of the Canadian Transport Commission (CTC).

It is recognized that there are cylinders that conform to the specification requirements of the DOT or the CTC, which are used in services beyond the jurisdiction of either of these authorities. In such cases it is recommended that state, provincial, local or other authorities having jurisdiction over these cylinders be guided by this Standard in determining adequate pressure-relief device requirements, provided that the cylinders are charged and maintained in accordance with DOT or CTC Regulations.

It is further recognized that there may be cylinders which are used in services beyond the jurisdiction of the DOT or CTC and which do not conform to the specification requirements of either authority. It is recommended that the authorities having jurisdiction over such cylinders be guided by this Standard in determining pressure-relief device requirements, provided that such cylinders are considered by the authority as having a construction at least equal to the equivalent DOT specification requirements and further provided that the cylinder shall be charged and maintained in accordance with DOT or CTC requirements.

For cylinders that come within the jurisdiction of state and local regulatory authorities, the user should check for compliance with all local regulations. A number of states and cities have pressure vessel laws and regulations which include requirements for pressure-relief devices. This Standard has been prepared specifically for compressed gas cylinders and the pressure-relief devices may not be acceptable unless special permission is obtained from the authority having jurisdiction.

For newly constructed cylinders that come within the jurisdiction of the DOT or CTC, pressure-relief devices must comply with requirements of this standard. This issue of the Standard (1979) is based on minimizing and optimizing the number of types of approved pressure-relief devices for each specific gas. It does not prejudice the continued use of previously approved and installed devices. If a pressure-relief device is replaced, the new device shall meet the requirements of these standards.

2. DEFINITIONS

For the purpose of this Standard the following terms are defined.

2.1. A *pressure-relief device* is a device designed to prevent rupture of a normally charged cylinder when it is placed in a fire as required by section 173.34 (d) of the DOT Regulations or paragraph 73.34 (d) of the CTC Regulations. The term "pressure-relief device" is synonymous with "safety-relief device" as used by DOT and CTC Regulations.

2.2. An *approach channel* is the passage or passages through which fluid must pass from the cylinder to reach the operating parts of the pressure-relief device.

2.3. A *discharge channel* is the passage or passages beyond the operating parts of the pressure-relief device through which fluid must pass to reach the atmosphere.

2.4. A *rupture disk device* is a nonreclosing pressure-relief device actuated by inlet static pressure and designed to function by the bursting of a pressure containing disk.

2.4.1. A *rupture disk* is the operating part of a pressure-relief device which, when installed in the device, is designed to burst at a predetermined pressure to permit the discharge of fluid. (Such disks, usually metal, are generally of flat, preformed, reinforced or grooved types.)

2.4.2. The *pressure opening* is the orifice against which the rupture disk functions.

2.4.3. The *rated burst pressure* of a rupture disk is the pressure for which the disk is designed to rupture when in contact with the pressure opening for which it was designed when tested as required in 6.3.

2.5. A *fusible plug device* is a nonreclosing pressure-relief device designed to function by the yielding or melting of a plug of suitable melting temperature material.

2.6. The *yield temperature* of a fusible plug is the temperature at which the fusible material becomes sufficiently soft to extrude from its holder to permit the discharge of fluid when tested in accordance with 6.2.

2.7. A *combination rupture disk-fusible plug* is a rupture disk in combination with a low-temperature melting material intended to prevent its bursting at its predetermined bursting pressure unless the temperature also is high enough to cause yielding or melting of the fusible material.

2.8. A *pressure-relief valve* is a pressure-relief device which is designed to reclose and prevent further flow of fluid after normal conditions have been restored.

2.9. A *pressure control valve* as used on a cryogenic cylinder vents only to maintain the proper working pressure of the cylinder.

2.10. The *set pressure* of a pressure-relief valve is the pressure marked on the valve and at which it is set to start-to-discharge (see 4.3.2).

2.11. The *start-to-discharge pressure* of a pressure-relief valve is the pressure at which the first bubble appears through a water seal of not over 4 in. (10.16 cm) water column on the outlet of the pressure-relief valve (see 6.5).

2.12. The *flow capacity* of a pressure-relief device is the capacity in cubic feet per minute of free air discharged at the required flow rating pressure.

2.13. *Flow rating pressure* is the inlet static pressure at which the relieving capacity of a pressure-relief device is measured for rating purposes.

2.14. A *nonliquefied compressed gas* is a gas, other than a gas in solution, which under the charging pressure is entirely gaseous at a temperature of 70 F (21.1 C).

2.15. A *liquefied compressed gas* is a gas which, under the charged pressure, is partially liquid at a temperature of 70 F (21.1 C).

2.16. A *compressed gas in solution* (acetylene) is a nonliquefied compressed gas which is dissolved in a solvent.

2.17. A *cryogenic liquid* is considered to be a liquid with a normal boiling point below -238 F (-150 C).

2.18. *Cylinders* refers to Specifications for Cylinders constructed under Subpart C of Part 178 of the DOT Regulations and similar cylinder specifications of the CTC Regulations.

2.19. The *test pressure of the cylinder* is the minimum pressure at which it must be tested as prescribed in the specifications for compressed gas cylinders by the DOT or CTC.

2.20. *Free air* or *free gas* is air or gas measured at a pressure of 14.7 psia (101.4 kPa) and a temperature of 60 F (15.6 C).

2.21. *DOT Regulations* refers to the Department of Transportation Regulations for the Transportation of Hazardous Materials under Code of Federal Regulations, Title 49, Parts 100 to 199.

2.22. *CTC Regulations* refers to regulations of the Canadian Transport Commission "Regulations for the Transportation of Dangerous Commodities by Rail."

3. TYPES OF PRESSURE-RELIEF DEVICES

Types of pressure-relief devices are designated as follows:

3.1. Type CG-1: Rupture disk.

3.2. Type CG-2: Fusible plug utilizing a fusible alloy with yield temperature not over 170 F (76.7 C) nor less than 157 F (69.4 C), 165 F nominal (73.9 C).

3.3. Type CG-3: Fusible plug utilizing a fusible alloy with yield temperature not over 220 F (104.4 C) nor less than 208 F (97.8 C), 212 F nominal (100 C).

3.4. Type CG-4: Combination rupture disk-fusible plug, utilizing a fusible alloy with yield temperature not over 170 F (76.7 C) nor less than 157 F (69.4 C), 165 F nominal (73.9 C).

3.5. Type CG-5: Combination rupture disk-fusible plug, utilizing a fusible alloy with yield temperature not over 220 F (104.4 C) nor less than 208 F (97.8 C), 212 F nominal (100 C).

3.6. Type CG-7: Pressure-relief valve.

4. APPLICATION REQUIREMENTS FOR PRESSURE-RELIEF DEVICES

4.1. General

4.1.1. Each cylinder charged with compressed gas, unless excepted in 4.1.1.1 must be equipped with one or more pressure-relief devices complying with this Standard. Cylinders that are found to be equipped with leaking or faulty pressure-relief devices shall have the contents of the cylinder removed, and the pressure-relief device corrected. Leak tests shall be made prior to shipment.

4.1.1.1. Exceptions to the requirement for a pressure-relief device on cylinders apply to Class A Poison (gases or liquids), or other gases designated by the DOT or CTC for shipment without pressure-relief devices.

4.1.2. The design, material and location of pressure-relief devices shall be suitable for the intended service. Consideration shall be given in the design and application of pressure-relief devices to the effect of the resultant thrust when the device functions.

4.1.3. When pressure-relief devices are required at both ends of a cylinder, each end shall have the required flow capacity.

4.1.4. When cylinders are not required to be equipped with pressure-relief devices at both ends, the flow capacity of the individual devices may be combined to meet the minimum total flow capacity requirement. This provision is limited to CG-1 and CG-7 pressure-relief device.

4.2. Rupture Disk Device

4.2.1. When a rupture disk device is used as a pressure-relief device on a compressed gas cylinder, the rated bursting pressure of the disk (when tested within the temperature range of 60 F to 160 F (15.6 to 71.1 C) in accordance with Section 6.3) shall not exceed the minimum required test pressure of the cylinder with which the disk is used, except as follows:

4.2.1.1. For DOT-3E or CTC-3E cylinders the rated bursting pressure of the disk shall not exceed 4500 psig.

4.2.1.2. For DOT-39 cylinders the burst pressure of the disk shall not exceed 80 percent of the minimum cylinder burst pressure and shall not be less than 105 percent of the cylinder test pressure.

4.3. Pressure-Relief Valves

4.3.1. The flow rating pressure shall not exceed the minimum required test pressure of the cylinder on which the pressure-relief valve is installed, and the reseating pressure shall not be less than the pressure in a normally charged cylinder at 130 F (54.5 C). The flow rating pressure for pressure-relief valves for DOT-39 cylinders shall not exceed 80 percent of the minimum required cylinder burst pressure. Pressure-relief valves for DOT-39 cylinders are not required to reseat.

4.3.1.1. A pressure-relief valve may incorporate a fusible element to relieve the total contents at a predetermined temperature. The minimum required flow capacity shall be satisfied by the pressure-relief valve.

4.3.2. The set pressure shall not be less than 75 percent nor more than 100 percent of the minimum required test pressure of the cylinder on which the pressure-relief valve is installed. For liquefied gases pressure-relief valve settings authorized for low-pressure cylinders for a particular gas shall be used on high-pressure (over 500 psi (3448 kPa) service pressure) cylinders for the same gas. For DOT-39 cylinders, the set pressure shall not exceed 80 percent of the minimum cylinder burst pressure and not less than 105 percent of the cylinder test pressure.

4.4. Piping of Pressure-Relief Devices

4.4.1. When fittings and piping are used on either the upstream or downstream side or both sides of a pressure-relief device or devices, the fittings and piping shall be so designed that the flow capcity of the pressure-relief device shall not be reduced below the capacity required for the cylinder on which the pressure-relief device assembly is installed, nor to the extent that the operation of the device could be impaired. Fittings, piping and method of attachment shall be designed to withstand normal handling and the pressures developed when the device or devices function.

4.4.2. A shutoff valve shall not be installed between the pressure-relief devices and the cylinder nor after the pressure-relief devices.

5. DESIGN AND CONSTRUCTION REQUIREMENTS FOR PRESSURE-RELIEF DEVICES

5.1. The design and material of pressure-relief devices shall be suitable for the intended service. In the design and application of pressure-relief devices consideration shall be given to the effect of the resultant thrust when the device functions.

5.2. The material, design and construction of a pressure-relief device shall be such that there will be no significant change in the functioning of the device and no serious corrosion or deterioration of the materials within the period between renewals.

5.3. In combination rupture disk-fusible plug devices, the fusible metal shall be on the discharge side of the rupture disk. The fusible metal shall not be used in lieu of a gasket to seal the disk against leakage around the edges. Gaskets, if used, shall be of a material that will not deteriorate rapidly at the maximum temperature range specified for the fusible metal.

5.4. The flow capacity of each design and modification thereof of all types of pressure-relief devices shall be determined by actual flow tests. Methods of conducting flow tests are given in 6.6.

5.5. For noninsulated cylinders for nonliquefied gas, the minimum required flow capacity of pressure-relief devices, except pressure-relief valves, shall be calculated using the following formula: (For pressure-relief valves refer to 5.7 and 5.8.)

$$Q_a = 0.154Wc$$

where

Q_a = flow capacity at 100 psia test pressure in cubic feet per minute of free air.

Wc = water capacity of the cylinder in pounds but not less than 25 lb.

Note: The above formula expresses flow capacity requirements equal to 70 percent of that which will discharge through a perfect orifice having a 0.00012 square inch area for each pound of water capacity of the cylinder.

5.6. For noninsulated cylinders for liquefied gas, the minimum required flow capacity of pressure-relief devices, except pressure-relief valves, shall be two times that required by the above formula in 5.5. (For pressure-relief valves refer to 5.7 and 5.8.)

5.7. For noninsulated cylinders for nonliquefied gas, the minimum required flow capacity of pressure-relief valves shall be calculated using the following formula:

$$Q_a = 0.00154PWc$$

where

Q_a = flow capacity in cubic feet per minute of free air.

P = flow rating pressure in pounds per square inch absolute.

Wc = water capacity of the cylinder in pounds, but not less than 12.5 lb.

5.8. For noninsulated cylinders for liquefied gas, the minimum required flow capacity of pressure-relief valves shall be two times that required by the formula in 5.7.

5.9. For Specification DOT-4L insulated cylinders containing cryogenic liquids listed in Table 3, the following requirements apply:

5.9.1. If all materials comprising a representative sample of the insulation system remain completely in place when subjected to 1200 F, the U value shall be as defined below and the

minimum required flow capacity of the pressure-relief device(s) shall be calculated using the following formula:

$$Q_a = G_i UA^{0.82}$$

where

U = total thermal conductance of cylinder insulating material Btu/hr ft^2 F when saturated with gaseous lading or air at atmospheric pressure, whichever is greater. Value of U is determined at 100 F except when 5.9.2 (b) and (c) apply. (total thermal conductance = thermal conductivity in Btu/hr ft^2 F/in. divided by insulation thickness in inches.)

A = total outside surface area of the cylinder in square feet.

Q_a = flow capacity in cubic feet per minute of free air at the rated burst pressure of the rupture disk.

G_i = gas factor for insulated containers obtained from Table 6 for the gas involved.

5.9.2. If any material comprising a representative sample of the insulation system deteriorates or only remains partly in place when subjected to 1200 F, one of the following procedures shall be used to determine the minimum flow capacity requirement of the pressure-relief device(s):

(a) Use the formula for uninsulated cylinders.
$$Q_a = G_u A^{0.82}$$
Q_a and A are as defined in 5.9.1.
G_u = gas factor for uninsulated containers obtained from Table 6 for the gas involved.

(b) Determine the total thermal conductance (U) for a representative sample of the insulation system with a 1200 F external test environment. This value of U shall then be used in the formula in 5.9.1 to determine the minimum required flow capacity of the pressure-relief device(s). The value of U shall be determined with the insulation saturated with gaseous lading or air at atmospheric pressure, whichever provides the greater thermal conductance.

(c) If the insulation system is equipped with a jacket that remains in place during fire conditions, the thermal conductance U shall be determined with no insulation and a 1200 F external test environment. The value of U shall be determined with gaseous lading or air at atmospheric pressure in the space between the jacket and cylinder, whichever provides the greater thermal conductance. This value of U shall then be used in the formula in 5.9.1 to determine the minimum required flow capacity of the pressure-relief device(s).

(d) An alternative procedure may be used to qualify a composite insulation, which when applied would consist of layers of several different insulations over the entire cylinder, by exposing a sample of the composite insulation to a temperature of 1600 F for 30 minutes and to use only the layer(s) of the insulation which is unaffected in determining the value of U to be used in the formula in 5.9.1 to calculate the minimum required flow capacity of the pressure-relief device(s). Such high-temperature insulation must be kept in place by a solid or mesh retainer (as required by the insulation) which will remain serviceable at 1600 F.

(e) Perform a fire test* on a full-scale cylinder, the results of which demonstrate that the pressure-relief devices are capable of preventing rupture of the normally charged cylinder.

5.9.3. For Specification DOT-4L cylinders a pressure control valve shall be provided and sized to provide adequate venting capacity with the insulation saturated with gaseous lading or air at atmospheric pressure, whichever provides the greater thermal conductance, as determined by:

$$Q_a = \frac{(130 - t) \; G_i UA}{4(1200 - t)}$$

*See Reference 7 for details on apparatus and procedure for Fire Testing of Cylinder/Safety Device Systems.

where

U = total thermal conductance Btu/hr ft^2 F, is determined at the average temperature of the insulation (alternatively the value of U at 100 F may be used).

Q_a = the flow capacity in cubic feet per minute of free air at a flow rating pressure of 120 percent of the set pressure of the pressure control valve.

A = the total outside surface area of the cylinder in square feet.

t = temperature in degrees F (Fahrenheit) of gas at pressure at flowing conditions.

G_i = gas factor for insulated containers obtained from Table 6 for the gas involved.

The pressure control valve shall have a set pressure not to exceed $1\frac{1}{4}$ times the marked service pressure of the DOT-4L cylinder, less 15 psi if vacuum insulation is used.

5.10. For acetylene cylinders, a fire test shall be used in determining pressure-relief device requirements (see Note F, Table 3).

6. TESTS

6.1. Test of Fusible Alloy

6.1.1. To determine the yield temperature, the following test on the alloy shall be conducted:

6.1.1.1. Select at random two sticks of the fusible alloy from each batch (heat).

6.1.1.2. A sample for test shall consist of a piece 2 in. (5.08 cm) long by approximately $\frac{1}{4}$ in. (0.635 cm) diameter cut from each stick. Each sample shall be placed horizontally on suitable supports spaced 1 in. (2.54 cm) apart and presenting knife edges to the sample so that the ends of the sample will overhang the knife edges $\frac{1}{2}$ in. (1.27 cm). The supported samples shall be immersed in a glycerine bath not closer than $\frac{1}{4}$ in. (0.635 cm) to the bottom of the container.

6.1.1.3. Two samples from the same stick shall be tested at one time. A temperature measuring device shall be inserted into the bath between and closely adjacent to the samples so that the sensor will be completely immersed at the same level as the samples. The bath temperature shall be raised at a rate not in excess of 5 F per minute.

6.1.1.4. The yield temperature shall be taken as that temperature at which the second of the four ends of the sample loses its rigidity and drops.

6.2. Tests of New or Reconditioned Fusible Plugs

6.2.1. Two representative samples shall be selected at random from each lot and subjected to the tests prescribed in 6.2.2 and 6.2.3. If both samples should fail to meet the requirements of 6.2.2 and 6.2.3 the lot shall be rejected. If one sample fails to meet the requirements of 6.2.2 and 6.2.3, four additional samples may be selected at random from the same lot and subjected to these tests. If any of these four additional samples fail to meet the requirements of 6.2.2 and 6.2.3 the lot shall be rejected. The production of either new or reconditioned fusible plugs by a manufacturer on any one day for any one range of minimum to maximum specified yield temperature, but in no case greater than 3000, shall constitute a lot.

6.2.2. A test to determine resistance to extrusion of the fusible alloy and leaks at a temperature of 130 F (54.5 C) or less, in a fusible plug shall be made as follows: Finished fusible plug shall be subjected to a controlled temperature of not less than 130 F (54.5 C) for 24 hours with a gas pressure of 500 psig (3447 kPa) on the end normally exposed to the contents of the cylinder. In order to pass this test no leakage nor visible extrusion of material shall be evident upon examination of the end exposed to atmospheric pressure.

6.2.3. A test for determining the yield temperature of a fusible plug shall be made as follows:

6.2.3.1. Subject plugs to an air pressure of not less than 3 pounds per square inch (20.7 kPa) applied to the end normally exposed to the contents of the cylinder. While subjected to this pressure, the plugs shall be immersed in a water bath or a glycerine-water bath at a temperature in the 5 F range immediately below the specified minimum yield temperature, and

held in that temperature range for a period of ten minutes. The temperature of the bath shall then be raised at a rate not in excess of 5 F per minute during which the pressure may be increased to not more than 50 pounds per square inch (345 kPa). When the temperature of the bath reaches the point where metal is exuded or spewed out sufficiently to produce leakage of air, the temperature of the bath shall be recorded as the yield temperature of the plugs. The yield temperature shall be within the temperature limits specified in Section 3 for that type of fusible plug.

6.2.3.2. As an alternate method, the plugs, after passing the test at a temperature of not less than 5 F below the specified minimum yield temperature may at once be immersed in another bath held at a temperature not exceeding the specified maximum yield temperature. If air leakage occurs within five minutes at that temperature, the requirements have been met.

6.2.3.3. Variation in temperature within the liquid bath in which the plug is immersed for either test in 6.2.3.1 or 6.2.3.2 shall be kept to a minimum by stirring while making these tests.

6.2.4. Fusible plugs to be used in chlorine service shall meet The Chlorine Institute, Inc.* requirements.

6.3. Tests of Rupture Disk Devices

6.3.1. The production of rupture disks shall be segregated into lots of not more than 3000 disks with appropriate control exercised to assure uniformity of production. Representative samples shall be selected at random for testing to verify rated bursting pressure. The number of samples selected shall be appropriate for the manufacturing procedures followed, but at least two samples shall be tested from each lot. Samples shall be mounted in a proper holder with a pressure opening having dimensions identical with that in the device in which it is to be used and submitted to a burst test at a temperature not lower than 60 F (15.6 C) nor higher than 160 F (71.1 C). The test pressure may be raised rapidly to 85 percent of the rated

*Chlorine Institute Inc., 342 Madison Avenue, New York, NY 10017.

burst pressure, held there for at least 30 seconds, and thereafter shall be raised at a rate not in excess of 100 psig (690 kPa) per minute, until the disk bursts. The actual burst pressure of the disk shall not be in excess of its rated burst pressure and not less than 90 percent of its rated burst pressure.

For rupture disks for DOT-39 cylinders see 4.2.1.2.

For DOT-4L cylinders, the actual burst pressure shall not exceed 105 percent of its rated burst pressure and not less than 90 percent of its rated burst pressure.

If the actual burst pressure is not within the limits prescribed above, the entire lot of rupture disks shall be rejected. If the manufacturer so desires, he may subject four more disks selected at random from the same lot to the same test. If all four additional disks meet the requirement, the lot may be used; otherwise the entire lot shall be rejected. Any elevated temperature determination may be arrived at by tests conducted at room temperature, provided that the relation of burst pressure to different temperatures is established by test for the type of material used.

6.3.2. The production of rupture disk holders (that part containing the pressure opening) of 3000 or less shall be considered a lot. Two representative holders selected at random from the lot shall be assembled with proper rupture disks from an acceptable lot as tested in 6.3.1 and subjected to the burst pressure test of 6.3.1. The actual burst pressure shall not be in excess of the rated burst pressure of the disk nor less than 85 percent of the rated burst pressure. For DOT 4L cylinders, the actual burst pressure of the disk shall not exceed 105 percent of its rated burst pressure and not less than 90 percent of its rated burst pressure. If the actual burst pressure at a temperature not less than 60 F (15.6 C) nor more than 160 F (71.1 C) is not within the above limits, the entire lot of rupture disk holders shall be rejected. If the manufacturer so desires, he may subject four more holders selected as above from the same lot to the same test. If all four holders meet the requirement, the lot may be used, otherwise the entire lot shall be rejected. Any elevated temperature determinations may be arrived at

by tests conducted at room temperature, provided that the relation of burst pressure to different temperatures is established by test for the type of material used.

6.3.3. Testing of the assembled rupture disk and holder for detail requirements specified in 6.3.1 and 6.3.2 in lieu of individual tests will be considered as complying with requirements of both 6.3.1 and 6.3.2.

6.4. Tests of Combination Rupture Disk-Fusible Plug Pressure-Relief Devices

6.4.1. The production of combination rupture disk-fusible plug devices of any one rated burst pressure and any one yield temperature during one shift, not to exceed 10 hours, shall be considered a lot. Two representative assembled devices shall be selected at random and submitted to a performance test conducted as follows:

6.4.1.1. Each assembled device shall be subjected to a pressure of 70 to 75 percent of rated burst pressure of the rupture disk used, and while under this pressure, shall be immersed in a liquid bath held at a temperature not less than 5 F below the minimum specified yield temperature of the fusible metal for at least ten minutes. The fusible metal shall not show evidence of melting. The temperature of the bath shall then be raised at a rate not in excess of 5 F per minute without material change in pressure. When the maximum specified yield temperature of the fusible metal is reached the fusible metal must have melted. There shall be no leakage.

6.4.1.2. The rupture disk shall then be tested in accordance with the requirements of 6.3.1. The device may be removed from the bath for this test.

6.4.1.3. As an alternate to tests in 6.4.1.1 and 6.4.1.2 the rupture disk and fusible metal may be tested separately to requirements 6.2.3 and 6.3.1, providing the design of the device is such as to allow for the separation of the parts and the separate tests.

6.4.1.4. If either of the devices fail to meet the requirements given in 6.4.1.1, 6.4.1.2 or 6.4.1.3 the entire lot shall be rejected. If the manufacturer so desires, he may subject four more such devices selected at random to the same test. If all four additional devices meet the requirements, the lot may be used.

6.5. Pressure Tests of Pressure-Relief Valves

6.5.1. Each pressure-relief valve, except those for DOT-39 cylinders, shall be subjected to an air or gas pressure test to determine that the start-to-discharge pressure at which the first bubble appears through a water seal of not over 4 in. on the outlet of the pressure-relief valve is not less than 75 percent nor more than 100 percent of the flow rating pressure for which the pressure-relief valve is marked. In any case, minimum required flow capacity must be achieved at flow rating pressure.

6.5.2. The production of pressure-relief valves for DOT-39 cylinders shall be subjected to an air or gas pressure test to determine the following:

6.5.2.1. Each pressure-relief valve shall be tested for leakage at cylinder test pressure, for a minimum of 30 seconds utilizing a water seal of not over 4 in. (10.16 cm) on the outlet of the pressure-relief valve or by any other method equally as sensitive. Any valve exhibiting leakage shall be rejected.

6.5.2.2. Two pressure-relief valves taken from each lot of 3000 valves or less shall be tested to determine that the start-to-discharge pressure at which the first bubble appears through a water seal of not over 4 in. (10.16 cm) on the outlet of the pressure-relief valve does not exceed 80 percent of the minimum cylinder burst pressure and is not less than 105 percent of the cylinder test pressure. If a failure occurs, the entire lot shall be rejected.

6.6. Flow Capacity Tests

6.6.1. The flow capacity of each design and modification thereof of all types of pressure-relief devices shall be determined by actual flow test. Three samples of each size of each device representative of standard production shall be tested. Each device shall be caused to operate either by pressure or by temperature, or by a combination of such effects, but not exceeding the maximum temperature and pressure for which it was designed.

6.6.1.1. After pressure testing and without cleaning, removal of parts or reconditioning, each pressure-relief device shall be subjected to an actual flow test wherein the amount of air or gas released by the device is measured. The rated flow capacity of the device shall be the average flow capacity of the three devices provided the individual flow capacities fall within 10 percent of the highest flow capacity recorded.

6.6.2. Acceptable methods of flow testing shall be one of the following:

6.6.2.1. Pressure-relief devices may be tested for flow capacity by testing with equipment conforming to the American Gas Association Gas Measurement Committee Report No. 3, "Orifice Metering of Natural Gas" as Reprinted with Revisions, 1969 (see Reference 1). Where this testing method is employed, such test may be made by the manufacturer of the pressure-relief device or a qualified test laboratory. The form Basis For Sizing of Pressure-Relief Device (see pages 88–89), showing the results of these tests, shall be completed and retained by the manufacturer.

6.6.2.2. Air or gas shall be supplied to the pressure-relief device through a supply pipe provided with a pressure gage and a temperature measuring device for indicating or recording the pressure and temperature of the supply. Observations shall be made and recorded after steady flow conditions have been established. Test conditions need not be the same as the conditions under which the device is expected to function in service, but the following limits must be met. The inlet pressure of the air or gas supplied to the pressure-relief device shall be not less than 100 psi absolute (690 kPa abs.), except that the flow test of a pressure-relief valve shall be made at the flow rating pressure, and the flow test of the rupture disk for the DOT-4L cylinders covered in 5.9 shall be made at the rated burst pressure of the rupture disk. Such test may be made by the manufacturer of the pressure-relief device or by a qualified test laboratory. The basis for sizing of pressure-relief device form (Appendix A), showing the results of these tests, shall be completed and retained by the manufacturer.

6.6.2.3. Where any other method of testing is used, a record of the accuracy of the test results prepared by a competent disinterested agency should be retained by the manufacturer.

6.7. Rejected Pressure-Relief Device

6.7.1. Rejected pressure-relief devices or components may be reworked provided they are subjected to such additional tests as are required to assure compliance with all the requirements of this Standard.

7. IDENTIFICATION

It is the purpose of this section to list certain safeguards or guides so that pressure-relief device performance may not be jeopardized by improper service practices. The aim in general, is to make it possible to identify the manufacturer of the device and to have the main replaceable parts so identified or coded that it may be readily determined, usually by reference to manufacturer's published data, whether parts are intended to function together, what operating pressure range or temperature range they will provide for, and whether they have adequate flow capacity for the cylinder with which they are to be employed. In particular, it is pointed out that rupture disks can be applied only against pressure openings for which they were specifically designed. Some manufacturers employ sharp pressure opening contours while others employ rounded or other shaped contours. Because of these contour variations, an interchange of the disks will give widely different burst pressures. In addition, variation in diameter for the pressure opening will give still wider variation in burst pressure if the disks are interchanged improperly.

7.1. Suitable marking shall be provided so that the manufacturer of the pressure-relief device may be determined.

7.2. When rupture disks and pressure opening parts are designed to be replaced as individual piece parts, they shall be marked to indicate the rated burst pressure (with the proper mating part), the flow capacity and the manufacturer. Suggested methods of marking are as follows:

7.2.1. Stamp with manufacturer's name or trademark and rated burst pressure or identifying part number on the part containing the pressure opening.

7.2.2. Ink or otherwise mark the number on the rupture disk or apply other code mark to facilitate determination of burst pressure range and proper mating part.

7.2.3. When rupture disk and pressure opening parts are combined in a factory assembled pressure-relief device designed to be replaced as a unit (CG-1, CG-4 or CG-5) the assembly shall be externally marked to indicate rated burst pressure, flow capacity, manufacturer and yield temperature if applicable.

7.3. Fusible metal pressure-relief devices (CG-2 or CG-3) shall be externally marked to indicate yield temperature and manufacturer.

7.4. Pressure-relief valves shall be marked to indicate:

(a) Manufacturer.

(b) The set pressure for which the valve is "set to start-to-discharge."

(c) The flow rating pressure in pounds per square inch gauge (psig) at which the flow capacity of the valve is determined.

(d) The flow capacity in cubic feet per minute of free air.

7.5. All markings required in 7.2 through 7.4 inclusive may be coded. Code designations shall be determinable from the manufacturer. Pressure-relief devices used on DOT-39 cylinders are exempt from marking requirements.

8. MAINTENANCE REQUIREMENTS FOR PRESSURE-RELIEF DEVICES

8.1. General Practices

8.1.1. As a precaution to keep cylinder pressure-relief devices in reliable operating condition, care shall be taken in the handling or storing of compressed gas cylinders to avoid damage. Care shall also be exercised to avoid plugging by paint or other dirt accumulation of pressure-relief device channels or other parts which could interfere with the functioning of the device. Only qualified personnel shall be allowed to service pressure-relief devices. Only assemblies or original manufacturer's parts shall be used in the repair of pressure-relief devices unless the interchange of parts has been proved by suitable tests.

8.2. Routine Checks When Filling Cylinders

8.2.1. Each time a compressed gas cylinder is received for refilling, all pressure-relief devices shall be examined externally for corrosion, damage, plugging of external pressure-relief device channels, and mechanical defects such as leakage or extrusion of fusible metal. This examination does not apply to DOT-4L cylinders. If there is any doubt regarding the suitability of the pressure-relief device for service, the cylinder shall not be filled until it is equipped with a suitable device.

9. REFERENCES

(1) Gas Measurement Committee Report No. 3, "Orifice Metering of Natural Gas." American Gas Association, 1515 Wilson Boulevard, Arlington, VA 22209. Reprinted with revisions 1969.
(2) Code of Federal Regulations, Title 49, Transportation, Parts 100 to 199. United States Government Printing Office, Washington, DC 20402.
(3) "Origin of Air Flow Capacity Requirements in Revised CGA Safety Device Schedule." Compressed Gas Association, Inc., 500 Fifth Avenue, New York, NY 10110.
(4) "Recommended Practice for the Manufacture of Fusible Plugs," Pamphlet S-4. Compressed Gas Association, Inc., 500 Fifth Avenue, New York, NY 10110.
(5) CGA Pamphlet V-1. American and Canadian Standard, "Compressed Gas Cylinder Valve Outlet and Inlet Connections." Compressed Gas Association, Inc., 500 Fifth Avenue, New York, NY 10110.
(6) Canadian Transport Commission, "Regulations for The Transportation of Dangerous Commodities by Rail." The Supervisor of Government Publications, Department of

Public Printing and Stationery, Ottawa, Canada K1A.
(7) CGA Pamphlet C-14. "Procedures for Fire Testing of DOT Cylinder/Safety Device Systems."
(8) CGA Pamphlet C-12. "Qualification Procedure for Acetylene Cylinder Design."

REQUIRED PRESSURE-RELIEF DEVICES

The types of pressure-relief devices listed in Table 1 are acceptable as indicated in Table 3 by a letter symbol or symbols for application on cylinders for various compressed gases and gas mixtures. In the event that a fire test is required, it shall be performed in accordance with CGA Pamphlets C-12 and C-14. A fire test shall be conducted when the flow capacity of a pressure-relief device is sized less than required by formula in this standard.

Requests for types and applications of pressure-relief devices other than those listed in Table 1 or Table 3 must be sent to the Compressed Gas Association, Inc. for assignment and be accompanied by test data as shown on form suggested in pages 88–89 of this Standard.

TABLE 1. TYPES OF PRESSURE-RELIEF DEVICES

CG-1	Rupture disk
CG-2	165 F (74 C) Fusible plug
CG-3	212 F (100 C) Fusible plug
CG-4	Rupture disk with 165 F (74 C) fusible alloy backing
CG-5	Rupture disk with 212 F (100 C) fusible alloy backing
CG-7	Pressure-relief valve

Note 1: When more than one type of device is listed in Table 3 for a particular gas, only one type is required.
Note 2: The letter codes used in Table 3 are defined on pages 82–83.
Note 3: Type CG-4 and CG-5 are not acceptable for 110 percent fill. See 173.302 (c) of 49 CFR.
Note 4: For certain gases, use of pressure-relief devices is not permitted. For such gases the pressure-relief device column is marked "Prohibited."

TABLE 2. FTSC NUMERICAL CODE FOR GAS CLASSIFICATION

1st Digit — Fire Potential

0	= inert
1	= supports combustion (oxidizing)
2	= flammable: Lower limit of flammility less than 13% or flammable range greater than 12%.
3	= pyrophoric
4	= highly oxidizing
5	= may decompose or polymerize and is flammable

2nd Digit — Toxicity

1	= non-toxic: over 500 ppm permitted for 8 hours exposure
2	= toxic: 50 to 500 ppm permitted (or very toxic with good warning properties) for 8 hours exposure.
3	= very toxic: less than 50 ppm permitted for 8 hours exposure.
4	= DOT poison A or others of similar toxicity
5	= DOT poison A or others of similar toxicity used in the electronic industry

3rd Digit — State of Gas: (in the cylinder @ 70 F) (21 C.)*

0	= non-cryogenic liquefied gas (less than 500 psi) (3500 kPa)**—gas withdrawal
1	= non-cryogenic liquefied gas (over 500 psi) (3500 kPa)—gas withdrawal
2	= liquefied gas (liquid withdrawal)***
3	= dissolved gas
4	= non-liquefied gas—or cryogenic gas withdrawal (less than 500 psi), (3500 kPa)
5	= Europe only
6	= non-liquefied gas between 500 and 3000 psi (3,500 & 20,000 kPa)
7	= non-liquefied gas above 3000 and below 10,000 psi (20,000 & 70,000 kPa)
8	= cryogenic gas (liquid withdrawal) above −240 C
9	= cryogenic gas (liquid withdrawal) below −240 C

4th Digit — Corrosiveness:

0	= non-corrosive
1	= non-halogen acid forming
2	= basic
3	= halogen acid forming

*The temperature of the cryogenic gases are always below 70 F (21 C).
**If pressure at 130 F (50 C) is over 600 psi (4000 kPa), use digit 1.
***When separate outlet for liquid withdrawal is specified.

TABLE 3. ALPHABETICAL LIST OF GASES AND DEVICES ASSIGNED

Cryogenic Liquids

FTSC	Name of Gas	CG-1 DISK	CG-2 165 F	CG-3 212 F	CG-4 165 F W/DISK	CG-5 212 F W/DISK	CG-7 RV
	ARGON	G					
	HELIUM	G					
	HYDROGEN	G					
	NEON	G					
	NITROGEN	G					
	OXYGEN	G					

Gases

FTSC	Name of Gas	CG-1 DISK	CG-2 165 F	CG-3 212 F	CG-4 165 F W/DISK	CG-5 212 F W/DISK	CG-7 RV
5130	Acetylene			F			
1160	Air	A		KB	B		K
2100	Allene		B				A
	Allylene (See Methylacetylene)						
0202	Ammonia, Anhydrous (Over 165# Weight)		E				
0303	Antimony Pentafluoride			Prohibited			
0160	Argon	A			B		K
2500	Arsine			Prohibited			
	Boron Chloride (See Boron Trichloride)						
	Boron Fluoride (See Boron Trifluoride)						
0203	*Boron Trichloride		L				
0263	Boron Trifluoride				B		
4303	*Bromine Pentafluoride			Prohibited			
4303	*Bromine Trifluoride			Prohibited			
0403	*Bromoacetone			Prohibited			
0100	*Bromochlorodifluoromethane (R12B1 or Halon 1211)	L					L
0100	*Bromochloromethane (Halon 1011)			None Required			
	Bromoethylene (See Vinyl Bromide)						
	Bromomethane (See Methyl Bromide)						
3100	Bromotrifluoroethylene (R113B1)	C					A
0100	Bromotrifluoromethane (R13B1) or (Halon 1301)	A					A
5100	1, 3 Butadiene, (Inhibited)						A
2100	Butane, Normal			M			A
2100	1-Butene						A
2100	2-Butene						A
0110	Carbon Dioxide	A			D		K
	Carbon Dioxide-Nitrous Oxide Mixture (Liquid)	A			D		K

*Not a compressed gas.

TABLE 3 (*Continued*)

FTSC	Name of Gas	CG-1 DISK	CG-2 165 F	CG-3 212 F	CG-4 165 F W/DISK	CG-5 212 F W/DISK	CG-7 RV
	Carbon Dioxide-Oxygen Mixture (Gas)	A			B		K
	Carbonic Acid (See Carbon Dioxide)						
2260	Carbon Monoxide				J	J	
	Carbon Oxysulfide (See Carbonyl Sulfide)						
	Carbon Tetrafluoride (See Tetrafluoromethane)						
	Carbonyl Chloride (See Phosgene)						
0413	Carbonyl Fluoride		Prohibited				
2301	Carbonyl Sulfide		B		BC		
1203	Chlorine (See Par. 6.2.4)		H				
4303	Chlorine Pentafluoride			Prohibited			
4303	Chlorine Trifluoride			Prohibited			
2100	1-Chloro-1, 1-Difluoroethane (R142B)						A
0100	Chlorodifluoromethane (R22)	A					A
	Chlorodifluoromethane Chloropentafluoroethane (Mixture) (R502)	A					A
	Chloroethane (See Ethyl Chloride)						
	Chloroethylene (See Vinyl Chloride)						
2100	Chlorofluoromethane (R31)						A
0100	Chloroheptafluorocyclobutane (R-C317)	A					A
	Chloromethane (See Methyl Chloride)						
0100	Chloropentafluoroethane (R115)	A					A
0100	1-Chloro-1,2,2,2-Tetrafluoroethane (R124)	A					A
0100	1-Chloro-2,2,2-Trifluoroethane (R133A)	A					A
5200	Chlorotrifluoroethylene (R1113)	C					A
0100	Chlorotrifluoromethane (R13)	A			P		
2400	Cyanogen			Prohibited			
0403	Cyanogen Chloride			Prohibited			
2100	Cyclobutane		E				A
2100	Cyclopropane		E		D		A
2160	Deuterium	N			J	J	
0213	Deuterium Chloride				B		
0203	*Deuterium Fluoride			None Required			
2500	Deuterium Selenide			Prohibited			
2301	Deuterium Sulfide		B		BC		
5360	Diborane				B	B	
1200	*Dibromodifluoroethane			None Required			
0200	*Dibromodifluoromethane (R12B2) or (Halon 1202)			None Required			
0100	*1,2 Dibromotetrafluoroethane (R114B2) (Halon 2402)	L					L
0100	*1,2 Dichlorodifluoroethylene			None Required			
0100	Dichlorodifluoromethane (R12)	A					A
	Dichlorodifluoromethane-Difluoroethane (Mixture) (R500)	A					A
0200	*1,2 Dichloroethylene (R1130)			None Required			

*Not a compressed gas.

TABLE 3 (Continued)

FTSC	Name of Gas	CG-1 DISK	CG-2 165 F	CG-3 212 F	CG-4 165 F W/DISK	CG-5 212 F W/DISK	CG-7 RV
0100	*Dichlorofluoromethane (R21)	L					L
0100	*1,2 Dichlorohexafluorocyclobutane (R-C316)			None Required			
2203	*Dichlorosilane				B		
0100	*1,1 Dichlorotetrafluoroethane (R114A)	L					L
0100	*Dichlorotetrafluoroethane (R114)	L					L
0100	*2,2 Dichloro-1,1,1-Trifluoroethane (R123)			None Required			
	Dicyan (See Cyanogen)						
3300	*Diethylzinc			Prohibited			
2100	1,1 Difluoroethane (R152A)		E				A
2110	1,1 Difluoroethylene (R1132A)	A			E		
	Difluoromethane (See Methylene Fluoride)						
2100	Difluoromonochloroethane (R142B)		E				A
2202	*Dimethylamine, Anhydrous			None Required			
2100	Dimethyl Ether						A
2100	*2,2 Dimethylpropane						L
0403	Diphosgene			Prohibited			
2110	Ethane	J				J	
2100	*Ethylacetylene						L
2100	*Ethyl Chloride						L
0403	Ethyldichloroarsine			Prohibited			
2160	Ethylene	J					
5200	*Ethylene Oxide			(See CFR 173.124)			
2100	*Ethyl Ether						L
2400	Ethyl Fluoride			Prohibited			
4343	Fluorine			Prohibited			
0100	Fluoroform (R23)	A			E		
2400	Germane			Prohibited			
0160	Helium	A			B		K
	Helium-Oxygen Mixture	A			B		K
2400	Heptafluorobutyronitrile			Prohibited			
0203	Hexafluoroacetone		B		B		
2400	Hexafluorocyclobutene			Prohibited			
0100	Hexafluoroethane (R116)	A			B		
0100	Hexafluoropropylene (R1216)	A					A
2160	Hydrogen	N			J	J	K
0203	Hydrogen Bromide				E		
0313	Hydrogen Chloride				B		
5301	Hydrogen Cyanide			Prohibited			
0203	*Hydrogen Fluoride			None Required			
0203	Hydrogen Iodide				B		
2500	Hydrogen Selenide			Prohibited			
2301	Hydrogen Sulfide		B		BC		

*Not a compressed gas.

TABLE 3 (*Continued*)

FTSC	Name of Gas	CG-1 DISK	CG-2 165 F	CG-3 212 F	CG-4 165 F W/DISK	CG-5 212 F W/DISK	CG-7 RV
4303	*Iodine Pentafluoride			None Required			
2100	Isobutane						
2100	Isobutylene						A
0160	Krypton	A			B		K
0403	Lewisite			Prohibited			
2160	Methane	N			J	J	K
2100	Methylacetylene						A
0300	*Methyl Bromide			None Required			
2100	*3-Methyl-1-Butene						L
2200	Methyl Chloride						A
0403	Methyldichloroarsine			Prohibited			
0110	Methylene Fluoride (R32)	A					A
2200	*Methyl Formate			None Required			
2201	Methyl Mercaptan			None Required			
2202	*Monoethylamine			None Required			
2202	Monomethylamine, Anhydrous			None Required			
0403	Mustard Gas			Prohibited			
2160	Natural Gas	N			J	J	K
0160	Neon	A			B	B	K
2400	*Nickel Carbonyl			Prohibited			
4361	Nitric Oxide			Prohibited			
0160	Nitrogen	A		KB	B		K
4401	*Nitrogen Dioxide }			Prohibited			
4401	*Nitrogen Tetroxide {						
4343	Nitrogen Trifluoride			B	B		
4301	Nitrogen Trioxide			Prohibited			
0203	Nitrosyl Chloride (Over 10 lbs. weight)			Not Required — 10 lb. weight and under			
0303	Nitrosyl Fluoride			Prohibited			
4110	Nitrous Oxide	A			D		
0303	Nitryl Fluoride			Prohibited			
0100	Octafluorocyclobutane (R-C318)						A
0100	Octafluoropropane (R218)	A					A
4160	Oxygen	A			B		K
4343	Oxygen Difluoride			Prohibited			
4330	Ozone (Dissolved in R13)			Prohibited			
3300	*Pentaborane			Prohibited			
2400	Pentafluoropropionitrile			Prohibited			
4303	Perchloryl Fluoride			Prohibited			
0100	Perfluorobutane		E				A
0200	*Perfluoro-2-Butene						L
0303	Phenylcarbylamine Chloride			Prohibited			

*Not a compressed gas.

TABLE 3 (Continued)

FTSC	Name of Gas	CG-1 DISK	CG-2 165 F	CG-3 212 F	CG-4 165 F W/DISK	CG-5 212 F W/DISK	CG-7 RV
0403	Phosgene			Prohibited			
3510	Phosphine			Prohibited			
0203	Phosphorous Pentafluoride				B		
0203	Phosphorous Trifluoride				B		
2100	Propane			M			A
2100	Propylene						A
3260	Silane				B		
0263	Silicon Tetrafluoride				B		
5300	Stibine			Prohibited			
0201	Sulfur Dioxide		E				
0100	Sulfur Hexafluoride	A				B	A
0203	Sulfur Tetrafluoride				B		
0300	Sulfuryl Fluoride		E				
5110	Tetrafluoroethylene-Inhibited (R1114)	A			B		
4343	Tetrafluorohydrazine			Prohibited			
0160	Tetrafluoromethane (R14)	A			B		K
2400	Tetramethyllead			Prohibited			
0100	*Trichlorofluoromethane (R11)	L					L
0100	*1,1,1 Trichlorotrifluoroethane (R113A)			None Required			
0100	*1,1,2 Trichlorotrifluoroethane (R113)			None Required			
3300	Triethylaluminum			Prohibited			
3300	Triethylborane			Prohibited			
2400	Trifluoroacetonitrile			Prohibited			
0303	Trifluoroacetyl Chloride			Prohibited			
2100	1,1,1 Trifluoroethane (R143A)		B				
4363	Trifluoromethyl Hypofluorite			Prohibited			
0200	Trifluoromethyl Iodide				B		
2202	*Trimethylamine			None Required			
3300	Trimethylstibine			Prohibited			
0303	*Tungsten Hexafluoride			Prohibited			
0303	*Uranium Hexafluoride			Prohibited			
5200	*Vinyl Bromide		L				L
5200	Vinyl Chloride		E				A
2100	Vinyl Fluoride				B		
5200	Vinyl Methyl Ether		E				A
0160	Xenon	A			B		K

*Not a compressed gas.

KEY TO SYMBOLS USED IN TABLE 3

A This device is required in one end of the cylinder only regardless of the length.

B When cylinders are over 65 in. (165 cm) long, exclusive of neck, this device is required at both ends. For shorter cylinders, the device is required in one end only.

C This device is permitted only in cylinders having a minimum required test pressure of 3000 psig (20,684 kPa) or higher, and is required in one end only. The bursting pressure of the disk shall be at least 75 percent of the minimum required test pressure of the cylinder.

D   This device is permitted only in cylinders which are in direct medical service. It is not to be used in cylinders which are in transfer service even though the gas itself is intended for medical purposes. When cylinders are over 55 in. (140 cm) long, exclusive of neck, the device is required in both ends.

E   When cylinders are over 30 in. (76 cm) long exclusive of neck, this device is required at both ends. For shorter cylinders, the device is required in one end only.

F   The number and location of pressure-relief devices for cylinders of any particular size shall be proved adequate as a result of the fire test and any change in style of cylinder, a filler or quantity of devices can only be approved if found adequate upon reapplication of the fire test. The fire test shall be conducted in accordance with CGA Pamphlet C-12, Qualification Procedure for Acetylene Cylinder Design (see Reference 8).

G   This device is required in one end of the cylinder only, regardless of length. A pressure controlling valve as required in 173.304 (b) (2) of DOT Regulation must also be used. This valve must be so sized and set as to limit the pressure in the cylinder to one and one-fourth times its marked service pressure less 15 psi if vacuum insulation is used. The insulation jacket shall be provided with a pressure-actuated device which will function at a pressure of not more than 25 psig (172 kPa) and provide a discharge area of 0.00012 square inch per pound water capacity of cylinder.

   An alternate pressure-relief valve, with a marked set pressure not to exceed 150 percent of the DOT Service Pressure may be used in lieu of the rupture disk device if the flow capacity required by 5.9 is provided at 120 percent of marked set pressure. Installation must provide for:

   (a)   Prevention of moisture accumulation at the seat by drainage away from that area.
   (b)   Periodic drainage of the vent piping.
   (c)   Avoidance of foreign material in the vent piping.

H   When cylinders are over 55 in. (140 cm) long, exclusive of neck, this device is required in both ends, except for cylinders purchased after October 1, 1944, which must contain no aperture other than that provided in the neck of the cylinder for attachment of a valve equipped with an approved pressure-relief device. (Chlorine cylinders do not generally exceed 55 in. (140 cm) in length, since DOT Regulations 173.304 (a) note 2, require that cylinders purchased after November 1, 1935 must not contain over 150 lb of chlorine.)

J   This device is required in only one end of cylinders having a length not exceeding 65 in. (165 cm), exclusive of neck. For cylinders over 65 in. (165 cm) long this device is required in both ends, and each device shall be arranged to discharge upwards and unobstructed to the open air in such a manner as to prevent any impingement of escaping gas upon the containers.

K   This device can be used up to 500 psig (3447 kPa) charging pressure.

L   This device recommended, but no pressure-relief device is required.

M   May be used in addition to CG-7.

N   For use only on cylinders over 65 in. (165 cm) long. This device is required on both ends and each device shall be arranged to discharge upwards and unobstructed to the open air in such a manner as to prevent any impingement of escaping gas upon the containers.

P   For use only on cylinders over 65 in. (165 cm) long. This device is required on both ends.

TABLE 4. TEMPERATURE CORRECTION FACTORS TO 60 F*

Degrees F	Factor	Degrees F	Factor	Degrees F	Factor
1	1.0621	51	1.0088	101	.9628
2	1.0609	52	1.0078	102	.9619
3	1.0598	53	1.0068	103	.9610
4	1.0586	54	1.0058	104	.9602
5	1.0575	55	1.0048	105	.9594
6	1.0564	56	1.0039	106	.9585
7	1.0552	57	1.0029	107	.9577
8	1.0541	58	1.0019	108	.9568
9	1.0530	59	1.0010	109	.9560
10	1.0518	60	1.0000	110	.9551
11	1.0507	61	.9990	111	.9543
12	1.0496	62	.9981	112	.9535
13	1.0485	63	.9971	113	.9526
14	1.0474	64	.9962	114	.9518
15	1.0463	65	.9952	115	.9510
16	1.0452	66	.9943	116	.9501
17	1.0441	67	.9933	117	.9493
18	1.0430	68	.9924	118	.9485
19	1.0419	69	.9915	119	.9477
20	1.0408	70	.9905	120	.9469
21	1.0398	71	.9896	121	.9460
22	1.0387	72	.9887	122	.9452
23	1.0376	73	.9877	123	.9444
24	1.0365	74	.9868	124	.9436
25	1.0355	75	.9859	125	.9428
26	1.0344	76	.9850	126	.9420
27	1.0333	77	.9840	127	.9412
28	1.0323	78	.9831	128	.9404
29	1.0312	79	.9822	129	.9396
30	1.0302	80	.9813	130	.9388
31	1.0291	81	.9804	131	.9380
32	1.0281	82	.9795	132	.9372
33	1.0270	83	.9786	133	.9364
34	1.0260	84	.9777	134	.9356
35	1.0249	85	.9768	135	.9349
36	1.0239	86	.9759	136	.9341
37	1.0229	87	.9750	137	.9333
38	1.0218	88	.9741	138	.9325
39	1.0208	89	.9732	139	.9317
40	1.0198	90	.9723	140	.9309
41	1.0188	91	.9715	141	.9302
42	1.0178	92	.9706	142	.9294
43	1.0168	93	.9697	143	.9286
44	1.0158	94	.9688	144	.9279
45	1.0147	95	.9680	145	.9271
46	1.0137	96	.9671	146	.9263
47	1.0127	97	.9662	147	.9256
48	1.0117	98	.9653	148	.9248
49	1.0108	99	.9645	149	.9240
50	1.0098	100	.9636	150	.9233

*From AGA Gas Measurement Report No. 3, "Orifice Metering of Natural Gas."

TABLE 5. BASIC ORIFICE FACTORS—FLANGE TAPS FOR FLOW PER MINUTE

Base Temperature 60 F
Base Pressure = 14.7 psia

Flow Temperature 60 F
Specific Gravity = 1.0

Orifice Diameter-Inches	Pipe Sizes — Extra Heavy, Schedule 80 Nominal and Published Inside Diameters (Inches)				
	2 1.939	3 2.900	4 3.826	6 5.761	8 7.981
.250	.2118	.2118*	.2115*	———	———
.375	.4740	.4730	.4726*	———	———
.500	.8431	.8386	.8372	.8364*	———
.625	1.3252	1.3114	1.3075	1.3049	———
.750	1.9270	1.8950	1.8858	1.8792	———
.875	2.6593	2.5902	2.5733	2.5605	2.5552
1.000	3.5412	3.4007	3.3700	3.3493	3.3398
1.125	4.6033	4.3325	4.2782	4.2453	4.2315
1.250	5.8930	5.3938	5.3005	5.2492	5.2297
1.375	7.4762	6.5967	6.4408	6.3617	6.3343
1.500		7.9560	7.7045	7.5838	7.5463
1.625		9.4942	9.0982	8.9172	8.8658
1.750		11.2407	10.631	10.363	10.293
1.875		13.2313	12.313	11.924	11.830
2.000		15.5108	14.157	13.602	13.475
2.125		18.1867	16.183	15.401	15.231
2.250			18.412	17.325	17.098
2.375			20.868	19.377	19.078
2.500			23.588	21.563	21.172
2.625			26.583	23.892	23.382
2.750			29.952	26.368	25.708
2.875				29.000	28.155
3.000				31.797	30.725
3.125				34.773	33.420
3.250				37.942	36.243
3.375				41.318	39.200
3.500				44.918	42.295
3.625				48.762	45.530
3.750				52.868	48.913
3.875				57.262	52.448
4.000				61.970	56.142
4.250				72.580	64.038
4.500					72.675
4.750					82.135
5.000					92.530
5.250					103.940
5.500					116.533
5.750					130.500
6.000					

*These orifices have diameter ratios lower than the minimum value for which the formulas used were derived and this size of plate should not be used unless it is understood that the accuracy of measurement will be relatively low.

(Data were taken from Gas Measurement Committee Report No. 3, "Orifice Metering of Natural Gas." (1956 Reprint), American Gas Association, and converted to calculations in cubic feet per minute.)

TABLE 6. VALUES OF G_i AND G_u FOR RATED BURST PRESSURES OF RUPTURE DISKS FOR DOT 4L CYLINDERS

Commodity	Rate Burst Pressure (psig) or Flow Rating Pressure	Value of G_i	Value of G_u
Argon, pressurized liquid	100	10.2	59.0
	200	11.8	69.5
	300	13.9	82.0
	400	17.9	108.0
Helium, pressurized liquid	200	52.5	
Hydrogen, liquefied	50	8.7	46.0
	100	10.6	56.0
Nitrogen, pressurized liquid	100	10.2	59.0
	200	11.8	69.5
	300	13.9	82.0
	400	17.9	108.0
Oxygen, pressurized liquid	100	10.2	59.0
	200	11.8	69.5
	300	13.9	82.0
	400	17.9	108.0
Neon, pressurized liquid	100	17.0	92.0
	200	20.8	113.5
	300	28.0	153.0

Note: When lower rated burst pressures than those shown are used, the values of G_i and G_u are on the safe side and may be used as shown or calculated as covered below. For higher rated burst pressures than shown, values of G_i and G_u must be calculated from the following formulas:

$$G_u = \frac{633,000}{LC} \sqrt{\frac{ZT}{M}} \quad \text{and} \quad G_i = \frac{73.4 \times (1200 - t)}{LC} \sqrt{\frac{ZT}{M}}$$

where

L = Latent heat at flowing conditions in Btu per pound.
C = Constant for gas or vapor related to ratio of specific heats ($k = C_p/C_v$) at 60 F and 14.7 psia (see Fig. 1).
Z = Compressibility factor at flowing conditions.
T = Temperature in degrees R (Rankin) of gas at pressure at flowing conditions ($t + 460$).
M = Molecular weight of gas.
t = Temperature in degrees F (Fahrenheit) of gas at pressure at flowing conditions.

When compressibility factor Z is not known, 1.0 is a safe value of Z to use. When gas constant C is not known, 315 is a safe value of C to use. For complete details concerning the basis and origin of these formulas, refer to "How to Size Safety Relief Devices," F. S. Heller, Phillips Petroleum Company, 1954.

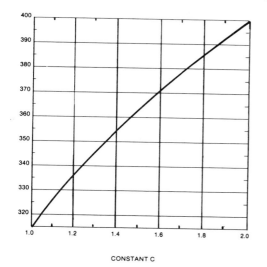

k	CON-STANT C	k	CON-STANT C	k	CON-STANT C
1.00	315	1.26	343	1.52	366
1.02	318	1.28	345	1.54	368
1.04	320	1.30	347	1.56	369
1.06	322	1.32	349	1.58	371
1.08	324	1.34	351	1.60	372
1.10	327	1.36	352	1.62	374
1.12	329	1.38	354	1.64	376
1.14	331	1.40	356	1.66	377
1.16	333	1.42	358	1.68	379
1.18	335	1.44	359	1.70	380
1.20	337	1.46	361	2.00	400
1.22	339	1.48	363	2.20	412
1.24	341	1.50	364		

CONSTANT C

Fig. 1. Constant C for gas or vapor related to ratio of specific heats ($k = C_p/C_v$) at 60 F and 14.7 psia. (Data from Figure UA-230, ASME Boiler and Pressure (Vessel Code, Section VIII, Division 1, Pressure Vessels.)

APPENDIX A

NOTE: This form is not suitable for acetylene cylinders. For further information contact the Compressed Gas Association, Inc., 500 Fifth Avenue, New York, New York 10036.

BASIS FOR SIZING
OF PRESSURE RELIEF DEVICE

Date_____

Manufacturer_____

Address_____

Catalog or Model No._____ _____

Dwg. No._____ Date of Dwg. and Latest Revision_____

Safety Relief Device Type CG—_____ (See Table 1 of these Standards)

Set Pressure_____psig. Flow Rating Pressure_____psia.

Yield Temperature _____ F. Rated Bursting Pressure_____psig.

Chemical Name of Gas_____Liquefied (   ) Non Liquefied (   )

Commercial Name of Gas_____

Percentage of Components for Mixed Gases_____

Specification and Service Pressure of DOT Cylinder(s) to be Used_____

Maximum Container Size for Which Approval is Requested _____
                                                        (pounds water capacity)

Minimum Required Flow-CFM of Air (See Pars. 4.5 to 4.9)_____

Actual Flow-CFM of Air at 60 F and Base Pressure of 14.7 psia. _____
                                                              (Item 16 of Test Data)

Test Conducted By: _____Title_____

    Company_____

    Signature_____Date_____

Test Requested By: _____Title_____

    Company_____

    Signature_____Date_____

NOTE: For safety relief devices on insulation jacket of DOT-4L cylinders, indicate:

Description of Device _____ Discharge Area_____ Set Pressure_____psig

(For Test Data, see next page)

APPENDIX A (*Continued*)

TEST DATA

This Form is suitable for Test Data using Orifice Meters.

Test Medium - Air_____or Name of Gas_____ Specific Gravity_____

Molecular Weight_____Ratio of Specific Heats (k)_____

ITEM		SAMPLES	1	2	3
1.	Start to Discharge Pressure—psig.				
2.	Resealing Pressure—psig.				
3.	Frangible Disc—Bursting Pressure—psig.				
4.	Fusible Plug—Yield Temperature—degrees F.				
5.	Flow Rating Pressure—psia. (psig + 14.7)				
6.	Orifice Diameter—Inches				
7.	Meter Pipe—Inside Diameter—Inches				
8.	Orifice Factor (For Flow in CFM.) See Table 3				
9.	Constant (Item 8 x $\sqrt{\text{Item 5}}$)				
10.	Differential Pressure— $\sqrt{\text{Inches Water}}$				
11.	Flow Temperature—degrees F.				
12.	Temperature Correction Factor. See Table 2.				
13.	Supercompressibility Factor (Air = 1.0)				
14.	Gas Constant Ratio(*).				
15.	Flow (Items 9 x 10 x 12 x 13 x 14)				
16.	AVERAGE FLOW AT 60 F and 14.7 psia				

(*) Gas constant ratio for air = 1.0; for other than air = 356/Gas Constant (C). See Figure. 1.

SECTION B

Pressure-Relief Device Standards—Part 2: Cargo and Portable Tanks for Compressed Gases

1. FOREWORD

This Part of the Pressure-Relief Device Standards represents the minimum requirements recommended by the Compressed Gas Association, Inc. for pressure-relief devices for use on cargo tanks and portable tanks for compressed gases having water capacity exceeding 1000 lb and which comply with the specifications and charging and maintenance regulations of the Department of Transportation (DOT) or the corresponding portable tank specifications and regulations of the Canadian Transport Commissioners (CTC).

It is recognized that there are cargo and portable tanks that conform with the specification requirements of the DOT and portable tanks that conform with the specification requirements of the CTC which are used in services that are beyond the jurisdiction of either of these authorities. In such cases it is recommended that state, provincial, local or other authorities having jurisdiction over these containers be guided by Part 2 of these Standards in determining adequate pressure-relief device requirements, provided that the cargo and portable tanks are charged with gas and maintained in accordance with the DOT or CTC Regulations that apply.

It is further recognized that there may be cargo and portable tanks which are used in services beyond the jurisdiction of the DOT or the CTC and which do not conform to the specification requirements of either authority. It is recommended that the authorities having jurisdiction over such cargo and portable tanks be guided by Part 2 of these Standards in determining pressure-relief device requirements, provided that such cargo and portable tanks are considered by the authority as having a construction at least equal to the equivalent DOT or CTC specification requirements and further provided that the cargo and portable tanks shall be charged with gas and maintained in accor-

dance with the DOT or CTC requirements that apply.

It is recommended that cargo and portable tanks fabricated after December 31, 1980 utilize pressure-relief devices which meet the requirements of this edition of Pamphlet S-1 Part 2 of the CGA Pressure-Relief Device Standards.

2. DEFINITIONS

For the purpose of these Standards the following terms are defined:

2.1. The term *cargo tank* means any container designed to be permanently attached to any motor vehicle or other highway vehicle and in which is to be transported any compressed gas. The term cargo tank shall not be construed to include any tank used solely for the purpose of supplying fuel for the propulsion of the vehicle or containers fabricated under specifications for cylinders.

2.2. The term *portable tank* means any container designed primarily to be temporarily attached to a motor vehicle, other vehicle, railroad car other than tank car, or marine vessel, and equipped with skids, mountings or accessories to facilitate handling of the container by mechanical means, in which is to be transported any compressed gas. The term portable tank shall not be construed to include any cargo tank, any tank car tank, or any tank of the DOT 106A and DOT 110-A-W type.

2.3. A *pressure-relief device* is a device designed to open at a specified value of pressure. It may be a safety relief valve, a nonreclosing pressure-relief device, or a nonreclosing pressure-relief device in combination with a safety relief valve.

2.4. A *safety relief valve* is a pressure-relief device characterized by rapid opening pop action or by opening generally proportional to the increase in pressure over the opening pressure.

2.5. The *set pressure* of a safety relief valve is the pressure marked on the valve and at which it is set to start-to-discharge.

2.6. The *start-to-discharge pressure* of a safety relief valve is the pressure measured at the valve inlet at which there is a measurable lift, or at which discharge becomes continuous as determined by seeing, feeling or hearing.

2.7. The *resealing pressure* is the value of decreasing inlet static pressure at which no further leakage is detected after closing. The method of detection may be a specified water seal on the outlet or other means appropriate for the application.

2.8. The *flow capacity* is the relieving capacity of a pressure-relief device determined at the flow rating pressure, expressed in cubic feet per minute of free air discharge.

2.9. The *flow rating pressure* is the inlet static pressure at which the relieving capacity of a pressure-relief device is determined for rating purposes.

2.10. Free air or *free gas* is air or gas measured at a pressure of 14.7 psia (101.4 kPa) and a temperature of 60 F (15.6 C).

2.11. A *nonreclosing pressure-relief device* is a pressure-relief device designed to remain open after operation. A manual resetting means may be provided.

2.11.1. A *rupture disk device* is a nonreclosing pressure-relief device actuated by inlet static pressure and designed to function by the bursting of a disk.

2.11.2. A *breaking pin device* is a nonreclosing pressure-relief device actuated by inlet static pressure and designed to function by the breakage of a load carrying section of a pin which supports a pressure containing member. It shall only be used in combination between a safety relief valve and the container.

2.11.3. A *fusible plug device* is a nonreclosing device designed to function by the yielding or melting of a plug.

2.12. A *combination pressure-relief device* is one of the following:

(a) A breaking pin device in combination with a safety relief valve.

(b) A rupture disk device in combination with a safety relief valve.

2.13. The *Code* as used in these Standards is defined as (1) Paragraph U-68, U-69, U-200 or U-201 of Section VIII of the Boiler and Pressure Vessel Code of the American Society of Mechanical Engineers, 1949 edition: or (2) Section VIII Division 1 of the Boiler and Pressure Vessel Code of the American Society of Mechanical Engineers, 1950 edition through the current edition, including addenda; or (3) The Code for Unfired Pressure Vessels for Petroleum Liquids and Gases of the American Petroleum Institute and the American Society of Mechanical Engineers (API-ASME),* 1951 edition.

2.14. The term *DOT design pressure* as used in this Standard is identical to the term *maximum allowable working pressure* as used in the *Code* and is the maximum gage pressure at the top of the tank in its operating position during normal operation.

Exception: For containers constructed in accordance with Paragraph U-68 or U-69 of Section VIII of the ASME Boiler and Pressure Vessel Code, 1949 edition, the maximum allowable working pressure for the purpose of these standards is considered to be 125 percent of the design pressure as provided in 173.315 of DOT or CTC Regulations.

2.15. DOT Regulations as used in these Standards refers to Code of Federal Regulations, Title 49, Parts 100 to 199, Subchapters A, B and C. The following list provides a comparison of terms used in the DOT Regulations and in this Standard.

DOT Regulations	CGA S-1.2
**Design pressure	**Design pressure
Safety relief device	Pressure-relief device
Safety relief valve	Safety relief valve
Frangible disk device	Rupture disk device
(not used)	Rupture disk
Fusible plug	Fusible plug device

Note: Terminology used in this Standard for pressure-relief devices is consistent with ANSI B95.1-1977 where possible.

*The API-ASME Code, as a joint publication and interpretation service, was discontinued as of December 31, 1956.

**Identical to "Maximum Allowable Working Pressure" in Section VIII Division 1 of ASME Code.

2.16. CTC Regulations as used in these Standards refers to Canadian Transport Commissioners, "Regulations for the Transportation of Dangerous Commodities by Rail."

3. TYPES OF PRESSURE-RELIEF DEVICES

Types of pressure-relief devices covered by this part are as follows:

3.1. Safety Relief Valve

3.2. Fusible plug device utilizing a fusible alloy with yield temperature not over 170 F (76.7 C) nor less than 157 F (69.4 C)(165 F nominal) (73.9 C nominal). (See Part 1 of these Standards. Type CG-2).

3.3. Rupture Disk Device

3.4. Rupture disk device in combination with safety relief valve.

3.5. Breaking pin device in combination with safety relief valve.

4. APPLICATION REQUIREMENTS FOR PRESSURE-RELIEF DEVICES

4.1. General

4.1.1. Each container shall be provided with one or more pressure-relief devices which unless otherwise specified, shall be safety relief valves of the spring-loaded type.

Safety relief valves shall meet the applicable requirements for design, materials, installation, set pressure tolerance markings and certification of capacity of the current edition of one of the following standards:

(a) "Safety Relief Valves for Anhydrous Ammonia and LP-Gas," UL 132 (Ref. 11).

(b) ASME Section VIII Division 1, UG-125 through UG-136 (Ref. 12).

(c) Appendix A of AAR Specifications for Tank Cars (Ref. 14).

Requirements for set pressure, flow capacity and overpressure are specified in Section 5 of this Standard.

4.1.2. Safety relief valves shall have a marked set pressure at the DOT design pressure of the container except as follows:

4.1.2.1. If an overdesigned container is used, the marked set pressure of the safety relief valve may be between the minimum required DOT design pressure for the lading and the DOT design pressure of the container used.

4.1.2.2. For sulfur dioxide containers, a minimum marked set pressure of 120 and 110 psig (827.4 and 758.4 kPa) is permitted for the 150 and 125 psig (1034 and 861.8 kPa) DOT design pressure containers, respectively. (See Table 1.)

4.1.2.3. For carbon dioxide (refrigerated), nitrous oxide (refrigerated), and the gases listed in 4.1.10, there shall be no minimum marked set pressure.

4.1.2.4. For butadiene, inhibited, and liquefied petroleum gas containers, a minimum marked set pressure of 90 percent of the minimum DOT design pressure permitted for these ladings may be used. (See Table 1.)

4.1.2.5. For containers constructed in accord with Paragraph U-68 or U-69 of the Code, 1949 edition, the set pressure marked on the safety relief valve may be 125 percent of the original DOT design pressure of the container.

4.1.3. The design, material and location of pressure-relief devices shall have been proved to be suitable for the intended service.

4.1.4. Pressure-relief devices shall have direct communication with the vapor space of the container.

4.1.5. Any portion of liquid piping or hose which at any time may be closed at each end must be provided with a means for pressure relief to operate at a safe pressure.

4.1.6. The following additional restrictions apply to pressure-relief devices on containers for carbon dioxide or nitrous oxide which are shipped in refrigerated and insulated containers:

4.1.6.1. The operating pressure in the container may be regulated by the use of one or more pressure-controlling devices, which devices shall not be in lieu of the safety relief valve required in paragraph 4.1.1.

4.1.6.2. All pressure-relief devices shall be so installed and located that the cooling effect of the contents will not prevent the effective operation of the device.

4.1.6.3. In addition to the safety relief valve required by paragraph 4.1.1 each container for carbon dioxide or nitrous oxide may be equipped with one or more rupture disk devices of suitable design with a stamped bursting pressure not exceeding two times the DOT design pressure of the container.

4.1.7. Subject to conditions of 173.315 (a) (1) of DOT Regulations for methyl chloride and sulfur dioxide optional portable tanks of 225 psig (1551 kPa) minimum DOT design pressure, one or more fusible plugs may be used in lieu of safety valves of the spring-loaded type. If the container is over 30 in. (76.2 cm) long a pressure-relief device having the total required flow capacity must be provided at each end.

4.1.8. When storage containers for liquefied petroleum gas are permitted to be shipped in accordance with 173.315 (j) of DOT Regulations they must be equipped with pressure-relief devices in compliance with the requirements for safety relief devices on above-ground containers as specified in the current edition of National Fire Protection Association Pamphlet No. 58 "Standard For The Storage And Handling of Liquefied Petroleum Gases."

4.1.9. When containers are filled by pumping equipment which has a discharge capacity in excess of the capacity of the container pressure-relief devices, and which is capable of producing pressures in excess of the DOT design pressure of the container, precautions should be taken to prevent the development of pressures in the container in excess of 120 percent of its DOT design pressure. This may be done by providing a bypass on the pump discharge, or by any other suitable method.

4.1.10. The following requirements apply to pressure-relief devices on containers for liquefied methane, ethane, ethylene, carbon monoxide (see 4.1.10.1.2.2) and hydrogen, helium (refrigerated), and pressurized liquid oxygen, nitrogen, argon and neon.

4.1.10.1. The liquid container shall be protected by pressure-relief devices consisting of one or more safety relief valves, and one or more rupture disk devices except as modified

by 4.1.10.1.2.1, so installed that the devices remain at ambient temperature during normal container operation.

4.1.10.1.1. The marked set pressure of the safety relief valve shall not exceed the DOT design pressure of the container. The minimum capacity of the safety relief valve shall be sized to provide adequate venting capacity at 120 percent of the container DOT design pressure for operational emergency contingencies, except fire, but including loss of vacuum with insulation saturated with gaseous lading or air at atmospheric pressure, whichever provides the greater thermal conductance. The minimum required capacity of the safety relief valve for the loss of vacuum condition is:

$$Q_a = \frac{(130 - t)}{4(1200 - t)} G_i UA$$

where the value of U at the average temperature of the insulation may be used (alternatively the value of U at 100 F may be used). For helium and hydrogen, the value of U must be chosen on the basis that the insulation space is filled with helium and hydrogen gas, respectively, at one atmosphere. (See Table 1 and 5.3.3 for nomenclature.)

4.1.10.1.2. The rupture disk device must have a stamped bursting pressure not to exceed 120 percent of the DOT design pressure at a coincident disk temperature not to exceed 800 F (426.7 C). In addition the room temperature rating for the lot of disks must be determined, and must be less than 150 percent of the DOT design pressure (plus 15 psi (103.4 kPa) if vacuum insulation is used) of the container.

4.1.10.1.2.1. An alternate safety relief valve, with a marked set pressure not to exceed 110 percent of DOT design pressure may be used in lieu of the rupture disk device. Installation must provide for:

(a) Prevention of moisture accumulation at the seat by drainage away from that area.

(b) Periodic drainage of the vent piping.

(c) Avoidance of foreign material in the vent piping.

4.1.10.1.2.2. Rupture disk devices are not permitted on containers in liquefied carbon monoxide service.

4.1.10.1.3. The combined capacity of the pressure-relief devices shall be adequate to relieve the vapor generated at 20 percent above the DOT design pressure when the container is exposed to fire or other unexpected source of external heat. This required flow capacity shall be determined in accordance with the provisions of 5.3.2, 5.3.3 or 5.3.4 when applicable.

4.1.10.1.3.1. When the flow capacity of the safety relief valve meets the requirements of 4.1.10.1.3, the supplemental pressure-relief device meeting the requirements of 4.1.10.1.2 or 4.1.10.1.2.1 shall have a marked set pressure or stamped bursting pressure not exceeding:

(a) For containers other than vacuum insulated:

(1) For safety relief valve in lieu of rupture disk device, the marked set pressure shall not exceed 136 percent of DOT design pressure.

(2) For rupture disk device the stamped bursting pressure at 70 F (21.1 C) shall not exceed 150 percent of DOT design pressure.

(b) For vacuum-insulated containers:

(1) For safety relief valve in lieu of rupture disk device, the marked set pressure shall not exceed 136 percent of (DOT design pressure + 15 psi)(103.4 kPa).

(2) For rupture disk device, the stamped bursting pressure at 70 F (21.1 C) shall not exceed 150 percent of (DOT design pressure + 15 psi)(103.4 kPa).

4.1.10.2. The outer shell of vacuum-insulated containers shall be protected with means for pressure relief as specified in 6(a) (13) of CGA-341, "Insulated Tank Truck Specification" (Ref. 10).

4.2. Piping of Safety Relief Devices

4.2.1. When fittings and piping are used on either the upstream or downstream side or both of a pressure-relief device or devices, the passages shall be so designed that the flow capacity of the pressure-relief device will not be reduced below the capacity required for the container on which the pressure-relief device assembly is installed.

4.2.2. Pressure-relief devices shall be arranged to discharge to the open air in such a manner as to prevent any impingement of escaping gas upon the container. Pressure-relief devices shall be arranged to discharge upward except this is not required for carbon dioxide, helium, neon, nitrous oxide and pressurized liquid argon, nitrogen and oxygen.

4.2.3. No shutoff valves shall be installed between the pressure-relief devices and the container except, in cases where two or more pressure-relief devices are installed in the same container, a shutoff valve may be used where the arrangement of the shutoff valve or valves is such as always to insure full required capacity flow through at least one pressure-relief device.

5. DESIGN AND CONSTRUCTION REQUIREMENTS FOR PRESSURE-RELIEF DEVICES

5.1. The material, design and construction of a pressure-relief device shall be such that there shall be no significant change in the functioning of the device and no serious corrosion or deterioration of the materials within the period between renewals, due to service conditions. The chemical and physical properties of the materials shall be uniform and suitable for the requirements of the part manufactured therefrom. Parts and components shall be suitably cleaned for the intended service.

5.2. Safety relief valves shall meet the applicable requirements for design, materials, installation, set pressure tolerance, markings and certification of flow capacity of one of the following standards:

(a) "Safety Relief Valves For Anhydrous Ammonia and LP-Gas," UL 132 (Ref. 11).

(b) ASME Section VIII Division 1, UG-125 through UG-136 (Ref. 12).

(c) Appendix A of AAR Specifications For Tank Cars (Ref. 14).

5.3. Pressure-relief devices shall have a total flow capacity as calculated by the applicable formulas in 5.3.2 or 5.3.3. These formulas are based on the principle of relieving the vapor in the container generated at 120 percent of the DOT design pressure of the container.

5.3.1. The flow capacity of safety relief devices of each design and modification thereof shall be determined as required by the applicable Standard (see 5.2 and Section 6).

5.3.2. For liquefied compressed gases in uninsulated containers and in insulated containers not meeting the requirements of 5.3.4, the minimum required flow capacity of the pressure-relief device(s) shall be calculated using the formula:

$$Q_a = G_u A^{0.82}$$

where

Q_a = Flow capacity in cubic feet per minute of free air.

G_u = Gas factor for uninsulated container obtained from Table 1 for the gas involved.

*A = Total outside surface area of the container in square feet.

Note: Graph of A versus $A^{0.82}$ is shown in Fig. 1.

5.3.3. For liquefied compressed gases in insulated containers where all materials comprising a representative sample of the insulation system remain completely in place when subjected to 1200 F, the U value shall be as defined below and the minimum required flow capacity of the pressure-relief device(s) shall be calculated using the formula:

$$Q_a = G_i U A^{0.82}$$

where

**U = Total thermal conductance of the container insulating material Btu/hr ft^2 F

*When the surface area is not stamped on the name plate or when the marking is not legible, the area can be calculated by using one of the following formulas:
(1) Cylindrical container with hemispherical heads:
Area = (overall length) × (outside diameter) × (3.1416).
(2) Cylindrical container with semi-ellipsoidal heads:
Area = (overall length + .3 outside diameter) × (outside diameter) × (3.1416).
(3) Spherical container:
Area = (outside diameter)^2 × (3.1416).

**Total thermal conductance = thermal conductivity in Btu/hr-ft^2 (F/in.) divided by thickness of the insulation in inches.

when saturated with gaseous lading or air at atmospheric pressure, whichever is greater. Value of U is determined at 100 F except when 5.3.4.2 and 5.3.4.3 apply. (Total thermal conductance = thermal conductivity in Btu/hr ft^2 (F/in.) divided by insulation thickness in inches.)

*A = Total outside surface area of the container in square feet.

G_i = Gas factor for insulated containers obtained from Table 1 for the gas involved.

Q_a = Flow capacity in cubic feet per minute of free air.

Note: Graph of A versus $A^{0.82}$ is shown in Fig. 1.

5.3.4. For liquefied compressed gases in insulated containers where any of the materials of a representative sample of the insulation system deteriorates when subjected to 1200 F, one of the following procedures shall be used to determine the minimum flow capacity requirement of the pressure-relief device(s):

5.3.4.1. Use the formula for uninsulated containers in 5.3.2.

5.3.4.2. Determine the total thermal conductance (U) for a representative sample of the insulation system with a 1200 F external test environment. This value of U shall then be used in the formula in 5.3.3 to determine the minimum required flow capacity of the pressure-relief device(s). The value of U shall be determined with the insulation saturated with gaseous lading or air at atmospheric pressure, whichever provides the greater thermal conductance.

5.3.4.3. If the insulation system is equipped with a jacket that remains in place during fire conditions, the thermal conductance U shall be determined with no insulation and a 1200 F external test environment. The value of U shall be determined with gaseous lading or air at atmospheric pressure in the space between the jacket and container, whichever provides the greater thermal conductance. This value of U shall then be used in the formula in 5.3.3 to determine the minimum required flow capacity of the pressure-relief device(s).

5.3.4.4. For insulated containers for pressurized oxygen, nitrogen and argon, the minimum required flow capacity of the pressure-relief devices may be calculated using the formula of 5.3.3 provided the flow capacity of the pressure-relief devices shall not be less than 0.004 cu ft/min of free air per pound of water capacity of the container at a flow rating pressure of 25 psig. The value of U shall be determined at 100 F with insulation saturated with gaseous nitrogen at atmospheric pressure. (Requirements for the insulation and jacket material are specified in CGA-341 (Ref. 10).

5.3.5. Values are given in Table 1 for G_i (for insulated containers) and G_u (for uninsulated containers) for use in formulas $Q_a = G_i UA^{0.82}$ and $Q_a = G_u A^{0.82}$. These values for G_i and G_u may be used in determining the required flow capacity at the flow rating pressure shown in Table 1, or below. Alternatively the G_i and G_u values may be calculated for the applicable flow rating pressure.

5.4. Safety Relief Valves

5.4.1. Safety relief valves shall be of the spring-loaded type. The inlet connection shall not be less than $\frac{3}{4}$ in. (1.9 cm) nominal pipe size with physical dimensions for the wall thickness not less than those of Schedule 80 pipe (extra heavy), except that safety relief valves or insulated containers for the gases listed in 4.1.10 shall have an inlet connection not less than $\frac{1}{2}$ in. (1.27 cm) nominal pipe size.

5.4.2. The minimum design pressures of containers required by the DOT and CTC for the various compressed gases are shown in Table 1.

5.4.3. Safety relief valves shall be designed so that the possibility of tampering will be minimized. If the pressure setting or adjustment is external, safety relief valves shall be provided with suitable means for sealing the adjustment.

5.4.4. If the design of a safety relief valve is such that liquid can collect on the discharge side, the valve shall be equipped with a drain at the lowest point where liquid can collect. Any discharge from the drain shall be directed to prevent impingement on the tank.

5.4.5. Seats or disks of cast iron shall not be used.

5.5. Rupture Disk Devices*

5.5.1. Where permitted in Section 4, a rupture disk device may be used as the sole pressure-relief device on the container, or as a supplemental device, or in a combination device. Rupture disk devices shall meet the requirements of Section VIII Division 1 of the ASME Code (Ref. 12).

5.6. Rupture Disk in Combination with a Safety Relief Valve

Where permitted in Section 4, a rupture disk device may be installed between a spring-loaded safety relief valve and the container provided such combination device meets the requirements of Section VIII Division 1 of the ASME Code (including UG-127 (a) (3) (b)).

5.6.1. The space between the rupture disk and the safety relief valve shall be provided with a telltale, try cock, needle valve or other suitable device to monitor or prevent accumulation of pressure. The device shall be open to the atmosphere during transportation except when prohibited by DOT or CTC.

Note: Users are warned that a rupture disk will not burst at its rated pressure if pressure builds up in the space between the disk and the safety relief valve which will occur should leakage develop in the disk due to corrosion or other cause.

5.7. Breaking Pin Devices

5.7.1. Breaking pin in combination with a safety relief valve. Where permitted in Section 4, a breaking pin device may be installed between a spring-loaded safety relief valve and the container provided such combination device meets the requirements of Section VIII Division 1 of the ASME Code (including UG-127 (b)).

5.7.1.1. The space between the breaking pin and the safety relief valve shall be provided with a telltale, try cock, needle valve or other

---

*Note: It is recommended that the user closely review rupture disk characteristics for the expected operating conditions to prevent failure of the rupture disk due to fatigue or creep. (Consult with the rupture disk manufacturer.)

suitable device to monitor or prevent accumulation of pressure. The device shall be open to the atmosphere during transportation except when prohibited by DOT or CTC.

Note: Users are warned that a breaking pin will not break at its rated pressure if back-pressure builds up in the space between the breaking pin and the safety relief valve.

5.7.2. Breaking pin devices shall not be used as single devices but only in combination with a safety relief valve.

5.8. Fusible Plug Devices

5.8.1. Where permitted in Section 4, fusible plug devices meeting the requirements of CGA S-1.1 may be used as a primary or supplemental device on containers.

6. TESTS OF PRESSURE-RELIEF DEVICES

6.1. Pressure Tests of Safety Relief Valves

6.1.1. Each safety relief valve shall be subject to an air or gas pressure test to determine the following:

6.1.1.1. That the start-to-discharge pressure setting is within tolerance of the set pressure marked on the valve as required by the applicable standard (see 5.2).

Note: In setting the valve, care must be taken that evidence of start-to-discharge is due to opening of the valve and not due to a defect.

6.1.1.2. That after the start-to-discharge pressure test, the resealing pressure is not less than 90 percent of the start-to-discharge pressure.

6.2. Flow Capacity Tests of Safety Relief Valves

6.2.1. The flow capacity of each design and modification thereof of a spring-loaded safety relief valve shall be determined at the flow rating pressure as required by the applicable standard (see 5.2).

6.2.2. Methods of flow testing shall be as required by the applicable standard (see 5.2).

6.3. Tests of Safety Relief Devices Other Than Spring-Loaded Safety Relief Valves

6.3.1. When fusible plug devices are used to satisfy the requirements of these Standards, the flow capacity at a pressure of 120 percent of the DOT design pressure of the container may be determined by calculation if the capacity at some other pressure has been determined by actual flow tests conducted in accordance with CGA S-1.1 (Ref. 7). All other test requirements of CGA S-1.1 shall apply.

6.3.2. When rupture disk devices or combination devices are used to satisfy the requirements of this Standard, the flow capacity at 120 percent of the DOT design pressure of the container shall be determined by the procedures of Section VIII Division 1 of the ASME Code.

6.4. Rejected material may be reworked providing the material is subject to such additional tests as are required to insure compliance with all requirements of these Standards.

7. IDENTIFICATION

7.1. Safety relief valves shall be marked as required by the applicable Standard (see 5.2), and shall include:

(a) Manufacturer's name or trademark and catalog number.

(b) The year of manufacture.

(c) The set pressure in psig.

(d) The flow capacity in cubic feet per minute of free air.

7.2. Fusible plug devices shall be marked as required by CGA S-1.1 (Ref. 7).

7.3. For pressure-relief devices other than spring-loaded safety relief valves and fusible plug devices, the requirements of Section VIII Division 1 of the ASME Code shall apply.

7.3.1. For rupture disk devices the 70 F bursting pressure shall be included as information on rupture disks having a marked coincident temperature other than 70 F (see 4.1.10.1.2).

8. MAINTENANCE REQUIREMENTS FOR PRESSURE-RELIEF DEVICES

8.1. Care shall be exercised to avoid damage to pressure-relief devices. Care shall also be exercised to avoid plugging by paint or other dirt accumulation of pressure-relief device channels or other parts which could interfere with the functioning of the device.

8.2. Repair work on safety relief valves involving machining, grinding, welding or other alterations or modifications can be performed only by the valve manufacturer or by the shipper with the manufacturer's permission except:

If a trained specialist is available, the seating surfaces of metal-to-meal seat valves may be lapped. The flat gasket face on a valve body mounting surface, or the gasket tongue, may be machined to remove nicks and burrs. However the tolerances on the gasket tongue must not be exceeded.

8.3. Repair work on rupture disk holders involving machining, grinding, welding or other alterations or modifications can be performed only by the rupture disk holder manufacturer or by the shipper with the manufacturer's permission.

8.4. Only replacement parts or assemblies provided and properly identified by the manufacturer of the pressure-relief device shall be used.

8.5. Routine Checks When Filling Containers

8.5.1. Pressure-relief devices periodically shall be examined externally for corrosion, damage, plugging of external pressure-relief device channels, mechanical defects and leakage. Valves equipped with secondary resilient seals shall have the seals inspected periodically. If there is any doubt regarding the suitability of the pressure-relief device for service the container shall not be filled until it is equipped with a suitable pressure-relief device.

9. REFERENCES

(1) Gas Measurement Committee Report No. 3, "Orifice Metering of Natural Gas," American Gas Association, 1515 Wilson Boulevard, Arlington, VA 22209. Reprinted with revisions 1956.

(2) Code of Federal Regulations, Title 49, Transportation Parts 100 to 199. United States Government Printing Office, Washington, DC 20402.

(3) "How to Size Safety Relief Devices," F. J. Heller, Phillips Petroleum Company, 1954.

(4) "Recommended Practice for the Manufacturer of Fusible Plugs," Pamphlet S-4, Compressed Gas Association, Inc., 500 Fifth Avenue, New York, NY 10110.

(5) CGA Pamphlet V-1. "American Standard— Canadian Standard Compressed Gas Cylinder Valve Outlet and Inlet Connections." Compressed Gas Association, Inc., 500 Fifth Avenue, New York, NY 10110.

(6) Canadian Transport Commissioners, "Regulations for the Transport of Dangerous Commodities by Rail." Available for nominal fee from the Supervisor of Government Publications, Department of Public Printing and Stationery, Ottawa, Canada.

(7) "Pressure Relief Device Standards—Part 1— Cylinders for Compressed Gases," Pamphlet S-1.1. Compressed Gas Association, Inc., 500 Fifth Avenue, New York, NY 10110.

(8) "Pressure Relief Device Standards—Part 3— Compressed Gas Storage Containers," Pamphlet S-1.3. Compressed Gas Association, Inc., 500 Fifth Avenue, New York, NY 10110.

(9) Safety and Relief Valves, Performance Test Codes, PTC 25.3-1976. American Society of Mechanical Engineers, 345 East 47th Street, New York, NY 10017.

(10) Pamphlet CGA-341. "Insulated Tank Truck Specification." Compressed Gas Association, Inc., 500 Fifth Avenue, New York, NY 10110.

(11) UL 132 "Safety Relief Valves For Anhydrous Ammonia And LP-Gas." Underwriters Laboratories, Inc., 207 East Ohio Street, Chicago, IL 60611.

(12) "ASME Boiler and Pressure Vessel Code Section VIII Division 1." American Society of Mechanical Engineers, 345 East 47th Street, New York, NY 10017.

(13) National Bureau Standards Monograph 111 "Technology of Liquid Helium." Available from Superintendent of Documents, U.S. Government Printing Office, Washington, DC 20402.

(14) "Specifications For Tank Cars," Standard M-1002, The Association of American Railroads, 1920 L Street, N.W., Washington, DC 20036.

TABLE 1. S-1.2: VALUES OF G_i AND G_u FOR MINIMUM DOT DESIGN PRESSURE CONTAINERS AND COMMONLY USED FLOW RATING PRESSURES (1)

Gas	(5) Minimum DOT Design Pressure of Container (psig)	Flow Rating Pressure (psig)	Value of G_i	Value of G_u
Anhydrous ammonia	265	318	2.80	22.1
Anhydrous dimethylamine	150	180	3.76	21.0
Anhydrous monomethylamine	150	180	3.55	29.4
Anhydrous trimethylamine	150	180	5.33	41.8
Argon, pressurized liquid (2)	—	100	10.2	59.0
	—	200	11.8	69.0
	—	300	13.8	82.0
	—	400	17.9	108.0
Butadiene, inhibited	100	120	4.17	35.8
Carbon dioxide (refrigerated)(See 173.315)	100	360	7.94	57.7
	100	390	6.46	47.4
Carbon monoxide, liquefied (2)	—	100	10.2	59.0
	—	200	11.8	69.0
	—	300	13.8	82.0
Chlorine	225	270	6.74	54.3
Dichlorodifluoromethane (R-12)	150	180	8.94	72.0
Dichlorodifluoromethane-difluoroethane mixture (R-500)	250	300	8.75	71.9
Dichlorodifluoromethane-dichlorotetrafluoroethane mixture R-12/R-114 mixture	150	180	9.34	81.0
Dichlorodifluoromethane-monofluorotrichloromethane mixture (R-12/R-11 mixture)	150	180	8.94	72.0
Difluoroethane (R-152A)	150	180	6.07	49.0
Difluoromonochloroethane	100	120	6.82	55.7
Ethylene, liquefied	—	100	5.42	36.8
Helium (3) (4)	—	200	52.5	—
Hydrogen, liquefied (3) (4)	—	50	8.6	45.8
	—	100	10.6	56.0
	—	150	17.3	93.0
Liquefied petroleum gas	See 173.315	300	6.56	53.6
Methyl chloride	150*	180	4.96	40.4
Methyl mercaptan	100	120	6.05	51.2
Monochlorodifluoromethane (R-22)	250	300	7.92	64.0
Neon, pressurized liquid (4)	—	100	17.0	92.0
	—	200	20.8	113.4
	—	300	28.0	153.0
Nitrogen, pressurized liquid	—	100	10.2	59.0
	—	200	11.8	69.0
	—	300	13.8	82.0
	—	400	17.9	108.0
Nitrous oxide (refrigerated) See 173.315	100	120	5.36	37.2
		420	6.20	46.0
Oxygen, pressurized liquid (2)	—	100	10.2	59.0
	—	200	11.8	69.0
	—	300	13.8	82.0
	—	400	17.9	108.8
Sulfur dioxide	*150**	180	4.84	40.0
Vinyl chloride	150	180	5.61	46.8

*See 4.1.7 for optional portable tank.

**For tanks over 1200 gallon water capacity, minimum DOT design pressure may be 125 psig.

Notes:

(1) Flow rating pressure shall not exceed 120 percent of DOT design pressure in determination of flow capacity. When lower flow rating pressures than those shown are used, the values of G_i and G_u are on the safe side and may be used as shown or calculated as covered below. For higher flow rating pressures than shown, values of G_i and G_u must be calculated from the following formulas:

$$G_u = \frac{633,000}{LC}\sqrt{\frac{ZT}{M}} \quad \text{and} \quad G_i = \frac{73.4 \times (1200 - t)}{LC}\sqrt{\frac{ZT}{M}}$$

where

L = Latent heat at flowing conditions in Btu per pound.
C = Constant for gas or vapor related to ratio of specific heats ($k = C_p/C_v$) at 60 F and 14.7 psia from Fig. 2.
Z = Compressibility factor at flowing conditions.
T = Temperature in degrees R (Rankin) of gas at pressure at flowing conditions ($t + 460$).
M = Molecular weight of gas.
t = Temperature in degrees F (Fahrenheit) of gas at pressure at flowing conditions.

When compressibility factor Z is not known, 1.0 is a safe value of Z to use. When gas constant C is not known, 315 is a safe value of C to use. For complete details concerning the basis and origin of these formulas, refer to Ref. 3.

(2) G_i and G_u values for carbon monoxide, oxygen and argon are based on nitrogen properties. For containers restricted to carbon monoxide only, oxygen only or to argon service only, G_i and G_u values may be calculated as above.

(3) For determination of G_i for supercritical helium, see Appendix A of Chapter 6 of Ref. 13. The same technique can be used for supercritical hydrogen.

(4) Depending on the specific insulation system used, it may be desirable to consider the effect of air condensation in sizing relief devices. See Chapter 6 of Ref. 13.

(5) Minimum design pressure recommended for noninsulated and nonrefrigerated containers; a review of applicable DOT or CTC regulations is recommended. Marked set pressure of safety relief valve shall not exceed design pressure of container.

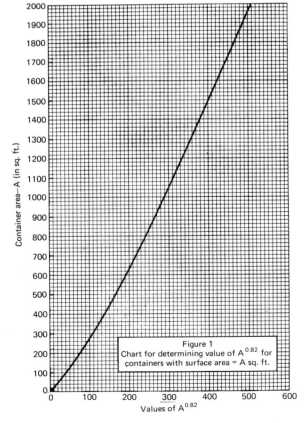

Figure 1
Chart for determining value of $A^{0.82}$ for containers with surface area = A sq. ft.

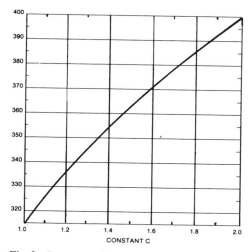

k	CON-STANT C	k	CON-STANT C	k	CON-STANT C
1.00	315	1.26	343	1.52	366
1.02	318	1.28	345	1.54	368
1.04	320	1.30	347	1.56	369
1.06	322	1.32	349	1.58	371
1.08	324	1.34	351	1.60	372
1.10	327	1.36	352	1.62	374
1.12	329	1.38	354	1.64	376
1.14	331	1.40	356	1.66	377
1.16	333	1.42	358	1.68	379
1.18	335	1.44	359	1.70	380
1.20	337	1.46	361	2.00	400
1.22	339	1.48	363	2.20	412
1.24	341	1.50	364		

Fig. 2. Constant C for gas or vapor related to ratio of specific heats ($k = C_p/C_v$) at 60 F and 14.7 psia. (Data from Figure UA-230, ASME Boiler and Pressure Vessel Code, Section VIII, Division 1, Pressure Vessels.)

SECTION C

Pressure-Relief Device Standards—Part 3: Compressed Gas Storage Containers

1. FOREWORD

This part of the Pressure-Relief Device Standards represents the minimum requirements recommended by the Compressed Gas Association, Inc., for pressure-relief devices for use on compressed gas storage containers constructed in accordance with the American Society of Mechanical Engineers (ASME) or the American Petroleum Institute (API)-ASME Codes,* or the equivalent. While these Standards are written specifically to cover "Compressed Gas" as defined in Par. 2.2, the procedures are applicable to storage containers with a maximum allowable working pressure exceeding 15 psig.

It is recommended that containers fabricated after Dec. 31, 1980 utilize pressure-relief devices which meet all requirements of Section VIII of the applicable edition of the ASME Code (see Par. 2.14), and the requirements of this edition of Pamphlet S-1 Part 3 of the CGA Pressure-Relief Device Standards, except that installa-

tions of storage containers for liquefied petroleum gas and anhydrous ammonia shall utilize pressure-relief devices meeting the requirements of the following existing standards. (1) Current edition of National Fire Protection Association Pamphlet No. 58, "Standard for the Storage and Handling of Liquefied Petroleum Gases." (2) Current edition of the American National Standard "Safety Requirements for the Storage and Handling of Anhydrous Ammonia," K6.1.**

Terminology employed in this Standard is consistent with ANSI B95.1-977*** where possible.

A number of states and cities have pressure vessel laws and regulations which include requirements for pressure-relief devices. Manufacturers and users are cautioned that in some instances the relief devices recommended in this

*The API-ASME Code, as a joint publication and interpretation service, was discontinued as of December 31, 1956.

**Obtainable for a nominal charge from the American National Standards Institute, 1430 Broadway, New York, NY 10018 and the Compressed Gas Association, Inc., 500 Fifth Avenue, New York, NY 10110.

***Obtainable for a nominal charge from the American National Standards Institute, 1430 Broadway, New York, NY 10018.

standard may not be acceptable unless special permission is obtained from the authority having jurisdiction.

2. DEFINITIONS

For the purpose of these Standards, the following terms are defined:

2.1. The term *storage container* means any container designed to be permanently mounted on a stationary foundation and in which is to be stored any compressed gas.

2.2. A *compressed gas* is defined as any material or mixture having in the container either an absolute pressure exceeding 40 pounds (275.8 kPa) per square inch at 70 F (21.1 C), or an absolute pressure exceeding 104 pounds per square inch at 130 F (717 kPa at 54 C), or both; or any liquid flammable material having Reid (Note 1) vapor pressure exceeding 40 pounds per square inch absolute at 100 F (275.8 kPa at 32.8 C).

Note 1: American Society for Testing and Materials Method of Test for Vapor Pressure of Petroleum Products (D-323).

2.3. A pressure-relief device is a device designed to open at a specified value of pressure. It may be a safety relief valve, a nonreclosing pressure-relief device or a nonreclosing pressure-relief device in combination with a safety relief valve.

2.4. A *safety relief valve* is a pressure-relief device characterized by rapid opening pop action or by opening generally proportional to the increase in pressure over the opening pressure.

2.5. The *set pressure* of a safety relief valve is the pressure marked on the valve and at which it is set to start-to-discharge.

2.6. The *start-to-discharge pressure* of a safety relief valve is the pressure, measured at the valve inlet, at which there is a measurable lift, or at which discharge becomes continuous as determined by seeing, feeling or hearing.

2.7. The *resealing pressure* is the value of decreasing inlet static pressure at which no further leakage is detected after closing. The method of detection may be a specified water seal on the outlet or other means appropriate for the application.

2.8. The *flow capacity* is the relieving capacity of a pressure-relief device determined at the flow rating pressure, expressed in cubic feet per minute of free air discharge.

2.9. The *flow rating pressure* is the inlet static pressure at which the relieving capacity of a pressure-relief device is determined for rating purposes.

2.10. *Free air* or *free gas* is air or gas measured at a pressure of 14.7 pounds per square inch absolute and a temperature of 60 F (101.4 kPa at 15.6 C).

2.11. A *nonreclosing pressure-relief device* is a pressure-relief device designed to remain open after operation. A manual resetting means may be provided.

2.11.1. A *rupture disk device* is a nonreclosing pressure-relief device actuated by inlet static pressure and designed to function by the bursting of a disk.

2.11.2. A *breaking pin device* is a nonreclosing pressure-relief device actuated by inlet static pressure and designed to function by the breakage of a load carrying section of a pin which supports a pressure containing member. It shall be used only in combination between a safety relief valve and the container.

2.12. A *combination pressure-relief device* is one of the following:

(a) A breaking pin device in combination with a safety relief valve.

(b) A rupture disk device in combination with a safety relief valve.

2.13. The *Code* as used in these Standards is defined as (1) Paragraph U-68, U-69, U-200 or U-201 of Section VIII of the Boiler and Pressure Vessel Code of the American Society of Mechanical Engineers, 1949 edition; or (2) Section VIII of the Boiler and Pressure Vessel Code of the American Society of Mechanical Engineers, 1950 edition through the current edition including addenda; or (3) the Code for Unfired Pressure Vessels for Petroleum Liquids and Gases of the American Petroleum Institute and the American Society of Mechanical Engineers (API-ASME)* 1951 edition.

*The API-ASME Code, as a joint publication and interpretation service, was discontinued as of December 31, 1956.

2.14. The term *maximum allowable working pressure* as used herein is identical to *maximum allowable working pressure* as used in the Code for the original design and construction of the container, except that for containers constructed in accord with Paragraph U-68 and U-69 of Section VIII of the Boiler and Pressure Vessel Code of the ASME, 1949 edition, the maximum allowable working pressure is 125 percent of the original ASME value provided the authority having jurisdiction approves such increased maximum allowable working pressure. The term *maximum allowable working pressure* as used herein applies to "design pressure" as used in Section VIII Division 2 of the ASME Code.

2.15. A *nonliquefied compressed gas* is a gas, other than a gas in solution which under the charging pressure, is entirely gaseous at a temperature of 70 F (21.1 C).

2.16. A *liquefied compressed gas* is a gas which under the charging pressure, is at least partially liquid at a temperature of 70 F (21.1 C). A flammable compressed gas which is normally nonliquefied at 70 F (21.1 C) but which is partially liquid under the charging pressure and temperature shall follow the requirements for liquefied compressed gas.

2.17. A *pressurized liquid compressed gas* is a compressed gas other than a compressed gas in solution, which cannot be liquefied at a temperature of 70 F (21.1 C) and which is maintained in the liquid state at a pressure not less than 40 psia (275.8 kPa abs.) by maintaining the gas at a temperature less than 70 F (21.1 C).

3. TYPES OF PRESSURE-RELIEF DEVICES

Types of pressure-relief devices covered by this part are as follows:

3.1. Safety Relief Valve

3.2. Rupture Disk Device

3.3. Rupture Disk Device in Combination with Safety Relief Valve

3.4. Breaking Pin Device in Combination with Safety Relief Valve

4. APPLICATION REQUIREMENTS FOR PRESSURE-RELIEF DEVICES

4.1. Each container shall be provided with one or more pressure-relief devices which unless otherwise specified shall be safety relief valves of the spring-loaded type.

Except as covered in 4.1.3 and 4.1.4, all pressure-relief devices shall meet the applicable requirements of UG-125 through UG-136 of Section VIII Division 1 or Part AR of Section VIII Division 2 of the ASME Code, including the requirements for installation, set pressure including tolerance, flow capacity, overpressure. markings and certification of capacity. Requirements for flow capacity are specified in Section 5 of this Standard.

The opening through all pipe and fittings between a container and its pressure-relief device(s) shall have at least the area of the pressure-relief device(s) inlet.

The size of the discharge lines shall be such that any pressure that may exist or develop will not reduce the relieving capacity of the pressure-relief device(s) below that required.

4.1.1. When a single pressure-relief device is used, its marked set pressure shall not exceed the maximum allowable working pressure of the container.

When additional pressure-relief devices are used, the marked set pressure shall meet the applicable requirement of Section VIII of the ASME Code.

The marked set pressure of the pressure-relief device having the lowest setting shall be as shown in Table 1 (MAWP) for those specific gases, except as follows:

4.1.1.1. If an overdesigned container is used, the marked set pressure of the pressure-relief device may be between the minimum required maximum allowable working pressure for the lading and the maximum allowable working pressure of the container used.

4.1.1.2. For carbon dioxide (refrigerated), nitrous oxide (refrigerated) and the gases covered by 4.9, there shall be no minimum marked set pressure.

4.1.1.3. For containers constructed in accord with Paragraph U-68 or U-69 of the Code, 1949 edition, the set pressure marked on the safety relief valve may be 125 percent of the original ASME maximum allowable working pressure of the container, with the concurrence of the authority having jurisdiction.

4.1.2. Nonreclosing pressure-relief devices or combination devices may be used in lieu of safety relief valves on containers containing substances that may render a safety relief valve inoperative, or where a loss of valuable material by leakage should be avoided, or contamination of the atmosphere by leakage of noxious gases must be avoided.

4.1.3. Storage containers for anhydrous ammonia shall be equipped with safety relief devices in compliance with the applicable edition of American National Standard K61.1 "Safety Requirements for the Storage and Handling of Anhydrous Ammonia."

4.1.4. Storage containers for liquefied petroleum gas shall be equipped with pressure-relief devices in compliance with the applicable edition of National Fire Protection Association Pamphlet No. 58 "Standard for the Storage and Handling of Liquefied Petroleum Gases."

4.2. The design, material and location of pressure-relief devices shall be suitable for the intended service.

4.3. Pressure-relief devices shall have direct communication with the vapor space of the container.

4.4. Pressure-relief devices shall be so installed and located that the cooling effect of the contents will not prevent the effective operation of the devices.

4.5. When fittings and piping are used on either the upstream or downstream side or both of a pressure-relief device or devices, the passages shall be so designed that the flow capacity of the pressure-relief device will not be reduced below the capacity required for the container on which the pressure-relief device assembly is installed.

4.6. Pressure-relief devices shall be arranged to discharge to the open air in such a manner as to prevent any impingement of escaping gas upon the container or upon operating personnel. Pressure-relief devices shall be arranged to discharge upward except this is not required for

carbon dioxide, nitrous oxide, helium and pressurized liquid argon, nitrogen and oxygen.

4.7. No shutoff valves shall be installed between the pressure-relief devices and the container except, in cases where two or more pressure-relief devices are installed on the same container, a shutoff valve may be used where the arrangements of the shutoff valve or valves is such as always to insure full required flow capacity through at least one pressure-relief device.

4.8. When storage containers are filled by pumping equipment which has a discharge capacity in excess of the capacity of the container pressure-relief devices, and which is capable of producing pressures in excess of the maximum allowable working pressure of the container, precautions should be taken to prevent the development of pressure in the container in excess of 116 percent of its maximum allowable working pressure. This may be done by providing additional capacity of the pressure-relief devices on the container, by providing a bypass on the pump discharge or by any other suitable method.

4.9. The following requirements apply to containers for liquefied methane, ethane, ethylene, carbon monoxide (see 4.9.1.2.2) and hydrogen, helium (refrigerated), and pressurized liquid oxygen, nitrogen, argon and neon.

4.9.1. The liquid container shall be protected by pressure-relief devices consisting of one or more safety relief valves, and one or more rupture disk devices except as modified by 4.9.1.2.1, so installed that the devices remain at ambient temperature during normal container operation.

4.9.1.1. The marked set pressure of the safety relief valve shall not exceed the maximum allowable working pressure of the container. The minimum capacity of the safety relief valve shall be sized to provide adequate venting capacity at 110 percent of maximum allowable working pressure for operational emergency contingencies, except fire, but including loss of vacuum with insulation saturated with gaseous lading or air at atmospheric pressure, whichever provides the greater thermal conductance. The minimum required capacity of the pressure-relief valve for the loss of vacuum condition is:

$$Q_a = \frac{(130 - t)}{4(1200 - t)} G_i UA$$

where the value of U at the average temperature of the insulation may be used (alternatively the value of U at 100 F may be used). For helium and hydrogen, the value of U must be chosen on the basis that the insulation space is filled with helium and hydrogen gas, respectively, at one atmosphere (see Table 1 and 5.3.5 for nomenclature).

If two or more pressure-relief valves are used to provide the above capacity, the additional valve(s) shall be set at 105 percent of maximum allowable working pressure and the combined capacity shall prevent the pressure from rising more than 16 percent above the maximum allowable working pressure.

4.9.1.2. The rupture disk device(s) must have a stamped bursting pressure not to exceed 110 percent of the maximum allowable working pressure at a coincident disk temperature not to exceed 800 F (427 C). In addition the room temperature rating for the lot of disks must be determined, and must be less than 150 percent of the maximum allowable working pressure (plus 15 psi if vacuum insulation is used) of the container.

4.9.1.2.1. An alternate safety relief valve with a marked set pressure not to exceed 110 percent of maximum allowable working pressure, may be used in lieu of the rupture disk device. Installation must provide for:

(a) Prevention of moisture accumulation at the seat by drainage away from that area.

(b) Periodic drainage of the vent piping.

(c) Avoidance of foreign material in the vent piping.

4.9.1.2.2. Rupture disk devices are not permitted on containers in liquefied carbon monoxide service.

4.9.1.3. The combined capacity of the pressure-relief devices shall be adequate to relieve the vapor generated at 21 percent above the maximum allowable working pressure when the container is exposed to fire or other unexpected source of external heat. This required flow capacity shall be determined in accordance with the provisions of 5.3.5 and 5.3.6.

4.9.1.3.1. When the flow capacity of the safety relief valve meets the requirements of 4.9.1.3 the supplemental pressure-relief device meeting the requirements of 4.9.1.2 or 4.9.1.2.1 shall have a marked set pressure or stamped burst pressure not exceeding:

(a) For containers other than vacuum insulated:

(1) For safety relief valve in lieu of rupture disk device, the marked set pressure shall not exceed 136 percent of maximum allowable working pressure.

(2) For rupture disk device, the stamped burst pressure at 70 F (21.1 C) shall not exceed 150 percent of maximum allowable working pressure.

(b) For vacuum insulated containers:

(1) For safety relief valve in lieu of rupture disk device, the marked set pressure shall not exceed 136 percent of (maximum allowable working pressure + 15 psi (103.4 kPa)).

(2) For rupture disk device, the stamped burst pressure at 70 F (21.1 C) shall not exceed 150 percent of (maximum allowable working pressure + 15 psi (103.4 kPa)).

4.9.2. The outer shell of vacuum-insulated containers shall be protected with means for pressure relief as specified in 6 (a)(13) of CGA-341 "Insulated Tank Truck Specification" (Ref. 10).

5. DESIGN AND CONSTRUCTION REQUIREMENTS FOR PRESSURE-RELIEF DEVICES

5.1. The material, design and construction of a pressure-relief device shall be such that there shall be no significant change in the functioning of the device and no serious corrosion or deterioration of the materials within the period between renewals, due to service conditions. The chemical and physical properties of the materials shall be uniform and suitable for the requirements of the part manufactured therefrom.

Parts and components shall be suitably cleaned for the intended service.

5.2. Safety relief valves shall meet the applicable requirements for design, materials, installation, set pressure including tolerance, markings and certification of capacity of ASME Section VIII Division 1 UG-125 through UG-136, or part AR of Section VIII Division 2 of the ASME Code (Ref. 8), except for those covered in 4.1.3 and 4.1.4 which shall meet the requirements of UL-132 (Ref. 9).

5.3. Pressure-relief devices shall have a total flow capacity as calculated by the applicable formulas in 5.3.2, 5.3.3 or 5.3.5 except as provided in 5.3.4. These formulas are based on the principle of preventing the pressure in the container from exceeding a maximum of 121 percent of the maximum allowable working pressure of the container.

5.3.2. For uninsulated containers for nonliquefied gases the minimum required flow capacity (see 6.3.2) of the pressure-relief device shall be calculated using the formula:

$$Q_a = 0.029 Wc$$

where

Q_a = Flow capacity in cubic feet per minute of free air.

Wc = Water capacity of the container in pounds.

5.3.3. For liquefied compressed gases in uninsulated containers and in insulated containers not meeting the requirements of 5.3.6 the minimum required flow capacity of the pressure-relief device(s) shall be calculated using the formula:

$$Q_a = G_u A^{0.82}$$

where

Q_a = Flow capacity in cubic feet per minute of free air.

G_u = Gas factor for uninsulated containers obtained from Table 1 for the gas involved.

*A = Total outside surface area of the container in square feet.

Note: Graph of A versus $A^{0.82}$ is shown in Fig. 1.

5.3.4. For containers complying with any one of the following, the flow capacity of the pressure-relief device(s) may be reduced to 30 percent of the capacity as determined in 5.3.3.

5.3.4.1. When the storage is underground (see also 5.3.5.1).

5.3.4.2. When the storage is used for nonflammable gas and is suitably isolated from possible envelopment in a fire.

5.3.4.3. When the storage container is used for nonflammable gas and is equipped with suitable water spray or fire extinguishing system.

5.3.5. For liquefied compressed gases in insulated containers where all materials comprising a representative sample of the insulation system remain completely in place when subjected to 1200 F (649 C) the U value shall be as defined below and the minimum required flow capacity of the pressure-relief device(s) shall be calculated using the formula:

$$Q_a = G_i UA^{0.82}$$

where

U = Total thermal conductance of the container insulating material Btu/hr ft² F when saturated with gaseous lading or air at atmospheric pressure, whichever is greater. Value of U is determined at 100 F except when 5.3.6.2 and 5.3.6.3 apply. Total thermal conductance = thermal conductivity in Btu/hr ft² (F/in.) divided by insulation thickness in inches.)

*A = Total outside surface area of the container in square feet.

G_i = Gas factor for insulated container obtained from Table 1 for the gas involved.

*When the surface area is not stamped on the name plate or when the marking is not legible, the area can be calculated by using one of the following formulas:
(1) Cylindrical container with hemispherical heads:
    Area = (overall length) × (outside diameter) × (3.1416).
(2) Cylindrical container with semi-ellipsoidal heads:
    Area = (overall length + .3 outside diameter) × (outside diameter) × (3.1416).
(3) Spherical container:
    Area = (outside diameter)² × (3.1416).

Q_a = Flow capacity in cubic feet per minute of free air.

Note: Graph of A versus $A^{0.82}$ is shown in Fig. 1.

5.3.5.1. The pressure-relief device capacity for underground containers may be determined in accordance with 5.3.5 by assigning a value of U for the minimum earth cover (see also 5.3.4.1).

5.3.6. For liquefied compressed gases in insulated containers where any of the materials of a representative sample of the insulation system deteriorates when subjected to 1200 F (649 C), one of the following procedures shall be used to determine the minimum flow capacity requirement of the pressure-relief device(s).

5.3.6.1. Use the formula for uninsulated containers in 5.3.3.

5.3.6.2. Determine the total thermal conductance (U) for a representative sample of the insulation system with a 1200 F (649 C) external test environment. This value of U shall then be used in the formula in 5.3.5 to determine the minimum required flow capacity of the pressure-relief device(s). The value of U shall be determined with the insulation saturated with gaseous lading or air at atmospheric pressure, whichever provides the greater thermal conductance.

5.3.6.3. If the insulation system is equipped with a jacket that remains in place during fire conditions, the thermal conductance U shall be determined with no insulation and a 1200 F (649 C) external test environment. The value of U shall be determined with gaseous lading or air at atmospheric pressure in the space between the jacket and container, whichever provides the greater thermal conductance. This value of U shall then be used in the formula in 5.3.5 to determine the minimum required flow capacity of the pressure-relief device(s).

5.3.7. Values are given in Table 1 below for G_i (for insulated containers) and G_u (for uninsulated containers) for use in formulas $Q_a = G_i U A^{0.82}$ and $Q_a = G_u A^{0.82}$. These values for G_i and G_u may be used in determining the required flow capacity at the flow rating pressure shown in Table 1, or below. Alternatively the G_i and G_u values may be calculated for the applicable flow rating pressure.

5.4. Safety Relief Valves

5.4.1. Safety relief valves shall be of the spring-loaded type. The inlet connection of the valve shall not be less than $\frac{3}{4}$ in. nominal pipe size with physical dimensions for the wall thickness not less than those of Schedule 80 pipe (extra heavy), except that safety relief valves on insulated containers meeting the requirements of 4.9 shall have an inlet connection not less than $\frac{1}{2}$ in. nominal pipe size.

5.4.2. Safety relief valves shall be designed so that the possibility of tampering will be minimized. If the pressure setting or adjustment is external the safety relief valves shall be provided with suitable means for sealing the adjustment.

5.4.3. If the design of a safety relief valve is such that liquid can collect on the discharge side, the valve shall be equipped with a drain at the lowest point where liquid can collect. Any discharge from the drain shall be directed to prevent impingement on the tank.

5.4.4. Seats or disks of cast iron shall not be used.

5.5. Rupture Disk Devices*

5.5.1. Where permitted in Section 4, a rupture disk device may be used as the sole pressure-relief device on the container, or as a supplemental device, or in a combination device. Rupture disk devices shall meet the requirements of Section VIII of the ASME Code (Ref. 8).

5.6. Rupture Disk in Combination With a Safety Relief Valve.

Where permitted in Section 4, a rupture disk device may be installed between a spring-loaded safety relief valve and the container provided

---

*It is recommended that the user closely review rupture disk characteristics for the expected operating conditions to prevent failure of the rupture disk due to fatigue or creep. (Consult with the rupture disk manufacturer.)

such combination device meets the requirements of Section VIII of the ASME Code.

5.6.1. The space between the rupture disk and the safety relief valve shall be provided with a telltale, try cock, needle valve or other suitable device to monitor or prevent accumulation of pressure.

Note: Users are warned that a rupture disk will not burst at its rated pressure if pressure builds up in the space between the disk and the safety relief valve which will occur should leakage develop in the disk due to corrosion or other cause.

5.7. Breaking Pin Devices

5.7.1. Breaking Pin in Combination with a Safety Relief Valve. Where permitted in Section 4, a breaking pin device may be installed between a spring-loaded safety relief valve and the container provided such combination device meets the requirements of Section VIII of the ASME Code.

5.7.1.1. The space between the breaking pin and the safety relief valve shall be provided with a telltale, try cock, needle valve or other suitable device or monitor or prevent accumulation of pressure.

Note: Users are warned that a breaking pin will not break at its rated pressure if backpressure builds up in the space between the breaking pin and safety relief valve.

5.7.2. Breaking pin devices shall not be used as single devices but only in combination with a safety relief valve.

6. TESTS OF PRESSURE-RELIEF DEVICES

6.1. Pressure Tests of Safety Relief Valves

6.1.1. Each safety relief valve shall be subject to an air or gas pressure test to determine the following:

6.1.1.1. That the start-to-discharge pressure setting is within tolerance of the set pressure marked on the valve as required by the applicable Standard (see 5.2).

Note: In setting the valve, care must be taken that evidence of start-to-discharge is due to opening of the valve and not due to a defect.

6.1.1.2. That after the start-to-discharge pressure test, the resealing pressure is not less than 90 percent of the start-to-discharge pressure.

6.2. Flow Capacity Tests of Safety Relief Valves

6.2.1. The flow capacity of each design and modification thereof of a spring-loaded safety relief valve shall be determined at the flow rating pressure as required by the applicable Standard (see 5.2).

6.2.2. Methods of flow testing shall be as required by the applicable Standard (see 5.2).

6.3. Tests of Safety Relief Devices Other Than Spring-Loaded Safety Relief Valves

6.3.1. When rupture disk devices or combination devices are used to satisfy the requirements of this Standard, the flow capacity at 121 percent of the maximum allowable working pressure of the container shall be determined by the procedures of Section VIII of the ASME Code, except as provided in 6.3.2.

6.3.2. Rupture disk devices used to provide the flow capacity required by 5.3.2 shall have a flow rating pressure of 100 psia (689.5 kPa abs.).

6.4. Rejected material may be reworked providing the material is subject to such additional tests as are required to insure compliance with all requirements of these Standards.

7. IDENTIFICATION

7.1. Safety relief valves shall be marked as required by the applicable Standard (see 5.2), and shall include:

(a) Manufacturer's name or trademark and catalog number.

(b) The year of manufacture.

(c) The set pressure in psig.

(d) The flow capacity in cubic feet per minute of free air.

7.2. For pressure-relief devices other than spring-loaded safety relief valves, the requirements of Section VIII of the ASME Code shall apply.

7.2.1. For rupture disk devices the 70 F (21.1 C) bursting pressure shall be included as information on rupture disks having a marked coincident temperature other than 70 F (21.1 C) (see 4.9.1.2).

8. MAINTENANCE REQUIREMENTS FOR PRESSURE-RELIEF DEVICES

8.1. Care shall be exercised to avoid damage to pressure-relief devices. Care shall also be exercised to avoid plugging by paint or other dirt accumulating of pressure-relief device channels or other parts which would interfere with the functioning of the device.

8.2. Repair work on safety relief valves involving machining, grinding, welding or other alterations or modifications can be performed only by the valve manufacturer or by the owner or user with the manufacturer's permission except:

If a trained specialist is available, the seating surfaces of metal-to-metal seat valves may be lapped. The flat gasket face on a valve body mounting surface, or the gasket tongue, may be machined to remove nicks and burrs. However, the tolerances on the gasket tongue must not be exceeded.

8.3. Repair work on rupture disk holders involving machining, grinding, welding or other alterations or modifications can be performed only by the rupture disk holder manufacturer or by the owner or user with the manufacturer's permission.

8.4. Only replacement parts of assemblies provided and properly identified by the manufacturer of the pressure-relief device shall be used.

8.5. Pressure-relief devices periodically shall be examined externally for corrosion, damage, plugging of external pressure-relief device channels, mechanical defects and leakage. Valves equipped with secondary resilient seals shall have the seals inspected periodically. The pressure-relief device shall be repaired or replaced, if there is any doubt regarding the suitability of the device for service.

9. REFERENCES

(1) Gas Measurement Committee Report No. 3, "Orifice Metering of Natural Gas," American Gas Association, 1515 Wilson Boulevard, Arlington, VA 22209. Reprinted with revisions, 1956.

(2) "How to Size Safety Relief Devices," F. J. Heller, Phillips Petroleum Company, 1954.

(3) "Recommended Practice For the Manufacture of Fusible Plugs," Pamphlet S-4, Compressed Gas Association, Inc., 500 Fifth Avenue, New York, NY 10110.

(4) "Pressure Relief Device Standards—Part 1—Cylinders For Compressed Gases," Pamphlet S-1.1, Compressed Gas Association, Inc., 500 Fifth Avenue, New York, NY 10110.

(5) "Pressure Relief Device Standards—Part 2—Cargo and Portable Tanks For Compressed Gases," Pamphlet S-1.2, Compressed Gas Association, Inc., 500 Fifth Avenue, New York, NY 10110.

(6) Safety and Relief Valves Power Test Codes PTC 25.3-1976, American Society of Mechanical Engineers, 345 East 42nd Street, New York, NY 10017.

(7) National Bureau of Standards Monograph 111 "Technology of Liquid Helium," available from Superintendent of Documents, U.S. Government Printing Office, Washington, DC 20402.

(8) "ASME Boiler and Pressure Vessel Code Section VIII Divisions 1 and 2." American Society of Mechanical Engineers, 345 East 47th Street, New York, NY 10017.

(9) UL-132 "Safety Relief Valves For Anhydrous Ammonia and LP-Gas." Underwriter Laboratories, 207 East Ohio Street, Chicago, IL 60611.

(10) Pamphlet CGA-341 "Insulated Tank Truck Specification," Compressed Gas Association, Inc., 500 Fifth Avenue, New York, NY 10110.

TABLE 1. S-1.3: VALUES OF G_i AND G_u FOR MINIMUM ALLOWABLE WORKING PRESSURES AND COMMONLY USED FLOW RATING PRESSURES (1)

Gas	(5) Minimum Allowable Working Pressure (psig)	Flow Rating Pressure (psig)	Value of G_i	Value of G_u
Anhydrous ammonia	250	300	2.80	22.1
Anhydrous dimethylamine	150	180	3.76	31.0
Anhydrous monomethylamine	150	180	3.55	29.4
Anhydrous trimethylamine	150	180	5.33	41.8
Argon pressurized liquid (2)	—	100	10.2	59.0
	—	200	11.8	69.0
	—	300	13.8	82.0
	—	400	17.9	108.0
Butadiene, inhibited	100	120	4.17	35.8
Carbon dioxide (refrigerated)	—	360	7.94	57.7
Carbon monoxide, liquefied (2)	—	100	10.2	59.0
	—	200	11.8	69.0
	—	300	13.8	82.0
Chlorine	225	270	6.74	54.3
Dichlorodifluoromethane (R-12)	150	180	8.94	72.0
Dichlorodifluoromethane-difluroethane mixtures (R-500)	250	300	8.75	71.9
Dichlorodifluoromethane-dichlorotetrafluoroethane mixture (R-12/R-114 mixture)	150	180	9.34	81.0
Dichlorodifluoromethane-monoflurotrichloromethane mixture (R-12/R-11 mixture)	150	180	8.94	72.0
Difluoroethane (R-152A)	150	180	6.07	49.0
Difluromonochloroethane	100	120	6.82	55.7
Ethylene, liquefied	—	100	5.42	36.8
Helium (3) (4)	—	200	52.5	—
Hydrogen, liquefied (3) (4)	—	50	8.6	45.8
	—	100	10.6	56.0
Liquefied petroleum gas	250	300	6.56	53.6
Methyle chloride	150	180	4.96	40.4
Methyl mercaptan	100	120	6.05	51.2
Monochlorodifluoromethane (R-22)	250	300	7.92	64.0
Neon, pressurized liquid (4)	—	100	17.0	92.0
Nitrogen, pressurized liquid	—	100	10.2	59.0
	—	200	11.8	69.0
	—	300	13.8	82.0
	—	400	17.9	108.0
Nitrous oxide (refrigerated)	—	120	5.36	37.2
	—	420	6.20	46.0
Oxygen, pressurized liquid	—	100	10.2	59.0
	—	200	11.8	69.0
	—	300	13.8	82.0
	—	400	17.9	108.0
Sulfur dioxide	150	180	4.84	40.0
Vinyl chloride	150	180	5.61	46.8

Notes:

(1) Flow rating pressure shall not exceed 121 percent of maximum allowable working pressure in determination of required flow capacity for fire exposure.

When lower flow rating pressures than those shown are used, the values of G_i and G_u are on the safe side and may be used as shown or calculated as covered below. For higher flow rating pressures than shown, values of G_i and G_u must be calculated from the following formulas:

$$G_u = \frac{633,000}{LC} \sqrt{\frac{ZT}{M}} \quad \text{and} \quad G_i = \frac{73.4 \times (1200 - t)}{LC} \sqrt{\frac{ZT}{M}}$$

where

L = Latent heat at flowing conditions in Btu per pound.
C = Constant for gas or vapor related to ratio of specific heats ($k = C_p/C_v$) at 60 F and 14.7 psia from Fig. 2.
Z = Compressibility factor at flowing conditions.
T = Temperature in degrees R (Rankin) of gas at pressure at flowing conditions ($t + 460$).
M = Molecular weight of gas.
t = Temperature in degrees F (Fahrenheit) of gas at pressure at flowing conditions.

When compressibility factor Z is not known, 1.0 is a safe value of Z to use. When gas constant C is not known, 315 is a safe value of C to use. For complete details concerning the basis and origin of these formulas, refer to Ref. 2.

(2) G_i and G_u values for carbon monoxide, oxygen and argon are based on nitrogen properties. For containers restricted to carbon monoxide only, oxygen only, or to argon service only, G_i and G_u values may be calculated as above.

(3) For determination of value of G_i for supercritical helium, see Appendix A of Chapter 6 of Ref. 7. The same technique can be used for supercritical hydrogen.

(4) Depending on the specific insulation system used, it may be desirable to consider the effect of air condensation in sizing relief devices; see Chapter 6 of Ref. 7.

(5) Minimum maximum allowable working pressure recommended for noninsulated and nonrefrigerated storage containers. Marked set pressure of safety relief valve shall not exceed maximum allowable working pressure of container.

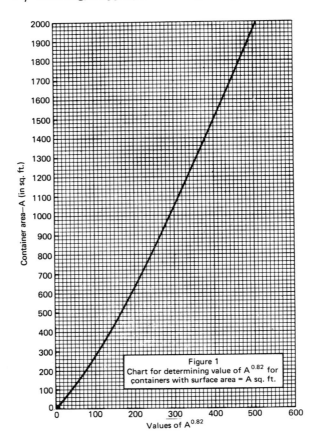

Figure 1
Chart for determining value of $A^{0.82}$ for containers with surface area = A sq. ft.

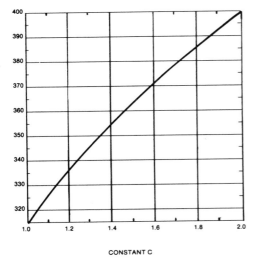

k	CON-STANT C	k	CON-STANT C	k	CON-STANT C
1.00	315	1.26	343	1.52	366
1.02	318	1.28	345	1.54	368
1.04	320	1.30	347	1.56	369
1.06	322	1.32	349	1.58	371
1.08	324	1.34	351	1.60	372
1.10	327	1.36	352	1.62	374
1.12	329	1.38	354	1.64	376
1.14	331	1.40	356	1.66	377
1.16	333	1.42	358	1.68	379
1.18	335	1.44	359	1.70	380
1.20	337	1.46	361	2.00	400
1.22	339	1.48	363	2.20	412
1.24	341	1.50	364		

CONSTANT C

Fig. 2. Constant for gas or vapor related to ratio of specific heats ($k = C_p/C_v$ at 60 F and 14.7 psia. (Data from Figure UA-230, ASME Boiler and Pressure Vessel Code, Section VIII, Division 1, Pressure Vessels.)

CHAPTER 6

Compressed Gas Cylinders:
Marking, Labeling,
Inspection, Testing,
Transfer and Disposition

Each section of this chapter presents a Compressed Gas Association standard that relates primarily to compressed gas cylinders. These are the CGA standards for:

Section A. Marking Portable Compressed Gas Containers to Identify the Material Contained

Section B. Precautionary Labeling and Marking of Compressed Gas Containers

Section C. Visual Inspection of Cylinders

Section D. Hydrostatic Testing of Cylinders

Section E. Disposition of Unserviceable Cylinders

In each section of this chapter where reference is made to cylinder specifications of the Department of Transportation (DOT), it applies equally to cylinders of the corresponding specifications of the Canadian Transport Commission (CTC) for Canada.

SECTION A

Marking Portable Compressed Gas Containers to Identify the Material Contained

This section presents the "American National Standard Method of Marking Portable Compressed Gas Containers to Identify the Material Contained," ANSI/CGA C-4-1978. It is one of a series of American industrial safety standards.

1. SCOPE

1.1. Requirements for marking portable compressed gas containers, not exceeding 1000 lb water capacity, to identify the material contained.

2. DEFINITIONS

2.1. A portable compressed gas container for the purposes of this standard, is any container that is constructed in accord with the specifications of a recognized authority (such as the specifications of the Department of Transportation, the Pressure Vessel Code of the American Society of Mechanical Engineers), and intended to contain compressed or liquefied gas, as defined in the Hazardous Materials Regulations of the Department of Transportation.

3. MARKING OF CONTAINERS

3.1. Compressed gas containers shall be legibly marked with at least the chemical name or a commonly accepted name of the material contained. Marking shall be by means of stenciling, stamping or labeling, and shall not be readily removable.

3.2. Wherever practical the marking shall be located at the valve end and off the cylindrical part of the body.

3.3. The lettering shall be of contrasting color to its background and shall be a minimum height of $\frac{3}{16}$ in. (0.476 cm.).

4. REGULATIONS

4.1. Compressed gas containers for use in international trade shall be legibly marked in a manner not readily removeable as required by this standard or in accordance with International Standards Organization (ISO) 448, Gas Cylinders for Industrial Use, Marking for Identification of Content. The latter standard calls for marking with the name of the material contained in the language of the country in which the container is filled and with the international chemical formula of the gas it contains.

4.2. None of the requirements contained in this standard are intended to conflict with or to supersede federal or state regulations such as the following:

Title 49, Transportation, Code of Federal Regulations Parts 100-199.

Note: When the material contained is specifically named on the label or the alternate mark-

ing authorized by Section 173.300 of the Hazardous Materials Regulations (DOT), the requirements of this standard have been met.

Regulations Governing the Transportation or Storage of Explosive or Other Dangerous Articles or Substances, and Combustible Liquids on Board Vessels, U.S. Coast Guard, Department of Transportation.

Regulations of the Food & Drug Administration, U.S. Department of Health, Education & Welfare.

Regulations of the Occupational Safety & Health Administration, U.S. Department of Labor.

APPENDIX A
AVAILABILITY OF PUBLICATIONS

Publications referenced in this Standard or otherwise of interest to the user may be obtained at nominal cost from the appropriate source as noted below:

ISO 448
International Standards Organization
1, Rue de Varembé
Case Postale 56
CH-1211 Genève 20
Switzerland/Suisse

49CFR Parts 100-199, Dept. of Transportation
29CFR Part 1910 (OSHA), Dept. of Labor
21CFR (Food & Drug),
    Dept. of Health, Education & Welfare

Superintendent of Documents
U.S. Government Printing Office
Washington, D.C. 20025

ASME Boiler & Pressure Vessel Code,
Section VIII
American Society of Mechanical Engineers
345 East 47th Street
New York, New York 10017

SECTION B

Precautionary Labeling and Marking of Compressed Gas Containers

1. INTRODUCTION

1.1. The compressed gas industry recognizes the value of precautionary labels on containers (cylinders, spheres and tubes) to warn of principal hazards. As a guide for the preparation of adequate container labels, the Compressed Gas Association, Inc., has prepared this statement defining general principles and giving illustrative labels for several types of gases.

1.2. The methods of preparing label precautionary information established by the Manufacturing Chemists' Association have been followed in this guide but modified where necessary to meet the specific labeling needs of the compressed gas industry.

1.3. Precautionary labeling should be applied to portable compressed gas containers, not exceeding 1000 lb water capacity for the purposes of identifying the container content, and to warn of principal hazards associated with the container and its content.

1.4. Precautionary labeling may not be necessary on those portable containers which, in transportation, storage and in use, are handled by and remain under the sole control of the supplier or his agent.

1.5. Precautionary labels shown in this section are examples prepared in accordance with the minimum requirements to warn of principal hazards involved.

1.5.1. DOT labels are the color-coded 4 in. diamond-shaped labels for labeling of explosives and other dangerous articles as described in Hazardous Materials Regulations of the Department of Transportation.

1.5.2. DOT markings permitted as an alternative to the DOT label as referenced in Appendix A, consist of color coded $1\frac{1}{4}$ in. diamond-shaped markings with associated DOT commodity list name.

1.6. Statutes or regulations such as those of the Federal Food and Drug Administration and the Department of Agriculture, may require that particular information be included on a label, or that a specific label be affixed to a container.

2. GENERAL PRINCIPLES

2.1. In preparing precautionary labels for compressed gas containers, the following general principles should serve as a guide.

2.1.1. "Compressed gas containers shall be legibly marked with at least the chemical name or a commonly accepted name of the material contained. Marking shall be by means of stenciling, stamping or labeling, and shall not be readily removable." (Quoted from Chapter 6 Section A.) Identification marking required to be in conformance with DOT regulations must use the product name as it appears in the DOT commodity list Part 172 of hazardous materials regulations of the Department of Transportation. For NOS listings, the chemical, or commonly accepted name of the material should appear.

2.1.2. For effectiveness all statements on labels should be brief, accurate and expressed in simple, easily understood terms.

2.1.3. Precautionary information should be used only where necessary. Unnecessary wording on labels may develop a disregard for the labels.

2.1.4. It is desirable to employ uniform wording in indicating the same hazards for different gases.

2.1.5. Mixtures of two or more gases may have properties that vary in kind or degree from those of the individual components. Any precautionary labels for mixtures should be based on the properties of the finished product.

2.1.6. Warning statements should be in easily legible type which is in contrast by typography, layout or color with other printed matter on the label. The label should be in a conspicuous place on the container.

2.2. In addition to the name of the gas (see 2.1.1) the following types of information, where necessary, should be considered for inclusion on the label:

2.2.1. Signal Word. The signal word is intended to draw attention to the presence and the degree of hazard. Recognized signal words are DANGER!, WARNING!, CAUTION.

2.2.1.1. The signal word DANGER should be used on labels of flammables, poisons and similar gases where the release of gas to the atmosphere would create an *immediate* hazard to health or property.

2.2.1.2. The signal word WARNING should be used on labels of gases such as oxygen, nitrous oxide, ammonia and the cryogenic liquids where a release of gas or liquid creates a *less than immediate* hazard but may be hazardous to health or property under certain conditions.

2.2.1.3. The signal word CAUTION should be used on labels of gases such as argon, helium, nitrogen and carbon dioxide where the release of gas creates *no immediate hazard* to health or property except that associated with a characteristic such as pressure or displacement of air.

2.2.2. Statement of Hazard. This statement gives notice of the hazards present in connection with the customary or reasonably anticipated handling or use of the product and should follow the signal word.

2.2.2.1. The following are examples of statement of hazards and associated signal words.

Statement of Hazard	Associated Signal Word
Flammable	
Poisonous	
Radioactive	
Extremely hazardous gas and/or liquid	DANGER
Harmful if inhaled	
Vigorously accelerates combustion	
Extremely cold liquid and/or gas under pressure	
Liquid causes burns	WARNING
Extremely irritating gas	
Corrosive gas or liquid	
High pressure	
Liquefied gas under pressure	
High concentrations in the atmosphere can cause immediate unconsciousness and asphyxiation	CAUTION

2.2.3. Precautionary Measures and Instructions in Case of Contact or Exposure. These instructions are intended to supplement, if necessary, the statement of hazard by briefly setting forth measurements to be taken to avoid injury or damage from stated hazards.

2.2.4. Federal Food and Drug Administration or Department of Agriculture regulations may require registration of labels with these agencies.

3. ILLUSTRATIVE PRECAUTIONARY LABELS

3.1. The following are examples of precautionary labels for compressed gas containers prepared in accordance with the general principles given above, for the categories listed:

(1) oxidizing;
(2) flammable;
(3) physiologically inert;
(4) physiologically corrosive-irritant;
(5) liquefied compressed gas;
(6) medical gas;
(7) poison;
(8) toxic;
(9) radioactive gas mixtures;

(10) low temperature and cryogenic liquefied gas;
(11) extreme pressure;
(12) pyrophoric.

3.1.1. Oxidizing.

(1) Name of cylinder content	**OXYGEN**
(2) Signal Word	**WARNING**
(3) Statement of Hazard	**VIGOROUSLY ACCELERATES COMBUSTION**
(4) Precautionary Measures	Keep oil and grease away. Use only with equipment conditioned for oxygen service.

3.1.2. Flammable.

ACETYLENE	**HYDROGEN**	**PROPANE**
DANGER	**DANGER**	**DANGER**
FLAMMABLE	**FLAMMABLE**	**FLAMMABLE**
	HIGH PRESSURE	
Keep away from heat, flame and sparks.	Keep away from heat, flame and sparks.	Keep away from heat, flame and sparks.
Close valve when not in use.	Close valve when not in use.	Close valve when not in use.

3.1.3. Physiologically Inert (Nonliquefied)

NITROGEN	**HELIUM**	**ARGON**
CAUTION	**CAUTION**	**CAUTION**
HIGH PRESSURE	**HIGH PRESSURE**	**HIGH PRESSURE**

3.1.4. Physiologically Corrosive-Irritant.

AMMONIA

WARNING

LIQUID CAUSES BURNS

GAS EXTREMELY IRRITATING

Do not breathe gas.
Do not get in eyes, on skin, on clothing.
In case of contact with eyes or skin, immediately flush with plenty of water for at least 15 minutes.
Call physician at once in case of exposure or burns, especially to eyes, nose and throat.

3.1.5. Liquefied Compressed Gas.

CARBON DIOXIDE	**DICHLORODIFLUOROMETHANE** (Refrigerant–12)
CAUTION	**CAUTION**
HIGH PRESSURE LIQUEFIED GAS	**LIQUEFIED GAS**

3.1.6. Medical Gas (See Par. 2.2.4).

NITROUS OXIDE (U.S.P.)

WARNING

VIGOROUSLY ACCELERATES COMBUSTION

HIGH PRESSURE LIQUEFIED GAS

Keep oil and grease away.

Use only with equipment conditioned for nitrous oxide service.

Store and use with adequate ventilation.

Administration of nitrous oxide may be hazardous or contraindicated. For use only by or under the supervision of a licensed practitioner who is experienced in the use and administration of nitrous oxide and is familiar with the indications, effects, dosages, methods, and frequency and duration of administration, and with the hazards, contraindications, and side effects, and the precautions to be taken.

CAUTION: Federal Law Prohibits Dispensing without Prescription.

3.1.7. Poison.

PHOSGENE

DANGER

MAY BE FATAL IF INHALED

Do not breathe gas.

Avoid exposure of skin and eyes to gas or liquid.

Use adequate ventilation.

No exertion should follow exposure. Obtain medical attention promptly.

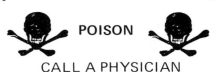

POISON

CALL A PHYSICIAN

(Add appropriate statement on antidote where necessary to comply with statutes.)

3.1.8. Toxic Gas.

CARBON MONOXIDE

DANGER

FLAMMABLE AND TOXIC

HARMFUL IF INHALED

Keep away from heat, flame and sparks.

Odorless. Gas cannot be detected by odor.

Close valve when not in use.

3.1.9. Radioactive Gas Mixture.

KRYPTON-85

Contents: Krypton-85

*Volume:

*Activity:

*Date:

*To be filled in as required.

WARNING

Use only as authorized by Atomic Energy Commission, and in conformance with state and local regulations.

3.1.10. Cryogenic Liquefied Gas.

LIQUID NITROGEN

WARNING

EXTREMELY COLD LIQUID

LIQUID MAY CAUSE BURNS

Use only by experienced personnel.

Keep container upright.

Avoid contact of liquid or cold gas with skin or eyes.

3.1.11. *Extreme Pressure.*

NITROGEN

CAUTION!

Pressure 6000 psi. Use only with equipment designated for 6000 psi or higher.

3.1.12. *Pyrophoric.*

SILANE

DANGER

FLAMMABLE

IGNITES ON CONTACT WITH AIR

Use only with equipment purged with inert gas prior to discharge from cylinder.

4. ADDITIONAL ILLUSTRATIONS

4.1. Many gases fall into more than one of the categories enumerated in 3.1. Below are illustrative labels which are appropriate for some of these gases.

4.1.1. *Oxidizing*   *Physiologically Corrosive*   *Toxic*

CHLORINE

DANGER

HAZARDOUS LIQUID AND GAS UNDER PRESSURE

Do not handle or use until safety precautions recommended by supplier have been read and
   understood.

Do not breathe air containing this gas.

Do not get in eyes or on skin.

Do not heat cylinders.

Have available emergency gas masks approved by U.S. Bureau of Mines for Chlorine service.

In case of exposure, move patient to fresh air, keep warm and quiet and call a physician.

4.1.2. *Highly Reactive*   *Physiologically Corrosive*

FLUORINE

DANGER

EXTREMELY HAZARDOUS GAS. CAUSES SEVERE BURNS.

CONTACT WITH ORGANIC OR SILICEOUS MATERIALS MAY CAUSE FIRE.

CONTACT WITH WATER MAY CAUSE VIOLENT REACTION.

Do not breathe vapor.

Do not get in eyes, or skin, or clothing.

Store out of sun and away from direct heat.

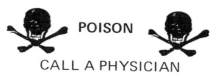

POISON

CALL A PHYSICIAN

Always have on hand a supply of magnesia paste (magnesium oxide and glycerine).
In case of contact or suspicion of contact, immediately flush skin with plenty of water (particularly under nails) until whiteness disappears.
Apply magnesia paste. Remove and wash clothing before reuse. Eyes—flush with water for at least 15 minutes.
GET MEDICAL ATTENTION.

4.1.3. Poison–Flammable

HYDROGEN SULFIDE

DANGER

POISONOUS LIQUID AND GAS

FLAMMABLE

Do not breathe gas.
Gas deadens the sense of smell. Do not depend on odor to detect presence of gas.
Keep away from heat flame and sparks.

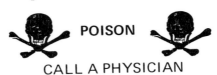 POISON

CALL A PHYSICIAN

No exertion should follow exposure. Obtain medical attention promptly.
(Add appropriate statement on antidote where necessary to comply with statutes.)

4.1.4. Cryogenic Liquefied Gas-Oxidizing

OXYGEN, Liquefied

WARNING

VIGOROUSLY ACCELERATES COMBUSTION.

MIXTURES OF COMBUSTIBLE MATERIALS AND LIQUID OXYGEN MAY EXPLODE ON IGNITION OR IMPACT.

EXTREMELY COLD LIQUID.

LIQUID MAY CAUSE BURNS.

Keep oil and grease away. Use only with equipment conditioned for oxygen service.
Avoid contact of liquid or cold gas with skin or eyes, or on clothing.
Avoid spills. Do not walk on, or roll equipment over spills.

5. ADDITIONAL PRECAUTIONARY STATEMENTS

5.1. The following statements, where applicable, may be considered for inclusion as additional label information:

    5.1.1. Return to supplier for recharging.

    5.1.2. Keep cylinder away from heat, flame and sparks.

    5.1.3. Keep away from combustible materials and other compressed gases.

    5.1.4. Do not drop.

    5.1.5. Keep valve closed when not in use.

    5.1.6. Store and use with adequate ventilation.

5.1.7. Gas does not support life.

5.1.8. Liquid may cause frostbite.

5.1.9. Liquid may cause burns.

5.1.10. Extremely flammable.

5.1.11. Liquefied gas under pressure.

5.1.12. Under pressure of ___ psig @70°F.
(This statement may be used in lieu of, rather than in addition to, statement "HIGH PRESSURE.")

5.1.13. Stand away from outlet when opening cylinder valve.

5.1.14. Do not open valve until cylinder is connected to utilization equipment.

5.1.15. Drain gas from regulator before opening cylinder valve.

5.1.16. Release regulator adjusting screw before opening cylinder valve.

5.1.17. For technical use only.

5.1.18. Not to be used for inhalation purposes.

5.1.19. Use only with equipment compatible with this product.

5.1.20. Keep oil and grease away.

The use of a signal word may be required with the above statements, if not already present on the label.

APPENDIX A ISSUED JANUARY 1976

The following Appendix was issued January 1976 to be consistent with new labeling requirements of DOT Hazardous Materials Regulations published in the Federal Register of April 15, 1976 under DOT Dockets HM-103 and HM-112. It replaced Appendix A and Appendix A-1 of the previous edition of this CGA standard.

APPENDIX A

CGA Marking System for Compressed Gas Cylinders

The CGA BASIC MARKING was developed, initially, to provide immediate identification of cylinder contents by both the commodity name and the hazard class within a single BASIC MARKING. Certain cylinders which bear this BASIC MARKING may be transported, without further labeling, under specific conditions set forth by the U.S. Department of Transportation in 49CFR172.400.

The CGA MARKING SYSTEM recognizes the need for providing additional useful information on cylinders, such as the name of the supplier, precautions to be observed in the handling, storage and use of the cylinder and/or its contents and other information of value to the consumer. The SYSTEM provides a method for combining this material with the BASIC MARKING. Thus, the SYSTEM may be used to provide the information for "Medical Use" or "In Shop Use" as may be required by other regulatory bodies.

The BASIC MARKING, illustrated in Fig. 1, shall consist of a diamond-shaped figure, denoting the hazard class of the contained gas, combined with a panel containing the proper shipping name of the contained gas. The panel shall be located to the left of the diamond. 49CFR172.101 indicates that certain gases, having an additional significant hazard, will require the application of two DOT labels. For such gases, the BASIC MARKING shall include a second diamond denoting the secondary hazard. The two diamonds shall be adjacent to one another but their adjoining points may overlap by not more than 3/8 in., as illustrated in Fig. 2.

The diamond figure shall measure 1-1/4 in. on each side and the corners shall have an in-

cluded angle of 90°. The hazard class words shall be imprinted across the center of the diamond in letters 3/16 in. in height. The color of the diamond and of the hazard class words shall be as specified for the appropriate label, as described in 49CFR172.407 through 172.450. Pictorial symbols are not used in the diamonds of the BASIC MARKING. For the oxygen marking, the word OXYGEN may be used in place of OXIDIZER as the hazard class, provided the letters are the same size as required for OXIDIZER.

The panel to the left of the diamond, or diamonds, shall be white and shall be imprinted with the commodity name of the contained gas in black letters not less than 3/16 in. in height. The panel shall measure not less than 1 in. from top to bottom but may vary in length to accommodate the name of the contained gas. When the contained gas is a mixture, the percentages of the components may be shown. The letters "USP"* or "NF,"** as required by the Federal Food and Drug Administration, may also be shown in this panel.

The BASIC MARKING may be incorporated into, or form a part of, other information to complete the CGA MARKING SYSTEM. When so used, the other information may appear above, below or beside the BASIC MARKING, provided that it shall in no way interfere with the ready recognition of the BASIC MARKING.

The complete CGA MARKING SYSTEM may be of any size or shape suitable for application to the cylinder on which it is to be used, sub-

*USP = United States Pharmacopeia
**NF = National Formulary

ject only to the restriction that the BASIC MARKING shall occupy a position of prominence. For suggested MARKING SYSTEMS, see pages 123 thru 125 of this addenda.

The BASIC MARK or BASIC MARKING SYSTEM shall be located (a) when space permits, on the shoulder of the cylinder, but not covering any permanent markings, or (b) on the side of the cylinder at a point approximately two thirds of the distance from the cylinder bottom to the top of the valve or cap.

The complete CGA MARKING SYSTEM may have, as a background color, any color which suits the user, with the exception that there must be a contrast between the background color and the colors which are required in the BASIC MARKING. When an identical or similar background color is desirable, this contrast may be accomplished by providing a border of contrasting color to separate the BASIC MARKING from the background. A similar border would be required where the BASIC MARKING was to be applied to a non-contrasting surface.

THE BASIC MARKING and/or the CGA MARKING SYSTEM shall be firmly affixed to the surface of the container and shall be of materials which are durable under conditions incident to transportation, storage and use and shall be maintained in legible condition.

The BASIC MARKING and/or the CGA MARKING SYSTEM shall remain affixed to the cylinder, full or empty, so long as it remains in the same gas service. The MARKING provides identification of the hazardous material contained in a filled cylinder. The MARKING is of equal value to the handlers of so-called "empty" cylinders, as it provides identification of any residual hazardous material which may be present in the cylinder. The removal or replacement of MARKINGS shall, therefore, be performed only by the supplier responsible for the filling of the cylinder.

Precautionary statements which are included in the information contained in the CGA MARKING SYSTEM are at the sole discretion of the supplier of the gas, except where federal or local regulations may require specific wording. The guidelines and cautionary statements, to be found elsewhere in this pamphlet, have been prepared to assist the supplier in developing his own labeling. Although suppliers may adopt these statements, they are not to be considered either mandatory or all-inclusive.

Fig. 1. The basic markings.

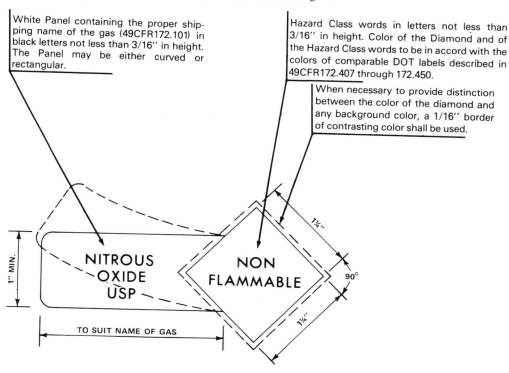

White Panel containing the proper shipping name of the gas (49CFR172.101) in black letters not less than 3/16″ in height. The Panel may be either curved or rectangular.

Hazard Class words in letters not less than 3/16″ in height. Color of the Diamond and of the Hazard Class words to be in accord with the colors of comparable DOT labels described in 49CFR172.407 through 172.450.

When necessary to provide distinction between the color of the diamond and any background color, a 1/16″ border of contrasting color shall be used.

NITROUS OXIDE USP

NON FLAMMABLE

1″ MIN.

TO SUIT NAME OF GAS

90°

Fig. 2. The BASIC MARKING in Fig. 2 is the same as in Fig. 1, except that a second diamond is added to denote the additional hazard when two labels are required by 49CFR for a flammable or nonflammable gas, except for fluorine, a nonflammable compressed gas, which must show a poison and an oxidizer label.

The colors of the individual Diamonds
may be different and shall be in accord
with colors prescribed in 49CFR172.407
through 172.450.

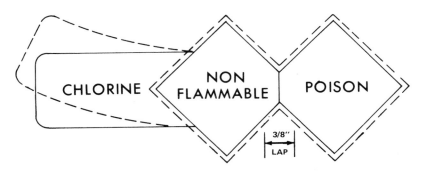

Scale: Full Size

EXAMPLES
CGA MARKING SYSTEM
(Not full size)

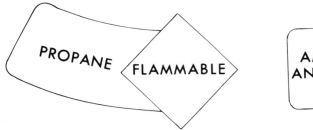

OXYGEN may replace OXIDIZER as the hazard class in which case it may be omitted from the left panel, except when USP markings are required by FDA, in which case the left panel must be as shown and the diamond can show either "OXYGEN" or "OXIDIZER".

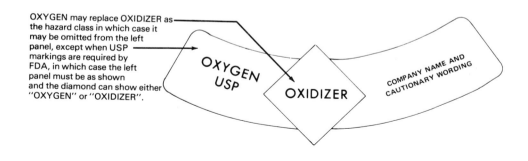

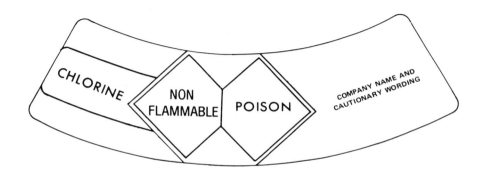

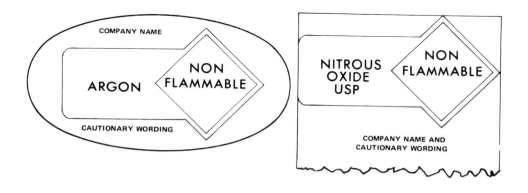

SECTION C

Visual Inspection of Cylinders

1. INTRODUCTION

1.1. Hazardous Materials Regulations (49 CFR Parts 170-179) of the Department of Transportation (DOT), formerly Interstate Commerce Commission Regulations, as well as the Regulations of the Canadian Transport Commission (CTC), require that a cylinder be condemned when it leaks, or when internal or external corrosion, denting, bulging, or evidence of rough usage exists to the extent that the cylinder is likely to be weakened appreciably.

Note 1: Under prescribed conditions of use a formal visual inspection has been authorized in lieu of the periodic hydrostatic retest for certain low-pressure cylinders used for noncorrosive gas services (Ref. DOT 173.34 (e) (10)).

Note 2: Wherever reference is made to DOT Regulations, similar requirements can be found in CTC Regulations.

1.2. This section has been prepared as a guide to cylinder users for establishing their own cylinder inspection procedures and standards. It is of necessity general in nature although some specific limits are recommended. It should be distinctly understood that it will not cover all circumstances for each individual cylinder type and condition of lading. Each cylinder user must expect to modify them to suit his own cylinder design or the conditions of use that may exist in his own service.

1.3. Experience in the inspection of cylinders is an important factor in determining the acceptability of a given cylinder for continued service. Users lacking this experience and having doubtful cylinders should return them to a manufacturer of the same type of cylinders or competent requalification agency for reinspection.

1.4. The suggestions contained in this standard do not apply to cylinders manufactured under specification DOT (ICC) 3HT. Because of the special provisions of this specification, separate recommendations covering service life and standards for visual inspection of these cylinders are contained in Pamphlet C-8, "Standard for Requalification of DOT (ICC) 3HT Cylinders."

1.5. Inspection procedures include preparation of cylinders for inspection, exterior inspection, interior inspection if required, nature and extent of damage to be looked for, some tests which indicate the conditions of the cylinder, etc. A sample inspection report form is appended which may be revised to suit user's requirements.

2. GENERAL

2.1. Inspection Equipment

2.1.1. Depth Gages, Scales, etc. Exterior corrosion, denting, bulging, gouges or digs are normally measured by simple direct measurement with scales or depth gages. In brief, a rigid straight edge of sufficient length is placed over the defect and a scale is used to measure the distance from the bottom of the straight edge to the bottom of the defect. There are also available commercial depth gages which are especially suitable for measuring the depth of small cuts or pits. It is important when measuring such defects to use a scale which spans the entire affected area. When measuring cuts, the upset metal should be removed or compensated for so that only actual depth of metal removed from the cylinder wall is measured.

2.1.2. Ultrasonic Devices. There are a variety of commercial ultrasonic devices available. These can be used to detect subsurface faults and to measure wall thickness.

2.1.3. Magnetic Particle Inspection. Magnetic particle inspection can be adapted for cylinder inspection to quickly locate surface faults not readily visible to the naked eye.

2.1.4. Penetrant Inspection Materials. Dye penetrant materials are available which show surface faults not readily visible to the naked eye.

2.2. Definitions

2.2.1. High- and Low-Pressure Cylinders.
For the purpose of this standard, high-pressure cylinders are those with a marked service pressure of 900 psi (6200 kPa) or greater; low-pressure cylinders are those with a marked service pressure less than 900 psi (6200 kPa).

2.2.2. Cylinder Disposition. The following definitions apply to terms used in this standard for disposition of cylinders failing visual inspection.

2.2.2.1. Condemned. Scrap, no longer fit for service.

2.2.2.2. Reject. Not fit for service in present condition. May be requalified by either additional testing to verify adequacy of cylinder for continued service, or by reheat treatment, repair or rebuilding to correct the defect as specified in 173.34 of DOT Regulations.

2.2.3. Minimum Allowable Wall Thickness.
For the purposes of this standard, the minimum allowable wall thickness is the minimum wall thickness required by the specification under which the cylinder was manufactured.

2.2.4. Dents. Dents in cylinders are deformations caused by its coming in contact with a blunt object in such a way that the thickness of metal is not materially impaired. A typical dent is shown in Fig. 1.

2.2.5. Cuts, Gouges or Digs. Cuts, gouges or digs in cylinders are deformations caused by contact with a sharp object in such a way as to cut into or upset the metal of the cylinder, decreasing the wall thickness at that point.

2.2.6. Corrosion or Pitting. Corrosion or pitting in cylinders involves the loss of wall

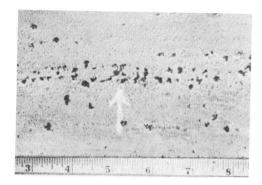

Fig. 2. Line corrosion.

thickness by corrosive media. There are several kinds of pitting or corrosion to be considered.

2.2.6.1. Isolated Pitting. Isolated pits of small diameter do not effectively weaken the cylinder.

2.2.6.2. Line Corrosion. When pits are connected to others in a narrow band or line, such a pattern is termed "line corrosion." This condition is more serious than isolated pitting. An example of line corrosion is shown in Fig. 2.

2.2.6.3. Crevice Corrosion. Corrosion which occurs in the area of the intersection of the footring or headring and the cylinders. Figure 3 is an example of crevice corrosion.

2.2.6.4. General Corrosion. General corrosion is that which covers considerable surface areas of the cylinder. It reduces the structural strength. It is often difficult to measure or estimate the depth of general corrosion because direct comparison with the original wall cannot

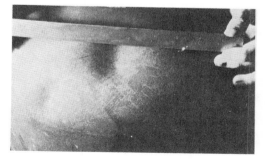

Fig. 1. Measuring the length of a typical dent.

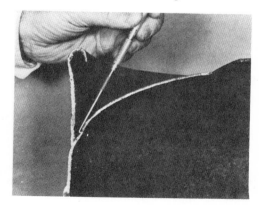

Fig. 3. Crevice corrosion near the cylinder footring.

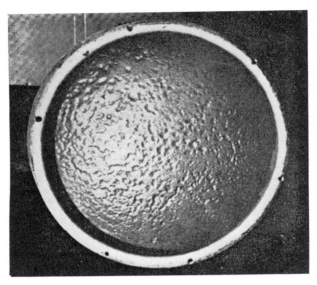

Fig. 4. General corrosion with pitting.

Fig. 5. General corrosion with pitting on cylinder wall.

always be made. General corrosion is often accompanied by pitting. This form of corrosion is shown in Figs. 4 and 5.

3. LOW-PRESSURE CYLINDERS EXEMPT FROM THE HYDROSTATIC TEST

This section covers cylinders exempt from hydrostatic retest requirements of the DOT by virtue of their exclusive use in certain noncorrosive gas service. They are not subject to internal corrosion and do not require internal shell inspection. If, due to unusual circumstances, cylinders of these types suffer internal corrosion they should be inspected in accordance with the procedure in Section 4.

3.1. Preparation for Inspection

Rust, scale, caked paint, etc. shall be removed from the exterior surface so that the surface can be adequately observed. Equipment should be provided to facilitate inspection of the bottom of the cylinder. This is important because experience has shown this area to be the most susceptible to corrosion.

3.2. Exterior Inspection

Cylinders shall be checked as outlined below for corrosion, general distortion or any other defect that might indicate a weakness which would render it unfit for service. Refer to Section 3.2.9 for rejection limits for aluminum cylinders.

3.2.1. Corrosion.

3.2.1.1. Corrosion Limits.
To fix corrosion limits for all types, designs and sizes of cylinders, and include them in this standard is not practicable. Failure to meet any of the following four general rules is cause for condemning a cylinder.

(a) A cylinder shall not be used when the tare weight is less than 95 percent of the original tare weight marked on the cylinder. When determining tare weight, be sure that the cylinder is empty.

(b) A cylinder shall be condemned when the remaining wall in an area having *isolated pitting* only is less than $\frac{1}{3}$ of the minimum allowable wall thickness.

(c) A cylinder shall be condemned when *line or crevice corrosion* on the cylinder is 3 in. (7.6 cm) in length or over and the remaining wall is less than $\frac{3}{4}$ of the minimum allowable wall thickness or when line or crevice corrosion is less than 3 in. (7.6 cm) in length and the remaining wall thickness is less than $\frac{1}{2}$ the minimum allowable wall thickness. Refer to Figs. 2 and 3.

(d) A cylinder shall be condemned when the remaining wall in an area of *general corrosion* is less than $\frac{1}{2}$ of the minimum allowable wall thickness. Refer to Figs. 4 and 5.

3.2.1.2. Representative Cylinder Wall Thickness.
To use the above criteria, it is necessary to know the original wall thickness of the cylinder or the minimum allowable wall thickness. Table 1 provides original minimum allowable wall thicknesses for a number of common size low-pressure cylinders which are likely to be governed by the criteria of this section.

3.2.1.3. Application of Limits.
To illustrate how to establish actual corrosion rejection limits, a cylinder of Specification 4B240 with $14\frac{1}{2}$ in. nominal diameter is used. From Table 1, the minimum allowable wall thickness is 0.125 in. Corrosion limits are determined as follows.

3.2.1.3.1. General Corrosion Accompanied by Pitting.

(a) When the actual wall thickness can be measured, the cylinder shall be condemned when the remaining wall is less than 0.063 in. (0.16 cm). See 3.2.1.1 (d).

(b) When the wall thickness cannot be measured and the original wall thickness is *not known*, the cylinder shall be condemned when the deepest pit in the general corrosion area exceeds $\frac{3}{64}$ or 0.047 in (0.12 cm). This is because general corrosion will already have removed 0.021 in. (0.053 cm) of original wall and the total pit depth is $0.021 + 0.042 = 0.063$ in. (0.160 cm). The remaining wall is then 0.062 in. (0.157 cm). See Notes 1 and 2 below.

(c) When the actual wall thickness cannot be measured but the original wall thickness *is known*, the cylinder shall be condemned when the original wall less $1\frac{1}{2}$ times the maximum pit depth is less than 0.063 in. (0.160 cm).

Note 1: Although general corrosion does not always follow a definite pattern, where there is appreciable pitting in areas of general corrosion, the pitted depth (Pd) usually is about twice the general corrosion loss (GC). See Fig. 6.

Note 2: Pitted depth may be measured by placing a straight edge across the pitted area and measuring the distance from the bottom of the straight edge to the bottom of the pit. Where this is not practical because of obstructions such as footrings, bands, etc., special curved measuring devices or even putty casts

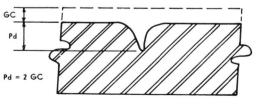

Fig. 6.

TABLE 1

Nominal Cylinder Diameter	DOT Specification Marking	Original Minimum Allowable Wall Thickness (*)
9" (23 cm)	DOT 4B240 DOT 4BA240, 4BW240, 4BA300, 4BW300 DOT 4E240, 4E300	.090" (.229 cm) .078" (.198 cm) .140" (.356 cm)
10" (25 cm)	DOT 4BA240, 4BW240, 4BA300, 4BW300 DOT 4E240	.078" (.198 cm) .140" (.356 cm)
12" (30 cm)	DOT 4B240 DOT 4BA240, 4BW240 DOT 4E240	.105" (.267 cm) .078" (.198 cm) .140" (.356 cm)
14½" (37 cm)	DOT 4B240 DOT 4BA, 4BW240 DOT 4AA480 DOT 3A480	.125" (.318 cm) .087" (.221 cm) .185" (.470 cm) .212" (.538 cm)
22" (56 cm)	DOT 4B240 DOT 4BA, 4BW240	.191" (.485 cm) .130" (.330 cm)
24" (61 cm)	DOT 4B240 DOT 4BA, 4BW240	.208" (.528 cm) .142" (.361 cm)
30" (76 cm)	DOT 4B240 DOT 4BA, 4BW240	.251" (.638 cm) .172" (.436 cm)

(*) Some cylinders will have thicker walls due to difference in manufacturing methods and inspection procedures. Values shown are for absolute minimums allowed by specifications. Higher values may be used if information showing thicker walls than those listed is obtained from the manufacturer of the cylinder.

may be prepared to be able to measure the depths of such pitting.

3.2.1.3.2. Isolated Pits Not in General Corrosion Area.

(a) When the actual wall thickness can be measured, the cylinder shall be condemned if its remaining wall is less than 0.047 in. (0.12 cm) thick.

(b) When the actual wall thickness cannot be measured and the original wall thickness *is not known*, the cylinder shall be condemned if the pit depths on the wall exceed 0.084 in. (0.21 cm).

(c) When the actual wall thickness cannot be measured and the original wall thickness *is known*, the cylinder shall be condemned if the remaining wall obtained by subtracting the maximum pit depth from the original wall is less than 0.047 in. (0.12 cm).

3.2.2. Dents. Dents are of concern where the metal deformation is sharp and confined, or where it is near a weld. Where metal deformation is not sharp, dents of larger magnitude can be tolerated.

3.2.2.1. Dents at Welds. Where denting occurs so that any part of the deformation includes a weld, the maximum allowable dent depth shall be ¼ in. (0.64 cm).

3.2.2.2. Dents Away from Welds. When denting occurs so that no part of the deformation includes a weld, the cylinder shall be rejected if the depth of the dent is greater than one-tenth of the greatest dimension of the dent.

3.2.3. Cuts, Gouges or Digs. Cuts, gouges or digs reduce the wall thickness of the cylinder and in addition are considered to be stress raisers. Depth limits are set in the next paragraphs; however, cylinders shall be condemned at one-half of the limit set whenever the length of the defect is 3 in. (7.6 cm) or more.

3.2.3.1. When the original wall thickness at manufacture *is not known*, and the actual wall thickness cannot be measured, a cylinder shall be condemned if the cut, gouge or dig exceeds one-half of the original minimum allowable wall thickness. (Example: In a $14\frac{1}{2}$ in. (36.8 cm) nominal diameter 4B240, 100 lb propane cylinder, the depth of the defect shall not exceed one-half of 0.125 in. or $\frac{1}{16}$ in. (0.15 cm). If the defect was 3 in. or more in length it shall not exceed one-half of $\frac{1}{16}$ in. or $\frac{1}{32}$ in. or 0.08 cm.

3.2.3.2. When the original wall thickness at manufacture *is known*, or the actual wall thickness is measured, a cylinder shall be condemned if the original wall thickness minus the depth of the defect is less than one-half of the original minimum allowable wall thickness.

3.2.4. Leaks. Leaks can originate from a number of sources, such as defects in a welded or brazed seam, defect at the threaded opening or from sharp dents, digs, gouges or pits. Any leakage is cause for rejection.

3.2.4.1. To check for leaks the cylinder shall be charged and carefully examined. All seams, pressure openings, sharp dents, digs, gouges and pits shall be coated with a soap or other suitable solution to detect the escape of gas. Any leakage is cause for rejection.

3.2.5. Fire Damage. Cylinders shall be carefully inspected for evidence of exposure to fire.

3.2.5.1. Inspection for Fire Damage. Common evidence of exposure to fire are (a) charring or burning of the paint or other protective coat, (b) burning or scarfing of the metal, (c) distortion of the cylinder, (d) melted out fuse plugs, (e) burning or melting of valve.

3.2.5.2. Evaluation of Fire Damage. DOT Regulations state that, "A cylinder which has been subjected to the action of fire must not again be placed in service until it has been properly reconditioned," in accordance with 173.34 (f). The general intent of this requirement is to remove from service cylinders which have been subject to the action of fire which has changed the metallurgical structure or the strength properties of the steel, or in the case of acetylene cylinders caused breakdown of porous filler. This is normally determined by visual examination as covered above with particular emphasis to the condition of the protective coating. If there is evidence that the protective coating has been burned off any portion of the cylinder surface, or if the cylinder body is burnt, warped or distorted, it is assumed that the cylinder has been overheated, and 173.34 (f) shall be compiled with. If, however, the protective coating is only smudged, discolored or blistered and is found by examination to be intact underneath, the cylinder shall not be considered affected within the scope of this requirement.

3.2.6. Bulges. Cylinders are manufactured with a reasonably symmetrical shape. Cylinders which have definite visible bulges shall be removed from service and evaluated.

3.2.6.1. Measurement. Bulges in cylinders can be measured as follows:

(a) Bulges on the cylinder sidewall can be measured by comparing a series of circumferential measurements.

(b) Bulges in the head, and also in some cases on the sidewall, can be measured by comparing a series of measurements of the peripheral distance between the valve spud and the center seam (if any) or an equivalent fixed location on the cylinder sidewall.

(c) Variations from normal cylinder contour can be measured directly by (1) measuring the height of a bulge with a scale; (2) comparing templates of bulged areas with similar areas not bulged.

3.2.6.2. Limits. Cylinders shall be condemned when a variation of 1 percent or more is found in the measured circumferences or in peripheral distances measured from the valve spud to the center seam (or equivalent fixed point). An example for a 15 in. outside diameter cylinder follows (metric equivalent not given for examples):

Normal cylinder outside diameter	15 in.
Cylinder circumference	47.12 in.
Maximum circumference	
$47.12 + 0.01 (47.12) = 47.59$ in.	

Variation in circumference   0.47 in.
Equivalent variation in diameter
                                   0.47 = 00.15 in.
If the bulge is uniform around the cylinder, the limiting height of the bulge would be 0.15 in./2 = 0.075 in.

3.2.7. Neck Defects. Cylinder, necks shall be examined for cracks, folds, flaws and distortion.

3.2.7.1. Neck cracks are normally detected by testing the neck during charging operations with a soap solution.

3.2.7.2. Cylinder neck threads shall be examined whenever the valve is removed from the cylinder. At manufacture, cylinders have a specified number of full threads of proper form as required in applicable thread standards. Cylinders shall be rejected if the required number of effective threads are materially reduced so that a gas tight seal cannot be obtained by reasonable valving methods. Common thread defects are worn or corroded crests, and broken or nicked threads.

3.2.8. Attachments. Attachments on cylinders may lose their intended function through service abuse. These attachments and the associated portion of the cylinder must receive careful inspection.

3.2.8.1. The footring and headring of cylinders may no longer perform their functions: (a) to cause the cylinder to remain stable and upright, (b) to protect the valve. Rings shall be examined for distortion, for looseness and for failure of welds. Appearances may often warrant removal of the cylinder from service.

3.2.8.2. When the cylinder bears a permanent attachment such as a footring, headring, double bottom or marking plate, which covers a portion of the cylinder surface proper, it must receive particular attention to assure the inspector that it is in the same relation to the cylinder as at the time of its attachment. The seal of the periphery of the part of the cylinder must be checked for possible entry of moisture. In the case of adhesive attachments, any evidence of a break in the seal is cause for removal of the attachment. Plastic materials must be checked carefully for gouges or splits, which if present, would also require their removal.

3.2.8.3. In the case of a marking plate not completely sealed and not removable for inspection of the cylinder wall, any evidence of corrosion between it and the wall shall require leak testing before the cylinder is returned to service.

Note 1: Repair rules for such cylinders are established in DOT Regulation 173.34, Par. (h) to (l).

3.2.9. Aluminum Cylinders. The exterior inspection requirements of Section 3.2 shall be met except as follows:

3.2.9.1. Aluminum cylinders shall be condemned when impairment to the surface (corrosion or mechanical defect) exceeds a depth where the remaining wall is less than $\frac{3}{4}$ of the minimum allowable wall thickness required by the specification under which the cylinder was manufactured.

3.2.9.2. Aluminum cylinders subject to the action of fire shall be condemned as required by 173.34 (f) of DOT Regulations.

3.3. Inspection Report Form

DOT Regulations (173.34 (e) (5)) require that results be recorded and a record kept by the owner or his authorized agent until either expiration of the retest period or until the cylinder is again reinspected or retested, whichever occurs first. A sample Visual Inspection Report Form is shown in Appendix A.

4. LOW-PRESSURE CYLINDERS SUBJECT TO HYDROSTATIC TESTING

Cylinders covered in this section are low-pressure cylinders other than those covered in Section 3. Cylinders not listed in DOT Regulations 173.34 (e) (10) differ essentially from such cylinders in that they require a periodic hydrostatic retest which includes an internal and external examination. Defect limits for the external examination are prescribed in Section 3. The additional procedures for internal inspection follow.

4.1. Preparation for Inspection

4.1.1. The provisions outlined in 3.1 shall be followed. Additionally, the interior of the cylinder shall be prepared for inspection by the removal of internal scale, or any other condition which would interfere with the inspection of the internal surface. Cylinders with interior coating shall be examined for defects in the coating. If the coating is defective, it shall be removed.

4.1.2. A good inspection light of sufficient intensity to clearly illuminate the interior walls is mandatory for internal inspection. Flammable gas cylinders shall be purged with inert gas or water before being examined with a light.

4.2. Internal Inspection

Cylinders shall be inspected internally at least every time the cylinder is periodically retested.

Note: The Chlorine Institute, 342 Madison Avenue, New York, NY 10017, Pamphlet No. 7 "Container Procedure for Chlorine Packaging" includes an internal examination of chlorine cylinders at the time of each filling.

4.2.1. General Corrosion. Interior corrosion is best evaluated by a hydrostatic test combined with careful visual inspection. Thickness measuring and flaw-detection devices of the ultrasonic type may be used to evaluate specific conditions. For basic corrosion limits refer to Section 3.2.1.

4.2.2. Defects Other Than Corrosion. Any cylinder shall be rejected when doubt exists as to its suitability for continued service. Where the bottom of the defect cannot be seen or where its extent cannot be measured by various inspection instruments, the cylinder shall be condemned. Examples of such internal defects are cuts, mechanical abrasions and fabrication irregularities.

5. HIGH-PRESSURE CYLINDERS

High-pressure cylinders are those with a marked service pressure of 900 psi (6200 kPa) or higher. They are seamless; no welding is permitted. The majority of such cylinders are DOT or CTC 3A or 3AA type.

5.1. Preparation for Inspection

Proper identification of the content of the cylinder shall be made. Cylinder content shall be released in a safe area by cracking the valve slowly. If the valve is damaged or obstructed, content may be released by slowly backing off the valve safety nut until content starts to escape.

Warning: Position cylinder to prevent safety nut from causing injury to personnel or damage to equipment if it becomes totally disconnected from valve. Remove valve from cylinder only after making certain that the cylinder is empty. Cylinders containing flammable or toxic gas shall be purged properly using inert gas or water in a safe area.

5.2. Hammer Test

The hammer test is a valuable indicator of internal corrosion and is a convenient test that can be made without removing the valve prior to each charging of the cylinder.

5.2.1. The hammer test consists of tapping the cylinder side wall with a light blow using a half pound ($\frac{1}{2}$) ball-peen hammer or equivalent. A cylinder will normally have a clear ring. A dull ring would indicate internal corrosion, liquid or accumulation of foreign material in the cylinder. Such cylinders shall be inspected internally in accordance with 5.4.

5.3. Exterior Inspection

5.3.1. Exterior of cylinder shall be cleaned to permit adequate visual inspection.

5.3.2. Corrosion Limits. To fix corrosion limits for all types, designs and sizes of cylinders, and include them in this standard is not practicable. Cylinders which do not meet any of the following general rules shall be condemned:

(a) The calculated wall stress, based on actual wall thickness measurements, in a corroded section shall not exceed the maximum wall stress limitation published in DOT Regulations, 173.302 (c) (3).

(b) The average wall stress, determined from the elastic expansion obtained in the hydrostatic test, shall not exceed the average wall stress limitation published in DOT Regulations, 173.302 (c) (3).

Note: This applies only to those cylinders which qualify for 110 percent filling. Where not qualified for 110 percent filling, refer to DOT Regulations 173.34 (e) (4).

(c) The remaining wall in a cylinder which has isolated pitting of small cross section only shall not be less than two-thirds the minimum allowable wall thickness.

5.3.2.1. The use of these criteria for a common size cylinder follows (metric equivalents not given in example):

Size (o.d. × length)	$9\text{-}\frac{1}{16}$ in. × 51 in.
Specification marking	3A2015
Minimum allowable wall at manufacture	0.243 in.
Limiting wall thickness in service	0.221 in.
Limiting elastic expansion in the hydrostatic test	180 cc

(a) Local Pitting or Corrosion or Line Corrosion.

(1) When the original wall thickness of the cylinder *is not known*, and the actual wall thickness cannot be measured, the cylinder shall be condemned if corrosion exceeds $\frac{1}{32}$ in. (0.079 cm) in depth. This is arrived at by subtracting from the minimum allowable wall at manufacture (0.243 in. or 0.617 cm), the limiting wall in service (0.221 in. or 0.561 cm), to give the maximum allowable corrosion limit of 0.022 in. (0.056 cm).

(2) When the wall thickness *is known*, or the actual wall thickness is measured, the difference between this known wall and the limiting value establishes the maximum corrosion figure. The normal hot forged cylinder of this size will have a measured wall thickness of about 0.250 in. (0.635 cm). Comparison of this with the limiting wall thickness shows that defects up to about $\frac{1}{16}$ in. (0.158 cm) are allowable provided, of course, that the actual wall is measured or is known.

(b) General Corrosion.

(1) Cylinders with general corrosion are evaluated by subjecting them to a hydrostatic test. Thus, a cylinder with an elastic expansion of 181 cc or greater would be removed from 110 percent filling service. If areas of pronounced pitting are included within the general corrosion, the depth of such pitting should also be measured (with the high spots of the actual surface as a reference plane) and the criteria established in the first example apply. Thus, the maximum corrosion limit would be 0.022 in. (0.056 cm) when the wall was not known.

(c) Isolated Pits.

(1) When the original or actual wall thickness of the cylinder *is not known*, and the actual wall thickness cannot be measured, the cylinder shall be condemned if the pit depth exceeds 0.081 in. (0.243 in. × $\frac{1}{3}$ or 0.206 cm).

(2) When the wall thickness *is known*, or the actual wall thickness is measured, this thickness less the pit depth shall not be less than 0.165 in. (0.243 in. × $\frac{2}{3}$ or 0.617 cm).

5.3.3. Cuts, Digs and Gouges.

5.3.3.1. Measurement. Cuts, digs or gouges may be measured with suitable depth gages (any upset metal shall be smoothed off to allow true measurements without causing further damage to parent metal).

5.3.3.2. Limits. Established by the stress considerations in 5.3.2(a).

5.3.3.3. General. Any defect of appreciable depth having a sharp bottom is a stress raiser and even though a cylinder may be acceptable from a stress standpoint, it is common practice to remove such defects. After any such conditioning operation, verification of the cylinder strength shall be made by wall thickness measurement followed by hydro-static test.

5.3.4. Dents. Dents can be tolerated when the cylinder wall is not deformed excessively or abruptly. Considerations of appearance play a major factor in evaluation of dents. In general, industry practice for a 9 in. diameter × 51 in. (22.9 cm diameter × 130 cm) long cylinder accepts dents up to $\frac{1}{16}$ in. (0.158 cm) depth when the major diameter of the dent is 2 in. (5 cm) or greater.

5.3.5. Arc and Torch Burns. Cylinders with arc or torch burns shall be rejected. Defects of this nature may be recognized by one of the following conditions:

(a) Removal of metal by scarfing or cratering.
(b) A scarfing or burning of the base metal.
(c) A hardened heat affected zone.

5.3.6. Bulges. Cylinders are manufactured with a reasonably symmetrical shape. Those with definite visible bulges shall be condemned.

5.3.7. Fire Damage. Cylinders shall be carefully inspected for evidences of exposure to fire.

5.3.7.1. Inspection for Fire Damage. Common evidence of exposure to fire are (a) charring or burning of the paint or other protective coat, (b) burning or scarfing of the metal, (c) distortion of the cylinder, (d) melted out fuse plugs, (e) burning or melting of valve.

5.3.7.2. Evaluation of Fire Damage. DOT Regulations state that, "A cylinder which has been subjected to the action of fire must not again be placed in service until it has been properly reconditioned," in accordance with 173.34 (f). The general intent of this requirement is to remove from service cylinders which have been subject to the action of fire which has changed the metallurgical structure or the strength properties of the steel. This is normally determined by visual examination as covered above with particular emphasis to the condition of the protective coating. If there is evidence that the protective coating has been burned off any portion of the cylinder surface, or if the cylinder body is burnt, warped or distorted, it is assumed that the cylinder has been

overheated, and 173.34 (f) shall be complied with. If, however, the protective coating is only smudged, discolored or blistered and is found by examination to be intact underneath, the cylinder shall not be considered affected within the scope of this requirement.

5.3.8. Neck Defects. Cylinder necks shall be examined for cracks, folds and other flaws.

5.3.8.1. Neck cracks are normally detected by testing the neck during charging operations with a soap solution.

5.3.8.2. Cylinder neck threads shall be examined whenever the valve is removed from the cylinder. At manufacture, cylinders have a specified number of full threads of proper form as required in applicable thread standards. Cylinders shall be rejected if the required number of effective threads are materially reduced so that a gas tight seal cannot be obtained by reasonable valving methods. Common thread defects are worn or corroded crests, and broken or nicked threads.

5.3.9. Attachments. Attachments on cylinders may lose their intended function through service abuse. These attachments and the associated portion of the cylinder shall receive careful inspection. Attachments may not be welded to a high-pressure cylinder surface.

5.3.9.1. The footring and neckring of cylinders may no longer perform their functions: (a) to cause the cylinder to remain stable and upright, (b) to provide proper attachment of valve protection cap. Rings shall be examined for distortion, looseness or condition of threads.

5.3.9.2. When the cylinder bears a permanent attachment such as a footring or neckring, which covers a portion of the cylinder surface proper, it must receive particular attention to assure the inspector that it is in the same relation to the cylinder as at the time of its attachment. The seal of the periphery of the part to the cylinder must be checked for possible entry of moisture. In the case of adhesive attachments, any evidence of a break in the seal is cause for removal of the attachment. Plastic materials must be checked carefully for gouges or splits, which if present, would also require their removal.

5.4. Internal Inspection

Cylinders shall be inspected internally at least every time the cylinder is periodically retested.

5.4.1. The interior of cylinders shall be prepared for inspection by the removal of dirt, scale or other condition as necessary to permit the inspection of the internal surface. Cylinders with interior coating shall be examined for defects in the coating. If the coating is defective, it shall be removed.

5.4.2. A good inspection light of sufficient intensity to clearly illuminate the interior walls is mandatory for internal inspection.

5.4.3. Corrosion and Pitting.

5.4.3.1. General Corrosion. Interior corrosion is best evaluated by a hydrostatic test combined with careful visual inspection. Thickness measuring and flaw-detection devices of the ultrasonic type may be used to evaluate specific conditions. For basic corrosion limits refer to Section 5.3.2.

5.4.4. Interior Defects. Any cylinder should be withdrawn from service for further evaluation when doubt exists as to its suitability for continued service. Examples of such internal defects are cuts, mechanical abrasions and fabrication irregularities.

APPENDIX A

The following is a sample visual inspection report form. It includes cylinder identification, protective coating, what cylinder inspected for and disposition.

SAMPLE VISUAL INSPECTION REPORT FORM

Reference: DOT Regulations 173.34(e)(10)

COMPANY _____

PLANT _____

DATE: _____ _____ _____
            Month   Year

RESPONSIBLE MANAGER _____

(Signature)

CYLINDER IDENTIFICATION					PROTECTIVE COATING		CYLINDERS INSPECTED FOR								DISPOSITION		
Serial No.	Identifying Symbol	ICC DOT Spec.	Mfg.	Date of Mfg.	Type	Condition	Corrosion and Pitting	Dents	Cuts, Digs and Gouges	Leaks	Fire Damage	Bulges	Neck Defects	Attachments	Disposition	Date Inspected	Inspector's Initials
90615	GAS INC	48A240	ABC.yl.	6-58	Paint	Good	✓	✓	✓	✓	✓	✓	✓	✓	OK	1-6-72	JHD
22 0196	ABC CYL	4BW240	AB Cyl.	10-70	Galvanized	FAIR	✓	SC	✓	✓	✓	✓	✓	✓	SC	1-6-72	JHD
109 640	OXYING	4AA280	AB Cyl.	5-60	Paint	Good	✓	✓	✓	✓	SC	SC	✓	R	R	1-6-72	JHD
180015	XYZ CO	4BA290	AB Cyl.	4-68	Paint	BURNED	✓	✓	✓	✓	SC	✓	✓	✓	SC	1-6-72	JHD
11225	ABCC.	4BA290	AB Cyl.	9-55	Paint	FAIR	✓	✓	✓	SC*	✓	✓	✓	✓	SC	1-6-72	JHD

Disposition Code: OK – Return to Service
  SC – Scrap (Condemned)
  R – Hold for Authorized Repair

*REMARKS: CYL. No. 11225 - Scrapped Due to Bullet Holes

137

SECTION D

Hydrostatic Testing of Cylinders

INTRODUCTION

Hydrostatic testing and retesting of compressed gas cylinders, as specified in DOT Regulations and Specifications, require testing by water jacket or other suitable method, operated so as to obtain accurate data. Except when modified by the regulations or specifications, the total expansion, permanent expansion and percent permanent expansion shall be determined. The test apparatus shall be approved as to type and operation, by the Bureau of Explosives. The following methods are in use.

(1) Water Jacket Volumetric Expansion Method. This method is applicable to all hydrostatic tests when volumetric expansion determinations are required. It consists of enclosing the cylinder in a vessel completely filled with water, measuring in a suitable device attached to the vessel the total and permanent volumetric expansion of the cylinder, by measuring the amount of water displaced by expansion of the cylinder when under pressure and after pressure is released.

(2) Direct Expansion Method. This method is applicable to all hydrostatic tests when volumetric expansion determinations are required. However, it has practical limitations in its use. It consists of forcing a measurable volume of water into a cylinder filled with a known weight of water at a known temperature, and measuring the volume of water expelled from the cylinder when the pressure is released. The permanent volumetric expansion of the cylinder is calculated by subtracting the volume of water expelled from the volume of water forced into the cylinder. The total volumetric expansion of the cylinder is calculated by subtracting the compressibility of the volume of water forced into the cylinder to raise pressure to desired test pressure. Although this also measures the elastic expansion, DOT Reg-

ulations do not allow this method of testing to qualify cylinders for charging to 10 percent in excess of marked service pressure.

(3) Pressure Recession Method. This method consists of subjecting the cylinder rapidly to hydrostatic test pressure, immediately cutting off pressure supply and observing recession of pressure in the cylinder due to permanent expansion. Since the expansion of the cylinder is not measured by burette, this method shall not be used to qualify cylinders for charging to 10 percent in excess of marked service pressure.

(4) Proof Pressure Method. This method is permitted where DOT Regulations and Specifications do not require the determination of total and permanent expansion. It consists of examining the cylinder under test for leaks and defects.

This section gives a brief description of the above four test methods. For detailed requirements for operation, equipment and use of the above four test methods see CGA Pamphlet C-1.

INSPECTION OF CYLINDERS

Regardless of the type of hydrostatic test method used, 173.34 (e) of DOT Regulations specifies that this periodic retest must require external and internal visual examination of the cylinder. Performance of these inspections prior to retest is recommended and all cylinders rejected by these inspections shall be condemned immediately.

(1) Careful internal inspection must be made with a suitable light. Scale and sludge deposits (if any) shall be removed and, if the cylinder is seriously corroded, it shall be condemned. (To avoid a possible flash, cylinders used in flammable gas service shall be thoroughly purged before the interior inspection light is inserted.)

(2) Exterior inspection of the cylinder, including the bottom, shall be carefully made for corrosion, dents, arc or torch burns and physi-

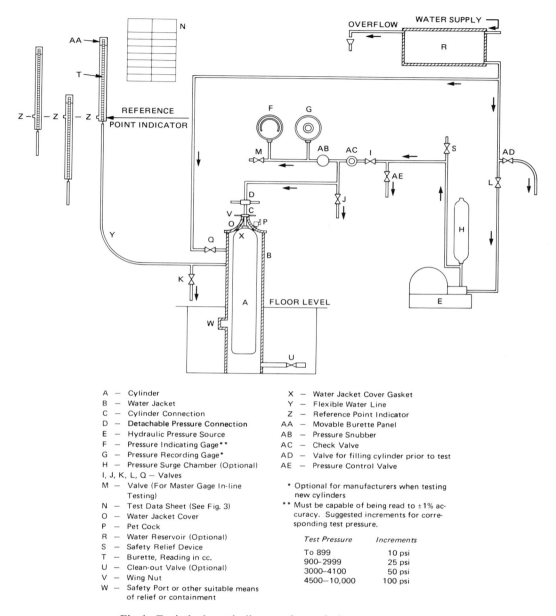

A	—	Cylinder	X —	Water Jacket Cover Gasket
B	—	Water Jacket	Y —	Flexible Water Line
C	—	Cylinder Connection	Z —	Reference Point Indicator
D	—	**Detachable Pressure Connection**	AA —	Movable Burette Panel
E	—	Hydraulic Pressure Source	AB —	Pressure Snubber
F	—	Pressure Indicating Gage**	AC —	Check Valve
G	—	Pressure Recording Gage*	AD —	Valve for filling cylinder prior to test
H	—	Pressure Surge Chamber (Optional)	AE —	Pressure Control Valve

I, J, K, L, Q — Valves
M — Valve (For Master Gage In-line Testing)
N — Test Data Sheet (See Fig. 3)
O — Water Jacket Cover
P — Pet Cock
R — Water Reservoir (Optional)
S — Safety Relief Device
T — Burette, Reading in cc.
U — Clean-out Valve (Optional)
V — Wing Nut
W — Safety Port or other suitable means of relief or containment

* Optional for manufacturers when testing new cylinders
** Must be capable of being read to ±1% accuracy. Suggested increments for corresponding test pressure.

Test Pressure	*Increments*
To 899	10 psi
900–2999	25 psi
3000–4100	50 psi
4500–10,000	100 psi

Fig. 1. Typical schematic diagram of water jacket test apparatus.

cal deformation. Seriously corroded cylinders shall be condemned. Cylinders dented, arc or torch burnt or physically damaged so as to weaken the cylinder appreciably shall be condemned.

Note 1: Each cylinder, standing alone, shall be tapped lightly with a $\frac{1}{2}$ lb metallic object, such as a machinist hammer, wrench or equivalent. Any cylinder which has a dull or dead ring shall be cleaned and if the dull or dead ring persists, the cylinder shall be condemned.

B.E. APPROVAL NO. 1234
ASSIGNED RETEST SYMBOL ZYX

XYZ OXYGEN CO.
THIS TOWN, THAT STATE 00000

PLANT CHART NO. 10
TEST DATE(S) 2-75

	Serial Number	Identifying Symbol	Size	DOT/ICC Rating	Test Pressure	Volumetric Expansion			Visual Insp.	Disposition*	Tested By	Remarks
						Total	Perm.	Elastic				
1	Calibrated	Cylinder			3360						SS	
2	X-1234	XYZ	9 × 51	3A2015	3360	157	1	156	OK	a		
3	X-5432	XYZ	9 × 51	3AA2015	3360	−	−	−	OK	b		Neck leaks
4	123456	ABC	9 × 51	3AA2015		−	−	−	C	c		Excessive corrosion
5	C-6789	CDE	9 × 51	3A2015	3360	195	21	174	OK	c		P.E. over 10 percent of T.E.
6	236567	GHI	9 × 51	3A2015	3360	188	2	186	OK	a		Not plus marked
7	K-1456	ABC	7 × 32	3A2015	3360	58	2	56	OK	a		
8	65432R	XYZ	7 × 43	3A2015	3360	71	0	71	OK	a		Passed 2nd test
9	X-5432	XYZ	9 × 51	3AA2015	3460	184	1	183	OK	a		
10	D-2345	ZYX	4 × 17	3AA2015	3360	12.7	.3	12.4	OK	a		
11	Calibrated	Cylinder			4000						NN	
12	Z-592	XYZ	9½ × 56	3AA2400	4000	225	1	224	OK	a		
13	Z-495	XYZ	9½ × 56	3AA2400	4000	227	2	225	OK	a		
14	Calibrated	Cylinder			3775							
15	Y10555	XYZ	9 × 51	3AA2265	3775	200	2	198	OK	a		
16	Y10554	XYZ	9 × 51	3AA2265	3775	207	0	207	OK	a		
17	K14999	ABC	7 × 32	3A2015					C	c		Failed hammer test

*Disposition code.
a Return to service.
b Set aside for further tests.
c Scrap.
d Set aside for heat treatment.
e Other (specify).

I hereby certify that all the above tests were made under my supervision and in accordance with DOT Regulations.

(Signed) _____.

Fig. 2. Typical water jacket test data.

Note 2: For additional guidelines on cylinder inspection, refer to Section C or CGA Pamphlet C-6, Standards for Visual Inspection of Compressed Gas Cylinders.

Note 3: DOT Regulations permit certain specification containers used exclusively in noncorrosive gas service to be retested by the proof pressure method (see 173.34 (e) (9)).

1. WATER JACKET METHOD

This is the standard method of testing high-pressure cylinders in the compressed gas industries. The water jacket leveling burette method of testing cylinders consists essentially of enclosing the cylinder, suspended in a jacket vessel provided with necessary connections and attachments and measuring the volume of water forced from the jacket upon the application of pressure to the interior of the cylinder, and the volume remaining displaced upon release of the pressure. These volumes represent the total and permanent expansions of the cylinder, respectively. This method is used also to determine

accurately the elastic expansion, which is directly related to the average wall thickness of a cylinder. In general, an increase in elastic expansion indicates reduction of average wall thickness.

A cylinder that is handled properly will retain its original condition unless physically damaged or attacked by corrosion.

A schematic diagram for the testing equipment is shown in Fig. 1. Design and details of apparatus to suit individual requirements may be adopted by individuals, but they should follow the safety recommendations outlined in Pamphlet C-1.

Data from a typical water jacket test are shown in Fig. 2.

2. DIRECT EXPANSION METHOD

This method determines the total expansion by measuring the amount of water forced into a cylinder to pressurize it to test pressure, and the permanent expansion by measuring the amount of water expelled from the cylinder

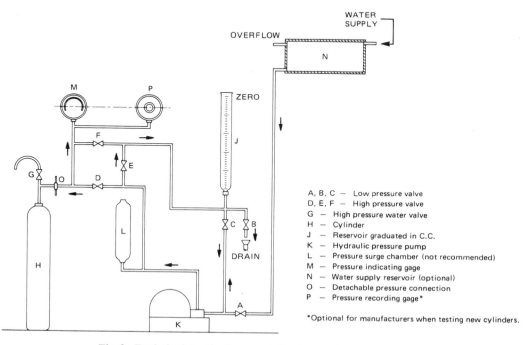

A, B, C — Low pressure valve
D, E, F — High pressure valve
G — High pressure water valve
H — Cylinder
J — Reservoir graduated in C.C.
K — Hydraulic pressure pump
L — Pressure surge chamber (not recommended)
M — Pressure indicating gage
N — Water supply reservoir (optional)
O — Detachable pressure connection
P — Pressure recording gage*

*Optional for manufacturers when testing new cylinders.

Fig. 3. Typical schematic diagram of direct expansion test apparatus.

APPROVAL NO. 4567
ASSIGNED RETEST SYMBOL ABC

XYZ OXYGEN CO.
THIS TOWN, THAT STATE 00000

TEST DATE 2-75
OPERATOR J.D.

	Serial Number	Identifying Symbol	DOT/ICC Rating	Test Pressure (psi)	Water Temp. (°F)	Wt. of H_2O W. (lb)	Vol. of H_2O to Test Press. cc	Water Expelled From Cyl. cc	Volumetric Expansion			Visual Insp.	Disposition*	Remarks
									Total	Perm.	Elastic			
1	L 456	XYZ	3A2015	3360	60	251	1745	1698	525	47	478	OK	a	
2	Z 123	XYZ	4B250	500	73	1657	2736	2667.5	1539.5	68.5	147.1	OK	a	

*Disposition code.
aReturn to service.
bSet aside for further tests.
cScrap.
dSet aside for heat treatment.
eOther (specify). _____

I hereby certify that all the above tests were made under my supervision and in accordance with DOT Regulations.

_____ (Signed) _____

Fig. 4. Typical direct expansion test data.

when pressure is released. Although the elastic expansion is also measured in this method, DOT Regulations forbid this method to be used to qualify cylinders that may be charged to 10 percent in excess of marked service pressure. A schematic diagram for testing equipment is shown in Fig. 3. Design and details of equipment to suit individual requirements can be adapted by the individual, but should follow the recommendations for accurate testing and safety outlined in Pamphlet C-1. Only properly trained personnel should be employed.

Data from a typical direct expansion test are shown in Fig. 4.

3. PRESSURE RECESSION METHOD

This method may be used for testing cylinders that require a test pressure of 2000 psi or more. The method consists essentially in subjecting the cylinder to the required hydrostatic test pressure, then immediately cutting off further pressure supply, and observing for at

least two minutes whether or not there occurs a recession of the pressure in the cylinder. If the pressure does not recede, this indicates that at the test pressure the cylinder does not show any permanent expansion.

A schematic diagram for testing equipment is shown in Fig. 5. Design and details of equipment to suit individual requirements can be adopted by the individual, but should follow the recommendations for accurate testing and safety outlined in Pamphlet C-1. Only properly trained personnel should be employed.

4. PROOF PRESSURE METHOD

This method may be used when DOT Regulations and Specifications do not require the determination of total and permanent volumetric expansions of the cylinder. The test consists essentially of applying an internal pressure equal to that stipulated for the volumetric expansion test and determining whether any weakness or leakage exist.

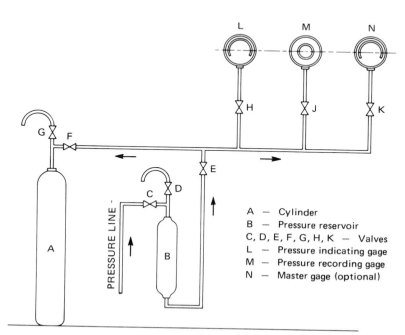

A — Cylinder
B — Pressure reservoir
C, D, E, F, G, H, K — Valves
L — Pressure indicating gage
M — Pressure recording gage
N — Master gage (optional)

Fig. 5. Typical schematic diagram of pressure recession test apparatus.

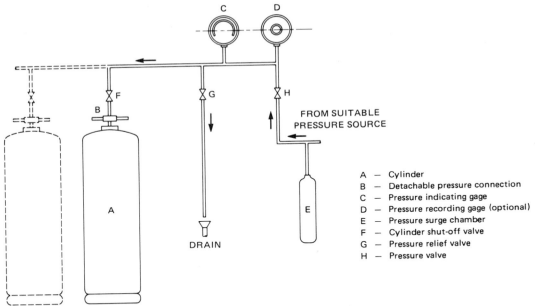

Fig. 6. Typical schematic diagram of proof pressure test apparatus.

A — Cylinder
B — Detachable pressure connection
C — Pressure indicating gage
D — Pressure recording gage (optional)
E — Pressure surge chamber
F — Cylinder shut-off valve
G — Pressure relief valve
H — Pressure valve

A schematic diagram for testing equipment is shown in Fig. 6. Design and details of equipment to suit individual requirements can be adopted by the individual, but should follow the recommendations for accurate testing and safety outlined in Pamphlet C-1.

SECTION E

Disposition of Unserviceable Cylinders

1. INTRODUCTION

1.1. From time to time the Compressed Gas Association, Inc., receives inquiries for recommendations for the disposal of compressed gas cylinders which, for one reason or another, are no longer considered to be serviceable. Some of these unserviceable cylinders failed to qualify for further use under the maintenance requirements of the Department of Transportation (DOT) or those of the Canadian Transport Commission (CTC). In other cases cylinders are occasionally found that appear to have been out of service for a long time, are inadequately marked and are considered to be unsafe for further use. In the latter case the cylinders may either be empty or charged with gas.

1.2. Nonreusable (nonrefillable) containers such as those made to DOT Specifications 39, 40 and 41 are by DOT and CTC Regulations considered unserviceable after one use and must not be refilled or reused for any purpose. Disposal of such cylinders should be in accordance with recommendations obtained from the suppliers who initially filled the cylinders. Such recommendations may sometimes be included in the labeling of the cylinders.

1.3. The proper safe disposition of unserviceable compressed gas cylinders is important, as a very substantial potential hazard may exist that must be recognized and evaluated by those who attempt to dispose of them. Where the content of the cylinder is unknown and there is no ready means for identifying its properties, the hazard is especially great. The disposal of unserviceable compressed gas cylinders is poten-

tially hazardous because they may contain:
- (a) gas under pressure;
- (b) flammable gas;
- (c) explosive mixtures;
- (d) poisonous or toxic materials;
- (e) corrosive, oxidizing or reactive materials.

1.4. In preparing these recommendations an attempt was made to anticipate practical considerations that might arise. These recommendations do not cover all the possible remedies. Questions concerned with the application of these recommendations should be directed to the gas supplier.

1.5. These recommendations are intended for use only by qualified personnel and are not intended for use by the general public.

1.6. Individuals or organizations such as scrap dealers, fire services, military organizations and others, who may have reason to dispose of an unserviceable compressed gas cylinder should acquaint themselves with the names and addresses of the nearest manufacturers or distributors of the type of compressed gas or gases for which the cylinder use was intended. The manufacturer or distributor should be requested to remove the cylinders for appropriate disposal. If this is not feasible, the manufacturer or distributor should be requested to provide appropriate instructions and supervision for the safe disposal and/or destruction of the cylinders.

2. DISPOSAL PROCEDURE

In general, the disposal of all unserviceable compressed gas cylinders requires the adherence to the following procedures:
- (a) identify contents (refer to Section 3);
- (b) safely purge contents (refer to Section 4);
- (c) remove valve;
- (d) obliterate all markings;
- (e) destroy the cylinder.

3. IDENTIFICATION OF CONTENTS

3.1. It is absolutely essential that cylinder contents be identified before steps for the disposition of the cylinder are taken.

3.2. The cylinder should be marked with the chemical name of the commodity contained by means of stenciling or a product label. If such marking is not on the cylinder or is illegible, do not place reliance upon the color of the cylinder or other color coding to determine the cylinder content identification. Instead, contact the supplier to arrange for the return of the cylinder. When a cylinder bears adequate product labeling, it is reasonable to rely upon the labels for the identity of the cylinder content.

3.2.1. Usually suppliers of compressed gases mark their cylinders with product names. These names may either be the proper chemical names or a commonly accepted name, such as a trade name. (Refer to American National Standard Method of Marking Compressed Gas Cylinders to Identify Content, Z48.1, CGA Pamphlet C-4.) Where such product labeling is used, it is presumed that it is not in conflict with the requirements of the DOT, the CTC or other authority that may have jurisdiction. Product labeling should give adequate information concerning the name of the firm which filled the cylinder with gas.

3.3. Cylinders bearing the marks ICC-8, ICC-8AL, CRC-8, CRC-8AL, BTC-8, BTC-8AL, DOT-8. DOT-8AL, CTC-8 or CTC-8AL are restricted to acetylene service. Cylinders marked ICC-3D, CRC-3D, BTC-3D, DOT-3D or CTC-3D are restricted to poison gas service. Such cylinders may be readily identified by these markings. Poison gas cylinders do not contain safety relief devices.

3.4. In order to determine whether there is gas in the cylinder under pressure, the valve should be opened slightly and immediately closed to determine by sound or soap suds whether any gas has escaped. This should only be done in a safe area and with the cylinder properly supported. This test should be made with caution and only a minimum amount of gas should be permitted to escape. When opening the valve, the outlet should be pointed away from anyone in the vicinity. Do not inhale the released gas for it can be poisonous, toxic or irritating. The absence of escaping gas does not necessarily mean the cylinder is empty as the valve may be inoperative or the cylinder may contain a low-pressure liquid.

3.5. It may be necessary to determine whether a cylinder contains gas in the liquid phase. It is not always possible to make such a determination by sound. However, with some thin-walled cylinders it is sometimes possible to hear the liquid in the cylinder when it is shaken. Cylinders for some liquefied gases are often marked with a tare weight. Weighing of such cylinders may therefore establish the presence of gas in the cylinder. A more exact determination can be made by inverting the cylinder with the valve down and opening the valve slightly and immediately closing it, noting the escape of the gas in the liquid form.

Caution: This should be done in a well-ventilated area or with the workmen wearing gas masks with an independent air supply. Care should be taken to be sure that the valve outlet is pointed away from the operator when the valve is opened.

3.6. Valve outlet thread dimensions have been standardized by the compressed gas industry. These dimensions are published by the Compressed Gas Association as "American National and Canadian Standard Compressed Gas Cylinder Valve Outlet and Inlet Connections," CGA V-1, ANSI-B57.1, CSA-B96. A comparison of the valve outlet connection on a cylinder valve with the recommended standard as contained in CGA V-1, ANSI-B57.1 or CSA-B96, may provide a guide to the determination of cylinder content. This method, however, is not conclusive, as the same dimensional valve outlet standard is often utilized for a number of gases. It is also possible that a cylinder may not be equipped with a valve having an outlet of the recommended standard.

3.7. While it is recommended that one should not rely upon safety relief devices as a means for checking cylinder content, it may sometimes be used as an aid in classifying the possible content of a cylinder. The Safety Relief Device Standards of the Compressed Gas Association contain a table which indicates the types of devices recommended for use on DOT and CTC specification containers. One should not rely entirely upon the type of safety relief device as an indication of the specific commodity contained in the cylinder. It is only a clue. The absence of a safety relief device could indicate that the cylinder has been in poisonous gas service.

4. DISPOSITION OF CYLINDER CONTENT

4.1. The contents of unserviceable cylinders should be disposed of by the appropriate methods for the contents. These methods should not be attempted unless someone experienced in a proper method of disposal is personally directing the procedure. If this cannot be done, all details on procedures and precautions to be employed must be obtained from a person so qualified. If the cylinder contents are:

(a) Flammable gases—dispose by burning by a suitable method.

(b) Toxic, reactive, poisonous or irritating gases—dispose by methods appropriate to the properties of the contents, i.e., incineration or chemical reaction.

4.2. If the content of a cylinder must be removed by some method other than through a properly operating valve, care must be employed to release the content slowly, so that the released energy does not cause the cylinder to rocket. Under such circumstances, it is recommended that the cylinder be firmly secured in a restraining device or by other suitable means, minimizing the danger of the cylinder being tossed about out of control. A needle valve should be attached to the discharge connection of the cylinder to adequately control and to prevent excessive discharge rates. If the cylinder contains an unknown or a chemically reactive material, no device should be so attached, as a rapid discharge of gas into the small trapping volume could initiate a dangerous explosion.

4.3. If a cylinder valve is damaged, preventing the discharge of the content in a normal manner, it may be possible to release the pressure in the cylinder through the safety relief device. However, only qualified personnel, familiar with gas cylinders and their safety relief devices should attempt this procedure. It should not be attempted where the gas content may be noxious or where the ejection of the safety relief device by high cylinder pressure could be hazardous, without proper protection or venting procedures.

4.4. Many of the compressed gases can be safely vented to the atmosphere. Obviously, the inert gases can be so vented without creating any undue hazard. If flammable gases or gases which may present a health hazard are vented into the outside atmosphere, it should be done very carefully and at an isolated location in such a manner and at such a rate as is necessary to assure safety from fire or contamination of the outside atmosphere.

4.5. As an extreme means of discharging the gas contained in an unserviceable cylinder, where the valve is damaged to such an extent that it cannot be used for the release of gas, it is possible to accomplish release of pressure by shearing the valve if it is made of brass or by perforating the cylinder with a high-powered rifle bullet. This method should be used only as a last resort. It must only be attempted by qualified individuals at a suitable isolated place, with the cylinder located wherever possible in a depression below surrounding grade level. The rifle must be fired from a distance not less than 50 yards from the cylinder through a protection shield that is adequate to protect personnel from any flying fragments. Wind direction and intensity must be considered.

4.6. After releasing the pressure and discharging the content from an unserviceable cylinder, the cylinder should be purged, if it previously contained a flammable, or reactive or toxic material, before any attempt is made to destroy it with a cutting torch. Purging can be accomplished with the use of inert gases, steam, or by filling the cylinder with water.

5. DISPOSITION OF UNSERVICEABLE EMPTY CYLINDERS
(other than acetylene)

Having removed the gas content of a cylinder and after purging in accordance with 4.6 the empty cylinder should be destroyed with a cutting torch, or by other appropriate means

that will make it unusable as a pressure vessel. It is further recommended that the specification markings on the cylinder be destroyed.

6. DISPOSITION OF UNSERVICEABLE ACETYLENE CYLINDERS

6.1. As noted in 3.3, cylinders marked ICC-8, DOT-8, CRC-8, BTC-8, CTC-8, ICC-8AL, DOT-8AL, CRC-8AL, BTC-8AL or CTC-8AL are authorized for acetylene only. These cylinders are filled with a porous mass which serves as an absorbent of a solvent that is utilized to retain acetylene in solution. An unserviceable acetylene cylinder, therefore, may retain varying quantities of solvent and acetylene. Before destroying one of these cylinders it is important that every precaution be taken to de-energize the cylinder. This work should be done by personnel completely familiar with these cylinders. The following procedures should be observed:

6.1.1. Cylinder Preparation. Drain the empty cylinder for a minimum of 24 hours with the valve open in an isolated location where escaping residual acetylene and solvent vapors will not present any hazard to personnel or property. The temperature must be above 45 F (7 C). Weigh the cylinder to be certain that it is empty of residual acetylene.

6.1.2. Cylinder Dismantling. The cylinder valve should be removed very carefully, making certain that all gas pressure has been released before completely unscrewing the valve. This operation should be performed with caution in the event of release of acetylene which may be retained in the cylinder if the valve happened to be clogged. Remove all fusible-metal plugs, or only the bottom fusible-metal plugs if water flushing is to be used in the next step. Remove the well (or core-hole) packing material, if it is used, from the well (or core-hole) in the filler. It should be noted that fusible-metal safety relief devices may be installed in both ends of the cylinder.

6.1.3. Acetylene and Solvent Removal. Allow cylinders to air for one week or expose to sun for two days in an isolated location. Then place the cylinder in an inverted position—inclined or vertical—and flush with water until water emerges from the bottom fusible plug openings. An alternate procedure is to allow cylinders to lie outside in an isolated location for four weeks during which time the ambient temperature must be above 32 F (0 C) for one week.

6.1.4. Cylinder Destruction. Remove top fusible-metal plugs, if not removed previously. Using a cutting torch cut out the cylinder valve spud and all markings on the cylinder, including registered symbol, serial number and other identification markings in such a manner as to make the cylinder unfit for future service. As an alternate method the markings must be obliterated and the pres-

sure shell must be pierced to render the cylinder incapable of further acetylene service.

6.1.5. Cylinder Disposal. Do not store scrapped cylinders or filler in a confined space because of possible accumulation of solvent vapors from the cylinder filler. Do not dispose of scrap cylinders by any method that would permit them to fall into the hands of uninformed persons.

6.1.6. Precautions. There may be some flame from burning solvent retained in the filler. Two men should be present during the cutting operations, and a fire extinguisher should be available for immediate use. Both men should wear goggles and welder's gloves. Some cylinders may be galvanized; therefore, particular care must be exercised to avoid breathing of fumes when cutting the cylinder.

CHAPTER 7

Compressed Gas Cylinder Valve Connection Systems

Accidentally connecting a compressed gas cylinder valve outlet with equipment not designed for the gas contained in the cylinder may result in serious hazards. Because of this, standard valve outlet connections have been established for valves used with cylinders containing the different gases. These standard connections are made so that the valve connection for one gas will not fit the connections prescribed for other incompatible gases. Standard valve outlet connections thus help to protect personnel and equipment.

Standards for gas cylinder valve connection systems developed by Compressed Gas Association, Inc., are used throughout the United States and Canada. The major CGA connection standards are summarized in this chapter as follows:

Section A. American–Canadian Standard Valve Outlet and Inlet Connections
Section B. Standard for 22 mm Anesthesia Breathing Circuit Connectors
Section C. Diameter-Index Safety System

SECTION A

American–Canadian Standard Valve Outlet and Inlet Connections

This section summarizes the uniform American-Canadian Standards for the outlet and inlet connections of compressed gas cylinder valves. Detailed drawings which give the full specifications for each standard connection and component in the system are published in Pamphlet V-1 of Compressed Gas Association, Inc., entitled "Compressed Gas Cylinder Valve Outlet and Inlet Connections," 1977 edition. The pamphlet presents American National Standard B57.1-1977 and Canadian Standard B96-1977. After Tables 1 and 2 identifying the standard outlet connections for gases and gas mixtures, this section presents (for illustrative purposes)

Pamphlet V-1's detailed drawing of Connection No. 540 (for oxygen), and V-1's section and table on cylinder valve inlet connections.

FOREWORD*

History

The first efforts to develop standards for compressed gas cylinder valve connections followed

*This Foreword is for general information and is not a part of the American and Canadian Standard Compressed Gas Cylinder Valve Outlet and Inlet Connections.

149

TABLE 1. ALPHABETICAL LIST OF GASES AND CONNECTIONS ASSIGNED

GAS	CHEMICAL SYMBOL	THREADED CONNECTIONS		YOKE CONNEC- TION STANDARD
		STANDARD	ALTERNATE	
Acetylene	C_2H_2	510 200*,520*	300, 410△	
Air - Up to 3000 psi		346	590,1310**	850, 950
3000 - 10,000 psi		†	677	
Cryogenic Liquid Withdrawal		440	295	
Allene	$CH_2:C:CH_2$	510		
Allylene: *See Methylacetylene*				
Ammonia	NH_3	240, 705		800, 845
Antimony Pentafluoride	SbF_5	330		
Argon - Up to 3000 psi	Ar	580		
3000 - 10,000 psi		†	677	
Cryogenic Liquid Withdrawal		295	440	
Arsine	AsH_3	350	660	
Bis (trifluoromethyl) Peroxide: *See Hexafluorodimethyl Peroxide*				
Boron Chloride: *See Boron Trichloride*				
Boron Fluoride: *See Boron Trifluoride*				
Boron Trichloride	BCl_3	330	660	
Boron Trifluoride	BF_3	330		
Bromine Pentafluoride	BrF_5	670		
Bromine Trifluoride	BrF_3	670		
Bromoacetone	$BrCH_2COCH_3$	330	660	
Bromochlorodifluoromethane (R12B1)	$CBrClF_2$	668 165*, 182*	660	
Bromochloromethane	CH_2BrCl	668 165*, 182*	660	
Bromoethylene: *See Vinyl Bromide*				
Bromomethane: *See Methyl Bromide*				
Bromotrifluoroethylene (R113B1)	$BrFC:CF_2$	510	660	
Bromotrifluoromethane (R13B1)	$CBrF_3$	668 165*, 182*	320, 660	
1, 3-Butadiene	$CH_2:CHCH:CH_2$	510		
Butane - Gas Withdrawal Liquid Withdrawal	$CH_3CH_2CH_2CH_3$	510 555	510	
1-Butene	$CH_3CH_2CH:CH_2$	510		
2-Butene	$CH_3CH:CHCH_3$	510		
α -Butylene: *See 1-Butene*				
β -Butylene: *See 2-Butene*				
1-Butyne: *See Ethylacetylene*				

△ Canada Only
* Limited Standard
† See 2.1.7 of Introduction
** Alternate yoke connection

TABLE 1 *(Continued)*

GAS	CHEMICAL SYMBOL	THREADED CONNECTIONS		YOKE CONNECTION STANDARD
		STANDARD	ALTERNATE	
Carbon Dioxide	CO_2	320		940
Carbonic Acid: *See Carbon Dioxide*				
Carbon Monoxide	CO	350		
Carbon Oxysulfide: *See Carbonyl Sulfide*				
Carbon Tetrafluoride: *See Tetrafluoromethane*				
Carbonyl Chloride: *See Phosgene*				
Carbonyl Fluoride	COF_2	750	660	
Carbonyl Sulfide	COS	330		
Chlorine	Cl_2		660	820, 840††
Chlorine Pentafluoride	ClF_5	670		
Chlorine Trifluoride	ClF_3	670		
1-Chloro-1, 1-difluoroethane (R142b)	CH_3CClF_2	510	660	
Chlorodifluoromethane (R22)	$CHClF_2$	668 165*, 182*	660	
Chloroethane: *See Ethyl Chloride* Chloroethylene: *See Vinyl Chloride*				
Chlorofluoromethane (R31)	CH_2ClF	510		
Chloroheptafluorocyclobutane (RC317)	C_4F_7Cl	668 165*, 182*	660	
Chloromethane: *See Methyl Chloride*				
Chloropentafluoroethane (R115)	C_2ClF_5	668 165*, 182*	660	
1-Chloro-1, 2, 2, 2-tetrafluoroethane (R124)	CF_3CHClF	668 165*, 182*	660	
1-Chloro-2, 2, 2-trifluoroethane (R133a)	CF_3CH_2Cl	668 165*, 182*	660	
Chlorotrifluoroethylene (R1113)	$CClF:CF_2$	510	660	
Chlorotrifluoromethane (R13)	$CClF_3$	668 165*, 182*	320, 660	
Cyanogen	C_2N_2	750	660	
Cyanogen Chloride	CNCl	750	660	
Cyclobutane	C_4H_8	510		
Cyclopropane	C_3H_6	510		920
Deuterium	D_2	350		
Deuterium Chloride	DCl	330		
Deuterium Fluoride	DF	330		
Deuterium Selenide	D_2Se	350	330	

†† Alternate Standard
 * Limited Standard

TABLE 1 (*Continued*)

| GAS | CHEMICAL SYMBOL | THREADED CONNECTIONS | | YOKE CONNEC-TION STANDARD |
		STANDARD	ALTERNATE	
Deuterium Sulfide	D_2S	330		
Diborane	B_2H_6	350		
Dibromodifluoroethane	$C_2H_2Br_2F_2$	668 165*, 182*	660	
Dibromodifluoromethane (R12B2)	CBr_2F_2	668 165*, 182*	660	
1, 2-Dibromotetrafluoroethane (R114B2)	$CBrF_2CBrF_2$	668 165*, 182*	660	
1, 2-Dichlorodifluoroethylene	ClFC:CClF	668 165*, 182*	660	
Dichlorodifluoromethane (R12)	CCl_2F_2	668 165*, 182*	660	
1, 2-Dichloroethylene (R1130)	ClCH:CHCl	668 165*, 182*	660	
Dichlorofluoromethane (R21)	$CHCl_2F$	668 165*, 182*	660	
1, 2-Dichlorohexafluorocyclobutane (RC316)	$C_4Cl_2F_6$	668 165*, 182*	660	
Dichlorosilane	H_2SiCl_2	330	510, 678	
1, 1-Dichlorotetrafluoroethane (R114a)	CF_3CCl_2F	668 165*, 182*	660	
1, 2-Dichlorotetrafluoroethane (R114)	$CClF_2CClF_2$	668 165*, 182*	660	
2, 2-Dichloro-1, 1, 1-trifluoroethane (R123)	$CHCl_2CF_3$	668 165*, 182*	660	
Dicyan: *See Cyanogen*				
Diethylzinc	$(C_2H_5)_2Zn$	750		
Difluorodibromoethane: *See Dibromodifluoroethane*				
Difluorodibromomethane *See Dibromodifluoromethane*				
1, 1-Difluoroethane (R152a)	CH_3CHF_2	510	660	
1, 1-Difluoroethylene (R1132a)	$CH_2:CF_2$	350	320	
Difluoromethane: *See Methylene Fluoride*				
Difluoromonochloroethane: *See Chlorodifluoroethane*				
Dimethylamine	$(CH_3)_2NH$	705	240	
Dimethyl Ether	CH_3OCH_3	510		
Dimethylhexafluoroperoxide: *See Hexafluorodimethyl Peroxide*				

* Limited Standard

TABLE 1 (*Continued*)

GAS	CHEMICAL SYMBOL	THREADED CONNECTIONS		YOKE CONNECTION STANDARD
		STANDARD	ALTERNATE	
2, 2-Dimethylpropane	$C(CH_3)_4$	510		
Dinitrogen Oxide: *See Nitrous Oxide*				
Dinitrogen Tetroxide: *See Nitrogen Dioxide*				
Dinitrogen Trioxide: *See Nitrogen Trioxide*				
Diphosgene	$ClCO_2CCl_3$	750	660	
Epoxyethane: *See Ethylene Oxide*				
Ethane	C_2H_6	350		
Ethene: *See Ethylene*				
Ethylacetylene	$CH_3CH_2C\colon CH$	510		
Ethylamine: *See Monoethylamine*				
Ethyl Chloride	CH_3CH_2Cl	510	300	
Ethyldichloroarsine	$C_2H_5AsCl_2$	750	660	
Ethylene	$CH_2\colon CH_2$	350		900
Ethylene dichloride: *See Dichloroethylene*				
Ethylene Oxide	C_2H_4O	510		
Ethyl Ether	$(C_2H_5)_2O$	510		
Ethyl Fluoride	C_2H_5F	750	660	
Ethylidene Fluoride: *See 1,1-Difluoroethane*				
Ethyl Methyl Ether: *See Methyl Ethyl Ether*				
Ethyne: *See Acetylene*				
Fluorine	F_2	679	670	
Fluoroethylene: *See Vinyl Fluoride*				
Fluoroform (R23)	CHF_3	668 165*, 182*	660, 320	
Fluoromethane: *See Methyl Fluoride*				
Germane	GeH_4	750	350, 510, 660	
Helium - Up to 3000 psi 3000 - 10,000 psi Cryogenic Liquid Withdrawal	He	580 † 792	677	930
Heptafluorobutyronitrile	C_4F_7N	750	660	
Hexafluoroacetone	C_3F_6O	330	660	
Hexafluorocyclobutene	C_4F_6	750	660	
Hexafluorodimethyl Peroxide	CF_3OOCF_3	755	660	
Hexafluoroethane (R116)	C_2F_6	668 165*, 182*	660, 320	
Hexafluoro-2-propanone *See Hexafluoroacetone*				

* Limited Standard
† See 2.1.7 of Introduction

TABLE 1 (*Continued*)

GAS	CHEMICAL SYMBOL	THREADED CONNECTIONS		YOKE CONNEC-TION STANDARD
		STANDARD	ALTERNATE	
Hexafluoropropylene	$CF_3CF:CF_2$	668 165*, 182*	660	
Hydriodic Acid, Anhydrous: *See Hydrogen Iodide*				
Hydrobromic Acid, Anhydrous *See Hydrogen Bromide*				
Hydrochloric Acid, Anhydrous *See Hydrogen Chloride*				
Hydrocyanic Acid, Anhydrous *See Hydrogen Cyanide*				
Hydrofluoric Acid, Anhydrous *See Hydrogen Fluoride*				
Hydrogen - Up to 3000 psi 3000 - 10,000 psi Cryogenic Liquid Withdrawal	H_2	350 † 795	677 792	
Hydrogen Bromide	HBr	330		
Hydrogen Chloride	HCl	330		
Hydrogen Cyanide	HCN	750	160	
Hydrogen Fluoride	HF	330	670, 660	
Hydrogen Iodide	HI	330	660	
Hydrogen Selenide	H_2Se	350	160, 660	
Hydrogen Sulfide	H_2S	330		
Industrial Gas Mixtures: *See Table 2, page 21*				
Iodine Pentafluoride	IF_5	670		
Isoamylene: *See 3-Methyl-1-butene*				
Isobutane	C_4H_{10}	510		
Isobutene: *See Isobutylene*				
Isobutylene	C_4H_8	510		
Isopropylethylene: *See 3-Methyl-1-butene*				
Krypton - Up to 3000 psi 3000 - 10,000 psi	Kr	580 †	677	
Laughing Gas: *See Nitrous Oxide*				
Lewisite [Dichloro (2-chlorovinyl) arsine]	$CICH:CHAsCl_2$	750	660	
Liquid Dioxide: *See Nitrogen Dioxide*				
Marsh Gas: *See Methane*				
Medical Gas Mixtures: *See Table 2, page 21*				
Methane - Up to 3000 psi 3000 - 10,000 psi Cryogenic Liquid Withdrawal	CH_4	350 † 450	677	

* Limited Standard
† See 2.1.7 of Introduction

TABLE 1 (Continued)

GAS	CHEMICAL SYMBOL	THREADED CONNECTIONS		YOKE CONNEC- TION STANDARD
		STANDARD	ALTERNATE	
Methanethiol: *See Methyl Mercaptan*				
Methoxyethylene: *See Vinyl Methyl Ether*				
Methylacetylene	$CH_3C\!:\!CH$	510		
Methylamine: *See Monomethylamine*				
Methyl Bromide	CH_3Br	330	660, 320 165, 182	
3-Methyl-1-butene	$(CH_3)_2CHCH\!:\!CH_2$	510		
Methyl Chloride	CH_3Cl	510	660, 165, 182	
Methyldichloroarsine	CH_3AsCl_2	750		
Methylene Fluoride (R32)	CH_2F_2	320		
Methyl Ether: *See Dimethyl Ether*				
Methyl Ethyl Ether	$CH_3OC_2H_5$	510		
Methyl Fluoride	CH_3F	350		
Methyl Formate	$HCOOCH_3$	510	660	
Methyl Mercaptan	CH_3SH	750	330	
2-Methylpropene: *See Isobutylene*				
Methyl Vinyl Ether: *See Vinyl Methyl Ether*				
Mixtures: *See Table 2, page 21*				
Monochlorodifluoromethane: *See Chlorodifluoromethane*				
Monochloropentafluoroethane: *See Chloropentafluoroethane*				
Monochlorotetrafluoroethane: *See Chlorotetrafluoroethane*				
Monochlorotrifluoromethane: *See Chlorotrifluoromethane*				
Monoethylamine	$CH_3CH_2NH_2$	705	240	
Monomethylamine	CH_3NH_2	705	240	
Mustard Gas [Bis (2-chloroethyl) Sulfide]	$S(C_2H_4Cl)_2$	750	350, 660	
Natural Gas - Up to 3000psi 3000 - 10,000 psi Cryogenic Liquid Withdrawal		350 † 450	677	
Neon - Up to 3000 psi 3000 - 10,000 psi Cryogenic Liquid Withdrawal	Ne	580 † 792	677	
Neopentane: *See 2, 2-Dimethylpropane*				
Nickel Carbonyl	$Ni(CO)_4$	750	320	

† See 2.1.7 of Introduction

TABLE 1 (*Continued*)

GAS	CHEMICAL SYMBOL	THREADED CONNECTIONS		YOKE CONNEC- TION STANDARD
		STANDARD	ALTERNATE	
Nickel Tetracarbonyl: *See Nickel Carbonyl*				
Nitric Oxide	NO	755	660	
Nitrogen - Up to 3000 psi	N_2	580	590, 555	960
3000 - 10,000 psi		†	677	
Cryogenic Liquid Withdrawal		295	440	
Nitrogen Dioxide	NO_2	755	160, 660	
Nitrogen Peroxide: *See Nitrogen Dioxide*				
Nitrogen Sesquioxide: *See Nitrogen Trioxide*				
Nitrogen Tetroxide: *See Nitrogen Dioxide*				
Nitrogen Trifluoride	NF_3	679		
Nitrogen Trioxide	N_2O_3	755	160, 660	
Nitrosyl Chloride	NOCl	330	660	
Nitrosyl Fluoride	NOF	330		
Nitrous Oxide	N_2O	326		910
Nitryl Fluoride	NO_2F	330		
Octafluorocyclobutane (RC318)	C_4F_8	668 165*, 182*	660	
Octafluoropropane	C_3F_8	668 165*, 182*	660	
Oxirane: *See Ethylene Oxide*				
Oxygen - Up to 3000 psi	O_2	540		870
Cryogenic Liquid Withdrawal		440	295	
Oxygen Difluoride	OF_2	679		
Ozone	O_3	755	660	
Pentaborane	B_5H_9	750	660, 350	
Pentachlorofluoroethane	CCl_3CCl_2F	668 165*, 182*	660	
Pentafluoroethane (R125)	CF_3CHF_2	668 165*, 182*	660	
Pentafluoroethyl Iodide	CF_3CF_2I	668 165*, 182*	660	
Pentafluoropropionitrile	CF_3CF_2CN	750	660	
Perchloryl Fluoride	ClO_3F	670		
Perfluoroacetone: *See Hexafluoroacetone*				
Perfluorobutane	C_4F_{10}	668 165*, 182*		
Perfluoro-2-butene	C_4F_8	668 165*, 182*	660	

* Limited Standard
† See 2.1.7 of Introduction

TABLE 1 (*Continued*)

GAS	CHEMICAL SYMBOL	THREADED CONNECTIONS		YOKE CONNECTION STANDARD
		STANDARD	ALTERNATE	
Perfluorocyclobutane: *See Octafluorocyclobutane*				
Perfluorodimethyl Peroxide: *See Hexafluorodimethyl Peroxide*				
Perfluoroethane: *See Hexafluoroethane*				
Perfluoropropane: *See Octafluoropropane*				
Phenylcarbylamine Chloride	$C_6H_5N:CCl_2$	330	660, 350	
Phosgene	$COCl_2$	750	160, 660	
Phosphine	PH_3	350	160, 660	
Phosphorous Pentafluoride	PF_5	330		
Phosphorous Trifluoride	PF_3	330		
Propadiene: *See Allene*				
Propane - Gas Withdrawal	C_3H_8	510		
- Liquid Withdrawal		555	510	
Propene: *See Propylene*				
Propylene	C_3H_6	510		
Propyne: *See Methylacetylene*				
"REFRIGERANTS" - Numerical Listing				
R11: *See Trichlorofluoromethane*				
R12: *See Dichlorodifluoromethane*				
R12BI: *See Bromochlorodifluoromethane*				
R12B2: *See Dibromodifluoromethane*				
R13: *See Chlorotrifluoromethane*				
R13BI: *See Bromotrifluoromethane*				
R14: *See Tetrafluoromethane*				
R21: *See Dichlorofluoromethane*				
R22: *See Chlorodifluoromethane*				
R23: *See Fluoroform*				
R31: *See Chlorofluoromethane*				
R32: *See Methylene Fluoride*				
R40: *See Methyl Chloride*				
R41: *See Methyl Fluoride*				
R50: *See Methane*				
R112: *See 1,1,2,2-Tetrachlorodifluoroethane*				
R112a: *See 1,1,1,2-Tetrachlorodifluoroethane*				
R113: *See 1,1,2-Trichlorotrifluoroethane*				
R113BI: *See Bromotrifluoroethylene*				

TABLE 1 (*Continued*)

GAS	CHEMICAL SYMBOL	THREADED CONNECTIONS		YOKE CONNECTION STANDARD
		STANDARD	ALTERNATE	
"REFRIGERANTS" - Numerical Listing (continued)				
R114: *See 1,2-Dichlorotetrafluoroethane*				
R114a: *See 1,1,-Dichlorotetrafluoroethane*				
R114B2: *See 1,2-Dibromotetrafluoroethane*				
R115: *See Chloropentafluoroethane*				
R116: *See Hexafluoroethane*				
R123: *See 2,2-Dichloro-1,1,1-trifluoroethane*				
R124: *See 1;Chloro-1,2,2,2-tetrafluoroethane*				
R125: *See Pentafluoroethane*				
R133a: *See 1-Chloro-2,2,2-trifluoroethane*				
R142b: *See 1-Chloro-1,1-difluoroethane*				
R143a: *See 1.1.1-Trifluoroethane*				
R152a: *See 1,1-Difluoroethane*				
R160: *See Ethyl Chloride*				
R170: *See Ethane*				
R218: *See Octafluoropropane*				
R290: *See Propane*				
RC316: *See Dichlorohexafluorocyclobutane*				
RC317: *See Chloroheptafluorocyclobutane*				
RC318: *See Octafluorocyclobutane*				
R600: *See Butane*				
R601: *See Isobutane*				
R630: *See Monomethylamine*				
R631: *See Monoethylamine*				
R717: *See Ammonia*				
R729: *See Air*				
R744: *See Carbon Dioxide*				
R744a: *See Nitrous Oxide*				
R764: *See Sulfur Dioxide*				
R1113: *See Chlorotrifluoroethylene*				
R1114: *See Tetrafluoroethylene*				
R1130: *See 1,2-Dichloroethylene*				
R1132a: *See 1,1-Difluoroethylene*				
R1140: *See Vinyl Chloride*				
R1141: *See Vinyl Fluoride*				
R1150: *See Ethylene*				
R1270: *See Propylene*				

TABLE 1 *(Continued)*

GAS	CHEMICAL SYMBOL	THREADED CONNECTIONS		YOKE CONNECTION STANDARD
		STANDARD	ALTERNATE	
Silane	SiH_4	350	510	
Silicon Tetrafluoride	SiF_4	330		
Silicon Tetrahydride: *See Silane*				
Stibine	SbH_3	350		
Sulfur Dioxide	SO_2	668	660 165, 182	
Sulfur Hexafluoride	SF_6	668	590,660	
Sulfur Tetrafluoride	SF_4	330		
Sulfuryl Fluoride	SO_2F_2	330	660	
1,1,1,2-Tetrachlorodifluoroethane (R112a)	$C_2Cl_4F_2$	668 165*, 182*	660	
1,1,2,2-Tetrachlorodifluoroethane (R112)	$C_2Cl_4F_2$	668 165*, 182*	660	
1,1,2,2-Tetrafluoro-1-chloroethane	C_2HClF_4	668 165*, 182*	660	
Tetrafluoroethylene	C_2F_4	350 165*, 182*	660	
Tetrafluorohydrazine	N_2F_4	679		
Tetrafluoromethane (R14)	CF_4	580	320,660	
Tetrafluorosilane: *See Silicon Tetrafluoride*				
Tetramethyllead	$(CH_3)_4Pb$	750	350	
Tetramethylmethane: *See 2,2-Dimethyl-propane*				
Trichlorofluoromethane (R11)	CCl_3F	668 165*, 182*	660	
Trichloromonofluoromethane: *See Trichlorofluoromethane*				
1,1,1-Trichlorotrifluoroethane	CF_3CCl_3	668 165*, 182*	660	
1,1,2-Trichlorotrifluoroethane (R113)	$CFCl_2CF_2Cl$	668 165*, 182*	660	
Triethylaluminum	$(C_2H_5)_3Al$	750	350	
Triethylborane	$(C_2H_5)_3B$	750	350	
Trifluoroacetonitrile	CF_3CN	750	350	
Trifluoroacetyl Chloride	CF_3COCl	330		
Trifluorobromomethane: *See Bromotri-fluoromethane*				
Trifluorochloroethylene: *See Chlorotri-fluoroethylene*				
1,1,1-Trifluoroethane (143a)	CH_3CF_3	510		

* Limited Standard

TABLE 1 (*Continued*)

GAS	CHEMICAL SYMBOL	THREADED CONNECTIONS		YOKE CONNEC-TION STANDARD
		STANDARD	ALTERNATE	
Trifluoroethylene	C_2F_3H	510		
Trifluoromethane: *See Fluoroform*				
Trifluoromethyl Chloride: *See Chlorotrifluoromethane*				
Trifluoromethyl Hypofluorite	CF_3OF	679		
Trifluoromethyl Iodide	CF_3I	668 165*, 182*	660	
Trifluorovinyl Bromide: *See Bromotrifluoroethylene*				
Trimethylamine	$(CH_3)_3N$	705	240	
Trimethylene: *See Cyclopropane*				
Trimethylmethane: *See Isobutane*				
Trimethylstibine	$(CH_3)_3Sb$	750	350	
Tungsten Hexafluoride	WF_6	330	670	
Uranium Hexafluoride	UF_6	330		
Vinyl Bromide	C_2H_3Br	510	290	
Vinyl Chloride	C_2H_3Cl	510	290	
Vinyl Fluoride	C_2H_3F	350	320	
Vinylidene Fluoride: *See 1,1-Difluoroethylene*				
Vinyl Methyl Ether	$C_2H_3OCH_3$	510	290	
Xenon - Up to 3000 psi 3000 - 10,000 psi	Xe	580 †	677	

* Limited Standard
† See 2.1.7 of Introduction

immediately after World War I, and were inspired by the difficulties encountered both by industry and the military services because of the multiplicity of connections then in use, and because of the danger from using the same connection for incompatible gases.

Through the activity of the Gas Cylinder Valve Thread Committee of the Compressed Gas Manufacturers' Association, Inc., substantial progress was made through the years that followed with the result that, when America became involved in World War II, the gas industries themselves had materially improved this situation. Several of the compressed gas industries themselves had achieved virtual standardization at tremendous cost for replacement of valve equipment. Their standards, however, were not completely formalized nor fully coordinated with other related standards. Much of the progress between World War I and World War II was the result of interest in this problem by the Federal Specifications Board.

The circumstances surrounding industrial and military users of compressed gases during World War II brought into clear focus the need for acceleration of the standardization project for cylinder valve threads. They created not only the necessity but also an opportunity for the compressed gas industry, the military services and other federal agencies to study cooperatively the standardization problems of valve outlet threads. These studies resulted in closer

TABLE 2. LIST OF MIXTURES AND CONNECTIONS ASSIGNED

MIXTURE	THREADED CONNECTIONS		YOKE CONNEC-TION
	STANDARD	ALTERNATE	STANDARD
Industrial Gas Mixtures for pressures up to 3,000 psi:			
Carbon Dioxide & Ethylene Oxide (**)	350		
Dichlorodifluoromethane (R12) & Difluoroethane (R152a) (**)	668 165*,182*	660	
Oxygen Mixtures (O$_2$ over 23%)	296		
Medical Gas Mixtures for pressures up to 3,000 psi:			
Carbon Dioxide & Oxygen (CO$_2$ not over 7.5%)	280		880
Carbon Dioxide & Oxygen (CO$_2$ over 7.5%)	500	320	940
Nonflammable, Noncorrosive Diagnostic & Medically Related Gas Mixtures	500		
Helium & Oxygen (He not over 80.5%)	280		890
Helium & Oxygen (He over 80.5%)	500	580	930
Nitrous Oxide & Oxygen (N$_2$O 47.5 to 52.5%)	280		965
Nitrogen & Oxygen (O$_2$ over 23.5%)	280		

* Limited Standard
** These mixtures are included in this standard because standards were shown in the 1965 edition. A system is being developed for assigning valve outlets to industrial gas mixtures.

definition and appreciation of each valve outlet and in a more balanced relationship between the many types and sizes.

When the Standards Associations representing Great Britain, Canada and the United States met in Ottawa in October 1945 to consider unification of screw threads, a fairly well-developed plan for standardization of compressed gas cylinder valve threads was presented to the Conference by the Valve Standardization Committee of the Compressed Gas Manufacturers' Association, Inc. These proposed standards represented the experience and knowledge of compressed gas manufacturers, valve manufacturers, and the needs and requirements of varied users of gas cylinder valves, including the military services and other federal agencies. Approval of these standards to the extent to which they were then developed was given by the U.S. Department of Commerce, the U.S. Army, and the U.S. Navy through the Interdepartmental Screw Thread Committee following a joint meeting

with the representatives of CGMA in August 1945. Much progress was made later in that year at the Canadian Section Meeting of CGMA tending to unify United States and Canadian practices. During January 1946, through conference between representatives of the CGMA Valve Thread Standardization Committee and the Interdepartmental Screw Thread Committee in Washington agreements were reached that resulted in final approval of considerable additional gas cylinder valve thread data for inclusion in the National Bureau of Standards Handbook H-28.

The Compressed Gas Manufacturers' Association, Inc., changed its name in January 1949 and its Valve Thread Standardization Committee became the Valve Standards Committee of the Compressed Gas Association, Inc. In 1971, the Valve Standards Committee became the Connections Standards Committee to recognize a broadened scope to cover all compressed gas connections. During the interval between Janu-

COMPRESSED GAS ASSOCIATION, INC.

CONNECTION NO. 540

.903-14NGO-RH-EXT

STANDARD CYLINDER VALVE OUTLET CONNECTION FOR
Pressures up to 3,000 psi for

Oxygen

WARNING — Do not use this thread for any other gas or for any gas mixture.

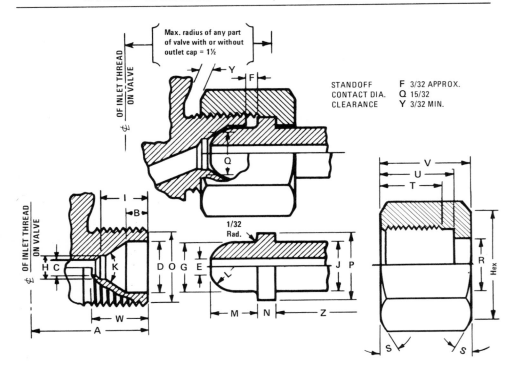

STANDOFF	F	3/32 APPROX.
CONTACT DIA.	Q	15/32
CLEARANCE	Y	3/32 MIN.

VALVE OUTLET		
THREAD		.903-14NGO-RH-EXT
MAJOR DIA.		.9030-.8980
PITCH DIA.		.8566-.8530
MINOR DIA.		.8154 MAX.
CHAMFER	O	45° x 51/64 DIA.
FULL THREAD	W	17/32 MIN.
BORE DEPTH	B	7/32 ± 1/64
DRILL	C	3/16 ± 1/16
BORE	D	.588-.598
ANGLE	K	70°
C'BORE DIA.	H	11/32 MAX.
C'BORE DEPTH	I	15/32 MIN.
LENGTH	A	1-5/16 MAX.

NIPPLE †		
DRILL	E	3/16 ± 1/16
NOSE DIA.	G	.562-.552
SHANK DIA.	J	.562-.557
SHANK LENGTH	Z	OPTIONAL
NOSE RADIUS	L	9/32
NOSE LENGTH	M	.510-.490
SHOULDER LENGTH	N	3/16 $^{+1/64}_{-0}$
SHOULDER DIA.	P	.752-.740

†Nipple may be made from 11/16 hex material.

HEXAGON NUT		
THREAD		.908-14NGO-RH-INT
MINOR DIA.		.8307-.8384
PITCH DIA.		.8616-.8652
MAJOR DIA.		.9080 MIN.
MAJ. DIA. NEW TAP		.915-.914
COUNTERSINK		90° x 59/64 DIA.
HEXAGON		1-1/8
HOLE	R	.567-.572
CHAMFER	S	30° x 1-1/8 DIA.
FULL THREAD	T	9/16 MIN.
BORE DEPTH	U	3/4
LENGTH	V	15/16 MIN.

ary 1946 and February 1949 this Committee developed its standards sufficiently to present them to the American Standards Association and the Canadian Standards Association. They were accepted as National Standards for Canada and the United States in 1949, accomplishing an objective established some 30 years before. Since that date, additional connections have been developed and have been included in subsequent editions of the standard. Similarly, alternate standard connections were removed as they became obsolete.

With the growth of the compressed gas industries and the introduction of many new gases, the need for classifying all existing gases became more acute. It was felt that a classification taking into account fire potential, toxicity, state of gas and corrosiveness of each gas, and grouping gases according to these properties would facilitate an orderly expansion of the system to cover the many new gases added in the 1974 edition. A system of classification was devised by CGA and all known gases grouped according to their properties. New connections were required and these were designed to fit into the existing system of noninterchangeable connections. New connections were tested and then assigned to various groups of gases as needed. Where NGO threads were shown for connections prior to 1977, they were retained. New connections added in the 1977 edition were designed with Unified threads since these were well established as American National Standards. The 1977 edition represents most of the results of this work.

Medical Gas Connections

As early as the spring of 1940, it was evident to various medical societies, as well as to the manufacturers of medical gases, that a system should be devised to prevent the interchangeability of medical gas cylinders equipped with flush-type valves when used with medical gas administering apparatus. Various means for accomplishing this were studied. The most difficult obstacle to overcome was that of devising a system that would permit the adjustment of existing apparatus without interfering with its use and without requiring that it be returned to the manufacturer for conversion. The system, known as the "Pin-Index Safety System" was incorporated in this standard in 1953.

The Pin-Index Safety System was submitted to the International Standards Organization (ISO) with the result that ISO Recommendation R407, "Yoke-Type Connections for Small Medical Gas Cylinders Used for Anesthetic and Resuscitation Purposes" was published in December 1964. The system provided ten noninterchangeable connections for medical gases, eight of which were assigned initially.

In the 1960s, it became apparent that additional noninterchangeable connections would be needed to cover new medical gases and mixtures. The Medical Division and Valve Standards Committee of CGA expanded the system to provide additional connections.

Suggestions for Changes

Suggestions for changes or additions to the standard are welcomed. Suppliers and users of gases or gas mixtures for which there is no standard outlet should contact the Compressed Gas Association, Inc., 500 Fifth Avenue, New York, NY 10110 for assistance in selecting the proper valve connection and initiating a proposal for its inclusion in the standard.

1. GENERAL

1.1. The Connection Standards Committee of the Compressed Gas Association, Inc., applying experience and knowledge of gas producers, valve manufacturers, military services, other federal agencies and gas consumers, established detailed dimensions for valve outlet and inlet connections.

1.2. These standards are based on a coordinated plan for the inclusion of future connections as they are required. Standard outlet connections for respective gases are fully defined and complete in themselves. The outlet connections are designed to minimize the possibility of hazardous misconnections.

1.3. Material specifications are not covered by this standard except in one case where it was essential for maintaining the integrity of noninterchangeability between connections intended for different gas services. The chemical and physical properties of the gases must be considered to determine the strength and suitability of materials used.

1.4. Some of the commodities covered by this standard do not meet the generally accepted definition of "compressed gas" as contained in the Hazardous Materials Regulations of the U.S. Department of Transportation. Even though it is not required for safe packaging, some of the products are handled in compressed gas cylinders. When they are, the cylinder shall have the standard outlet to minimize hazardous misconnections.

1.5. Except where a connection standard is specified for liquid withdrawal, standards for any of the liquefied gases are understood to be suitable for both liquid and gas withdrawal.

1.6. Except where otherwise noted, all dimensions are in inches.

2. OUTLET CONNECTIONS

2.1. Basic Divisions

2.1.1. The threaded outlets are separated into four basic divisions—internal, external, right hand and left hand. Further separation is made within each division by varying the diameter of the threads. The diameters within each division are so spaced that adjoining sizes will not engage. Still further separation is made by varying the size and shape of seats and nipples for any given thread size.

2.1.2. With the exception of outlets having taper pipe threads which seal at the threads, each outlet provides for screw threads which do not seal but merely hold the nipple against its seat or against a washer. For the purpose of this standard, nipples modified to incorporate soft tips or O-rings, and handwheels to permit hand-tight connections are considered in compliance if noninterchangeability is maintained.

2.1.3. For the purpose of this standard, flow passages modified to incorporate filters, flow restrictions, check valves or other features are considered in compliance if the integrity of the joint and noninterchangeability are maintained.

2.1.4. There is a separate group of pin-indexed yoke connections for medical gases (Nos. 870-965) where pins in the yoke and mating holes in the valve body preclude unintended connections. Some gases and gas mixtures for which pin-indexed positions are not assigned are currently being packaged in cylinders with non-pin-indexed yoke connections. Recent development of new pin-indexed connections will make possible additional assignments. Future assignment of these newly available connections should eventually make it possible to discontinue the use of non-pin-indexed connections.

2.1.5. Past practice has firmly established many outlet connections for specific gases or groups of gases and in many cases, these connections were retained. By adhering to existing outlets where practicable, it was possible to put the new standard system into effect without the inconvenience and expense of a cumbersome and costly changeover.

2.1.6. Alternate and limited standards have been established for some gases. A standard connection is the connection recommended for a particular gas. A limited standard is a connection used at present which is not expected to be discontinued and is considered safe. Its use is limited to the particular application mentioned in the standard. An alternate standard is a connection used at present which is expected to be phased out in favor of the standard within five years unless otherwise stated.

2.1.7. Connection 677 was established as an alternate for high-pressure gases while suitable standards are being developed. When new standards are adopted a date will be established for obsolescence of Connection 677 as an alternate standard.

2.2. Numbering System

2.2.1. Prior to 1977, the numbering system provided for separate numbers for the complete outlet connection, valve outlet, mating assembly, nipple, nut and washer. For example, Connection 540 indicated the complete outlet connection and Part 541 was the valve outlet, 543 was the nipple and 544 was the nut. This system was discontinued with the 1977 edition to provide additional numbers for new valve connections.

2.2.2. A single number is now used to describe the complete outlet connection as well as its parts. For example, Connection 540 describes the complete outlet connection. Other parts would be described as valve outlet 540, nipple 540, nut 540, etc.

2.3. Outlet Threads

2.3.1. Outlet threads are of two systems: National Gas Outlet (NGO) and Unified (UN). Both systems are similar in form and kind and use 60° threads.

2.3.2. NGO threads served to consolidate the practice of the industry in 1945. Where NGO threads were used for connections prior to 1977, they are retained because of the satisfactory experience with them, both as to production and performance and because tools and gages are well provided.

2.3.3. Unified threads came into being in 1948 and 1949. Threads selected for new connections added in the 1977 edition are Unified threads because they meet the demands of valve connections and are widely accepted as the American National Standard. Many of these new threads, before being selected for CGA connections, were satisfactorily used in compressed gas service for small cylinders, regulator and hose connections, manifold fittings, refrigeration fittings, cryogenic transfer lines and high-pressure fittings.

2.4. NGO Thread Details

2.4.1. The full designation for NGO threads describes the major diameter of the valve outlet in inches, the number of threads per inch, the identifying letters NGO, the letters RH or LH to denote the thread direction and the letters

EXT or INT to denote an external or internal thread. For example:

> .903-14NGO-RH-EXT
> .965-14NGO-LH-INT

2.5. Unified Thread Details

2.5.1. The full designation for Unified threads includes the major diameter of the thread in inches, the number of threads per inch, the identifying letters UN with any additional letters (C, F, EF, S) to denote its series as listed in 2.5.2, the class—such as 2A or 2B—where A denotes an external thread and B an internal thread. For example:

> .750-14UNF-2A

For American National Standard Unified Screw Threads, B-1.1, all threads are understood to be right hand unless LH is added. In this valve outlet standard, RH, LH, INT and EXT are added to the UN designations for the sake of uniformity with NGO practice. The above UN designation therefore appears in this standard as:

> .750-14UNF-2A-RH-EXT

2.5.2. Unified threads fall into several series:

> UNC = Unified coarse
> UNF = Unified fine
> UNEF = Unified extra fine
> UN = Unified constant pitch
> UNS = Unified special, where the diameter or its pitch do not fall into the standard diameter-pitch combinations.

2.5.3. Class 2 tolerances have been selected except minor internal diameters have been specified with 3B limits. Internal thread designations therefore are shown as modified (MOD).

2.6. Other Threads

2.6.1. Beside Unified and NGO threads, there are two more threads used for outlet connections:

> NPSL = Straight pipe thread for locknut connections (see Connection 240).
> NGT = National Gas Taper pipe threads.

2.7. Reference Literature for Threads

Thread Series	Detailed Dimensions	Gages and Gaging
NGO	H28(II), Table IX.4, p. 76	H28(I), §VI, p. 107
UNC, UNF, UNEF, UN, UNS	H28(I) or ANSI-B1.1	H28(I), pp. 39 to 68 or ANSI-B1.2
NPSL	H28(II), Table VII.7, p. 10	H28(II), §VII, ¶8, p. 11 or ANSI-B2.1
NGT	H28(II), Table IX.5, pp. 76-78 or this standard, pp. 74-75.	H28(II), pp. 79-90

Notes: "H28" refers to National Bureau of Standards Handbook H28 (1957) Screw-Thread Standards for Federal Services. "(I)" and "(II)" indicate Parts I and II of H28.

"ANSI-B1.1" refers to American National Standard Unified Inch Screw Threads (UN and UNR Thread Form), B1.1-1974.

"ANSI-B1.2" refers to American National Standard Gages and Gaging for Unified Screw Threads, B1.2-1966.

"ANSI-B2.1" refers to American National Standard Pipe Threads (except dryseal), B2.1-1968.

2.8. Valve Clearance

2.8.1. The maximum radius of any part of the valve from its centerline has been specified to insure clearance for the cylinder valve protecting cap or collar.

2.9. Valves for Small Cylinders*

2.9.1. Nearly all gases, besides being available in commercial cylinders of the more conventional size, are likewise supplied in small cylinders which may incorporate a valve different from the one on the larger cylinders. Small cylinders known as "lecture bottles" used Connections 110 and 170 for all gases prior to 1974. These connections are intended to be phased out in favor of Connection 180 which may be used as a limited standard for all gases (except acetylene) in small cylinders. Connection 180 provides for a swivel connection so that the equipment may be oriented in relation to the cylinder. It is understood that standards shown in Table 1 may be used for the gases specified for all sizes of cylinders.

2.9.2. Acetylene Connection 200 (normally having a $\frac{3}{8}$ in. NGT inlet thread) is a limited standard restricted to cylinders holding approximately 10 cu ft of acetylene. Acetylene Connection 520 (normally having a $\frac{3}{8}$ in. NGT inlet thread) is a limited standard restricted to cylinders containing between 35 and 75 cu ft of acetylene.

3. INLET THREADS

3.1. Inlet threads on the valve and in the cylinder neck have also been standardized and are included at the end of this section.

4. ADAPTERS

4.1. In the standardization of compressed gas cylinder valve outlet connections, more than one outlet is specified for some gases. To provide interchangeability of user's equipment *for the same gas*, adapters may be required. CGA recognizes adapters designed to connect a cylinder valve outlet to a regulator, charging connection or other mating part having a different connection for the same gas. CGA adapters are permitted to make connections between standard, alternate standard, or limited standard connections for the same gas, unless the use of such adapters could result in undesirable connections. Other adapters may be used by suppliers of gases or related equipment.

---

*For the purpose of this standard, "small cylinders" are understood to mean refillable cylinders having a maximum capacity of 110 cu in. (4 lb of water).

COMPRESSED GAS CYLINDER VALVE INLET CONNECTIONS

The threads on the inlet, or valve to cylinder connection and on some outlet connections are:
A. National Gas Taper Threads, denoted by the symbol NGT or by NGT (Cl) for chlorine.
B. Special Gas Taper Threads, denoted by the symbol SGT.
C. National Gas Straight Threads, denoted by the symbol NGS.*

A. NGT Threads

The NGT threads are based on the American Standard for taper pipe threads, but are longer to provide fresh threads if further tightening is necessary. Should there be an unintentional difference in taper at the pitch elements of the valve and of the cylinder threads, it is preferred to have the greater tightness at the bottom of

---

*Straight threads other than NGS are used satisfactorily. Some of these threads are being studied for inclusion in a future revision of this standard.

the valve. The threads also have their well-defined details as given in Table 3 and have their own tolerances which require gages developed specifically for these threads.

Limits on Size. Final inspection limits on size (pitch diameter) of both the external and internal threads are ±1 turn from basic; although the preferred working limits are $\pm\frac{1}{2}$ turn from basic.

Limits on Taper.

(a) The taper on the pitch elements of external threads shall be $\frac{3}{4}$ in. per foot on diameter, with a minus tolerance of 1 turn, but with no plus tolerance in gaging to assure a taper on the pitch elements not more than basic.

(b) The taper on the pitch elements of internal threads shall be $\frac{3}{4}$ in. per foot on diameter, with a plus tolerance of 1 turn, but with no minus tolerance in gaging to assure a taper on the pitch elements not less than basic.

Limits on Crest and Root. Same as for American National Standard for taper pipe threads.

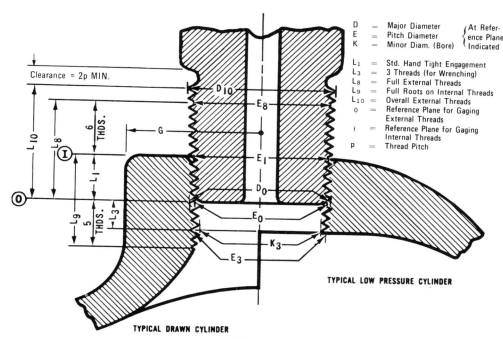

D = Major Diameter	{At Reference Plane Indicated}
E = Pitch Diameter	
K = Minor Diam. (Bore)	
L_1 = Std. Hand Tight Engagement	
L_3 = 3 Threads (for Wrenching)	
L_8 = Full External Threads	
L_9 = Full Roots on Internal Threads	
L_{10} = Overall External Threads	
o = Reference Plane for Gaging External Threads	
i = Reference Plane for Gaging Internal Threads	
p = Thread Pitch	

Clearance = 2p MIN.

TYPICAL LOW PRESSURE CYLINDER

TYPICAL DRAWN CYLINDER

Fig. 1.

TABLE 3. NATIONAL GAS TAPER (NGT) THREADS[6]

		EXTERNAL									INTERNAL				
		SMALL END			FULL THREADS		LARGE END				C'SINK	BORE	FULL THREADS		
SYMBOL (Designation of Thread) (1)	HAND-TIGHT ENGAGEMENT L_1 (3)	MAJOR DIAM. D_0	PITCH DIAM. E_0	CHAMFER 45°×MIN. DIAM. GG	PITCH DIAM. E_8	LENGTH L_8 (4)	MAJOR DIAM. APPROX. D_{10}	OVERALL LENGTH APPROX. L_{10}	NECK RADIUS MIN. G	PITCH DIAM. AT FACE E_1	C'SINK 90°×MAX. DIAM. KK	BORE MAX. K_3	PITCH DIAM. E_3	LENGTH L_1+L_3	LENGTH OF FULL ROOT MIN. L_9 (5)
1	2	3	4	5	6	7	8	9	10	11	12	13	14	15	16
1/8–27NGT	0.1615	0.3931	0.3635	21/64	0.3875	0.3837	0.4204	7/16	9/32	0.3736	13/32	.3269	0.3566	0.2726	0.3467
1/4–18NGT	0.2278	0.5218	0.4774	27/64	0.5125	0.5611	0.5530	5/8	3/8	0.4916	9/16	.4225	0.4670	0.3944	0.5056
3/8–18NGT	0.2400	0.6564	0.6120	9/16	0.6479	0.5733	0.6915	11/16	7/16	0.6270	11/16	.5572	0.6016	0.4067	0.5178
1/2–14NGT	0.3200	0.8156	0.7584	11/16	0.8052	0.7486	0.8625	13/16	9/16	0.7784	7/8	.6879	0.7450	0.5343	0.6771
3/4–14NGT	0.3390	1.0248	0.9677	29/32	1.0157	0.7676	1.0795	7/8	11/16	0.9889	1-1/16	.8972	0.9543	0.5533	0.6961
3/4–14NGT(Cl)-1	0.3390	1.0248	0.9677	29/32	1.0268	0.9461	1.0951	1-1/8							
3/4–14NGT(Cl)-2	0.3390	1.0427	0.9856	59/64	1.0447	0.9461	1.1130	1-1/8	11/16	0.9889	1-1/16	.8972	0.9543	0.5533	0.9461
3/4–14NGT(Cl)-3	0.3390	1.0628	1.0057	15/16	1.0648	0.9461	1.1331	1-1/8							
3/4–14NGT(Cl)-4	0.3390	1.0873	1.0302	31/32	1.0893	0.9461	1.1576	1-1/8							
1-11½–NGT	0.4000	1.2832	1.2136	1-1/8	1.2712	0.9217	1.3457	1	13/16	1.2386	1-5/16	1.1278	1.1973	0.6609	0.8348
1-1/4-11½–NGT	0.4200	1.6267	1.5571	1-15/32	1.6160	0.9417	1.6931	1-1/16	1	1.5834	1-43/64	1.4713	1.5408	0.6809	0.8548
1-1/2-11½–NGT	0.4200	1.8657	1.7961	1-45/64	1.8550	0.9417	1.9360	1-1/8	1-5/32	1.8223	1-29/32	1.7102	1.7798	0.6809	0.8548
3/4–14SGT (2)	0.4008	1.047	0.9852	59/64	1.0731	0.7030	1.1564	7/8	11/16	1.0353	1-7/64	.8556	0.9474	0.5714	0.7030

All dimensions are basic and are given in inches. All NGT threads are right hand.

(1) **Symbol (Designation of Thread)**

Oversize valves — For uses other than chlorine, oversize threads for revolving are generally but not always at 4 or 7 turns oversize. For chlorine, the ¾–14NGT(Cl)–1 is not oversize; the –2 is 4 turns oversize; the –3 is 8½ turns oversize; and the –4 is 14 turns oversize.

(2) **¾ – 14 SGT**

The ¾ –14SGT (Special Gas Taper) Thread is a standard having a taper of 1½" per foot on diameter with a 60° thread normal to the axis and 0.0618" deep. For this thread Col. 13, 14 and 15 are based on gages 0.7030" long. Cylinders are held to final inspection limits from basic to plus or minus 1 turn.

(3) **Handtight Engagement**

The basic condition of fit is that the External Thread with a pitch diameter of E_0 at the end (reference plane for gaging External Thread) shall enter by hand engagement to a distance L_1 into the Internal Thread with a pitch diameter of E_1 at the opening (reference plane for gaging Internal Thread).

(4) **Length**

External Threads shall be threaded the approximate length L_{10} but gaged up to L_8. Dimension L_8 is equal to L_1 plus six (6) threads for all NGT threads and L_1 plus eight and a half (8½) threads for the NGT (Cl) threads. Dimension E_8 is measured at distance L_8 from E_0, and dimension D_{10} is measured at distance L_{10} from E_0. These longer External Threads are desirable if further tightening should be necessary. To facilitate gaging, provision should be made to allow the L_8 ring gage to advance a distance of 2 full threads beyond the L_8 length (one turn for allowable variation in pitch diameter and one turn for allowable variation in taper).

(5) **Length of Full Root Min.**

Full Internal Threads at the crests and roots shall extend throughout lengths L_1 plus L_3 (L_3 = 3 threads). This dimension determines the minimum metal on the inside of the neck to produce maximum bore K_3. Any metal below L_3 shall have tapped threads with full roots to a minimum length L_9 (L_1 + 5 threads for all NGT threads and L_1 + 8½ threads for the NGT (Cl) threads).

(6) **Gaging NGT Threads**

Because of their length and more rigid requirements for sealing compressed gases against leaks, NGT threads require special gages. One such type of gage has been developed and described by Federal Services for inspecting NGT threads. Typical gages are fully described in Military Supply Procedures Manual 8310 IGMS-5008. Any other method of gaging which will give the required results can be used.

B. SGT Threads

The SGT threads are similar to NGT threads except that the taper is $1\frac{1}{2}$ in. per foot on diameter instead of $\frac{3}{4}$ in. per foot on diameter.

C. NGS Threads*

Whenever straight threads are specified they shall be National Gas Straight Threads, denoted by the symbol NGS. The diameters and the form for both the external and internal threads shall conform to American Standard Straight Pipe Threads for Mechanical Joints (NPSM), and the length of engagement shall be $L_1 + L_3$ as shown in Table 3 for NGT Threads. The seal for tightness shall be at or close to the end face of the cylinder whether it incorporates the external or the internal threads.

*Straight threads other than NGS are used satisfactorily. Some of these threads are being studied for inclusion in a future revision of this standard.

SECTION B

Standard for 22 mm Anesthesia Breathing Circuit Connectors

1. SCOPE

1.1. This standard concerns the dimensions and profile of male and female of nominal 22 mm size adult anesthesia breathing circuit connectors intended for use in nonload bearing applications.

1.2. This standard is intended to define a practice for marking anesthesia breathing circuit components with regard to direction of gas flow for those components where direction of flow is important for the safety of the patient and/or the efficient operation of the component or the anesthesia breathing circuit.

1.3. It is recognized that other kinds of non-anesthesia breathing circuits may use similar connectors, but the general application of those specified herein could in some instances present a hazard to the patient by making possible one or more dangerous misconnections in the breathing circuit; the application for these uses is not recommended.

2. GENERAL OBJECTIVES

2.1. The aim of this standard is to provide dependability and interchangeability of like components of breathing circuits for anesthesia apparatus.

2.2. Such apparatus shall be designed primarily for the safety of the patient and only on a secondary basis for the convenience of the anesthesiologist or manufacturer.

2.3. Connections between the components of the anesthesia breathing circuit shall be reasonably leak proof and have sufficient stability to prevent accidental disconnection during ordinary usage.

2.4. Mating connectors shall provide a broad area of mutual contact and shall not mate solely at the rim, end or edge of either.

3. TERMINOLOGY

3.1. Breathing Circuit

The assembled components from or through which the patient breathes.

3.2. Breathing Circuit Connection

The conduit formed by the mating of the appropriate connectors of two breathing circuit components.

3.3. Breathing Circuit Connector

Those details of a breathing circuit component intended to mate with their counterparts of other components.

3.4. Gas-Tight Seal

A connection which allows no bubbling when immersed in water and subjected to a specified differential pressure shall be deemed gas tight.

4. CONFIGURATION AND DIMENSIONS

4.1. Dimensions are specified in millimeters with inch equivalents for reference.

4.2. Connectors forming an integral or removable part of standard components of adult size anesthesia breathing equipment whose purpose is to permit attachment of these components to breathing bags, breathing tubes or masks, shall be male. Adult size breathing tube, mask and breathing bag connectors shall be female.

4.3. Male connectors of adult size breathing circuit components shall be of rigid construction.

4.4. Male connectors of adult size shall be of one of the three following configurations:

(1) Adult male breathing circuit connector as illustrated in Fig. 1.
(2) Adult male high-retention breathing circuit connector illustrated in Fig. 2.
(3) Mask connector, 22 mm male as shown in Fig. 3.

4.4.1. The adult male breathing circuit connector and the adult male high-retention breathing circuit connector may be used interchangeably. The degree of retention offered by each of these male connections is specified in Sections 6.3.1 and 6.4.1, respectively.

4.4.2. The male connector illustrated in Fig. 3, is intended for use in making mask connections; it embodies a coaxial 15 mm female connector for accommodation of standard tracheal tube adapters.*

---

*The configuration of the 15 mm female conductor is defined in American National Standards Institute Specification Z-79.2-1976 entitled "American Standard Specifications for Anesthetic Equipment: Endotracheal Tube Connectors and Adapters."

4.5. Connectors of standard adult size breathing tubes, breathing bags and face masks shall be female, of resilient construction and shall fit securely to and provide a gas-tight seal with a male plug gage as shown in Fig. 4.

5. MATERIALS

5.1. Because of the continuing rapid development of new materials, no strict standards relative to material are provided. However, in general, the following characteristics are desirable:

5.2. Male Connectors. Materials having a Rockwell "B" scale hardness of 97 or greater shall be considered rigid for the purposes of this standard.

5.2.1. The material used in the fabrication of rigid connectors shall be light in weight, resistant to deformation or actual breakage, and sufficiently elastic to resume original shape after reasonable deforming forces are removed.

5.2.2. The material shall be reasonably resistant to agents used in chemical cleansing and sterilizing, to anesthetic gases and vapors, and to deterioration by autoclaving, unless part of a disposable component.

5.2.3. The material shall meet the electrical conductivity standards of the appropriate** authority, when applicable.

5.3. Female Connectors. Material having a Durometer "A" scale hardness of less than 97 shall be considered nonrigid for purposes of this standard.

5.3.1. The material shall be reasonably resistant to agents used in chemical cleansing and sterilizing, to anesthetic gases and vapors and to deterioration by autoclaving, unless part of a disposable component.

5.3.2. The material shall meet the electrical conductivity standards of the appropriate** authority, when applicable.

**NFPA No. 56A, Standard for the Use of Inhalation Anesthetics-1971, may be obtained from the National Fire Protection Association, 60 Batterymarch Street, Boston, MA 02110.

6. CONNECTORS

6.1. Nonrigid Female Connectors

6.1.1. Tensile Test. Any nonrigid female connector, when dipped in distilled water and engaged with a plug gage (described in 4.5) shall be capable of withstanding a minimum tensile load of 1.5 kg applied in any direction for a period of one minute. If any movement of the connector in the direction of disengagement is detectable the connector is unacceptable. The connector shall permit engagement to the shoulder of the gage.

6.1.2. Pressure Test. Any nonrigid female connector when dipped in distilled water and engaged with a plug gage (described in 4.5) shall be capable of maintaining a gas-tight seal when an internal pressure of 150 mm Hg above ambient is applied. The connector shall permit engagement to the shoulder of the gage.

6.2. Adult Male Breathing Circuit Connectors

6.2.1. Tensile Test. Any connection formed by a nonrigid female connector (qualified by the test outlined in 6.1.1) dipped in distilled water, and an adult male breathing circuit connector, shall be capable of withstanding a minimum tensile load of 1.5 kg applied in any direction for a period of one minute. If any movement of the female in the direction of disengagement is detectable, the male connector is unacceptable. The connector shall permit engagement of at least 21 mm.

6.2.2. Pressure Test. Any connection formed by a nonrigid female connector (qualified by test outlined in 6.1.1 and 6.1.2) dipped in distilled water, and an adult male breathing circuit connector, shall be capable of maintaining a gas-tight seal when an internal pressure of 150 mm Hg above ambient is applied.

6.3. Adult Male High-Retention Breathing Circuit Connectors

6.3.1. Tensile Test. Any connection formed by a nonrigid female connector (qualified by the test outlined in 6.1.1) dipped in distilled water, and an adult male high-retention breathing circuit connector shall be capable of withstanding a minimum tensile load of 2.27 kg applied in any direction for a period of one minute. If any movement of the female in the direction of disengagement is detectable, the male connector is unacceptable. The connector shall permit engagement of at least 24 mm.

6.3.2. Pressure Test. Any connection formed by a nonrigid female connector (qualified by the tests outlined in 6.1.1 and 6.1.2) dipped in distilled water, and an adult male high-retention breathing circuit connector shall be capable of maintaining a gas-tight seal when an internal pressure of 200 mm Hg above ambient is applied.

6.4. Mask Connector, 22 mm Male

6.4.1. Tensile Test. Any connection formed by a nonrigid female connector (qualified by the test outlined in 6.1.1), and dipped in distilled water, and a mask connector, 22 mm male, shall be capable of withstanding a minimum tensile load of 1.5 kg applied in any direction for a period of one minute. If any movement of the female in the direction of disengagement is detectable, the male connector is unacceptable. The connector shall permit engagement of not more than 21 mm.

6.4.2. Pressure Test. Any connection formed by a nonrigid female connector (qualified by the tests outlined in 6.1.1 and 6.1.2), and dipped in distilled water, and a mask connector, 22 mm male, shall be capable of maintaining a gas-tight seal when an internal pressure of 150 mm Hg above ambient is applied.

7. MARKING

7.1. All anesthesia breathing circuit components in which the direction of gas flow is critical shall be marked in such a way that the intended direction of gas flow is immediately apparent to the operator.

7.2. For the purpose of this standard all anesthesia breathing circuit components that contain a valve or valves, the purpose of which is to establish the direction of gas flow, are considered to be flow critical and shall be marked as described in Section 8. Examples include inhalation check valves, exhalation check valves, and nonrebreathing valves.

7.3. Breathing circuit components which are designed to accommodate gas flow in a specific direction, but which do not contain valves, the purpose of which are to establish gas flow direction, shall be marked as specified in Section 8. Examples may include vaporizers, humidifiers, moisture traps, carbon dioxide absorbers, and

filters. The manufacturer shall be responsible for determining if the direction of flow through such components is related to the safety of the patient or the efficiency of the breathing circuit.

7.4. Breathing circuit components in which the direction of flow is not critical need not be marked to indicate specific flow direction. Examples include masks, breathing tubes, and rebreathing bags.

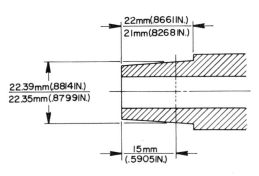

TAPER PER UNIT ON DIA.
.0240 TO .0260

Fig. 1. Adult male breathing circuit connector.

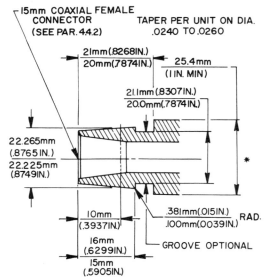

* OTHER MECHANICAL MEANS OF PREVENTING INSERTION OF THE MASK CONNECTOR INTO A MASK TO A DEPTH GREATER THAN 21 mm (.8268IN.) SHALL BE CONSIDERED TO BE SATISFACTORY ALTERNATIVES.

Fig. 3. Mask connector, 22 mm male.

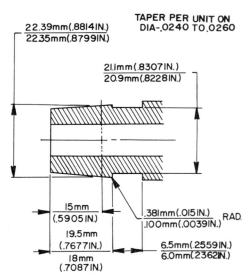

Fig. 2. Adult male high-retention breathing circuit connector.

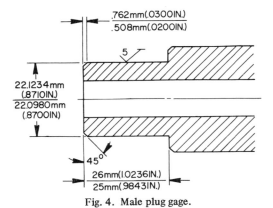

Fig. 4. Male plug gage.

8. METHOD OF MARKING

8.1. Where markings are applied to breathing circuit components in order to indicate the direction of gas flow, the minimum acceptable marking shall consist of at least one headed arrow permanently affixed to the component. Where a component contains more than one check valve, the minimum acceptable marking shall consist of at least one headed arrow indicating the direction of flow through each check valve.

8.2. At the discretion of the manufacturer, additional permanent markings can be applied, grouped as follows:

Group	Group 2	Group 3	Group 4
Inlet	In	Inhalation	Inhalation
Outlet	Out	Exhalation	Exhalation
Patient	Patient	Patient	Patient

All such markings used on a given component must be chosen from a single group. However, all markings of a particular group need not be used.

8.3. All markings which are applied to breathing circuit components for purposes of indicating direction of gas flow are to be located, if possible, so that they will fall in the normal field of view of the operator when the equipment is in use.

SECTION C

Diameter-Index Safety System

1. INTRODUCTION

1.1. In any field of human endeavor there are hazards which may be created by preoccupation, mental lapse, carelessness and the like. The medical gas industry, organized within the Compressed Gas Association, Inc. (CGA), has long recognized its responsibility to the medical profession and to the general public by working, to the extent of mechanical practicability, toward eliminating the hazards of inadvertent substitution of the wrong medical gases by users of anesthetic, resuscitation and therapeutic administering equipment.

1.2. The success achieved by the Compressed Gas Association in developing the Pin-Index Safety System for flush-type medical gas cylinder valves, has been evidenced by its adoption as the American National* and Canadian Standard as well as an International Standard. This system was developed to reduce the hazard of accidental substitution of the wrong gases on equipment utilizing yoke-type connections.

1.3. Another important step taken by the Association was standardization of threaded outlets for cylinder valves.* This action, taken at considerable expense to the industry, required the removal from service of all cylinder valves with threaded outlets which did not conform.

2. SCOPE

2.1. First approved in 1959, the Diameter-Index Safety System was developed by the Compressed Gas Association, Inc., to meet the need for a standard to provide noninterchangeable connections where removable exposed threaded connections are employed in conjunction with individual gas lines of medical gas administering equipment, at pressures of 200 psig (1400 kPa, gage) or less, such as outlets from medical gas regulators and connectors for anesthesia, resuscitation and therapy apparatus. Removable threaded connections are those which are commonly and readily engaged or disengaged in routine use and service.

2.2. The Diameter-Index Safety System supplements but does not replace:

*See Section A of this chapter.

(a) any of the means for medical gas identification now in use;

(b) the Pin-Index Safety System;

(c) the existing threaded outlet standards for cylinder valves; or

(d) automatic quick coupler valves which also provide noninterchangeable connections for medical gases, and suction equipment.

2.3. The Compressed Gas Association, Inc., does not presume to designate specifically where the Diameter-Index Safety System should find application on medical equipment.

2.4. The system presents a concept in design for low-pressure medical gas connections. For all medical gases and suction connections (except oxygen), noninterchangeable indexing is achieved by a series of increasing and decreasing diameters (see Fig. 1, Dimensions A and B) in the component parts of the connections. The long established $\frac{9}{16}$ in.-18 thread connection has been retained for oxygen (see Fig. 7).

2.5. The system recognizes the established use of suction (vacuum) in the immediate areas where connections to that equipment could inadvertently be interchanged with medical gas equipment. Therefore, suction has been assigned an indexed position in this system. Provision has been made for the larger bore required for suction service (see Fig. 6).

2.6. The direction of flow through connections of the system is optional unless otherwise specified for a particular connection assignment.

3. MAJOR CHANGES AND ADDITIONS IN THE 1978 REVISION

3.1. The Diameter-Index Safety System issued as a CGA standard in 1959 included 12 connections: 11 assigned for low-pressure medical gases and one for suction (vacuum). Except for suction and oxygen these were all based on the 1000 Series.

3.2. In addition to the 1000 Series included in the 1959 standard, this 1978 revision incorporates a Co-Standard 1000-A Series which utilizes the same body as the 1000 Series and the suction nut (Connection No. 1220) for all connections except oxygen (Connection No. 1240).

The shank diameter of corresponding 1000-A Series nipples is increased from $\frac{5}{16}$ in. to $\frac{17}{32}$ in. to mate with the suction nut as well as to allow for heavier duty application. Figures 1 and 2 depict typical mating sets for the Standard 1000 Series and Co-Standard 1000-A Series connections.

3.3. Some applications have required greater flow rates than were available through the connections of the 1959 standard, thus the drill size through the body and nipple have been increased. It is recommended that connections be manufactured with the maximum hole diameter unless a specific application requires otherwise.

3.4. The addition of eight new connections (1500 Series) is made in the 1978 revision. This 1500 Series has been designed and shown (Figs. 8 and 9) in metric dimensions in accordance with the International System of Units (SI) recommendations. One of these, Connection No. 1570, has been assigned to mixtures of oxygen and nitrous oxide where the concentration of nitrous oxide is between 47.5 and 52.5 percent.

3.5. To avoid inadvertent insertion of existing Standard 1000 Series and Co-Standard 1000-A Series nipples in the bodies of new Standard 1500 Series and further to provide protection of the nipple against damage, a nut stop is now required on *all* connections, except on Connection No. 1240, to limit the distance the nut can move away from the nipple shoulder (see Figs. 1 and 2).

3.6. Requirements for optional check valves and secondary seals have been added in Section 4.

3.7. Connection No. 1120 has been assigned to nitrogen.

3.8. Connection No. 1020 has been assigned to be used with special gas mixtures when a connection is needed for limited experimental applications.

4. OPTIONAL FEATURES

4.1. Check Valves. The addition of a check valve in the body portion of connections is permissible provided:

(a) A corresponding nut and nipple, meeting the connection specifications, will open the valve.

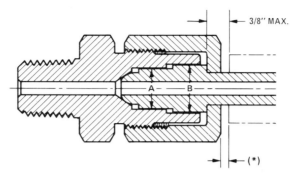

Fig. 1. Typical standard connection, Series No. 1000. (*) Nut stop to be contained in this area by use of any of the following or combination of hose ferrule, washer, retaining ring, body of mating device.

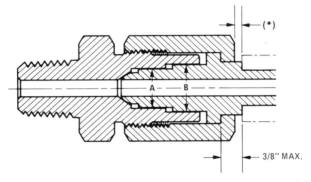

Fig. 2. Typical co-standard connection, Series No. 1000-A. (*) Nut stop to be contained in this area by use of any of the following or combination of hose ferrule, washer, retaining ring, body of mating device.

(b) No gas will flow unless the threads are engaged at least $\frac{3}{4}$ turn.

(c) Strength, noninterchangeability and nose to conical seat sealing are maintained.

Manufacturers should recognize that for some applications, flow capacity may be critical and the design of a valve should minimize flow restriction through the connection.

4.2. Secondary Seals. A secondary elastomer seal is permissible on the large indexing diameter and/or on the nose radius of the nipple provided:

(a) The strength and noninterchangeability of the connection are maintained.

(b) Omission of the secondary seal will not affect the seal or basic integrity of the connection.

4.2.1. On Connection No. 1220, which has only one diameter, the secondary seal, if used, must be on that portion of the nipple equivalent to the large indexing diameter and/or on the nose radius.

4.2.2. On Connection No. 1240, a secondary seal, if used, must meet the criteria noted above in 4.2(a) and (b).

5. IMPLEMENTATION OF NEW CONNECTIONS

The Co-Standard 1000-A Series and new Standard 1500 Series connections, along with the assignment of Connection No. 1120 for nitrogen and Connection No. 1020 for special gas mixtures, became effective in 1978. Other designated changes became mandatory for all connections produced beginning March 1, 1979.

TABLE 1. ASSIGNED GASES AND GAS MIXTURES CONNECTION
NOS. 1020–1200, 1020-A–1200-A, 1220, 1240 AND 1570

Connection No.	Gas Name	Gas Symbol
1020 1020-A	Special mixtures. For limited experimental applications. The word "SPECIAL" is permissible.	
1040 1040-A	Nitrous oxide	N_2O
1060 1060-A	Helium, helium-oxygen mixtures (helium over 80.5 percent)	He $He-O_2$ mixture
1080 1080-A	Carbon dioxide, carbon dioxide-oxygen mixtures (CO_2 over 7.5 percent)	CO_2 CO_2-O_2 mixture
1100 1100-A	Cyclopropane	C_3H_6
1120 1120-A	Nitrogen	N_2
1140 1140-A	Ethylene	C_2H_4
1160 1160-A	Air	
1180 1180-A	Oxygen-helium mixtures (helium not over 80.5 percent)	O_2-He mixture
1200 1200-A	Oxygen-carbon dioxide mixtures (CO_2 not over 7.5 percent)	O_2-CO_2 mixture
1220	Suction (vacuum)	
1240	Oxygen	O_2
1570	Nitrous oxide-oxygen mixtures (N_2O 47.5 to 52.5 percent)	N_2O-O_2 mixture

6. CONNECTION ASSIGNMENTS

Assignments of various gases and gas mixtures to the different connections in this standard are shown in Table 1.

Suppliers and users of the Diameter-Index Safety System who require a connection assignment for a gas or gas mixture or who need assistance in selecting the proper connection, should contact the Compressed Gas Association, Inc., 500 Fifth Avenue, New York, NY 10110.

7. DESCRIPTION OF THE DIAMETER-INDEX SAFETY SYSTEM

Each connection of the Diameter-Index Safety System consists of a body, nipple and nut. This system is based on having two concentric and specific bores in the body, and two concentric and mating diameters on the nipple, except for Connection Nos. 1220 and 1240. To achieve noninterchangeability between different connections, the two diameters on each part vary in opposite directions so that as one diameter increases, the other decreases. In this way, only the properly mated and intended parts fit together to allow thread engagement.

7.1. Connections 1020 thru 1200 (1000 Series). Attempts to bring together unmated parts result in interference, either at the large diameter or at the small diameter, preventing thread engagement. The critical diameters from one size to the next vary in basic steps of 0.012 in. Manufacturing tolerances on mating connections provide a clearance from 0.003 in. to

0.009 in. at each of the diameters. The interference at either the small or the large diameter of nonmating connections is 0.003 in. minimum.

Diameters A and B as shown in Figs. 1 and 2 are the variable diameters controlling the Diameter-Index Safety System. The small bore in the body mates with the small diameter of the nipple, and the large bore in the body mates with the large diameter of the nipple.

7.2. Connections 1020-A thru 1200-A (1000-A Series). Co-Standard Connections 1020-A thru 1200-A have indexing diameters identical to respective Connections 1020 thru 1200, but have an added shoulder at the rear of the large indexing diameter. This shoulder is identical to the shoulder for the suction Connection No. 1220, and utilizes the suction nut. This Co-Standard permits a larger shank diameter on the nipple allowing for heavier duty applications.

Note: Nut and nipple of the 1000 and 1000-A series connections will both fit the same body.

7.3. Connection 1220–Suction (Vacuum). On Connection No. 1220 (as shown in Fig. 6) the critical diameters become equal, forming single mating diameters for the bore and for the nipple. The nipple of this connection has the largest nose diameter and is therefore designed with an extra large bore and assigned to suction.

For mechanical reasons the design of the suction nut is $\frac{1}{8}$ in. longer than the nut used with Standard Connections 1020 thru 1200. This extra length makes it possible to have thread engagement between nonmating parts if the suction nut is inadvertently used in place of the shorter nut on Standard Connections 1020 thru 1200. To avoid this, the hole in the suction nut is enlarged so that the nipples for Standard Connections 1020 thru 1200 will pass through it.

7.4. Connection 1240–Oxygen. Standard Connection No. 1240 (as shown in Fig. 7) instead of having increasing and decreasing diameters, retains the long established standard oxygen connection with the $\frac{9}{16}$ in.-18 thread. This connection is interchangeable with Connection No. 022 in CGA Pamphlet E-1, Standard Connections for Regulator Outlets, Torches and Fitted Hose for Welding and Cutting Equipment.

7.5. Connections 1500 thru 1580 (1500 Series). The 1978 revision provides for eight additional Diameter-Index Safety System connections with indexing principles very similar to that of the Standard 1000 Series connections. The nut on the 1500 Series connections has a larger thread diameter and will slip over the body thread on the 1000 Series arrangement. The step arrangement for indexing is similar to that used previously with slightly greater differences between the large and small indexing diameters. These 1500 Series connections have been designed to metric dimensions in accordance with the International System of Units (SI) recommendations.

When establishing the eight new connections of the 1500 Series connections, certain nipples from the 1000 and 1000-A Series connections may enter the body of certain 1500 Series connections. These connections cannot be completed, since the thread diameter of the 1000 and 1000-A Series connections is too small to go over the body of the 1500 Series connections. In addition, the 1978 revision requires a nut stop, such as a retaining ring, to limit the distance which the nut can move away from the shoulder. With such a nut stop the 1000 and 1000-A Series nipples cannot enter the 1500 Series bodies (see Figs. 1 and 2).

As an added safety precaution, the hole in the nut for the 1500 Series connections is made larger to permit the 1000 and 1000-A Series nipples to fall through the nut should an attempt be made to make up a mismatched set.

7.6. Detailed Drawings. The pages that follow contain detailed drawings illustrating the various connections comprising the Diameter-Index Safety System as developed by the Compressed Gas Association, Inc. Figures 3, 4, and 8 show the basic dimensions which are common for their respective series. The dimensions which vary from one connection to the next are shown in table form on the pages following basic dimension drawings. Separate and individual drawings for Connection No. 1220 (Fig. 6) for suction and No. 1240 (Fig. 7) for oxygen are complete in themselves.

STANDARD 1000 SERIES LOW-PRESSURE CONNECTIONS FOR

Medical Gases

BASIC DIMENSIONS FOR CONNECTION NUMBERS 1020 THRU 1200
(Dimensions in Inches)

Gas or mixture name or symbol
must be marked on or at body

STANDOFF	F	5/64 APPROX.
CONTACT DIA.	Q	.240
CLEARANCE	Y	3/32 APPROX.

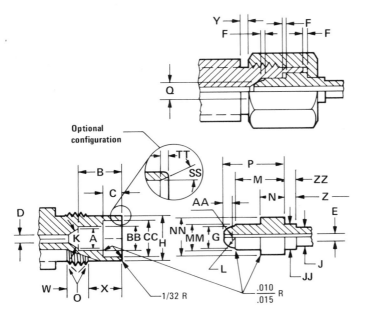

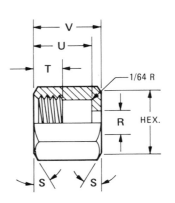

BODY			NIPPLE			HEXAGON NUT		
THREAD		.750-16UNF-2A-RH	DRILL	E	.161 MAX.	THREAD		.750-16UNF-2B-RH (MOD)
MAJOR DIA.		.7485-.7391	*NOSE DIA.	G	.296-.293	MINOR DIA.		.690-.696 (MOD)
PITCH DIA.		.7079-.7029	NOSE RADIUS	L	.1480-.1465	PITCH DIA.		.7094-.7159
MINOR DIA.		.6718 REF.	*SM. INDEX DIA.	MM	†	MAJOR DIA.		.7500 MIN.
*SEAT DIA.	A	.299-.302	LENGTH	M	.625 ± .005	COUNTERSINK		90° x 49/64 DIA.
BORE DEPTH	B	.625 ± .005	*LG INDEX DIA.	NN	†	HEXAGON		7/8
*BORE DIA.	BB	†	LENGTH	N	.312 ± .005	HOLE	R	.323-.333
C'BORE DEPTH	C	.312 ± .005	HEAD LENGTH	P	.781 ± .005	CHAMFER	S	30° x 7/8 DIA.
*C'BORE DIA.	CC	†	STEP DIA.	JJ	.317 ± .307	FULL THREAD	T	3/8 MIN.
DRILL	D	.161 MAX.	SHANK DIA.	J	OPTIONAL	BORE DEPTH	U	.745-.755
BODY DIA.	H	.656-.650	STEP LENGTH	ZZ	3/16 to 3/8	LENGTH	V	7/8 MIN.
ANGLE	K	70°	SHANK LENGTH	Z	OPTIONAL			
CHAMFER	O	45° x DIA. H	RADIUS DIST.	AA	.120 ± .003			
COUNTERSINK	SS	25°						
COUNTERSINK	TT	.031						
FULL THREAD	W	1/4 MIN.						
LENGTH	X	.437 ± .005						

*Body diameters A, BB and CC as well as nipple diameters G, MM and NN should be concentric within .002 Full Indicator Movement (FIM). These are critical dimensions for safety that must be adhered to on final product whether plated or not.

†See Fig. 5 for these dimensions.

Fig. 3.

CO-STANDARD 1000-A SERIES LOW-PRESSURE CONNECTIONS FOR
Medical Gases
BASIC DIMENSIONS FOR CONNECTION NUMBERS 1020-A THRU 1200-A
(Dimensions in Inches)

Gas or mixture name or symbol
must be marked on or at body

STANDOFF	F	5/64 APPROX.
CONTACT DIA.	Q	.240
CLEARANCE	Y	3/32 APPROX.

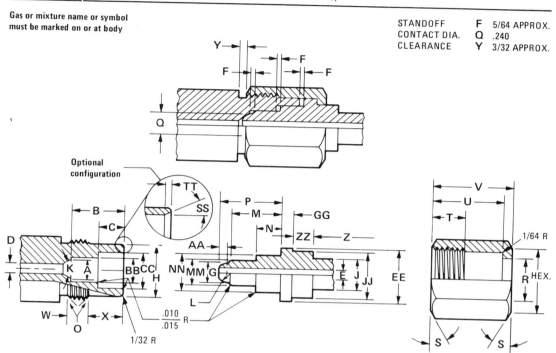

BODY		
THREAD		.750-16UNF-2A-RH
MAJOR DIA.		.7485-.7391
PITCH DIA.		.7079-.7029
MINOR DIA.		.6718 REF.
*SEAT DIA.	A	.299-.302
BORE DEPTH	B	.625 ± .005
*BORE DIA.	BB	†
C'BORE DEPTH	C	.312 ± .005
*C'BORE DIA.	CC	†
DRILL	D	.161 MAX.
BODY DIA.	H	.656-.650
ANGLE	K	70°
CHAMFER	O	45° x DIA. H
COUNTERSINK	SS	25°
COUNTERSINK	TT	.031
FULL THREAD	W	1/4 MIN.
LENGTH	X	.437 ± .005

NIPPLE		
DRILL	E	.161 MAX.
SHOULDER DIA.	EE	.672 ± .005
*NOSE DIA.	G	.296-.293
SHOULDER LGTH	GG	.125 ± .005
NOSE RADIUS	L	.1480-.1465
*SM INDEX DIA.	MM	†
LENGTH	M	.625 ± .005
LG INDEX DIA.	NN	†
LENGTH	N	.312 ± .005
HEAD LENGTH	P	.781 ± .005
STEP DIA.	JJ	.536-.531
SHANK DIA.	J	OPTIONAL
STEP LENGTH	ZZ	3/16 to 3/8
SHANK LENGTH	Z	OPTIONAL
RADIUS DIST.	AA	.120 ± .003

HEXAGON NUT		
THREAD		.750-16UNF-2B-RH (MOD)
MINOR DIA.		.690-.696 (MOD)
PITCH DIA.		.7094-.7159
MAJOR DIA.		.7500 MIN.
COUNTERSINK		90° x 49/64
HEXAGON		7/8
HOLE	R	.546-.551
CHAMFER	S	30° x 7/8 DIA.
FULL THREAD	T	3/8 MIN.
BORE DEPTH	U	.870-.880
LENGTH	V	1.000 MIN.

*Body diameters A, BB and CC as well as nipple diameters G, MM and NN should be concentric within .002 Full Indicator Movement (FIM). These are critical dimensions for safety that must be adhered to on final product whether plated or not.
†See **Fig. 5** for these dimensions.

Fig. 4.

STANDARD 1000 AND 1000-A SERIES DIAMETER-INDEX DIMENSIONS FOR
Medical Gases

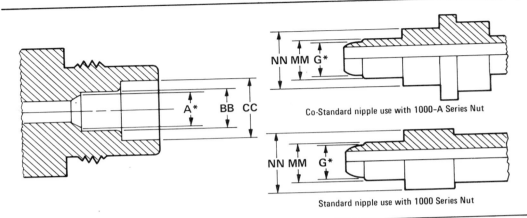

Co-Standard nipple use with 1000-A Series Nut

Standard nipple use with 1000 Series Nut

GAS NAME	GAS SYMBOL	CONN. NO.	DIAMETERS (INCHES)			
			*BB	*CC	*MM	*NN
Special Mixtures. For Limited Experimental Applications. The word "SPECIAL" is permissible.		1020 1020-A	.299-.302	.539-.542	.296-.293	.536-.533
Nitrous Oxide	N_2O	1040 1040-A	.311-.314	.527-.530	.308-.305	.524-.521
Helium, Helium-Oxygen Mixtures (Helium over 80.5%)	He, He-O_2 Mixture	1060 1060-A	.323-.326	.515-.518	.320-.317	.512-.509
Carbon Dioxide, Carbon Dioxide-Oxygen Mixtures (CO_2 over 7.5%)	CO_2, CO_2-O_2 Mixture	1080 1080-A	.335-.338	.503-.506	.332-.329	.500-.497
Cyclopropane	C_3H_6	1100 1100-A	.347-.350	.491-.494	.344-.341	.488-.485
Nitrogen	N_2	1120 1120-A	.359-.362	.479-.482	.356-.353	.476-.473
Ethylene	C_2H_4	1140 1140-A	.371-.374	.467-.470	.368-.365	.464-.461
Air		1160 1160-A	.383-.386	.455-.458	.380-.377	.452-.449
Oxygen-Helium Mixtures (Helium not over 80.5%)	O_2-He Mixture	1180 1180-A	.395-.398	.443-.446	.392-.389	.440-.437
Oxygen-Carbon Dioxide Mixture (CO_2 not over 7.5%)	O_2-CO_2 Mixture	1200 1200-A	.407-.410	.431-.434	.404-.401	.428-.425

* Body diameters A, BB and CC as well as nipple diameters G, MM and NN should be concentric within .002 Full Indicator Movement (FIM). These are critical dimensions for safety that must be adhered to on final product whether plated or not.

Fig. 5.

STANDARD CONNECTION FOR
Suction (Vacuum)
(Dimensions in Inches)

Suction or vacuum must
be marked on or at body

STANDOFF   F   3/32 APPROX.
CONTACT DIA.   Q   .340
CLEARANCE   Y   1/8 APPROX.

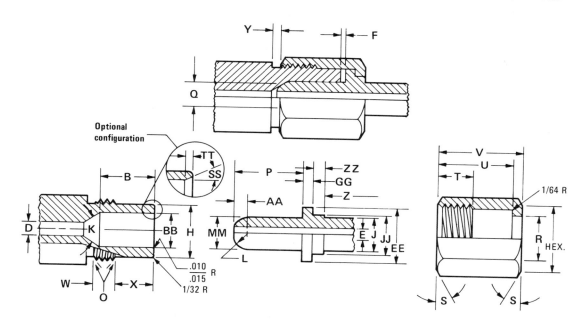

BODY		
THREAD		.750-16UNF-2A-RH
MAJOR DIA.		.7485-.7391
PITCH DIA.		.7079-.7029
MINOR DIA.		.6718 REF.
BORE DEPTH	B	.625 ± .005
*BORE DIA.	BB	.419-.422
DRILL	D	.255 MAX.
BODY DIA.	H	.656-.650
ANGLE	K	70°
CHAMFER	O	45° x DIA. H
COUNTERSINK	SS	25°
COUNTERSINK	TT	.031
FULL THREAD	W	1/4 MIN.
LENGTH	X	.437 ± .005

NIPPLE		
DRILL	E	.255 MAX.
SHOULDER DIA.	EE	.672 ± .005
NOSE RADIUS	L	.2080-.2065
*NOSE DIA.	MM	.416-.413
SHOULDER LGTH	GG	.125 ± .005
NOSE LENGTH	P	.812 ± .005
STEP DIA.	JJ	.536-.531
STEP LENGTH	J	OPTIONAL
SHANK DIA.	ZZ	3/16 to 3/8
SHANK LENGTH	Z	OPTIONAL
RADIUS DIST.	AA	.159 ± .003

HEXAGON NUT		
THREAD		.750-16UNF-2B-RH (MOD)
MINOR DIA.		.690-.696 (MOD)
PITCH DIA.		.7094-.7159
MAJOR DIA.		.7500 MIN.
COUNTERSINK		90° x 49/64 DIA.
HEXAGON		7/8
HOLE	R	.546-.551
CHAMFER	S	30° x 7/8 DIA.
FULL THREAD	T	3/8 MIN.
BORE DEPTH	U	.870-.880
LENGTH	V	1.000 MIN.

*Critical dimensions for safety that must be adhered to on final product whether plated or not.

Fig. 6.

STANDARD LOW-PRESSURE CONNECTION FOR

Oxygen

(Dimensions in Inches)

Note: Interchangeable with CGA connection No. 022 (CGA Pamphlet E-1)

FACE TO BACK	F	13/64
CONTACT DIA.	Q	.340
THREADS ENGAGED	Y	4

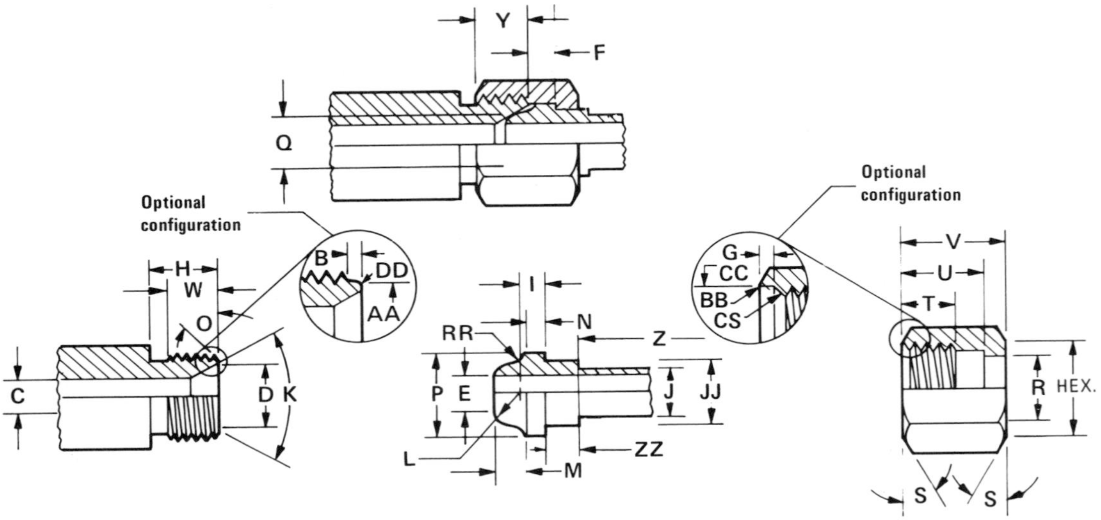

BODY			NIPPLE			HEXAGON NUT		
THREAD		.5625-18UNF-2A-RH	DRILL	E	.255 MAX.	THREAD		.5625-18UNF-2B-RH
MAJOR DIA.		.5611-.5524	RADIUS DIST.	I	.175 ± .005	MINOR DIA.		.502-.515
PITCH DIA.		.5250-.5205	NOSE RADIUS	L	.196	PITCH DIA.		.5264-.5323
MINOR DIA.		.4929 REF.	NOSE LENGTH	M	3/16	MAJOR DIA.		.5625 MIN.
DIA.	AA	.470	SHOULDER LGTH	N	1/8	COUNTERSINK		90° x 37/64 DIA.
LENGTH	B	.060	SHOULDER DIA.	P	.500-.496	HEXAGON		11/16
DRILL	C	.255 MAX.	BLEND RADIUS	RR	3/64	RADIUS	BB	.010
SEAT DIA.	D	.428-.438	STEP DIA.	JJ	.430-.425	C'BORE DIA.	CC	.590
RADIUS	DD	.010	SHANK DIA.	J	OPTIONAL	COUNTERSINK	CS	45° x .590
LENGTH TO			STEP LENGTH	ZZ	9/64 MIN.	LENGTH	G	.070
SHOULDER	H	13/32 MIN.	SHANK LENGTH	Z	OPTIONAL	HOLE	R	.4375-.4425
SEAT ANGLE	K	60°				CHAMFER	S	30° x 11/16 DIA.
CHAMFER	O	45° x 15/32 DIA.				FULL THREAD	T	5/16
FULL THREAD	W	5/16				BORE DEPTH	U	1/2
						LENGTH	V	5/8

Note: All dimensions must be adhered to on final product whether plated or not.

Fig. 7.

METRIC STANDARD 1500 SERIES
LOW-PRESSURE CONNECTIONS FOR
Medical Gases
BASIC DIMENSIONS FOR CONNECTION NUMBERS 1510 THRU 1580 (Dimensions in Millimeters)

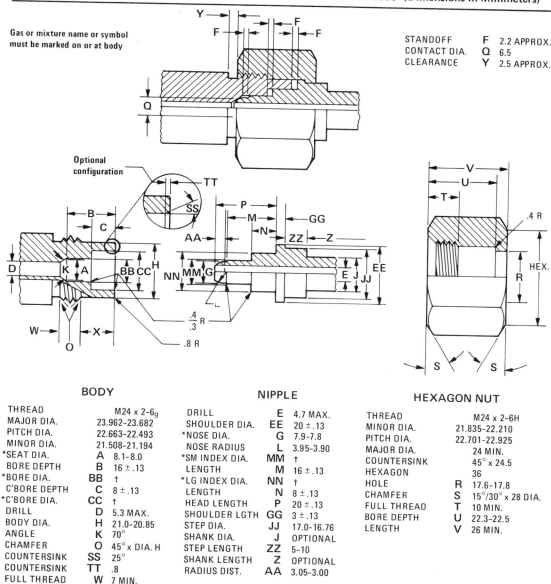

Gas or mixture name or symbol
must be marked on or at body

STANDOFF	F	2.2 APPROX.
CONTACT DIA.	Q	6.5
CLEARANCE	Y	2.5 APPROX.

BODY

THREAD		M24 x 2-6$_g$
MAJOR DIA.		23.962-23.682
PITCH DIA.		22.663-22.493
MINOR DIA.		21.508-21.194
*SEAT DIA.	A	8.1-8.0
BORE DEPTH	B	16 ±.13
*BORE DIA.	BB	†
C'BORE DEPTH	C	8 ±.13
*C'BORE DIA.	CC	†
DRILL	D	5.3 MAX.
BODY DIA.	H	21.0-20.85
ANGLE	K	70°
CHAMFER	O	45° x DIA. H
COUNTERSINK	SS	25°
COUNTERSINK	TT	.8
FULL THREAD	W	7 MIN.
LENGTH	X	11 ±.13

NIPPLE

DRILL	E	4.7 MAX.
SHOULDER DIA.	EE	20 ±.13
*NOSE DIA.	G	7.9-7.8
NOSE RADIUS	L	3.95-3.90
*SM INDEX DIA.	MM	†
LENGTH	M	16 ±.13
*LG INDEX DIA.	NN	†
LENGTH	N	8 ±.13
HEAD LENGTH	P	20 ±.13
SHOULDER LGTH	GG	3 ±.13
STEP DIA.	JJ	17.0-16.76
SHANK DIA.	J	OPTIONAL
STEP LENGTH	ZZ	5-10
SHANK LENGTH	Z	OPTIONAL
RADIUS DIST.	AA	3.05-3.00

HEXAGON NUT

THREAD		M24 x 2-6H
MINOR DIA.		21.835-22.210
PITCH DIA.		22.701-22.925
MAJOR DIA.		24 MIN.
COUNTERSINK		45° x 24.5
HEXAGON		36
HOLE	R	17.6-17.8
CHAMFER	S	15°/30° x 28 DIA.
FULL THREAD	T	10 MIN.
BORE DEPTH	U	22.3-22.5
LENGTH	V	26 MIN.

*Body diameters A, BB and CC as well as nipple diameters G, MM and NN should be concentric within 0.05 Full Indicator Movement (FIM). These are critical dimensions for safety that must be adhered to on final product whether plated or not.
†See Fig. 9 for these dimensions.

Fig. 8.

STANDARD 1500 SERIES
METRIC DIAMETER-INDEX DIMENSIONS FOR
Medical Gases

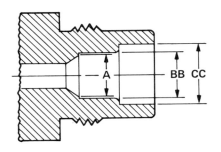
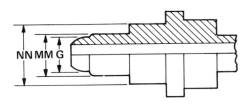

GAS NAME	GAS SYMBOL	CONN. NO.	DIAMETERS (MILLIMETERS)			
			*BB	*CC	*MM	*NN
†		1510	8.0–8.1	13.6–13.7	7.9–7.8	13.5–13.4
†		1520	8.4–8.5	13.2–13.3	8.3–8.2	13.1–13.0
†		1530	8.8–8.9	12.8–12.9	8.7–8.6	12.7–12.6
†		1540	9.2–9.3	12.4–12.5	9.1–9.0	12.3–12.2
†		1550	9.6–9.7	12.0–12.1	9.5–9.4	11.9–11.8
†		1560	10.0–10.1	11.6–11.7	9.9–9.8	11.5–11.4
Nitrous Oxide-Oxygen Mixtures (N_2O 47.5 to 52.5%)	N_2O–O_2 Mixture	1570	10.4–10.5	11.2–11.3	10.3–10.2	11.1–11.0
†		1580	10.8–10.9	10.8–10.9	10.7–10.6	10.7–10.6

* Body diameters A, BB, and CC as well as nipple diameters G, MM, and NN should be concentric within 0.05 Full Indicator Movement (FIM). These are critical dimensions for safety that must be adhered to on final product whether plated or not.

† Not assigned. SUPPLIERS AND USERS OF THE DIAMETER-INDEX SAFETY SYSTEM WHO REQUIRE A CONNECTION ASSIGN-MENT FOR A GAS OR GAS MIXTURE OR WHO NEED ASSISTANCE IN SELECTING THE PROPER CONNECTION, SHOULD CON-TACT THE COMPRESSED GAS ASSOCIATION, INC., 500 FIFTH AVENUE, NEW YORK, N.Y. 10036.

Fig. 9.

CHAPTER 8

How to Receive and Unload Liquefied Compressed Gases

Liquefied compressed gases commonly shipped to users in bulk by rail or highway are: anhydrous ammonia, liquefied petroleum gas (LP-gas), butadiene, chlorine, methyl chloride, methyl mercaptan, sulfur dioxide, vinyl chloride, and broadly, liquefied fluorinated hydrocarbons. Carbon dioxide, hydrogen chloride, nitrous oxide and vinyl fluoride are shipped as liquefied compressed gases at low temperatures. However, due to special considerations, these gases are not covered in this chapter.

All subject gases covered in this chapter are shipped in transport vehicles such as single-unit tank cars, multi-unit (TMU) tank cars and cargo tanks except that multi-unit tank cars are not authorized for butadiene. Safe unloading of single-unit tank cars, safe removal of containers from multi-unit tank cars, unloading of these containers at point of use and safe unloading of cargo tank trucks is a matter of applying known safety procedures which will be covered in some detail later.

Liquefied compressed gases can be safely handled and unloaded only:

(1) when their physical and chemical properties are understood;

(2) when equipment specifically designed for these physical and chemical properties are used;

(3) when regulations and standards governing their handling are complied with fully;

(4) when steps are taken to minimize accidents which are results of human failure.

Compressed Gas Association, Inc., which is and has been active for over 65 years in developing procedures for safe and efficient transportation, storage and handling of all compressed gases, as well as the related standards, takes the position that one must observe proper safety precautions at all times during any stage of the loading or unloading operations. Safe procedures are spelled out in regulations, standards, the literature of technical associations and in the instructions of compressed gas manufacturers.

DEFINITIONS

Cargo tank is defined by DOT an any tank permanently attached to or forming a part of any motor vehicle; or any compressed gas packaging not permanently attached but which by reason of its size, construction or attachment to a motor vehicle, is loaded or unloaded without being removed from the motor vehicle. The capacity of cargo tanks range up to approximately 10,800 gal (40.882 m^3). A *motor vehicle* includes a vehicle, machine, tractor, trailer, or semitrailer or any combination thereof, propelled or drawn by mechanical power and used on the highway. A *transport vehicle* means a motor vehicle or rail car used for the transportation of cargo by any mode.

PERSONNEL AND TRAINING

Start with careful selection of the personnel charged with the responsibility for loading or unloading. Choose reliable, intelligent people with a high sense of responsibility and arrange to train them thoroughly. Remember that employees may be promoted, transferred or may leave the company and therefore the training program must be continuous. Replacements must be thoroughly trained before assuming their duties. If possible include a medical officer, safety supervisor and fire protection engineer in the training program.

The compressed gas manufacturer is the logical source of technical data describing the physical and chemical properties of the compressed gas in question. He can provide knowledge concerning its characteristics, information about protective equipment, storage and handling procedures to take to minimize accidents and the corrective and first-aid steps in case of an accident. Make it a must for the trainees to study these data until they are completely versed in the contents. Some compressed gas manufacturers furnish qualified technical representatives or training aids to train personnel in every step of the loading and/or unloading procedure. Take advantage of this service where it is available.

The training program must impart a knowledge of all applicable rules and regulations, especially those found in Title 49 of the Code of Federal Regulations, Parts 107, 171, 172, 173, 174, 177, 178 and 179. Personnel should be made aware of the civil penalties for which they are liable. These are found in 49 CFR, Part 107. There may also be local, state or provincial regulations which govern. Pertinent hazardous materials regulations of the Department of Transportation are copied and republished in R. M. Graziano's Tariff. In Canada, the regulations of the Board of Transport Commis-

sioners for Canada (BTC) govern and these generally parallel the regulations of the DOT.

GENERAL PRECAUTIONS

Phone or wire the company personel responsible for maintaining the transport vehicle fleet; the compressed gas supplier or manufacturer; and/or the vehicle owner for assistance and instructions as needed if:

(1) A tank car is received in bad order. Meanwhile do not attempt to load it. A loaded car with a bad order card should be unloaded if it can be done safely. Check with safety personnel in the absence of an established company procedure. Empty bad order cars should be moved to repair facilities.

(2) There are failures of fittings or a leak that cannot be readily repaired by simple adjustment or tightening of the fitting. Isolate the transport vehicle and permit only properly instructed and protected personnel to enter the area.

(3) The transport vehicle cannot be unloaded after following all instructions.

(4) There is an accident of any kind.

Do not attempt to correct leaks at unions or other fittings in a line with a wrench while the line is under pressure. Never break a hose coupling under pressure.

Loading and unloading personnel should wear protective gloves to prevent contact with liquids. On release at normal temperatures and atmospheric pressure the liquids vaporize rapidly, and in contact with skin surfaces this evaporation causes severe burns (frostbite). When large quantities of the liquid are involved, deep and severe freezing of the area in contact with the liquid results. Large-lensed safety spectacles or goggles should be worn to prevent liquid from contacting the eyes. Gas masks, protective clothing, fire extinguishers and related equipment of approved design and in good condition should be stored in areas that are accessible in case of a leak. Personnel should be thoroughly instructed in their use and continuously trained by occasional "dry runs."

Frequently inspect loading and unloading hoses, coupling and flexible connectors for wear, deterioration and abuse.

TANK CARS

Details about compressed gas shipping containers are found in Chapter 5. Single- unit tank cars for transporting flammable compressed gases may be of the Class 105, 112 or 114 type and after December 31, 1980 the Class 112 and 114 tank cars must be equipped with thermal protection and tank head puncture resistance systems such as head shields (see 179.105 of the DOT Regulations). Single-unit tank cars for transporting anhydrous ammonia may be of the Class 105, 112 or 114 type and after December 31, 1979 the Class 112 and 114 tank cars must have head shields. These shields can be items separate from the tank, located outside each head, or they can be integral with the jacket head in the case of a jacketed tank. The head shield is made part of the jacket by using $\frac{1}{2}$ in. (1.27 cm) thick material for at least the lower half of the jacket head.

The DOT Class 105 specification is being reviewed by DOT and the AAR Tank Car Committee and some changes are imminent. The DOT is also studying tank car packaging requirements on a product by product basis. Therefore, CGA strongly recommends that shippers refer to the latest DOT regulations for information on proper shipping containers. The latest issue of the AAR Specifications for Tank Cars should also be reviewed because sometimes the AAR requirements for tank cars are more stringent than the DOT Regulations.

Tank cars are never filled liquid full but have a vapor space, termed outage, the volume of which is calculated for each specific gas and temperature condition.

Arrangement of most compressed gas tank car valves and fittings is similar (Fig. 1). These are located on a manhole cover plate usually on top of the car, within a protective housing. These are sometimes referred to as "dome fittings"

A   Liquid eduction valve

B   Vapor valve

C   Safety relief valve

D   Gaging device
1. Gaging pointer
2. Gage rod lock
3. Gage rod valve
4. Gage rod
5. Gage rod brake
6. Packing gland nut
7. Protective housing
8. Gasket
9. Gaging rod shield vent holes
10. Lubricator assembly

E   Sample valve

F   Thermometer well

G   Excess flow valves

H   Liquid eduction pipe

I   Sample line

J   Screen

K   4-in. insulation

L   Liquid level

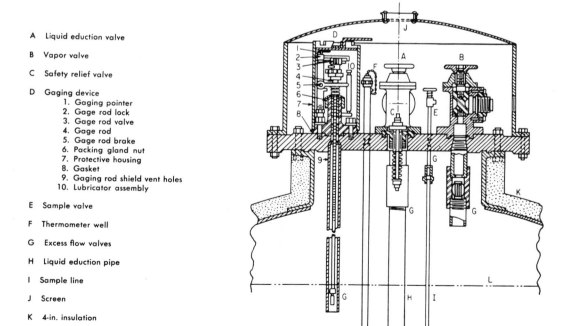

Fig. 1. Tank cars for liquefied compressed gases have identical valve arrangement.

although the current design tank has no dome as such. A few tank cars built to the 114 specification have been equipped with bottom outlets. Since they are few in number and are found in special service, we will confine our discussion to tank cars with top fittings.

Usually there are two valves in line with the track which are connected to liquid eduction pipes that extend to the bottom of the cars. These are referred to as "liquid valves" or "eduction valves." One valve is mounted toward the side of the car, terminating in the tank's vapor space. Sometimes an additional valve is located on the opposite side. Such valves are referred to as "vapor valves."

Additional cover plate fittings may include a liquid level gage, thermometer well, sample line and safety relief valve. Excess flow check valves are installed on the liquid eduction lines of most single-unit tank cars handling compressed gases. These excess flow valves are intended only to contain the tank contents in transit, or in a wreck environment in event of valve wipeoff.

Many unloading systems have hose or piping smaller than the tank car piping system and are not capable of the flow rates required to close the tank car excess flow valves. To prevent loss of product in the event of a hose break in the unloading system an additional, smaller, excess flow valve should be a part of the line connected to the tank car and located on the tank car side of the hose. (*Caution:* Check state and local regulations to be certain an excess flow valve is considered proper protection.) The design closing flow of this excess flow valve should be about 50 percent greater than the line flow during unloading. The unloading line should never be a pipe size smaller than that of the excess flow valve installed in the line. An excess flow valve in a loading line is not practical because it would close on excess flow into the tank car but would not protect against loss of vapor out of the tank car. Additional protection can be obtained by installing a backflow check valve in the loading line at the tank car end of any hose or swivel-type piping.

Another method of preventing loss of contents from the tank car is to install an emergency shutoff valve in the loading or unloading line near the tank car valve. In fact, NFPA Standard 58, Paragraph 4055 states: "When a hose or swivel-type piping is used for loading or unloading railroad tank cars, an emergency shutoff valve complying with 2343 shall be used at the tank car end of the hose or swivel-type piping." Paragraph 2343 requires the valve to have automatic shutoff through thermal actuation, manual shutoff from a remote location and manual shutoff at the installed location. NFPA 58 does not apply to marine terminals, pipeline terminals, natural gas processing plants, refineries, tank farms, utility gas plants or chemical plants.

Occasionally the operator will open the tank car liquid valves too quickly and the excess flow valves will slam shut. Never hammer tank car fittings in an attempt to open the check valves. To open a closed excess flow valve, close the tank car liquid angle valves and the pressure will equalize and the check valve will reopen with an audible click. The tank car liquid angle valves can then be opened slowly without further premature closing of the excess flow valve.

Single-unit tank cars are protected from excessive internal pressure by a spring-loaded pressure-relief valve located on the manhole cover plate. The valve setting depends upon the specification car used. Loading and unloading procedures must be such that the pressure developed does not cause the pressure-relief valve to open.

INITIAL PRECAUTIONS AT RAIL LOADING AND UNLOADING RACKS

Rail track at loading and unloading spots should be essentially level so the cars can be properly gaged and completely unloaded. Brakes must be set and wheels blocked on all cars. Caution signs as prescribed in DOT Regulations, Part 174, must be so placed on the track or cars to give necessary warning to persons approaching the cars from the open end(s) of the siding and must be left up until after the liquid transfer is complete and the cars' loading or unloading lines are disconnected. Once caution signs are posted do not allow the tank cars so protected to be coupled or moved. Do not permit other cars to be placed on the same track except after notifying the person who placed the signs, and provided there is a standard derail properly set and locked in the derailing position between the cars being protected and the cars to be set on the same track.

Training aids for instructing personnel about pretrip inspections are available from CGA. Before connecting the loading lines, the inspection should include but is not limited to the following:

(1) Is the tank car authorized for the product to be shipped?

(2) If the car has a product label or marking different from the product to be loaded, make certain the product residue in the tank is compatible with the product to be loaded. Also change markings to correspond with new product.

(3) Check condition of undercarriage.

(4) Check safety appliances (handrails, grab irons, etc.).

(5) Check condition of tank paint and stenciling, marking or labeling.

(6) Check condition of manway bolts and gaskets. Gasket material should be compatible with products and gaskets should be maintained or replaced regularly.

(7) Check the test due date for tank and safety valve. Do not load if either test is overdue or if the pressure-relief valve is leaking or inoperable.

(8) Check condition of internal and external valves and fittings.

(9) Remove steam coil closures and bottom outlet valve closures during loading. If leakage occurs at these fittings during loading, the loading must stop and repairs made or the car must be shopped for repairs.

(10) Check condition of tank.

After disconnecting the loading lines, the inspection should continue and include the following:

(1) If a product marking is not required on the tank, product markings which do not refer to the loaded product should be removed.

(2) If a product marking is required, the tank should be marked with the name of the loaded product in accordance with Appendix C of the AAR Specifications for Tank Cars, and DOT Regulations, Part 172, Subpart D.

(3) All valves should be closed and tightened and not leaking. It is recommended that valve handles be removed or adequately secured before shipment if it is not covered by a protective housing.

(4) Caps or plugs on sample valves, loading and unloading valves and gaging device valves should be installed and wrench tightened. Gaging device must be secure in place and not leaking.

(5) Closures for bottom outlet valves and heater coil connections should be installed and tightened.

(6) The protective housing cover must be secure, pinned and proper seals should be in place.

(7) Tank cars should be placarded in accordance with DOT Regulations, Part 172, Subpart F.

Remember that it is illegal to ship a defective or leaking tank car. If a car cannot be put in suitable shipping condition, notify the proper company personnel and request instructions. In some cases, the car owner is the proper person to contact for assistance.

UNLOADING TANK CARS

With single-unit tank cars and multi-unit tank cars, full responsibility for unloading them safely rests with the consignee. Proper equipment and well-trained personnel are a must at unloading spots.

Transfer lines and their flexible hoses must be of materials suitable for the product being unloaded, as recommended by the compressed gas manufacturer or industry standards. When not in use, plug or cap the ends of the lines to prevent accumulation of moisture and dirt. Use only Class 1, Group D (explosion proof) lights, switches, motors and other electrical appliances in the vicinity of racks where flammable gas is being unloaded. In the case of flammable gases, establish a firm "no smoking" rule for all personnel in the unloading rack area. Check the area for all open flames and extinguish them before starting operations.

In preparation for unloading, break the seal on the tank car protective housing cover and other seals as required. Make the car ready for unloading by carefully and cautiously removing the plugs from the car's liquid and vapor valves. If gaging the tank car is necessary and a slip tube gage is provided, do not place your head or body directly over the tank car gaging device when releasing the hold down latch. Tank pressure may force the slip tube up rapidly and with considerable force. Never place your head or face over the relief-valve opening.

After connecting the unloading lines and, when all connections are tight, pressurize the system and check for any leaks. When the system is ready for unloading, remember that throughout the entire period of unloading and while the car is connected to the unloading device the car must be monitored by the unloader. Tank cars should not be allowed to stand with unloading connections attached after the unloading is completed.

Different methods of unloading may require different valve types and related equipment. In all cases the liquid unloading line should be equipped with a pressure gage and shutoff valve. As mentioned earlier, the unloading lines should be equipped with properly sized excess flow valves or emergency shutoff valves to protect against loss of product in the event of hose failure.

Liquefied compressed gases are usually transferred from single-unit tank cars by pressure differential. All equipment must be designed for the particular gas being transferred. Methods of obtaining the pressure differential required and the gases to which the method may be applied are:

(1) By compressor: anhydrous ammonia, LP-gas, butadiene, methyl chloride, vinyl chloride, sulfur dioxide, and liquefied fluorinated hydrocarbons.
(2) By vaporizer: butane.
(3) By gas repressuring: LP-gas, butadiene, vinyl chloride.
(4) By direct-acting liquid pump: LP-gas, anhydrous ammonia, vinyl chloride.
(5) By air padding: chlorine and sulfur dioxide.

UNLOADING WITH A COMPRESSOR

A typical compressor unloading setup is shown in Fig. 2. The suction side of the compressor is connected to the vapor line of the storage tank

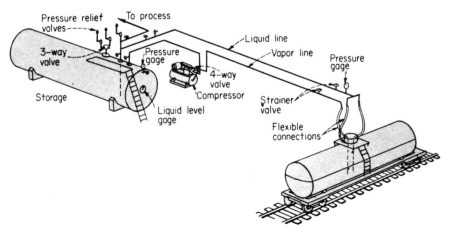

Fig. 2. Using a compressor to unload a single-unit tank car.

and the discharge side is connected to the vapor valve on the tank car. (*Note:* Oil-lubricated compressors should be equipped with an oil mist extractor on the compressor discharge to prevent contamination.) One or both liquid eduction lines on the tank car are connected to the storage tank with appropriate piping. The compressor withdraws and compresses vapor from the storage tank and transfers it to the tank car's vapor space. In this manner, the desired pressure differential is created to transfer liquid to storage from the tank car.

To use the compressor unloading system after the piping hookup is completed, open both liquid eduction valves on the tank car slowly but completely. Then open all other valves in the liquid line, working from the tank car to the storage tank. Open the storage tank filling valves slowly. If the tank car pressure is higher than that of the storage tank, do not open the valves in the vapor line or operate the compressor. When the rate of liquid flow drops to an unsatisfactory rate, open the vapor valves between the tank car and the storage tank, then start the compressor.

The sight glass, when one is installed in the liquid discharge line, pressure gages in the transfer lines and the sample line on the tank car can be used to determine when all the liquid is removed from the tank car. When this is established, prepare the car for disconnecting (with some exceptions such as chlorine) by closing down liquid transfer lines first. Start at the storage tank and close all valves up to and including the tank car's liquid valve.

To withdraw vapors from the tank car after it is empty of liquid, shut down the compressor, close the liquid valves starting at the storage tank and working to the tank car; then turn the compressor's reversing valve. Start the compressor and withdraw most of the vapors in the tank car, returning them to storage. The amount of vapors withdrawn is largely based on the operating costs of the compressor versus value of the vapors or safety considerations.

In shutting down the vapor transfer lines, close the tank car's vapor valve first, and then successively close the other valves by working from the tank car to the source of pressure.

After bleeding off all transfer lines to atmospheric pressure, disconnect from the tank car.

UNLOADING WITH VAPORIZER

Liquid from storage is charged into a vaporizer sometimes called an evaporator and its vapor section is connected to one or both vapor valves on the tank car. When heat is turned into the vaporizer, the vapor pressure generated exerts pressure on top of the liquid in the tank car causing it to flow to storage. By this method, all of the liquid can be transferred to storage but the vapor remains in the car. A steam vaporizer method of tank car unloading is illustrated in Fig. 3.

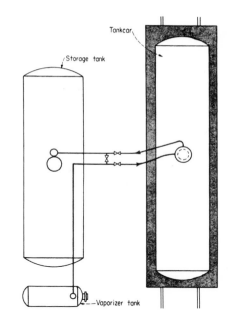

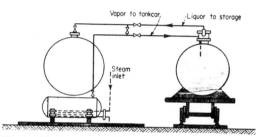

Fig. 3. Unloading a single-unit tank car by steam vaporizer.

UNLOADING WITH A PUMP

A tank car can be unloaded with a liquid pump but it is impossible to remove any of the vapors in the tank car. For this reason, the compressor is almost always used. The basic procedure is the same as that followed when using the compressor except that the pump is connected in the liquid line from the tank car to the storage tank. A vapor equalizing line is connected to each tank. The vapor line between the storage tank and the tank car must be large enough so that the tank car pressure never drops appreciably below that of the storage tank. Otherwise, the pump may vapor lock and unloading will be difficult if not impossible.

If at the start of unloading the tank car pressure is above the storage tank pressure, do not open the valves in the vapor line. The higher tank car pressure will help the pump and unloading will be faster. Open the vapor valves only when the storage tank pressure is equal to, or higher, than the tank car pressure.

UNLOADING BY GAS PRESSURE

This method or any method of unloading that leaves the car filled with noncondensible gas should not be used without permission of the tank car owner or lessee. The gas contaminates the next load and makes loading the car very difficult and may cause premature operation of the pressure-relief valve. Therefore, all noncondensible gases will have to be vented before the cars are reloaded. The loading rack must be equipped to handle this situation. If gas pressuring must be used to unload a tank car, the gas used must be clean, dry and noncorrosive. For some products an inert gas is required. A dryer may be needed to remove moisture from the gas. The pressure supplied must not be high enough to actuate the tank car pressure-relief valve.

To unload, connect the liquid hoses to the tank car liquid eduction valves and storage tank manifold lines. Connect the pressuring hose to the tank car vapor valve and to the gas supply line. This line must have an approved check valve in it to prevent the flow of gas from the tank car into the gas supply system. Open both tank car liquid eduction valves slowly but completely. Then open all the other valves in the liquid line, working from the tank car to the storage tank. Open the storage tank filling valve slowly. Be careful not to open this valve too far if the tank car pressure is above the storage tank pressure or the tank car excess flow valves may close. If the tank car pressure is higher than the storage tank pressure, do not open the tank car vapor valve or the valves in the pressuring gas supply line. When the liquid flow drops to an unsatisfactory rate with the storage tank fill valve wide open, open the pressuring gas supply valve and then slowly open the tank car vapor valve. Maintain the tank car pressure 5 to 10 lb above the storage tank pressure by controlling the pressuring gas supply to the tank car.

A flow of gas instead of liquid through the sight flow glass in the unloading line indicates that the tank car is empty. This should be checked by opening the tank car sample valve. When the tank car is empty, shut off the repressuring gas supply, then close all valves in the liquid line working from the storage tank to the tank car. Close all the valves on the tank car. After bleeding the pressure from all the hoses, disconnect the hoses and replace the plugs in the tank car valves.

UNLOADING WITH AIR PADDING

When air padding is used it is imperative that clean, oil-free, cooled, dry compressed air be introduced into the tank car's vapor space through its vapor valve to transfer the car's liquid. The setup consists of an air compressor and its tank equipped with a pressure regulator, pressure-relief filter and an air dryer with appropriate valves and gages similar to those as shown in Fig. 4. A dependable check valve must be incorporated in the air-padding line as also shown. Never use a plant air system for air padding since the possibility exists that vapors can be drawn into the plant air system.

HANDLING MULTI-UNIT TANK CARS

Multi-unit tank car tanks are commonly known as "TMU" tanks, "ton tanks" or "ton

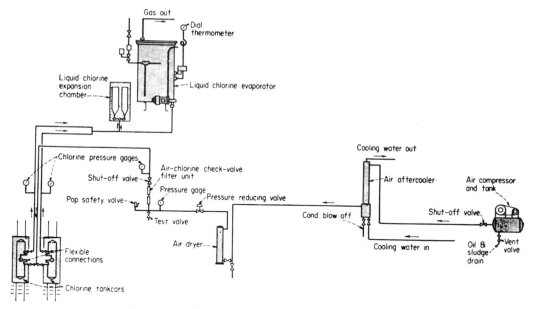

Fig. 4. Unloading two single-unit tank cars by air padding.

containers." This name comes from their capacity in terms of chlorine. They are shipped on a car underframe (shipments usually consisting of 15 containers). Fewer than 15 units may be shipped as a TMU car but this arrangement is impractical since freight charge is figured for the full 15 containers at prevailing carload rates.

Before removing containers from the car frame, it is necessary to observe the same DOT Regulations concerning personnel, setting brakes, blocking wheels of the car and posting caution placards as in unloading single-unit cars. Full containers are removed from the underframe by a lifting hook in combination with a hoist on a trolley or a jib boom. Never use a lifting magnet, or a rope or chain sling, to unload or otherwise handle these containers. Containers must be protected from shock which might damage valves, fusible plugs or the container itself. Keep valve protecting hoods in place at all times when containers are being moved. When containers are trucked from the rail siding into the storage area, they must be placed on saddles on the truck. It is preferable to clamp them down to prevent shifting and rolling.

Each container has two valves, each equipped with a $\frac{1}{2}$ in. eduction pipe. When the container

is positioned horizontally with a slight downward pitch and one valve is directly above the other, the top eduction pipe ends in the vapor phase and the eduction pipe of the bottom valve in the liquid phase. Either liquid or vapor can be withdrawn by connecting to the appropriate valve. These two valves are protected by hoods. Keep hoods in place at all times except when the container is connected for withdrawing its content.

Some containers may be protected by fusible plugs in each head. In some cases, the fusible metal is designed to soften or melt at temperatures as low as 157 F (69.4 C). Do not tamper with the fusible plugs under any circumstances.

DOT Regulations provide that multi-unit containers may be transported under certain conditions on trucks or semitrailers. They must be chocked or clamped on the truck to prevent shifting. You must have adequate facilities available when transfer intransit is necessary. TMU containers may be removed from the car underframe when the car is spotted on carrier tracks provided the DOT Regulations are complied with. The carrier may give permission for the unloading only if a private siding is not available within a reasonable trucking distance of the final destination and the consignee must furnish an adequately strong mechanical hoist to lift containers from the car and deposit them directly on vehicles furnished by the consignee. The containers are then transported by truck, as previously described, to the user's plant where similar handling equipment must be available. Cautions concerning the handling of these containers as earlier described must be strictly observed.

Ton containers should be stored on the user's premises in a cool dry place and protected against heat sources. A convenient storage rack can be made by supporting the containers at each end on a railroad rail or an I-beam. When containers are not being used the valve protective hoods should be kept in place at all times. Store full and empty containers in different places to avoid confusion in handling. It is good practice to tag empties. Do not store containers near elevators or gangways or in locations where heavy objects may fall and strike them. Never store containers near combustible or flammable materials. When containers hold flammable gases, keep them away from all ignition sources. Keep the storage room well ventilated and so arranged that any container can be removed with a minimum of handling of other containers. When practical the storage room should be fireproof. Avoid storage in subsurface locations. Make certain containers stored outdoors are free of debris, tall grass and away from public access. They should be kept clean and inspected regularly.

Containers storing a flammable gas or a gas that affects the respiratory system should not be placed where fumes can enter a ventilating system or where wind can carry fumes to populated areas.

Except when recommended by the compressed gas manufacturer, TMU containers should not be manifolded to withdraw contents from two or more containers simultaneously. To discharge the contents of a container, place it so that it is in a nearly horizontal position with a slight downward pitch (about 1-in. or 2.54 cm overall) toward the valves. Assembly for withdrawing a container's content usually consists of a transfer line equipped with a pressure-reducing valve, rotometer, control valve and diffuser. In chlorine service the system may include only pressure and control valves.

For all subject gases, test TMU container connections and transfer piping for leaks as later described under handling for each specific gas. Sometimes it is necessary to use heat to help the flow of either gas or liquid from sulfur dioxide or a liquefied fluorinated hydrocarbon ton container. When heat is used, the method approved by the compressed gas manufacturer must be followed. Refer to CGA Pamphlet G-3 for handling sulfur dioxide. Exercise great care as fusible plugs in ton containers may melt. Fusible plugs must be vapor tight at not less than 130 F (54.5 C), therefore, we recommend never allowing containers to reach a temperature above 125 F (51.7 C). Never apply blow torches or steam hoses or use an open flame from any source to heat containers.

CARGO TANKS

DOT requirements for carriage by public highway are found in 49 CFR, Part 177. In part, they provide that while loading or unloading a flammable gas there must be no smoking, lighting matches or carrying any flame or lighted cigar, pipe or cigarette. No hazardous material shall be loaded or unloaded unless the handbrake is securely set and other reasonable precautions taken to prevent motion of the vehicle. The truck should be parked so that the cargo tank is level for loading or gaging. For unloading, park so that the outlet opening is in the lowest part of the tank.

No flammable gas shall be loaded or unloaded with the motor vehicle engine running unless the engine is used for the operation of the transfer pump of the vehicle. Unless the delivery hose is equipped with a shutoff valve at its discharge end, the engine of the motor vehicle shall be stopped at the finish of the unloading operation while the discharge connection is disconnected.

No tank motor vehicle shall be moved, coupled or uncoupled when the loading connections are attached to the vehicle, nor shall any semitrailer or trailer be left without the power unit unless chocked or equivalent means are provided to prevent motion.

Most liquefied compressed gases are, or may be, shipped in cargo tanks conforming to DOT Specifications MC-330 or MC-331 (see CFR, Title 49, Section 173.315). The transport vehicle must be marked and placarded in accordance with DOT Regulations, Part 172. Chlorine cargo tanks may be shipped only if the contents are to be unloaded at one point. MC-330 and MC-331 cargo tanks constructed of quenched and tempered steel can be used to transport anhydrous ammonia if it has minimum water content of 0.2 percent by weight, and to transport LP-gases determined to be noncorrosive.

For a product being loaded, the vapor pressure (psig) at 115 F (46.1 C) must not exceed the design pressure of the cargo tank. Also, cargo tanks must not be filled liquid full as space must be provided for the liquid to expand as the temperature rises. Permissible filling density for each product is found in 49 CFR, Part 173. Cargo tanks may be filled by weight or volume unless otherwise specified for a particular product. Highway load limits must also be observed.

The actual loading operation will vary depending on the design of the cargo tank and the method of measurement used at that particular loading point. To load properly, the loader must usually know the specific gravity at the temperature of the liquid when loaded. Always be sure that the product last shipped is compatible with the product to be shipped unless the tank has been cleaned. For example, do not load LP-gas into a tank which previously contained anhydrous ammonia.

During the unloading of a cargo tank, the motor carrier who transports hazardous materials must ensure that a qualified person is in attendance at all times. However, the carrier's obligation to monitor unloading ceases when:

(1) the carrier's obligation for transporting the materials is fulfilled;
(2) the cargo tank has been placed on the consignee's premises; and

(3) the motive power has been removed from the cargo tank and removed from the premises.

When the carrier is not responsible, the consignee is responsible for the safe transfer of the product and must have a qualified person in attendance at all times. Definition of "attends" and "qualified person" can be found in 49 CFR, Part 177, Subpart B, Section 177.834 (i) (2).

Where responsibility for unloading cargo tank trucks rests with the common carrier or transport owner, his operator should be trained in all phases of the operation. He should be thoroughly familiar with the regulations, both federal and local.

Anhydrous ammonia is unloaded from cargo tanks by a compressor or a liquid pump; LP-gas by a compressor or liquid pump; and methyl chloride and sulfur dioxide by a compressor. Liquefied fluorinated hydrocarbons are usually unloaded from the cargo tank trucks by a turbine pump, although a compressor may be used. DOT Regulations for unloading cargo tank trucks and the precautions stated are to be observed. Procedures are generally those described in unloading single-unit tank cars. Even where responsibility for unloading cargo tank trucks rests with the carrier or transport owner, the consignee should be as familiar with the entire unloading procedure as is the transport's operator in charge of transfer.

RETURN OF EMPTY TRANSPORT EQUIPMENT

Consignee Becomes Shipper

After the unloading operation is completed, the consignee must prepare the transport vehicle for return to the supplier. Thus, the consignee now becomes the shipper.

After unloading a single-unit tank car and valves are closed with plugs or caps in place, close the protective housing cover and latch it in place. Valves on TMU containers should be closed and the protecting hoods in place when ready for return. Load the containers on the multi-unit tank car and secure them in place. The placards must be removed from the empty tank car and replaced or reversed to show an "empty" placard in accordance with DOT Regulations, 49 CFR, Part 172. The return billing paper must show "Empty," or "Empty: Last contained," followed by the name of the material last contained in the tank, the hazard class of the material and the word, "Placarded."

After a cargo tank is emptied of a hazardous material, it must remain placarded for the returned shipment unless it is:

(1) reloaded with a material not subject to DOT Regulation; or
(2) sufficiently cleaned and purged of vapors to remove any potential hazard.

The return shipping paper for a cargo tank containing the residue of a hazardous material may contain the words, "Empty" or "Empty: Last contained," followed by the name of the material.

PROBLEMS WITH SPECIFIC GASES

Butadiene Inhibited

Butadiene is a flammable gas which must be inhibited when offered for transportation because of polymerization hazards. In high concentration it is an anesthetic that can cause respiratory paralysis and death. Lower concentrations may produce slight irritation of eyes, nose and throat. Butadiene is mildly aromatic and blurring of vision and nausea are characteristic symptoms of exposure. The liquid will cause a freezing burn of the skin. Refer to the Butadiene Chemical Safety Data Sheet SD-55 issued by the Chemical Manufacturers Association, Inc., for recommended personal protective equipment and respiratory protection.

Large butadiene fires are difficult to extinguish. Control the fire by shutting off the source of the fuel and keep all adjacent occupancies wetted with water spray. Small fires can be extinguished with CO_2 or dry chemical first-aid extinguishers.

Air pressure should never be used for unloading. Butadiene should be unloaded through a closed system using a vapor return line and compressor or inert gas (such as nitrogen) pressure. When all the liquid is removed from the transport vehicle, valves should be closed and no air permitted to enter the tank. Butadiene transport vehicles should not be used for any other product unless the tank has been cleaned, gas freed and purged of air.

Vinyl Chloride

Vinyl chloride monomer is a flammable gas which is easily ignited producing hazardous combustion gases largely composed of hydrogen chloride and carbon monoxide. The odor of vinyl chloride vapors is pleasant and when inhaled acts as an anesthetic. Full details on the chemical and physiological properties, personal protective equipment and respiratory protection recommendations can be found in the Vinyl Chloride Chemical Safety Data Sheet SD-56 issued by the Chemical Manufacturers Association, Inc.

An explosion hazard can exist when drawing samples or venting to the atmosphere. Fire involving large quantities of liquid are difficult to extinguish since vinyl chloride is not miscible with water and is lighter than water. Most small fires can be extinguished with carbon dioxide or dry chemical agents if properly applied. Vinyl chloride can be unloaded by pump, compressor or inert gas pressure. Air must not be permitted to enter the tank.

Anhydrous Ammonia

Anhydrous ammonia is classified by the DOT as a nonflammable gas.

In transferring ammonia from containers, including single-unit tank cars, TMU containers and cargo tank trucks, never use compressed air as it will contaminate ammonia.

The continuous presence of the sharp, irritating odor of ammonia is evidence of a leak. Leaks can be located in ammonia by allowing fumes from an open bottle of hydrochloric acid (from a squeeze bottle of sulfuric acid or from a sulfur dioxide aerosol container) to come in contact with ammonia vapor. This produces a dense fog. Leaks may also be detected with moist phenolphthalein or litmus paper. Sulfur tapers for detecting ammonia leaks are not recommended.

When there is a leak around an ammonia container valve stem, it usually can be corrected by tightening the packing gland nut which has a left-hand thread.

When a leak occurs in a congested area where atmospheric dissipation is not feasible, absorb the ammonia in water. Its high solubility in water may be utilized to control escape of ammonia vapor. Applying a large volume of water from a fog or spray nozzle lessens vaporization, as the vapor pressure of ammonia in water is much less than that of liquid ammonia.

Do not neutralize liquid ammonia with acid— heat generated by the reaction may increase the fumes.

Only an authorized person should attempt to stop a leak, and if there is any question as to the seriousness of the leak, a gas mask of the type approved by the U.S. Bureau of Mines for use with ammonia must be worn. Have all persons not equipped with such masks leave the affected area until the leak is stopped.

Also, provide personnel subject to exposure to ammonia with a hat, gloves, suit and boots, all garments of rubber. Garments worn beneath rubber outer clothing should be of cotton. Some protection to the skin may be obtained by applying protecting oils before exposure to ammonia. Supply approved eye goggles if the eyes are not protected by a full face mask.

Although ammonia is flammable only within the narrow limits of 16 to 25 percent by volume, the mixture of oil with ammonia broadens this range. Therefore, take every precaution to keep sources of flame or sparks from areas that have ammonia storage or use.

In the event a fire does break out in an area containing ammonia, make every effort to remove portable containers from the premises. If they cannot be removed, inform the firemen of their location.

For data concerning the physiological effects of ammonia, protective equipment and first-aid measures, obtain a copy of "Pamphlet G-2, Anhydrous Ammonia" published by Compressed Gas Association, Inc.

LP-Gases

All LP-gases are classified by the DOT as flammable gases and must be treated as such. Obtain a copy of the National Fire Protection Association Pamphlet 58, "Standard for the Storage and Handling of Liquefied Petroleum Gases."

Liquid LP-gas leaks in transfer piping are indicated by frost at the point of leakage due to the low boiling point of the material.

Extremely small leaks and leaks in vapor transfer piping can be detected by applying soap suds or a similar material to the suspected area.

Under no circumstances should a flame be used to detect a leak.

The most important safety considerations in unloading LP-gas are: avoid unnecessary release of the product; keep any open flames and other sources of ignition away from the unloading area; and make sure suitable first-aid type fire extinguishers are available (dry chemical or carbon dioxide types are suitable for LP-gas fires).

It is important for personnel to be familiar with the characteristics of LP-gas. You must realize the importance of not extinguishing a fire unless by doing so you stop the source of the leakage (as by closing a valve). This is so because if you extinguish the fire and leakage is allowed to continue, unburned vapor could accumulate and possibly result in a more serious hazard than if you allowed the escaping gas to burn.

If a fire is in progress and an LP-gas storage tank is exposed, it is of prime importance to keep the container cool by applying hose streams of water until the fire is properly extinguished. By the same token, if fire threatens a single-unit tank car, a TMU car or a cargo transport truck, immediately remove these from the area. If this is not possible, apply hose streams of water, as in protecting storage tanks.

Chlorine

In both its liquid and gaseous form, chlorine is neither flammable nor explosive. It is classified as a nonflammable gas by the DOT. Its principal hazard arises from inhalation from

Tank truck for LP-gas service, above, contains its own unloading equipment.

leaks. For data describing its physiological effects, handling, employee training and protection, and its chemical characteristics and physical properties, obtain a copy of The Chlorine Institute's manual.

Also, consult with the chlorine manufacturer concerning these matters, as well as the equipment required for handling chlorine, specifications and maintenance of the equipment and the special safety precautions and safety equipment required.

All chlorine containers, including single-unit tank cars, are intended for use by the customer to deliver chlorine direct to process. Thus, differing from the other liquefied compressed gases, there is no need for intermediate storage tanks between the shipping container and the process. Moreover, the use of chlorine storage tanks between the container and process is frequently hazardous.

Usually, chlorine tank cars are filled at low temperature and pressure. Normally, the inherent pressure of the vapor in the tank car is sufficient to accomplish withdrawal of liquid chlorine to process, but sometimes, especially during the winter, the car is air padded to accomplish liquid withdrawal.

In Fig. 4 a typical arrangement for unloading liquid chlorine from tank cars is shown. Liquid flow is through the tank car's liquid eduction valve to a liquid evaporator within battery limits, then direct to process. Whenever liquid chlorine can be trapped between two valves, the line must be protected by a heated expansion chamber.

When air padding is required, introduce only clean, oil-free, cooled, dry compressed air into the tank car through its vapor valve.

Absorbing Chlorine in a Liquid. If chlorine is to be absorbed in a liquid, there is a tendency for the liquid to suck back into the container when the container becomes empty due to the creation of a partial vacuum. Avoid this as it has resulted in numerous accidents.

As soon as the container is empty and its pressure has dropped to zero, shut the container valve. Then, vent air into the line leading from the container, after the valve has been shut off to prevent liquid from "sucking back" into the line. For this purpose, install a "vacuum break" valve or loop on the chlorinator well line.

Rules for Unloading. In general, rules for unloading all chlorine containers apply to the tank car operation. These include:

Safety devices must never be tampered with.

Open container valves slowly.

Make sure that threaded connections are the same as those on the container valve outlets. Never force connections that do not fit. The outlet threads on valves of TMU containers are not tapered pipe threads.

Containers or valves should never be altered or repaired by unauthorized personnel.

Gas leaks around the valve stem may usually be checked by tightening the packing nut.

Test for chlorine leaks by attaching a cloth to one end of a stick, soak the cloth with strong ammonia water, and apply to the suspected area. A white cloud of ammonium chloride results if there is any chlorine gas leakage. Do not use the usual household ammonia—it is not strong enough.

Use only reducing valves and gages designed for chlorine. Consult the chlorine producer for details.

In addition to complying with DOT Regulations governing unloading tank cars, the following special rules are recommended where chlorine is being handled in tank car quantities:

Switch-tracks on which chlorine tank cars are placed for unloading should be devoted solely to this purpose.

When chlorine tank cars are located on a dead-end siding, protect the cars on the switch end by a locked derail; if on an open-end siding, at both ends. Keys should be in the charge of a designated responsible person.

Unloading single-unit chlorine tank cars should be done through flexible metal connections that compensate for the vertical rise of the tank car as its springs decompress during unloading. For all details relating to the unloading line, consult the chlorine producer. However, do not depend upon flexible connections as a safety factor in case of a bump by switching operations during unloading.

Place derails as stated previously and use a blue lantern, suitably placed, if unloading at night.

Perform unloading, connecting and disconnecting operations only in well-lighted places by reliable persons who are properly instructed, responsible for operations and equipped with chlorine gas masks.

In shutting off the flow of liquid chlorine, be careful not to leave the line full of liquid chlorine with valves closed at both ends. Otherwise, the liquid remaining in the line may warm up and the resulting expansion of the liquid burst the pipeline by the hydrostatic pressure thus developed.

Close the tank car valve first and then allow the discharge line to empty by continuing to use the chlorine remaining in the line. After this, close both ends of the line.

Never apply heat directly to tank cars. It may, however, be desirable in some locations subject to low winter temperatures, to unload the tank cars in a shed maintained at 70 to 75 F (21.1 to 23.9 C).

In certain cases, particularly in winter weather with its low outside temperature, the pressure in single-unit cars may not be adequate and it is then necessary to increase it by adding dried compressed air to secure the desired rate of discharge.

Use only thoroughly dried clean air for this purpose. Have clean air supplied by a separate air compressor used only for this service. Never connect an existing compressed air system to the chlorine tank car.

The air dryer must be of suitable capacity to supply the full requirements. It must be maintained in proper operating condition. Consult the chlorine producer for all details relating to the application of dry air to single-unit tank cars. A specification for adding dry air to single-unit tank cars can also be obtained from The Chlorine Institute, Inc.

If chlorine is to be used at an elevation substantially above that of the tank car, high pressure is required in the car to force the liquid to that higher elevation. It is much safer to locate a chlorine evaporator at the elevation of the tank car and pipe the gaseous chlorine to the higher elevation.

With single-unit tank cars used for all other liquefied compressed gases, the content is usually withdrawn to storage. It is desirable to unload them as quickly as possible. However, the rate of withdrawal of liquid chlorine from tank cars is governed by the requirements of the process, and withdrawal may not be continuous. Accordingly, it may be desirable to unload through only one liquid valve. Open this valve, as is the case when two are used, slowly but completely.

When air padding is necessary, connection of the compressed air line is made to one or both vapor valves. When a tank car has been "padded" with air, the pressure may become excessively high if the car is allowed to warm up when no chlorine is being withdrawn. Consequently, check the tank car pressure periodically during prolonged shutdowns so that excessive pressure may be vented carefully into caustic soda solution or into milk of lime in a large tank before the safety valve releases.

Chlorine can be withdrawn in gaseous form by unloading through the vapor valve or valves. However, this procedure is quite unusual.

The pipe connected to a tank car discharge line should be no longer than 18 in. (45.7 cm) and should be screwed in with a pipe wrench not larger than an 18 in. size. Use of longer connecting lines and larger wrenches may result in tilting the tank car valve on its gasket, causing a chlorine leak.

Extra-heavy black iron or steel pipe is recommended for chlorine service. All threads should be clean and sharp, preferably cut with new dies.

If night operations are necessary, adequately light the area where the tank car is hooked up.

Gas Masks. Canister gas masks of a type approved by the U.S. Bureau of Mines for chlorine service should always be readily available when chlorine is unloaded, stored, transported or used. Locate gas masks outside the probable area of contamination so that it is possible to reach them in case of emergency.

Canister-type gas masks do not supply oxygen; they absorb the chlorine content present in the air leaving clean air to breathe. Where the chlorine content is greater than 1 percent, use a self-contained oxygen breathing apparatus or fresh air hose mask. Each chlorine consumer should have at least one such device available in his plant. All personnel who may be required to use gas masks should be properly instructed in their application and use. Obtain a copy of National Safety Council Pamphlet No. 64, "Respiratory Protective Equipment."

A poor gas mask is worse than no gas mask at all. Since the active materials in a canister become inactive when exposed to chlorine or air, keep canisters sealed. Renew them after each use.

Emergency Measures. In an emergency, telephone the chlorine manufacturer or supplier.

In case of fire, remove chlorine containers from the fire zone immediately.

As soon as there is any indication of the presence of chlorine in the air, take steps to correct the condition. Chlorine leaks never get better; they always get worse if not corrected promptly.

Keep on the windward side of the leak and higher than the leak. Since gaseous chlorine is approximately $2\frac{1}{2}$ times as heavy as air, it tends to lie close to the ground.

When a chlorine leak occurs, authorized, trained personnel equipped with gas masks should investigate. All other persons should be kept away from the affected area until the cause of the leak is discovered and corrected. If the leak is extensive, warn all persons in the path of the fumes.

Do not spray water on a chlorine leak. To do so makes the leak worse because of the corrosive action of wet chlorine. Heat supplied by even cold water causes liquid chlorine to vaporize at a faster rate.

When a leak occurs in equipment in which chlorine is being used, immediately close the chlorine container valve.

If a chlorine container is leaking in such a position that chlorine is escaping as a liquid, turn the container so that chlorine gas escapes.

The quantity of chlorine escaping from a gas leak is about $\frac{1}{15}$ the amount that escapes from a liquid leak through the same size hole.

If a chlorine leak occurs in transit in a congested area, it is recommended that the vehicle keep moving, if possible, until it reaches an open area where the escaping gas will be less hazardous. If a chlorine leak occurs in transit and the conveying vehicle is wrecked, the container or containers should be shifted so that gaseous chlorine, rather than liquid, is escaping. If possible, transfer the container to a suitable vehicle and take it to open country.

Leaks at valve stems are often stopped by tightening the valve packing nuts or closing the valve.

The severity of a chlorine leak can be lessened by reducing the pressure on the leaking container. This may be done by absorbing chlorine gas, from the container, in caustic soda solution. Evaporation of some of the liquid chlorine cools the remaining liquid, reducing its pressure.

At regular points of storage and use, make emergency preparations for disposing of chlorine from leaking cylinders or ton containers. Chlorine may be absorbed in caustic soda, soda ash or hydrated lime solution. Caustic soda solution is preferred as it absorbs chlorine most readily.

A suitable container to hold the solution should be provided in a convenient location. Chlorine may be passed into the solution through an iron pipe or rubber hose properly weighted to hold it under the surface. Do not immerse the container in the solution.

Methyl Chloride

Methyl chloride is classified as a flammable gas. It burns feebly but forms explosive mixtures with air. The end product of high-temperature decomposition may be toxic. For data concerning the chemical and physical properties of methyl chloride, its physiological effects, protective equipment, etc., obtain a copy of the "Methyl Chloride Data Sheet" published by the Chemical Manufacturers Association, Inc.

Throughout the entire single-unit tank car unloading operation, particularly while connecting and disconnecting, take great caution to make sure the working area is free of heated surfaces, flames, static electricity, railroad locomotives, gasoline tractors and all other sources of ignition. Tank cars and TMU containers should be grounded electrically.

Under no circumstances should water or other materials be introduced into tank cars which contain, or have contained methyl chloride.

In testing for leaks, use soapy water; in freezing weather or around very cold pipes or equipment, use glycerine. Never test for leaks with an open flame. This is prohibited, and it applies not only to tank cars but also to all containers holding methyl chloride.

When unloading single-unit tank cars after all of the liquid has been transferred, the greater part (but not all) of the methyl chloride vapors may be recovered by creating a slight pressure differential. Observe caution in this operation, as a slight residual pressure must always remain on the tank car so that no air is drawn into it to form explosive mixtures when the pipes are disconnected.

When unloading tank cars, if a leak occurs which cannot be readily repaired by simple adjustment or tightening of the fittings, telephone the methyl chloride manufacturer at once for instructions, and evacuate the area around the car immediately. Permit only properly protected and instructed personnel to enter the contaminated area.

In case of TMU container leaks, all sources of ignition must be removed from the area at once, and if the leak cannot be stopped, transfer the methyl chloride to another container. As to the procedure, obtain and follow the detailed instructions of the methyl chloride manufacturer or supplier.

Fire and Explosion Hazards. Avoid all sources of ignition, of whatever nature, when unloading single-unit tank cars and TMU containers holding methyl chloride.

Methyl chloride fires are gas fires. The most

effective method of extinguishing them is to shut off the flow of vapor by closing the valves.

Carbon dioxide or dry chemical may be used to extinguish the flame to permit access to shut-off valves. If the valve is in the area of the fire, attack it if possible, close it, and then attack and extinguish the secondary fire which consists of other burning material ignited by the gas fire.

Circumstances may make it impossible to attack the valve, and in such a case, the flame may be allowed to continue burning while the surrounding area and objects are cooled with water spray.

Provide employees engaged in extinguishing fires with gas masks to protect them from methyl chloride vapors and the toxic combustion products formed.

Sulfur Dioxide

Sulfur dioxide is classified by the DOT as a nonflammable gas. In both its gaseous and liquid form it is neither flammable nor explosive. It is a respiratory and a skin and eye irritant. Obtain a copy of the "Sulfur Dioxide" pamphlet published by Compressed Gas Association, Inc., for data concerning its chemical and physical properties, physiological effects, protective equipment, etc.

The pressure differential necessary to unload sulfur dioxide from a single-unit tank car into a storage tank may be controlled in several ways. In some instances it is only necessary to release the pressure on the storage tank by actually using sulfur dioxide from the storage tank in a regular way while unloading the tank car.

Where you have two or more storage tanks, the car may be unloaded by building the pressure up in one of the storage tanks to 25 or 30 psig (172.4 or 206.8 kPa) higher than the pressure in the tank to be filled. This higher pressure is then applied to the tank car to force the liquid into the storage tank.

Probably the best method of unloading a single-unit tank car is by installing a compressor with the suction side connected to the top of the storage tank and the discharge side connected to one of the gas valves of the tank car.

The least desirable method of unloading is to apply air pressure to the tank car through the vapor valve. If this method is used, exercise great care to insure that the supply of air is clean and dry and at a sufficiently high pressure to prevent backflow of sulfur dioxide into the air line.

Feeding SO$_2$ into a Solution. Where it is desirable to feed sulfur dioxide gas into a solution, the pipe leading from the valve on the container will frost on the outside and may even cause the solution to freeze at the point where the sulfur dioxide is being absorbed. If this happens you may correct the trouble by steam jacketing a small section of the feed line.

When gas is being fed to a solution, the evaporation of the liquid in the container refrigerates the entire content. As a result, the pressure might be reduced to a point where there is little or no flow of gas. In this case it is necessary to apply heat to the container or to connect a number of containers in parallel. This retards the rate of evaporation in each one.

Because of the corrosive nature of sulfurous acid, exercise great care to prevent the solution from drawing back into the upper valve chambers when the feed valve is shut off at the container or storage tank. To prevent this from happening, it is imperative that the feed line is vented or that a stainless steel ball check or check valve is installed in the line.

Leaks. The occurrence of sulfur dioxide leaks are indicated by the pungent odor of the gas. Locate by using a wad of cloth soaked in ammonium hydroxide. This produces white fumes near the point of the leak.

Leaks which might develop are ordinarily not serious and can be readily controlled. Where leaks do occur, shut off the supply of sulfur dioxide by closing the appropriate valve. Leaks at unions or other fittings are often eliminated by tightening the connection.

Do not work on a line while it is under pressure. If corrosion is indicated, take care to empty the lines before working on them. A broken fitting might lead to a serious loss of sulfur dioxide before the supply valves can be shut off.

Although serious leaks rarely occur, careless handling sometimes results in this condition. Tank car and TMU container valves are made of brass and if struck by a heavy object, may be broken.

Leaks may also occur from carelessness in heating TMU containers, causing the fusible plug to melt and discharge the contents. Care in handling eliminates those dangers. Occasionally, leaks may develop in the valve packing, but these can be checked by tightening the packing nut.

In the event of a leaking container, wherever possible, move the container to an open area where the hazard due to escaping sulfur dioxide is minimized.

If a container is discharging too freely to permit movement, it should be arranged, if possible, in such a position that the leak is at the top, thus discharging gaseous sulfur dioxide and not liquid.

If large quantities of gas can be withdrawn rapidly into equipment, or satisfactorily vented, the evaporation often lowers the pressure in the container to such a point that you can move it to the open without difficulty; or, possibly, the leak can be repaired by persons provided with suitable masks.

If the leak still prevents removing the container to an open area, gaseous, or liquid sulfur dioxide can be vented into a solution of lime, caustic soda or other alkaline material. One lb of sulfur dioxide is equivalent to about 2 lb (0.454 kg) of lime or $1\frac{1}{2}$ lb (0.907 or 0.567 kg) of caustic soda.

When a leak does occur, only an authorized employee should attempt to stop it, and if there is any question as to the seriousness of a leak, a suitable gas mask should be worn. In general, where leaks are serious, the employee even when equipped with a suitable gas mask, should remain in the contaminated area only long enough to make emergency adjustments.

Provisions for an Emergency. All employees handling sulfur dioxide should be impressed with the potential danger it represents and should be trained in its safe handling. In addition, provide them with personal protective equipment for use in an emergency, and drill them until they are familiar with its use.

This protective equipment should include a gas mask of a type approved by the Bureau of Mines for sulfur dioxide service. Take care to assure that masks are kept in proper working order and that they are stored so as to be readily available in case of need.

Canister-type masks are unsafe for high concentrations of sulfur dioxide. Warn employees of possible failure of this type of mask in the event of a really serious leak. Self-contained oxygen breathing apparatus or a mask with a long air hose and outside source of air may be required under extreme conditions.

Other protective equipment provided should include goggles or large-lensed spectacles to eliminate the possibility of liquid sulfur dioxide coming in contact with the eyes and causing possible injury.

If sulfur dioxide should be released, the irritating effect of the gas will force personnel to leave the area before they have long been exposed to dangerous concentrations. To facilitate their rapid evacuation, provide sufficient, well-marked, easily accessible exits.

Since sulfur dioxide neither burns nor supports combustion, there is no danger of fire or explosion due to igniting gas or liquid.

If a fire breaks out due to some other cause in an area containing sulfur dioxide, make every effort to remove the containers from the area to prevent overheating which would lead to melting of the fuse plugs. If they cannot be removed, inform the firemen of their location.

Fluorocarbons

Liquefied fluorinated hydrocarbons are classified as nonflammable gases by the DOT. Those presently shipped in single-unit tank cars, TMU containers and cargo tank trucks are: dichlorodifluoromethane, trichloromonofluoromethane, monochlorodifluoromethane, dichlorotetrafluoromethane and trichlorotrifluoromethane.

These gases are odorless, and leaks cannot be detected by sense of smell. Frosting is evidence of a large leak while smaller leaks may be located by means of a halide torch.

Avoid contact with the liquid and excessive inhalation of vapor. In case of a severe leak, persons entering area of dense concentration of vapor should wear an air gas mask of the type approved by the U.S. Bureau of Mines for liquefied fluorinated hydrocarbon service.

Check Regulations and Industry Standards

DOT regulations, and such local regulations as may exist in your area as well as industry standards are the "Bible" for handling and unloading compressed gases. Statements made herein are intended only to complement these regulations and standards and not to supplant them.

Obviously, within the limits of this explanation, it is not possible to set forth all the precautions to be taken when handling a particular gas. It is possible only to mention some of the more important. In all cases, the compressed gas manufacturer must be consulted concerning specific handling and transfer problems, and he will advise in full concerning them.

REFERENCES

(1) Tariff BOE-6000 Publishing "Hazardous Materials Regulations of the Department of Transportation by Air, Rail, Highway, Water, and Military Explosives by Water, including Specifications for Shipping Containers," Bureau of Explosives, 1920 L St., N.W., Washington, DC 20036. Nominal charge.

(2) "Standard 1202 (Feb. 1960), Measuring, Sampling and Calculating Tank Car Quantities and Calibrating Tank Car Tanks (Pressure Type Tank Cars)," American Petroleum Institute, 2101 L St., N.W., Washington, DC 20037. Nominal charge.

(3) "Anhydrous Ammonia (Pamphlet G-2)," Compressed Gas Association, Inc., 500 Fifth Avenue, New York, NY 10110. Nominal charge.

(4) "Standard for the Storage and Handling of Liquefied Petroleum Gases, NFPA Pamphlet No. 58," National Fire Protection Association 470 Atlantic Ave., Boston, MA 02210. Nominal charge.

(5) "Chlorine Manual," The Chlorine Institute, Inc., 342 Madison Ave., New York, NY, 10017. Nominal charge.

(6) "Respiratory Protective Equipment," Nation Association, 470 Atlantic Ave., Boston, MA 02210. Nominal charge.

(7) "Methyl Chloride" Safety Data Sheet SD-40, Chemical Manufacturers Association, 1825 Connecticut Ave., N.W., Washington, DC 20009. Nominal charge.

(8) "Sulfur Dioxide" (Pamphlet G-3), Compressed Gas Association, Inc. Nominal charge.

(9) "Butadiene" Safety Data Sheet SD-55, Chemical Manufacturers Association.

(10) "Vinyl Chloride" Safety Data Sheet SD-56, Chemical Manufacturers Association.

CHAPTER 9

Safe Handling and Use of
Cryogenic Liquids

This chapter provides general information regarding the safe handling and use of the cryogenic liquids that are more commonly needed in the work of industrial and institutional organizations. The safe handling of these liquefied gases is largely a matter of knowing their specific properties and compatability with materials, and using common-sense procedures based on that knowledge. This information does not supplant, but is intended to complement national, state, provincial, municipal and insurance company safety requirements.

The information in this chapter is intended for the use of cryogenic liquid consumers, shippers, carriers, distributors, equipment designers or installers, safety administrators and others desiring an introductory knowledge of the subject. Anyone requiring more detailed or specialized information should consult his cryogenic liquid supplier.

This chapter is broken down into the following sections:

Section A. General Introduction.
Section B. General Safety Practices.
Section C. Fire Prevention and Fire Fighting.
Section D. Safe Handling Recommendations.
Section E. Bulk Storage Site Considerations.
References

SECTION A

General Introduction

A cryogenic liquid is considered to be a liquid with a normal boiling point below -238 F (-150 C) as defined in the National Bureau of Standards Handbook 44.

The most commonly used industrial gases that are transported, handled and stored in the liquid state at cryogenic temperatures are oxygen, nitrogen, argon, hydrogen and helium. Three rare atmospheric gases, neon, krypton and xenon, are also used as cryogenic liquids in industry. Liquefied natural gas (LNG) and/or liquid methane and carbon monoxide are also handled as cryogenic liquids, although they are not usually classified as industrial gases. Liquefied ethylene, carbon dioxide and nitrous oxide are not classified as cryogenic liquids because they have normal boiling points above -238 F (-150 C). Although they are transported and stored in their liquid form, they are not discussed here.

The large-scale transportation and storage of cryogenic liquids is usually based on lower transportation and handling costs than for compressed gases. Sometimes the gas is used in the liquid state, as in liquid nitrogen refrigeration systems, but usually the liquid is vaporized and used as a gas. Examples are the uses of oxygen in hospitals, welding and steelmaking.

The handling of cryogenic liquids in large volumes is not new. Liquid oxygen was first shipped by tank truck in 1932. Today it is common to see portable liquid containers, cryogenic trailers and trucks, and railroad tank cars hauling large quantities of liquefied gases across the country. Cryogenic tanker ships transport LNG overseas. Air transport also is commonly used to transport some of these liquefied gases, especially liquid helium.

Industry has maintained a superior safety record based on a thorough understanding of the risks and hazards involved in handling each cryogenic liquid. This information is available to users, and strict adherence to safety precautions should be routinely observed. Many of the safety precautions observed for gases in the gaseous state, also apply to the same gases in the liquid state. However, certain additional precautions are necessary because of the special properties exhibited by fluids at cryogenic temperatures. Although most of these precautions are common to all cryogenic fluids, some are limited to only one or two of them.

All cryogenic fluids are extremely cold. Cryogenic liquids and their cold "boil-off" vapors can rapidly freeze human tissue, and cause many common materials such as carbon steel, plastics and rubber to become brittle, or even fracture under stress. Liquids in poorly insulated or uninsulated containers and piping at temperatures at or below the boiling point of liquefied air (-318 F, -194 C) can actually condense the surrounding air to a liquid. This liquid air is oxygen-rich and should be treated as liquid oxygen. The extremely cold liquefied gases (LHe, LH$_2$, LNe) can even solidify directly exposed air or other gases.

All cryogenic liquids also produce large volumes of gas when they vaporize. For example, one volume of liquid nitrogen at its boiling temperature at 1 atm vaporizes to 696.5 volumes of nitrogen gas when warmed at room

temperature (70 F or 21 C) at 1 atm. The volume expansion ratio of oxygen is 860.6 to 1. Liquid neon has the highest expansion ratio of any industrial gas at 1445 to 1. If these liquids are vaporized in a sealed container, they will produce enormous pressures. For example, one volume of liquid helium at 1 atm when vaporized and warmed to room temperature in a totally enclosed container has the potential to generate a pressure of over 14,500 psig. For this reason, pressurized cryogenic containers are protected with multiple devices for pressure relief, usually a pressure relief valve for primary protection, and a frangible disc for secondary protection.

Most cryogenic liquids are odorless, colorless and tasteless when vaporized to the gaseous state. Most of them have no color as a liquid, although liquid oxygen is light blue. However, the extremely cold liquid and vapor has a built-in warning property that appears whenever they are exposed to the atmosphere. The cold boil-off gases condense the moisture in the air, creating a highly visible fog. The fog normally extends over a larger area than the vaporizing gas.

All of the gases except oxygen can cause asphyxiation by displacing breathable air in an enclosed workspace. However, a worker cannot detect the presence of such gases without instrumentation. Therefore, he can be asphyxiated before he realizes that a problem exists. Carbon monoxide presents the added hazards of high toxicity and flammability.

Liquid nitrogen and argon are inert, chemically inactive and noncorrosive at cryogenic temperatures. Liquid oxygen is an oxidizer and will vigorously support the combustion of other materials, but it will not burn by itself. Liquid carbon monoxide, natural gas and liquid hydrogen are classed (and must be handled) as flammable gases.

STANDARDS AND SPECIFICATIONS

Many industry and government standards are available to guide the user in the proper selection of the type of equipment and installation to be used in specific operations. A partial list of these Standards and Specifications includes:

(1) Code of Federal Regulations, Title 49, Transportation, Parts 100 to 199, 179.400/179.401-	
DOT Spec. 113A60W	Liquefied hydrogen tank car
DOT Specification 4L, 178.57	Welded cylinders insulated
(2) ASME Section VIII, Division 1	Pressure vessels
(3, 4, 5) CGA S-1.1, S-1.2, S-1.3	Relief devices
(8) AAR 204W	Tank cars
(6) CGA 341	Tank trucks
(9) NFPA 50	Bulk oxygen systems at consumer sites
(10) NFPA 50B	Liquefied hydrogen system at consumer sites
(12) NFPA 56F	Nonflammable medical gas
(13) NFPA 59A	Storage and handling of liquefied natural gas
(16) IATA Restricted Articles Regulations	Air shipment of cryogens
(17) Air Transport Restricted Articles Tariff No. 6-D	Air shipment of cryogens

A more complete list of standards, specifications and texts will be found under "References" at the end of this chapter. In some applications there may be additional requirements specified by municipal, state or federal authorities, and the user should check any such requirements in preparing for his specific operations. Parenthetic numbers of the sources listed above correspond to the numbers in the "References" list.

SECTION B

General Safety Practices

General safety practices for handling, transporting, and using the common cryogenic liquids are covered first. Special precautions are then detailed to cover those few liquids such as carbon monoxide, hydrogen, LNG and oxygen which require special handling.

Physical Properties Important for Safe Handling. Table I is a useful guide to the properties of cryogenic liquids that can affect their safe handling and use. These liquids have been listed by order of decreasing boiling point. Although xenon boils above the -238 F (-150 C) level, it also has been included. Natural gas is a mixture of methane and other hydrocarbons with a boiling point depending upon its composition. However, natural gas is primarily methane, and methane data is listed in Table 1. Data on the flammable gases, e.g., methane, carbon monoxide and hydrogen, include flammability limits in air and oxygen, spontaneous ignition temperatures in air at one atmosphere, and other useful information that will help to determine safe-handling procedures. None of the gases listed are corrosive at ambient temperatures, and only carbon monoxide is toxic.

AVOID CONTACT

Always handle cryogenic liquids carefully. At their extremely low temperatures, they can produce frostbite on skin and exposed eye tissue. When spilled, they tend to cover a surface completely, cooling a large area. The vapors issuing from these liquids are also extremely cold. Delicate tissues, such as those of

TABLE 1. PHYSICAL PROPERTIES OF CRYOGENIC FLUIDS
(Also see Table 3)

	Xenon Xe	Kryp-ton Kr	Methane CH_4	Oxygen O_2	Argon Ar	Carbon Mon-oxide CO	Nitro-gen N_2	Neon Ne	Hydro-gen H_2	Helium He
Boiling point, 1 atm										
°F	−163	−244	−259	−297	−303	−313	−321	−411	−423	−452
°C	−108	−153	−161	−183	−186	−192	−196	−246	−253	−268
Melting point, 1 atm										
°F	−169	−251	−296	−362	−309	−341	−346	−416	−435	−[1]
°C	−112	−157	−182	−219	−189	−207	−210	−249	−259	−
Density at boiling point and 1 atm lb/cu ft	191	151	26	71	87	49	50	75	4.4	7.8
Heat of vaporization at boiling point Btu/lb	41	46	219	92	70	98	85	37	193	10
Volume expansion ratio, liquid at 1 atm and boiling point to gas at 70°F and 1 atm	559	693	625	860	842	680	696	1445	860	755
Flammable	No	No	Yes	No[2]	No	Yes	No	No	Yes	No

[1] Helium does not solidify at 1 atm pressure.

[2] Oxygen does not burn, but supports and accelerates combustion; however, high oxygen atmospheres substantially increase combustion rates of other materials, and many form explosive mixtures with other combustibles. Flame temperatures in oxygen are higher than those in air.

the eyes, can be damaged by exposure to these cold gases, even when the contact is too brief to affect the skin of the hands or face.

Stand clear of boiling or splashing liquid and its vapors. Boiling and splashing always occur when charging a warm container, or when inserting warm objects into a liquid. Always perform these operations slowly to minimize boiling and splashing. If cold liquid or vapor contacts the skin or eyes, follow the first-aid recommendations in the next section.

Never allow any unprotected part of the body to touch uninsulated pipes or vessels which contain cryogenic fluids. The extremely cold metal will cause the flesh to stick fast and tear when one attempts to withdraw from it. Even non-metallic materials are dangerous to touch at low temperatures. Use tongs to withdraw objects immersed in a cryogenic liquid. In addition to the hazards of frostbite or flesh sticking to cold materials, objects that are soft and pliable at room temperature, such as rubber or plastics, are easily broken because they become hard and brittle at these extremely low temperatures. Carbon steels also become brittle at low temperatures and will easily fracture.

TREATING COLD CONTACT BURNS

Workers will rarely ever come in contact with a cryogenic liquid if proper handling procedures are used, but in the unlikely event of contact with a liquid or cold gas, a cold-contact "burn" may occur. Actually, the skin or eye tissue freezes. Following are the recommended emergency treatments for a cold-contact burn while awaiting medical assistance:

Remove any clothing that may restrict the circulation to the frozen area. Do not rub frozen parts as tissue damage may result. Obtain medical assistance as soon as possible.

As soon as practical, place the affected part of the body in a warm-water bath, which has a temperature of not less than 105 F or more than 115 F (40 C to 46 C). Never use dry heat. The victim also should be in a warm room if possible.

If there has been massive exposure so that the general body temperature is depressed, the patient must be rewarmed by total immersion into a warm-water bath. Supportive treatment for shock should be provided.

Frozen tissues are painless and appear waxy with a possible yellow color. They will become swollen, painful and prone to infection when thawed. Do not rewarm rapidly if the accident occurs in the field and the patient cannot be transported to medical attention immediately. Thawing may require from 15 to 60 minutes and should be continued until the pale blue tint of the skin turns pink or red. Narcotics, such as morphine or tranquilizers may be required to control the pain during thawing and should be administered under professional medical supervision.

If the frozen part of the body has thawed by the time medical attention has been obtained, cover the area with dry sterile dressings with a large bulky protective covering.

Alcoholic beverages and smoking decrease blood flow to the frozen tissues and should not be used. Warm drinks and food may be administered to a conscious victim.

PROTECTIVE CLOTHING AND EQUIPMENT

Eye and hand protection for handling cryogenic liquids should be used by all workers with any chance of exposure to liquids or boil-off vapor. Adequate eye and hand protection serves primarily to protect workers against splashing and possible cold-contact burns.

Safety glasses are recommended during transfer and normal handling of cryogens. If severe spraying or splashing may occur, a face shield or chemical goggles should be worn for additional protection.

Insulated gloves should always be worn when handling anything that comes in contact with cold liquids and vapor. Gloves should be loose fitting so that they can be removed quickly if liquids are spilled into them.

Trousers should be left outside of boots or work shoes.

SPECIAL OXYGEN PRECAUTIONS

Keep all combustible materials, especially oil or grease, away from oxygen. Do not permit smoking or open flames in any area where liquid oxygen is stored or handled. Post "No Smoking" signs conspicuously in such areas.

Oxygen is nonflammable but it vigorously accelerates and supports combustion. Substances that burn in air will burn much more vigorously in oxygen. The upper flammable limit for a flammable gas in air is raised in oxygen-enriched air atmospheres, which means that fire or explosion is possible over a wider range of gas mixtures.

Do not permit liquid oxygen or oxygen-rich air atmospheres to come in contact with organic materials or flammable substances of any kind. Some of the organic materials that can react violently with oxygen when ignited even by a hot spark are oil, grease, asphalt, kerosene, cloth, tar and dirt that may contain oil or grease. If liquid oxygen spills on asphalt or other surfaces contaminated with combustibles, (e.g., oil-soaked concrete or gravel) do not walk on or roll equipment over the area of the spill. Keep sources of ignition away for at least 30 minutes after all frost or fog has disappeared.

Always operate valves in oxygen slowly. Abruptly starting and stopping oxygen flow may ignite any contaminants that might be in the system.

Any clothing that has been splashed or soaked with liquid oxygen should be removed immediately and aired for at least an hour until it is completely free of excess oxygen. If workers are exposed to high-oxygen atmospheres, they should leave the area, avoid all sources of ignition, particularly smoking, and wait for a half hour until clothing and the exposed area are both completely ventilated. Clothing saturated with oxygen is readily ignitable and will burn vigorously.

SPECIAL INERT GAS PRECAUTIONS

Rupture of containers, pipelines or systems, and asphyxiation are the primary hazards of inert gas systems. A cryogen cannot be indefinitely maintained as a liquid even in well-insulated containers. Any liquid or even cold vapor trapped between valves has the potential to cause an excessive pressure buildup to the point of violent rupture of a container or piping, hence use of reliable pressure-relief devices is mandatory.

Loss of vacuum in vacuum-jacketed tanks containing cryogenic liquids will cause increased evaporation within the system. This may cause the relief devices to function and result in product venting. The vented gases should be routed to a safe outdoor location. If outdoor venting is not done, the user must assure himself that adequate ventilation is maintained.

SPECIAL FLAMMABLE GAS PRECAUTIONS

Do not permit smoking or open flames in any area where flammable fluids are stored or handled. Post "No Smoking—Flammable Gas" signs in such areas. It may be advisable to wear work clothes that minimize ignition sources (such as static electricity) in atmospheres containing possible hazardous concentrations of flammable gases.

All major stationary equipment should be properly grounded. Stationary and mobile equipment should be electrically bonded together before loading and unloading any flammable gas. All electrical equipment used in or near flammable gas loading and unloading areas, or in atmospheres that might contain explosive mixtures should be in accordance

with the NFPA Publications 50B (Liquid Hydrogen) (10) and 59A (Liquefied Natural Gas) (13) and/or the National Electrical Code, Article 500, for other flammable gas hazardous locations.

Flammable cryogenic liquids and gases shall only be handled inside rooms or buildings with adequate positive mechanical ventilation electrical equipment and wiring in accordance with Article 501 of the NEC; NFPA No. 70 (15), and possible need for monitoring for the presence of flammable and/or explosive venting as described in NFPA 68, "Explosion Venting Guide" (14).

Flashoff gas from closed liquid hydrogen containers used or stored inside buildings should be piped to a laboratory hood and vented outside the building or vented by other means to a safe location. If hydrogen is vented by any of these means into ductwork, this ventilation system should be independent of other systems and sources of ignition must be eliminated at the exit of the exhaust system. Liquid hydrogen cylinders should be stored in accordance with NFPA 50B. Consideration should be given to equipping such vent systems with inert gas (nitrogen) purge capability to prevent electrostatic ignition of hydrogen and air mixtures at the start of venting.

SPECIAL CARBON MONOXIDE PRECAUTIONS

Carbon monoxide (CO) is odorless, colorless, toxic and flammable. Because of these characteristics, all leaks must be eliminated before a system is put into operation, or returned to operation after long shutdowns. Liquid nitrogen must be used to test and purge vessels, pipelines, vaporizers, and controls at pressures and temperatures near actual operating conditions. This allows detection and correction of all leaks to assure that the system is safe for liquid carbon monoxide.

At least two men should be assigned to any operation involving liquid carbon monoxide. Refer to Section D on "Safe Handling Recommendations" for specific safety recommendations.

A suitable supply of oxygen should be available in the general vicinity to administer oxygen to personnel who may be exposed to excessive carbon monoxide concentrations. The American Conference of Industrial and Government Hygienists lists 50 ppm by volume as the maximum allowable concentration for an eight-hour day exposure to carbon monoxide. While 400 to 500 ppm carbon monoxide by volume can be inhaled for one hour without appreciable effect, no leaks should be tolerated in view of the dangerous potential of CO. Concentrations of 4000 ppm and above may be fatal in exposure of less than one hour. Exposure to very high concentrations can cause death almost immediately. A sensitive monitoring system capable of measuring 10 to 200 ppm levels should be maintained continuously.

ASPHYXIATION

All gases should be used and stored in well ventilated areas. Oxygen is the only gas that will support life. High concentrations of all other gases reduce the breathable oxygen in the air below a safe level.

Asphyxiation can occur suddenly or develop slowly without the worker being aware that he is in trouble. Refer to CGA Safety Bulletin SB-2 (7) for additional information. The problem is easily avoided unless large quantities of inert gas are present, simply by using proper ventilation at all times. When it is absolutely necessary to enter a work area that may have an oxygen content below 19 percent, by volume, portable air packs or a hose mask connected to a breathing-air source must be used. An absorptive gas mask will *not* prevent asphyxiation.

Asphyxia develops slowly as the oxygen content of the air is gradually reduced from 21 percent and the victim will not be aware of a problem, as he generally will not recognize the symptoms of gradual asphyxia. These symptoms are described below, with decreasing oxygen levels:

(1) The normal oxygen concentration in air is about 21 percent by volume and provides a safe working environment with respect to oxygen required to support life.

(2) Depletion of oxygen in a given volume

of air by combustion, displacement with inert gas or by increased elevation is a potential hazard to personnel. The true measure of oxygen availability is the oxygen partial pressure. While the percent of oxygen in the atmosphere is always 20.9 percent at all elevations, the partial pressure of oxygen in the atmosphere varies with the ratio of the atmospheric pressure at the elevation being considered, to the atmospheric pressure at sea level. In Denver, at an elevation of 5280 ft, there is only about 82 percent as much oxygen available in each volume of air as at sea level. Consideration must be given to the role of elevation since the oxygen concentrations referred to in this chapter are sea level values.

(3) When the oxygen content of air is reduced to about 15 to 16 percent the flame of ordinary combustible materials, including those commonly used as fuel for heat or light, will be extinguished. Somewhat below this concentration an individual breathing the air is mentally incapable of diagnosing the situation as the symptoms of sleepiness, fatigue, lassitude, loss of coordination, errors in judgment and confusion will be masked by a state of "euphoria" giving the victim a false sense of security and well being.

(4) Human exposure to atmospheres containing 12 percent or less oxygen will bring about unconsciousness without warning and so quickly that the individual cannot help or protect himself. This is true if the condition is reached by immediate change of environment or by gradual depletion of oxygen. The victim's condition and degree of activity will have an appreciable effect on signs and symptoms at various oxygen levels. In some instances, prolonged reduction of oxygen may cause brain damage if the individual survives.

Areas where it is possible to have low-oxygen content must be well ventilated. Nitrogen vents shall be piped outside of buildings to safe areas. Where low-oxygen atmospheres are possible, installation of analyzers with alarms should be used. Constant monitors, sniffers and other precautions must be used to check atmospheres and maintain surveillance of the atmosphere when personnel enter such enclosed areas or vessels. When there is any doubt of maintaining safe breathing atmospheres, self-contained breathing apparatus or approved air lines and masks must be used.

Most personnel working in or around oxygen-deficient atmospheres rely on the buddy system for protection—but the buddy is equally liable to asphyxiation if he enters the area to rescue his unconscious partner unless he is equipped with a portable air supply. The best protection is obtained by providing both the worker and his buddy with a portable supply of respirable air. Life lines are acceptable only if the area is free of obstructions and the buddy is capable of lifting his partner's weight rapidly and without straining himself. In practice this has seldom been possible. Also use more than one person if required to remove a worker in any emergency.

WORK PERMITS

Most applications of cryogenic liquids are quite safe, involving few if any hazards to workers. The liquid usually is transferred, transported or stored in closed pressure vessels, and piping systems designed for use at low temperatures. The liquid may be vaporized into pipelines for use in various processes. The liquid may flow directly into refrigeration or process equipment. In most applications workers probably will never see or contact the cryogenic liquid.

However, maintenance work occasionally must be performed on a pressure vessel, a ship's hold, a pipeline, storage vessel or tank that previously contained a flammable or potentially suffocating cryogenic liquid. It is recommended that such work be covered by work permits issued by the responsible supervisor after a detailed review of the specific project. These permits detail the allowable working times, conditions and procedures for the job. They specify who may enter or do the work, and what special safety precautions are required.

The American Welding Society, American Gas Association, American Petroleum Institute and National Fire Prevention Association have detailed literature on purging, testing and working on potentially hazardous jobs.

TRAINING

The best single investment in safety is trained personnel. Some personnel need only detailed training in a particular type of equipment, cryogen or repair operation. Others require broader training in safe handling practices for a variety of cryogenic fluids.

The following subjects should be explained to all persons who will work with cryogenic liquids, including handling, storage and transfer operations:

(1) nature and properties of the cryogen in both the liquid and gaseous phases;
(2) specific instructions on the equipment to be used;
(3) approved materials that are compatible with the cryogen;
(4) use and care of protective equipment and clothing;
(5) safety, first aid, and self-aid when first aid and/or medical treatment is not immediately available;
(6) handling emergency situations such as fire, leaks and spills;
(7) good housekeeping practices.

HOUSEKEEPING

Good housekeeping is essential for the safety of personnel. Few cryogenic liquids are spontaneously hazardous, but the liquefied gases under consideration in this chapter have different degrees of potential hazard, whether in gaseous or liquid form.

Liquid oxygen with fuels, oils and/or grease may form mixtures that are shock sensitive. Solids such as asphalt or wood that have a porous structure can become saturated with oxygen and also become shock sensitive. Ignition is more likely to occur with weaker sparks and lower temperatures than would be the case in air.

The flammable gases such as hydrogen and methane are lighter than air at normal temperatures and will rise. At temperatures which occur just after evaporation from the liquid, the saturated vapor is heavier than air and tends to fall. Wind and/or forced ventilation will affect the direction of the released gases. The provisions for disposal of any leaking fluid should be considered within the satisfactory "housekeeping" requirements.

The location and maintenance of safety and fire-fighting equipment in good operating condition is important. Subcontractors and other outside personnel must be informed of all necessary safeguards before entering a potentially hazardous area. Good housekeeping rules also minimize negligence by demanding a high level of worker conduct everywhere in the facility.

SECTION C

Fire Prevention and Fire Fighting

The only cryogenic liquids discussed in this Chapter that present a direct fire hazard are methane or liquefied natural gas (LNG), carbon monoxide, hydrogen and (although not flammable) oxygen. All other gases are capable of blanketing and extinguishing fires.

Prevention of fires is best accomplished by eliminating all leakage of flammable gases, and any potential ignition sources. Experience has shown that flammable cryogenic liquids that leak and vaporize are easily ignited by various means. Therefore, the primary effort should be containment of the fluid, detection of any leaks, and ventilation of the area.

Outbreak of a fire in the vicinity of cryogenic liquids or gases in a plant, storage vessel or transport vehicles creates considerable potential danger. Everything feasible should be done to prevent a fire from starting. If a fire starts, call the fire department and start to fight the fire immediately in an accepted manner. The key elements are personnel training, conduct of fire drills, upkeep of all fire equipment and fire-fighting materials, and liaison with technical staff and local fire departments when required.

A fire plan should contain sketch maps and diagrams locating all fire alarms, fire-brigade call points, emergency exits, escape routes, rescue and safety equipment, hydrants, fire-fighting equipment and plant controls featured in the plan. Special or unusual risks should be defined and explained.

One of the special problems with cryogenic liquids that make fire fighting more difficult is their potential to rapidly freeze water. The indiscriminate use of water on surfaces of cryogenic containers or piping can lead to heavy icing and possible blockage of pressure relief devices. Valve stems may be iced over so badly that they cannot be operated. Meanwhile, the relatively warm water will cause the liquefied gas to vaporize more rapidly. Increasing the boil-off rate of a flammable cryogenic liquid will, of course, generate even more gas to feed the fire.

There are no general safety objections to using carbon dioxide extinguishers to fight smaller cryogenic liquid fires. However, the operation of CO_2 extinguishers may create electrostatic discharges of sufficient magnitude to ignite some hydrogen/air mixtures. They are ineffective with oxygen-rich fires.

After large volumes of cryogenic substances have been released into the atmosphere, a fog will form from water vapor condensing in the surrounding air. This fog may severely reduce visibility. Escape routes in potentially vulnerable areas should include blind reference points such as hand rails (about 4 ft high), and simple solid objects such as curbs to permit location and identification by touch alone.

FIGHTING FIRES

It is not possible to outline specific fire-fighting techniques that will cover all types of fires involving cryogenic liquids. Such measures depend upon the quantity and nature of the cryogen involved, the location of the fire with respect to adjacent areas and their occupants and other factors. The following general procedures, however, are applicable to all fires involving cryogenic liquids:

(1) Everyone not actively engaged in fighting the fire should leave the area. If a flammable cryogenic liquid is involved, the flammable-mixture zone, under unusual atmospheric conditions, may extend beyond the normal fog cloud produced by condensing water vapor in the air. People should be evacuated well outside the fog area.

(2) The single best fire-fighting technique is simply shut off the flow of liquid or gaseous cryogen if a liquefied flammable gas is involved.

(3) If electrical equipment is involved in the fire, be sure that the power supply is disconnected before using water for fire fighting, or else use carbon dioxide or dry chemical extinguishers.

(4) When using water, use large quantities, preferably in spray form, to cool equipment in areas surrounding the fire. Use the spray to cool any burning material below its ignition temperature. If possible, do not spray cold areas of equipment, or direct water onto the cryogenic fluid. Fire hoses with stream-to-spray nozzles should be available where large quantities of flammable cryogenic liquids are handled.

(5) Depending upon the circumstances, it is not usually advisable to extinguish a flammable cryogen in a confined area. If the flammable gas supply cannot be shut off, the continued escape of unburned gas can create an explosive mixture in the air. The mixture may be reignited by other burning material or hot surfaces. It is usually better to allow the gas to burn itself out in a confined area and keep adjacent objects cool with water, rather than risk exposing personnel at the site of a potential explosion.

(6) If an inert cryogenic liquid is involved, judgment should be used in deciding whether to allow the gas to escape, with possible risk of asphyxiation of fire fighters, or to cut off the gas flow. Where asphyxiation is not a factor, it is preferable to reduce the pressure at the source of the inert cryogenic liquid.

(7) Observe the special precautions listed hereafter for fire-fighting equipment, and for fighting oxygen, carbon monoxide, liquefied natural gas and hydrogen fires.

Table 2 is a general guide to specific fire-extinguishing agents for different cryogenic liquids. Table 3 provides a guide to the relative explosion and fire risks involved with the common cryogenic liquids. The next subsection lists special precautions for several specific cryogenic fluids.

TABLE 2. FIRE EXTINGUISHING AGENTS

Extinguishing Agent	Cryogenic Fluid in Fire	
	Oxygen	Hydrogen, Methane, Carbon Monoxide
Water	Preferred	Used to protect adjacent equipment or property and to spray personnel. Not to be applied directly onto burning vapor or cryogenic liquid since the water will evaporate additional flammable material.
Soda acid	No effect	Unacceptable
Carbon dioxide	No effect	Fair. Apply at base of flame.
Dry powder	No effect	Good. Apply at base of flame.
Methyl bromide	Unacceptable	Not normally used unless authorized and supplied for individual premises or equipment.

TABLE 3. EXPLOSIVE AND FIRE HAZARDS OF COMMON CRYOGENIC FLUIDS

	Oxygen	Nitrogen	Argon	Helium	Krypton	Xenon	Neon	Methane	Hydrogen	Carbon Monoxide
Explosive hazard with combustible materials	Yes	No	No	No	No	No	No	No	No	No
Explosive hazard with oxygen or air	—	No	No	No	No	No	No	Yes	Yes	Yes
								(within flammable limits)		
Pressure rupture if liquid or cold vapor is trapped	Yes	Yes	Yes	Yes	Yes	Yes	Yes	Yes	Yes	Yes
Fire hazard type Combustible	Nil	Nil	Nil	Nil	Nil	Nil	Nil	Yes	Yes	Yes
Promotes ignition	Yes	No	No	No	No	No	No	Yes	Yes	Yes
Condenses air and expands flammable range	No	Yes	No	Yes	No	No	Yes	No	Yes	Yes
Flammable limits in air, percent by volume	—	—	—	—	—	—	—	5-15	4-74	13-74
Spontaneous ignition temperature in air at 1 atm										
°F	—	—	—	—	—	—	—	1000	1085	1204
°C	—	—	—	—	—	—	—	538	585	650
Minimum ignition energy, mJ	—	—	—	—	—	—	—	0.30[2]	0.02	
Flame temperature in air										
°F	—	—	—	—	—	—	—	3434	3722	3812[1]
°C	—	—	—	—	—	—	—	1080	2050	2100
Flame velocity in air										
ft/sec	—	—	—	—	—	—	—	1.28	6.9	1.08
cm/sec	—	—	—	—	—	—	—	39	271	33
Limiting oxygen index volume[3]	—	—	—	—	—	—	—	12	5	

[1]The maximum flame temperature for carbon monoxide occurs at about the stoichiometric mixture (66.67% carbon monoxide-33.3% oxygen). The range of flame temperatures is 1560°C to 2100°C.

[2]Some components of LNG, such as ethylene have ignition energies as low as 0.08 mJ.

[3]Minimum oxygen concentration to support flame propagation when stoichiometric fuel-air mixture is diluted with nitrogen.

FIGHTING FIRES WITH OXYGEN PRESENT

Oxygen does not burn, so there can be no fire unless combustible materials also are present. If a fire involves gaseous or liquid oxygen, the oxygen plays the same part as oxygen from the air does in an ordinary fire. The difference is that the presence of additional oxygen will make combustible materials burn much faster and more violently, or explode. Shut off the oxygen supply if at all possible.

Neither liquid nor gaseous oxygen can be effectively blanketed by such agents as carbon dioxide, dry chemical or foam. It is necessary to cool combustible materials below their ignition temperatures to stop the fire. Use large quantities of water in spray form.

The insulation on cryogenic equipment greatly retards heat transfer into the liquid container. The insulation helps protect the liquid container under fire conditions and allows the use of smaller safety relief devices which prevents overpressure if the tank should be exposed to fire conditions. Hose streams should be played on the insulation jacket to keep it cool and in place under fire conditions. If any of the insulation is lost, exposing the bare liquid container to fire, the area should be evacuated. This is recommended because of the remote possibility that the tank might fail under these conditions, due to localized heat input to the tank shell and its lower strength at the higher metal temperatures.

If a fire is supported by liquid oxygen flowing into large quantities of combustible materials, shut off the flow of oxygen. After that has been done, put out the fire with water, fog or foam.

If a fire is supported by fuel flowing into large quantities of liquid oxygen, shut off the flow of fuel and allow the fire to burn out. If other combustible material in the area is burning, water streams or fogs may be used to control the fires. Foams should not be applied to liquid oxygen.

If electrical equipment is involved in the fire, use carbon dioxide or dry chemical extinguishers. If possible, turn off the electric current at a main switch, circuit breaker or fuse box. Do not use water.

FIGHTING HYDROGEN FIRES

Small hydrogen flames are invisible. They also radiate less heat than ordinary fires so they give little warning of their presence either by sight or heat. The invisible hydrogen flames may be long and shift quickly with the slightest breeze. Personnel fighting hydrogen fires should be alert to these dangers.

Fight hydrogen fires from a safe distance upwind of the blaze. Also, keep in mind that liquid hydrogen's low temperature is capable of condensing air, which may create an explosive mixture of hydrogen and oxygen-enriched air.

The usual fire fighting practice for hydrogen fires is first to prevent the fire from spreading. Let it burn until the hydrogen is consumed. The only effective way to fight this type of fire is to shut off the flow of hydrogen gas. Carbon dioxide extinguishers, modified by sawing off the fog nozzle, have been used to blow out the fire rather than smother it; however, if the fire is extinguished without stopping the flow of gas, an explosive mixture may form in a confined area, creating a more serious hazard than the fire itself.

Dry powder fire extinguishers should be available in the area around any hydrogen storage facility. Adequate water supply should be available to keep surrounding equipment cool in the event of a hydrogen fire. The local fire department should be advised of the flammable characteristics of hydrogen and the best known methods for combating such fires.

FIGHTING NATURAL GAS FIRES

See Chapter 9 of NFPA 59A entitled "Standard For The Production, Storage and Handling of Liquefied Natural Gas (LNG)."

FIGHTING CARBON MONOXIDE FIRES

Carbon monoxide fires should be treated similarly to hydrogen fires in the sense that both gases are flammable and can combine with oxygen to form explosive mixtures. However, the toxic nature of carbon monoxide adds a special problem. It is imperative to evacuate the area until the fire subsides and it is determined that the toxic vapor level, measured by instrumentation, is no longer hazardous.

It is preferable to cut off the source of supply and allow a carbon monoxide fire to burn itself out. When the source of supply cannot be isolated, carefully consider possible exposure to personnel by carbon monoxide gas with consequences more serious than if the fire were permitted to continue to burn.

Water spray, carbon dioxide or dry powder are effective carbon monoxide fire-extinguishing agents. Only self-contained breathing apparatus should be used where a toxic vapor level might exist.

FIGHTING FIRES INVOLVING INERT GASES

Nitrogen, argon and the other inert gases are noncombustible and inert. They do not present any fire hazard. However, if a system should rupture due to a fire, these gases may displace air to the point where there is not enough oxygen to support life.

Excessive pressure buildup in piping and storage vessels can be prevented through proper selection of pressure-relief devices and flow-control equipment.

SECTION D

Safe Handling Recommendations

Cryogenic liquids are stored and transported in a wide range of containers from small dewars to railroads tank cars. Use only equipment and containers designed for the intended product and service pressure and temperature. The Code of Federal Regulations 49, Transportation Parts 100 to 199 (1), lists some of the requirements for various containers, by the individual gases or cryogenic liquid. If there is any question as to correct handling or transport procedure, or compatibility of materials with a given cryogenic liquid, consult your gas supplier.

Cryogenic liquids ordinarily should not be handled in open containers unless they are specifically designed for that purpose and for the specific product. Containers must be clean, especially those used to hold oxygen. Cryogenic containers must be made from materials suitable for cryogenic temperatures such as austenitic stainless steels, certain nickel bearing steels, copper or certain aluminum alloys. Most other materials become extremely brittle at cryogenic temperatures.

Personnel should have proper training and instructions for the specific cryogen and equipment involved. Transfer of cryogens into warm lines or containers should be done slowly to prevent thermal shock to the piping and container and possible excessive pressure buildup within the system.

When transferring liquid from one container to another, the receiving container should be cooled gradually to prevent thermal shock and to avoid splashing. High concentrations of escaping gas should be vented so as not to collect in an enclosed area.

Do not drop solids or liquids into cryogenic liquids. Violent boiling can splash liquid onto personnel and equipment.

Avoid breathing vapor from any cryogenic liquid source, except for liquid oxygen equipment specifically designed to supply breathing oxygen. When discharging cryogenic liquids from drain valves or blowdown lines, open the valves slowly to avoid being splashed by cold liquid.

Do not smoke, or permit smoking or open flames in any area where flammable liquids or gases, or liquid oxygen are stored, handled or used, or where they are loaded or unloaded.

Post "No Smoking" signs conspicuously in all such areas.

LIQUID DEWARS, LIQUID CYLINDERS AND LABORATORY DEWARS

Three types of portable liquid storage vessels are generally used to hold and dispense liquid or gaseous product. They are described as liquid dewars or flasks, pressurized liquid cylinders and laboratory dewars.

Liquid dewars are open-mouthed nonpressurized vacuum-jacketed vessels used to hold cryogenic fluids (usually liquid argon, nitrogen or oxygen). Some dewars are designed for specific lightweight liquids and maximum holding times, and their internal support system cannot hold some of the heavier cryogenic liquid. Be sure that no ice accumulates in the neck or on the cover which can cause a blockage and subsequent pressure buildup within the dewar.

Liquid cylinders are pressurized containers, usually vertical vessels, designed and fabricated in accordance with DOT-4L specifications. Liquid cylinders designed specifically for liquid helium and hydrogen are also available. There are three major types of liquid cylinders: one used for the dispensing of liquid or gas, one for gas withdrawal only, and one for liquid withdrawal only.

Each type of liquid cylinder has appropriate valves for filling and dispensing product and is adequately protected by a pressure-control valve and frangible disk against cylinder overpressurization. All valve outlet connections should be in accordance with industry standards for the product stored.

Liquid cylinders of various sizes can be handled by hand but it is preferable to move them using portable handcarts. A strap should be used to secure the cylinder to the handcart to keep the cylinder from slipping off.

Laboratory liquid dewars with wide-mouthed openings have neither lids nor covers to protect the liquid. Most are made of metal but some smaller units are made of glass. These smaller glass containers are often used in laboratories.

A cold outside jacket on a cryogenic vessel indicates some loss of insulating vacuum. The vessel should be drained of liquid, removed from service and set aside for repair. Repairs involving vacuum deterioration should be handled by the manufacturer of an authorized company with personnel qualified in fabrication and repair of the equipment.

All vessels should be precooled gradually before filling to avoid thermal shock as well as to reduce excessive flashing and loss of product. This is especially true for liquid hydrogen or liquid helium vessels. Cryogenic dewars, containers and cylinders must be handled very carefully. They should not be dropped or tipped over on their sides.

TRANSFER LINES

Many types of filling or transfer lines are used to handle the flow of cryogenic fluids from one point to another, including small, noninsulated copper or stainless steel lines, large diameter rigid lines, or flexible hose systems, vacuum-jacketed lines or other insulated systems.

Liquid product can be transferred by three methods, the simplest being by gravity transfer in which the height of stored liquid serves as the transfer medium. Pressurized transfer uses the vapor pressure of the product, or pressure from an external source as the prime mover of the liquid to the lower pressure receiving container.

Various types of cryogenic pumps are also used to transfer cryogenic liquids. Flow rates may vary from less than one gallon per minute to several hundred gallons per minute for industrial systems.

It is desirable to keep the product being transferred in totally liquid form within the transfer line. Any vaporization of liquid within the system may cause excessive pressure drop and two-phase flow. Two-phase flow is detrimental to the operation of cryogenic pumps.

Short transfer lines used for intermittent service are normally not insulated. Lines that are used for continuous transfer of cryogens are usually insulated. All liquid transfer hoses should be provided with dust caps.

Vacuum-jacketed transfer lines are required

for liquid hydrogen and liquid helium transfer. This is because of the extremely cold temperatures and low heats of vaporization of both liquids. Occasionally vacuum-jacketed lines are used for in-plant transfer of other cryogenic fluids when it is justified in terms of cost savings and/or much reduced line and flash losses.

BULK LIQUID STORAGE SYSTEMS

Bulk liquid storage systems usually consist of a storage tank, vaporizer (where the user wants gas instead of liquid), associated valves and piping and cryogenic pump (when needed). Storage capacities may range from several hundred to millions of gallons of liquid product.

The tanks used vary in physical size and shape. There are spherical tanks, vertical cylindrical tanks, horizontal cylindrical tanks and flat-bottom cylindrical tanks. Operating pressures usually range from atmospheric to several hundred pounds per square inch. Personnel should learn the characteristics of the particular installation.

No smoking or open flames should be permitted in the immediate vicinity of liquid oxygen, carbon monoxide, hydrogen or LNG storage tanks. Combustible materials should not be stored in the immediate vicinity of tanks holding any of these liquids, nor near their vent stack outlets.

On vacuum-jacketed containers, loss of vacuum in the annular space, or continual venting of the relief device warrants investigation. Visual signs of vacuum deterioration include "cold (frost) spots" on the outer jacket or rapid pressure rise observed on the tank pressure gage. Notify the supplier immediately.

All storage tanks containing liquefied flammable compressed gases must be grounded and the integrity of the grounds periodically checked. Delivery vehicles should also be grounded to the tank prior to transfer of the product.

Excessive ice formation or other foreign matter should not be allowed to accumulate beneath the vaporizer or tank. A sufficiently thick ice buildup underneath or on the vaporizer might impede air flow, resulting in discharging excessively cold gas.

Only trained personnel should be permitted to perform transfer operations or operate cryogenic systems.

Never overfill a storage tank. Fill it only to the full trycock level per the tank manufacturer's instructions. A minimum vapor space must be maintained in all tanks to accommodate expansion of the liquid. This vapor space may vary from about 3 percent of the tank's capacity on newer tanks to as much as ten percent depending upon the vessel design.

Operators of LNG storage facilities should have a full understanding of the potential differences in densities and temperatures for LNG of varying compositions. If LNG of a different density from that already in the container is added, there is a possibility of stratification and a possible fast release of large amounts of vapor. It is generally best to insure that a new product added to a tank with an existing product is well mixed with the existing product so that stratification will not take place.

SAFETY VALVES AND RUPTURE DISKS

Relief devices should function only during abnormal operation and/or for emergencies. If they operate, check the system for loss of vacuum or leaks. Do not tamper with safety valve settings. Report leaking or improperly set relief valves to the gas supplier so steps can be taken to have them replaced or reset by qualified personnel. Report all safety valves with broken seals to a supervisor. Report frosting, ice formation or excessive corrosion on safety valves, as such conditions may render the valves inoperative.

LIQUID TANKER TRANSFERS

Do not allow unauthorized personnel to congregate near the loading and unloading operation. Keep spectators away. Make sure that trailer wheels are blocked against unintended rolling.

During transfer operations, personnel should wear gloves and safety glasses. Additional eye protection may be required for certain transfer operations. When transferring liquid oxygen, clothing should be free of oil and grease.

There should be no smoking or open flames in the immediate vicinity when transferring oxygen, hydrogen, natural gas or carbon monoxide.

Oxygen transfer is prohibited near the simultaneous transfer of gaseous or liquid fuels. Unloading or loading operations should not commence in any area or on any surface considered unsafe. A concrete trailer apron under the rear of the vehicle is recommended. Asphalt or blacktop should never be used.

Hot air, steam or hot water should be used to thaw frozen equipment. Frozen valves should be thawed by directing water on the valve bonnet and extension below the packing nut. Never direct water onto the packing nut itself. The nut will freeze and render the valve inoperative.

Hose connections should be compatible with the product being transferred. Transfer hoses must be drained prior to disconnecting the hose after transfer. Fill connections should be capped at all times when not in use.

Prior to starting the tractor engine, the driver should make a complete walk around of the trailer to ensure that transfer and ground lines are disconnected, that compartment doors are closed and latched and that all wheel chocks have been removed.

LIQUID HYDROGEN AND HELIUM TRANSFER

All the above suggestions for transferring cryogenic liquids also apply to liquid helium and hydrogen trailers. However, because of the extremely cold temperature of these liquids and their volatility and the flammable nature of hydrogen, these additional requirements should be observed.

All transfer hoses in liquid hydrogen service must be purged with the hydrogen gas before the transfer operation begins. Only helium gas should be used to purge a liquid hydrogen line when the piping is at or below the dew point of nitrogen. All other gases will freeze at the temperature of liquid hydrogen. Liquid helium transfer lines should only be purged with helium.

Hydrogen tankers must be adequately grounded to the storage tank prior to loading and unloading and remain grounded throughout the entire transfer operation.

LIQUID CARBON MONOXIDE TRANSFER

Carbon monoxide is odorless, colorless, toxic and flammable. Therefore, in addition to the general requirements outlined in the previous two sections, the following procedures should be followed:

(1) At least two trained men should be assigned to a liquid carbon monoxide transfer operation. Under both normal and emergency conditions, one man should be assigned and equipped to monitor the area for leaks, leaving the other man free to transfer the liquid. Monitoring should be done on a continuous basis.

(2) Loading or unloading operations should never be left unattended. The two men should always be there.

(3) All personnel should be required to wear self-contained breathing apparatus when making or breaking transfer hose connections and when operating loading or unloading valves. Personnel should also be required to wear this equipment if any leakage occurs and the transfer should be stopped immediately.

Where worker mobility is required, self-contained breathing apparatus is preferable to hose masks. Hose masks supplying pure, filtered breathing air from a cylinder bank or other clean source have a longer uninterrupted supply potential and may be desired for fixed locations such as at unloading operations. Absorptive-type canister units for removing carbon monoxide from the air, and rebreathing respiratory equipment are *not* suitable for use when transferring toxic gases.

A flame should never be used to detect carbon monoxide leaks. Analyzers or detectors capable of measuring the presence of carbon monoxide in the 10 to 200 ppm range should be used and maintained in good operating condition. These analyzers should be calibrated periodically to ensure safe operation. No one should be allowed to enter an area where there is a known or suspected toxic gas leak without wearing a self-contained breathing apparatus.

Large quantities of carbon monoxide gas may be vented in the event that safety valves blow or when tanks are cooled down without nitrogen precooling. Provision should be made either to recover the venting gas, burn it in a flare stack, or vent it to a safe location out of operating areas and where there will be no exposure to personnel.

Carbon monoxide forms a combustible mixture with air over a wide range, although personnel exposure to carbon monoxide is the primary hazard. Process equipment and piping including transfer hoses, storage and transport vessels should be suitably purged with an inert gas before liquid carbon monoxide is introduced.

CONDENSED AIR PROBLEMS

Atmospheric air that comes in contact with the outside of a pipe or other metal at or below liquid nitrogen temperature (-320 F or -196 C) may condense into a liquid dependent upon surrounding conditions. The condensation of air is normally prevented due to the frost accumulation which acts as an insulator. Since oxygen has a higher boiling point (-297 F or -183 C) than the nitrogen in the air, the condensed air that might drip off a pipe may actually be enriched to 35 to 40 percent oxygen concentration. Most materials that burn in air will burn much faster in this oxygen-enriched air.

Atmospheric air will condense on the surface of any exposed line which contains liquid hydrogen or liquid helium. See NFPA 50B (10) for guidance on requirements for insulating materials for liquid hydrogen systems.

HANDLING AIR-SOLIDIFYING PROBLEMS

Liquid hydrogen and helium present special problems because they are so cold that they can solidify air and all other gases. Use containers specifically designed to hold liquid hydrogen or liquid helium. These containers are made from materials that can withstand the rapid changes and extreme differences in temperature encountered with these extremely cold cryogenic fluids.

All containers should be vented or protected by safety devices that permit vapor to escape while excluding air. The vent should be checked at regular intervals to ensure that it does not become plugged with ice which can cause excessive gas pressure that may damage or burst the container.

Before the initial filling of a warm container with liquid hydrogen or helium, it is advantageous to precool the container to minimize flashoff. Liquid nitrogen is frequently used for precooling (consult the container manufacturer for maximum weight of liquid nitrogen that the container can carry). Precooling with liquid nitrogen will purge the air from the container. It is most important however, that the precooling gas be purged by pure helium gas before the container is filled with liquid helium. Before filling containers with liquid hydrogen, either helium or hydrogen purge gas can be used. If an inert gas is in the container, as would be the case if liquid nitrogen precooling were used, it is important to completely remove the liquid nitrogen to prevent plugging problems which could result from its later solidification.

If liquid nitrogen is not readily available, the container may be purged with gaseous nitrogen to make sure that all air is removed, after which the helium or hydrogen purge described above may be used.

Purging is not required prior to the refilling of liquid containers that previously held liquid hydrogen or liquid helium if a positive pressure of the lading has been maintained. If the container has become contaminated with air or other harmful impurities, the remaining liquid should be drained from the container. The container should then be allowed to warm sufficiently to vaporize any collected impurities and then purged with nitrogen gas.

LIQUID HYDROGEN PRECAUTIONS

Liquid hydrogen should be stored, transferred and handled out-of-doors, unless a specific review for safe handling indoors has been made (see Section B, "Special Flammable Gas Precautions," in this chapter).

Use transfer equipment that has been designed specifically for liquid hydrogen service. Liquid hydrogen should not be poured from one container to another or transferred in an air atmosphere. If this is done, oxygen from the air will condense in the liquid hydrogen, presenting a possible explosion hazard. Dewars or other equipment made of glass are not recommended for liquid hydrogen service. The possibility of explosion makes breakage too hazardous to risk.

Cold hydrogen gas released from boiling liquid has about the same density as air, although it diffuses rapidly in all directions. Tests conducted by the U.S. Bureau of Mines with liquid hydrogen spills in a block house show that the hydrogen concentration after about 25 sec was roughly the same throughout the test building.

Every effort must be made to avoid spills, regardless of the rate of ventilation, because it is impossible to avoid a flammable vapor cloud from being created.

LIQUID HELIUM PRECAUTIONS

A good reference on liquid helium is the National Bureau of Standards Monograph No. 111, "Technology of Liquid Helium" (21). Liquid helium's extremely low latent heat of vaporization (8.8 Btu/lb or 20.5 kJ/kg) makes it important to use handling and transfer practices that minimize heat input into the liquid.

A technique often used in dewars to minimize losses is to use the sensible heat of helium flash-off gas from the liquid to absorb incoming heat through the insulation, piping and supports. This principle is very effective with helium since its sensible heat from normal boiling point to 70 F (21 C) is 640 Btu/lb (1489 kJ/kg). That is some 75 times greater than helium's latent heat of vaporization at 1 atm. The use of sensible heat is accomplished by having the boil-off gas flow out along the walls of the long-necked tube that supports the inner container of the helium dewar.

Since liquid helium is normally handled only in closed systems, such systems must be pressurized to greater than atmospheric pressure to

prevent the backflow of air into the equipment. Relief valves also must be designed to prevent back leakage of air into the system.

The problem of backflow or diffusion of air into liquid helium and hydrogen vessels and equipment must be emphasized. It is the major safety hazard with these cryogenic fluids, and if it occurs, blockage of the openings may result, leading to rupture of the container. Another form of the problem exists where liquid nitrogen is used to shield liquid helium equipment (in which case nitrogen can be the contaminating element rather than air if a leak develops between the two vessels). Proper equipment design should eliminate this problem.

The problem of air backflow often is caused by variations in atmospheric pressure. A substantial drop in atmospheric pressure can cause a vessel to vent and blow down, if regulated by a conventional pressure-relief valve. Upon subsequent rise in atmospheric pressure, the vessel may be subjected to air backflow if the atmospheric pressure exceeds the helium pressure in the vessel, and the relief device does not provide a perfect seal. This problem is especially common in air cargo shipments of liquid helium. The changes in atmospheric pressure between ground and high altitude even in pressurized cabins can create the problem. Ordinary relief devices can freeze or plug up following rapid ejection of cold gas during altitude changes.

The most satisfactory solution to this problem is the use of absolute pressure-relief valves. With such devices a vessel full of liquid helium can be air transported or trucked by land over mountainous terrain without changing the dewar's internal liquid saturation conditions or its normal venting pattern. However, the helium supplier, and the user should discuss the problem when unusual changes in elevation are expected during truck or rail shipment or when air shipment is planned.

When air shipment is required for liquid helium, it is suggested that the shipper make advance arrangements with the local cargo manager of the appropriate airline to avoid any question or unnecessary delay when the container is presented for shipment.

Each liquid helium container must be packaged in its proper shipping crate or frame to prevent unloading from aircraft in any position other than upright. Details are provided in the manufacturer's instruction manual.

The following identification notices must also be prominently displayed on the container:

THIS SIDE UP (at 120° intervals with arrows)

DO NOT DROP

HELIUM, LIQUID, LOW PRESSURE (for shipments at pressures which do not exceed 40 psia)

CARGO AIRCRAFT ONLY (label for shipments over 60 liters)

Detailed air shipment data including the requirements for suitable pressure-relief devices are provided in the following publications.

For Domestic Shipment: "Official Air Transport-Restricted Articles Tariff No. 6-D" (17). Obtain this document from:

Airline Tariff Publishers, Inc.
1825 K Street, N.W.
Washington, DC 20006

For International Air Shipment: "IATA Restricted Articles Regulations" (16). Obtain this document from:

International Air Transport Association
P.O. Box 315
1215 Geneva 15 Airport, Switzerland

Note: Regulations for domestic and international air shipment are identical.

Upon arrival, every liquid helium container should be examined for abnormal relief valve condition and also to ensure that no neck plug has formed in the dewar. Each dewar also should be periodically checked in the user's plant or laboratory to make sure that the condition of relief devices is normal and that no neck plug has formed.

ELIMINATING PLUGS

Plugs of ice or foreign material can develop in cryogenic container vents and openings. If the proper action is not immediately taken, the vessel can rupture. Prevent plugs by following the supplier's operating procedures. If a plug does develop, contact the supplier immediately. Do not attempt to remove the plug without consulting the supplier. Remove the vessel to a remote location if possible, in accordance with the supplier's recommendations.

SECTION E

Bulk Storage Site Considerations

Some general principles are summarized here for the installation of bulk storage units for cryogenic liquid on industrial and institutional consumer premises where the cryogenic liquid is delivered to the site by mobile equipment.

The primary considerations in locating the cryogenic storage vessel, regardless of the product stored, are to minimize exposure to potential fire and ignition sources and to ensure adequate ventilation to avoid excessive concentrations of vapor or liquid in the event of leakage of the product.

Some special requirements and precautions for the bulk storage of the various cryogenic liquids are also mentioned in this section, including the applicable tank fabrication specifications, distances from various types of structures in locating the tanks and precautions in the event that the cryogenic liquid is inert, oxidizing, flammable or toxic.

BULK STORAGE VESSEL SPECIFICATIONS

Stationary vessels installed at the consumer site having an operating pressure in excess of 15 psig (103 kPa, gage) should meet all requirements of Section VIII of the ASME Boiler and Pressure Vessel Code, including the required stamping and nameplate.

In applications governed by laws or regulations issued by municipal, state, provincial or federal authorities covering pressure vessels, those laws or regulations should be reviewed to determine size or service limitations which may be more restrictive than cited references or specifications in this chapter.

Some vessels are used for a combination of transportation and dispensing of the product at the consumer site. Such vessels are fabricated to a U.S. Department of Transportation Specification and are subject to applicable DOT regulations.

The pressure-relief devices of stationary vessels operating above 15 psig normally consist of a pressure-relief valve and rupture disk piped in parallel in accordance with CGA Pamphlet S-1.3, "Safety Relief Device Standards, Compressed Gas Storage Containers" (also see Chapter 5, Section C of this book). An outer vessel-relief device must also be provided.

Vessels installed at consumer sites having an operating pressure not in excess of 15 psig may be fabricated to meet the requirements of Section VIII of the ASME Code, or the applicable provisions of API-620, "Recommended Rules for Design and Construction of Large, Welded, Low-Pressure Storage Tanks."

INSTALLATION

Bulk storage tanks and vessels should be installed above ground and out-of-doors so that they are readily accessible to mobile supply equipment at ground level and to authorized personnel. Bulk storage vessels, piping valves, regulating equipment, vaporizers and other accessories should be suitably protected against physical damage. During the installation of the vessel, all spectators and other unauthorized personnel should be kept from the immediate vicinity.

Permanently installed vessels should be provided with substantial noncombustible supports securely anchored on firm noncombustible foundations such as concrete. Permanent bulk liquid oxygen tanks should be installed on concrete pad foundations. A concrete apron should be provided to accommodate the rear end of the delivery vehicle. Macadam, wood, asphalt or any other oil-base surfaces should never be used

due to possible reaction with oxygen which might be spilled.

For vessels containing flammable cryogens and supported by steel columns in excess of 18 in. (46 cm) in height should be coated with protective coatings having a 2-hour fire-resistance rating.

Permanent wheel stops should be provided at each installation where a tanker truck is backed close to the tank to deliver a product. Protective guards or corner posts should be installed when appropriate to prevent vehicular traffic from hitting the tank or its projecting piping.

Electrical equipment should be installed in accordance with the applicable provisions of the National Electrical Code (NFPA No. 70, ANSI C-1). The general purpose of weatherproof types of electrical wiring and equipment are acceptable depending upon whether the wiring is indoors or outdoors for installations in oxygen, or inert (nitrogen, argon, helium, neon, xenon) liquid service.

For flammable cryogenic hydrogen installations, electrical wiring and equipment located within 3 ft (about 1 m) of a point where connections are regularly made and disconnected should be in accordance with Article 501 of NFPA 70 for Class I, Division 1 locations, and if within 3 to 25 ft (0.9 to 7.6 m) of a point where connections are regularly made and disconnected or within 25 ft (7.6 m) of a flammable liquid storage vessel should be in accordance with Article 501 for Class I, Group B, Division 2 locations. For other flammable cryogenic liquid installations, refer to the National Electrical Code, Article 501, NFPA No. 70.

A vessel and associated piping containing a flammable cryogen should be electrically bonded and grounded. The mobile supply unit should also be electrically grounded to the grounded bulk storage vessel prior to the start of transfer of a product and should remain grounded throughout the entire transfer operation.

The possibility of leakage from connections to the vessel should be considered in the vessel installation and a slight slope should be provided away from the vessel toward a suitable drainage.

It is advisable to locate liquid oxygen vessels on ground higher than any nearby flammable or combustible liquid storage site. If the bulk oxygen system is lower, suitable means should be taken (such as by diking, diversion curbs or grading) to prevent accumulations of flammable liquids under the bulk oxygen system. Similar steps should be taken to prevent the accumula-tion of liquid oxygen under vessels containing flammable cryogenic fluids.

All cryogenic vessels should be placarded to indicate the product. NO SMOKING–NO FLAMES warning signs should be affixed to vessels containing liquid oxygen in accordance with NFPA 50. FLAMMABLE GAS–NO SMOKING–NO OPEN FLAMES warning signs

TABLE 4. DISTANCE BETWEEN BULK OXYGEN SYSTEMS AND EXPOSURES*

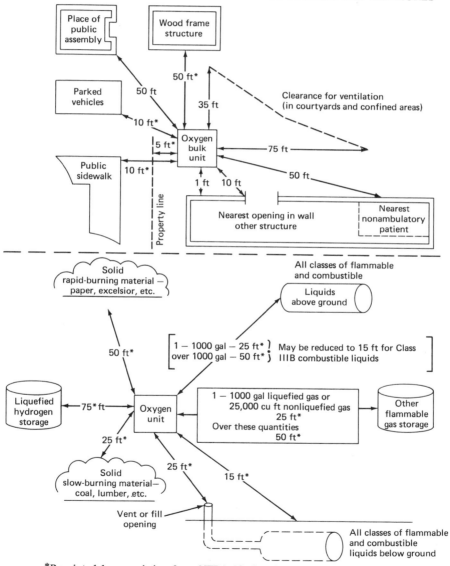

*Reprinted by permission from NFPA 50, Standard for Bulk Oxygen Systems at Consumer Sites, Copyright © 1979, National Fire Protection Association, Boston, MA.

should be affixed to vessels containing flammable cryogens. Tanks should also be properly marked with product decals visible from the unloading station to indicate the name of the contained product.

For installations which require any operation of the equipment by the user, legible instructions should be provided at operating locations.

SPECIAL LIQUID OXYGEN REQUIREMENTS

Bulk oxygen storage systems are probably the most numerous of the cryogenic storage systems in use today and are found at most hospitals, at many industrial plants and at many other locations. These systems are considered to be very safe provided that the requirements and safeguards of NFPA Pamphlet No. 50,

"Bulk Oxygen Systems" are observed. NFPA No. 50 now serves as the official guide for most fire department approvals of such installations.

Table 4 is a pictorial display of the requirements of NFPA No. 50.

CGA Pamphlet G-4.4, "Industrial Practices for Gaseous Oxygen Transmission Systems," presents information on the current practices used in gaseous oxygen transmission and distribution piping systems. Such piping systems are frequently supplied with vaporized oxygen from a bulk oxygen storage vessel.

SPECIAL LIQUID HYDROGEN REQUIREMENTS

Bulk hydrogen storage systems require careful consideration of the location and installation of

TABLE 5. MINIMUM DISTANCE IN FEET FROM LIQUID HYDROGEN SYSTEMS TO EXPOSURE**

Type of Exposure	Liquefied Hydrogen Storage (Capacity in Gallons)		
	39.63 (150 liters) to 3,500	3,501 to 15,000	15,000 to 30,000
1. Fire-resistive Building*	5	5	5
2. Noncombustible/Limited-Combustible Building*	25	50	75
3. Building Construction Types other than 1 and 2*	50	75	100
4. Wall Openings, Air Compressor Intakes, Inlets for Air-conditioning or Ventilating Equipment	75	75	75
5. Class I, II and IIIA Flammable and Combustible Liquids (Aboveground and Vent or Fill Openings if Below Ground)(See 514)	50	75	100
6. Between Stationary Liquefied Hydrogen Containers	5	5	5
7. Flammable Gas Storage (other than hydrogen)	50	75	75
8. Liquid Oxygen Storage and other Oxidizers (See 514)	75	75	75
9. Combustible Solids	50	75	100
10. Open Flames, Smoking and Welding	50	50	50
11. Places of Public Assembly	75	75	75
12. Public Ways, Railroads, and Property Lines	25	50	75
13. Protective Structures (Notes 1 and 2)	5	5	5

*Refer to NFPA No. 220, Standard Types of Building Construction, for definitions of construction types.

Note 1: The distances in Nos. 2, 3, 5, 7, 8, 9 and 12, in Table 2, may be reduced where protective structures, such as fire walls equal to height of top of the container, to safeguard the liquefied hydrogen storage system, are located between the liquefied hydrogen storage installation and the exposure. (See definition of Outdoor Location.)

Note 2: Where protective structures are provided, ventilation and confinement of product shall be considered. The 5-foot distance in Nos. 1 and 13 facilitates maintenance and enhances ventilation.

**Reprinted by permission from NFPA 50B, Standard for Liquefied Hydrogen Systems at Consumer Sites, Copyright © 1978, National Fire Protection, Association, Boston, MA.

such facilities due to hydrogen's wide flammable range. The requirements of NFPA 50B, "Liquefied Hydrogen Systems," should be met. Table 5 provides a summary of the minimum distances between the bulk storage vessel and the specified exposure, from NFPA 50B.

Storage Area Restrictions. Bulk liquid hydrogen systems should be installed to prevent access by unauthorized personnel and to prevent smoking or other ignition sources within the area.

REFERENCES

(1) Code of Federal Regulations 49, Transportation Parts 100 to 199, 1975. Superintendent of Documents, U.S. Government Printing Office, Washington, DC 20402.

(2) Section VIII, Division 1 of ASME Boiler and Pressure Vessel Code, The American Society of Mechanical Engineers, United Engineering Center, 345 E. 47th Street, New York, NY 10017.

The following CGA Pamphlets are available from the Compressed Gas Association, Inc., 500 Fifth Avenue, New York, NY 10110.

(3) S-1.1 Pressure Relief Device Standards—Cylinders for Compressed Gases.

(4) S-1.2 Pressure Relief Device Standards—Cargo and Portable Tanks for Compressed Gases.

(5) S-1.3 Pressure Relief Device Standards—Compressed Gas Storage Containers.

(6) 341 Standard Insulated Tank Truck Specification.

(7) SB-2 Oxygen Deficient Atmospheres.

(8) Specifications for Tank Cars, The Association of American Railroads, 1920 L Street N.W., Washington, DC 20036.

The following are available from the National Fire Protection Association, 470 Atlantic Avenue, Boston, MA 02210:

(9) NFPA 50 Bulk Oxygen Systems.

(10) NFPA 50B Liquefied Hydrogen Systems.

(11) NFPA 51A Acetylene Cylinder Charging Plants.

(12) NFPA 56F Nonflammable Medical Gas Systems.

(13) NFPA 59A Storage and Handling Liquefied Natural Gas.

(14) NFPA 68 Explosion Venting.

(15) NFPA 70 National Electrical Code.

The following are available as indicated:

(16) IATA Restricted Articles Regulations, International Air Transport Association, P.O. Box 315 Geneva 15 Airport, Switzerland.

(17) Air Transport Restricted Articles Tariff No. 6-D, Airline Tariff Publishers Company, Agent, Dulles International Airport, P.O. Box 17415, Washington, DC 20041.

(18) M. Zabetakis, U.S. Bureau of Mines, *Safety with Cryogenic Fluids*, 1967, Plenum Press, 277 W. 17th Street, New York, NY 10011.

(19) Safety Panel of British Cryogenics Council, *Cryogenics Safety Manual*, 1970, British Cryogenics Council, 16 Belgrave Square, London SW1.

(20) R. Scott, *Cryogenic Engineering*, 1959, Van Nostrand Reinhold, Inc., New York, NY.

(21) R. H. Kropschot, B. W. Birmingham, and D. B. Mann, Ed., *Technology of Liquid Helium*, National Bureau of Standards Monograph 111, Superintendent of Documents, U.S. Government Printing Office, Washington, DC 20402.

(22) National Bureau of Standards Handbook No. 44, Specifications, *Tolerances and Other Technical Requirements for Commercial Weighing and Measuring Devices*, U.S. Superintendent of Documents, U.S. Government Printing Office, Washington, DC 20402.

In addition to the foregoing specific references, the following companies and safety organizations contributed safety literature and information which was quite helpful in the preparation of this chapter:

Air Products and Chemicals, Inc.
Airco, Inc.
British Cryogenic Council
Chemetron Corp.
Liquid Carbonic Co.
Union Carbide Corp., Linde Division

CHAPTER 10

Equipment Cleaned for Oxygen Service

1. SCOPE

The cleaning methods described in this chapter are intended for cleaning equipment used in the production, storage, distribution and use of liquid and gaseous oxygen. Examples of such equipment (illustrative of the primary intent of this chapter) are: stationary storage tanks, trucks and tank cars; pressure vessels such as heat exchangers and rectification columns; and associated piping, valves and instrumentation. The cleaning methods, however, are not limited to the above equipment and with proper consideration or modification may be utilized in cleaning other oxygen service equipment such as cylinders, cylinder valves, regulators, welding torches, pipelines, compressors and pumps.*

2. OBJECTIVES

Oxygen equipment and systems, including all components and parts thereof, must be adequately cleaned to remove harmful contamination prior to the introduction of oxygen. Harmful contamination would include both organic and inorganic materials such as oils, greases, paper, fiber, rags, wood pieces, solvents, weld

*CGA Pamphlet C-10, "Recommendations for Changes of Service for Compressed Gas Cylinders," includes the conversion of cylinders to oxygen service. Where the requirements for this conversion differ from the recommendations of CGA Pamphlet G4.1, the recommendations of CGA C-10 should take precedence.

slag dirt and sand which if not removed could cause a combustion reaction in an oxygen atmosphere or result in an unacceptable product purity.

This chapter presents methods for cleaning oxygen service equipment. When properly used, these cleaning methods and subsequent inspections will result in the degree of cleanliness required for the safe operation of oxygen service equipment and the necessary product purity required in CGA Commodity Specification for Oxygen, Pamphlet G-4.3. Suggested maximum levels of contamination and ways of determining if a component or system is sufficiently clean to be used in oxygen service are given along with procedures for keeping such equipment clean before being placed in service.

Cleaning a component or system for oxygen service involves the removal of combustible contaminants including the surface residue from manufacturing, hot work, and assembly operations, as well as the removal of all cleaning agents and the prevention of recontamination before final assembly, installation and use. These cleaning agents and contaminants include solvents, acids, alkalies, water, moisture, corrosion products, noncompatible thread lubricants, filings, dirt, scale, slag, weld splatter, organic material (such as oil, grease, crayon and paint) lint and other foreign materials.

The removal of injurious contaminants can be accomplished by precleaning all parts and maintaining this condition during construction, by completely cleaning the system after construction or by a combination of the two.

3. PLANNING REQUIREMENTS

3.1. Supervision

An individual skilled in the techniques required for oxygen service cleaning shall be responsible for monitoring the cleaning operation and determining if a component or system is clean so that it can function in an oxygen environment. Where piping systems with multiple branches are involved it is of paramount importance that the cleaning procedures be well established, suitably integrated with the sequence of construction operations and precisely followed since it may be neither practical nor possible to inspect such a system completely for cleanliness after construction and final cleaning.

3.2. Selecting Procedures

3.2.1. In order to decide on the most practical method of cleaning, inspecting and testing, it is first desirable to estimate the nature, possible location and degree of contamination. In addition, the arrangement of passages must be studied so that cleaning, washing or draining practices can be adjusted to make sure that deadend passages and possible traps are adequately cleaned.

3.2.2. The cleaning procedure selected, which includes removal of the cleaning agent, will depend on several factors such as the following:

(a) the nature of the contaminants;
(b) the location and degree of contamination;
(c) the arrangement of passages with respect to their ability to be flushed and drained;
(d) the effectiveness of the cleaning agent in removing the contaminants;
(e) the compatibility of the cleaning agent with the contaminants, metals and material involved:
(f) the availability and cost of cleaning agents and cleaning methods, the availability of

personnel experienced in handling these materials;
(g) the speed and effectiveness of cleaning and the desired level of cleanliness.

3.2.3. A list of typical cleaning procedures would include:

(a) steam cleaning (including hot water and detergents);
(b) vapor degreasing;
(c) solvent washing (including ultrasonics);
(d) alkaline (caustic) washing;
(e) acid cleaning;
(f) mechanical cleaning (blast cleaning, wire brushing, etc.);
(g) purging.

3.2.4. A detailed cleaning procedure in accordance with the instructions of the manufacturer of the cleaning agent should be specified to the satisfaction of both the manufacturer and the purchaser of the oxygen equipment and followed throughout the project.

4. PRECLEANING

Prior to cleaning, component material not compatible with the cleaning agent shall be removed or isolated. Gross amounts of foreign material such as scale, dirt, grit, solid objects and hydrocarbons shall be removed. Removal may be accomplished by grinding, wire brushing, blast cleaning, sweeping, vacuuming, swabbing, etc.

5. STEAM OR HOT WATER CLEANING

Steam or hot water cleaning may be described as the use of steam or hot water propelled through a nozzle or sprayhead and usually assisted by a detergent to remove contaminants such as dirt, oil and loose scale.

5.1. Materials

The steam or hot water should be clean and oil free. In most steam or hot water cleaning operations, a detergent solution is combined with the steam or hot water to provide an acceptable level of final cleanliness. The detergent selected, shall be suitable for the contaminants involved and shall also be compatible with the surfaces being cleaned.

5.2. Steam Cleaning

5.2.1. Equipment. The equipment used may consist of a steam and water supply, a length of hose, and a steam lance with or without a spray nozzle.

5.2.2. Steam Cleaning Procedure. Either plant steam or steam from a portable steam generator can be used. If a steam lance is used, the detergent solution may enter the steam gun by venturi action and mix with the steam. Steam removes oils, greases and soaps by first "thinning" them with heat. Dispersion and emulsification of the oils then occur followed by dilution with the condensed steam. The system should provide control over the steam, water and detergent flows so that full effect of the detergent's chemical action, the heat effect of the steam and the "abrasive" action of the pressure jet is attained for maximum cleaning efficiency.

If the steam is clean and free of organic material, a secondary cleaning operation with a solvent or alkaline degreaser may not be required in cases where the initial contamination is not heavy or is readily removed with steam.

5.3. Hot Water Cleaning

5.3.1. Equipment. Cleaning with a hot detergent solution may utilize a spray system or a cleaning vat with suitable agitation of the solution or the parts.

5.3.2. Hot Water Cleaning Procedure. Hot detergent solution cleaning can be used where a steam temperature is not necessary to free and fluidize contaminants. Proper consideration shall be given to the size, shape and the number of parts to be cleaned so as to assure adequate contact between the surfaces to be cleaned and the solution. The solution temperature should

be in accordance with the recommendation of the manufacturer of the cleaning agent.

5.4. Removal of Cleaning Agents

Most detergents are water soluble and are best removed by prompt flushing with sufficient quantities of hot or cold clean water as appropriate before the cleaning agents have time to precipitate. The equipment is then dried by purging with dry oil-free air or nitrogen which may be heated to shorten the drying time.

6. CAUSTIC CLEANING

Caustic cleaning is described as cleaning with solutions of high alkalinity for the removal of heavy or tenacious surface contamination followed by a rinsing operation.

6.1. Materials

There are many effective cleaning materials available for caustic washing. They are basically alkalies and are water soluble, nonflammable and may be harmful in contact with the skin or if swallowed. The cleaning agents should be chosen so that they do not react chemically with the metal being cleaned.

The water that is used for rinsing should be free of oil and other hydrocarbons and should contain no particles larger than those acceptable on the cleaned surface.

Filtration may be required. It may be desirable to analyze the water to determine the type and quantity of impurities. Some impurities may cause undesirable products or reactions with the particular caustic cleaner used.

6.2. Caustic Washing Procedure

The cleaning solution can be applied by spraying, immersion flushing or hand swabbing. Spraying works well, but requires a method whereby the cleaning solution reaches all areas of the surface. It is also desirable to have provisions for draining the solution faster than it is introduced to avoid accumulation.

Immersion or flushing should be total rather than partial as the solution tends to dry on the surface that is exposed to air.

Hand-swabbed surfaces should be rinsed before the cleaning solution dries.

Generally, cleaning solutions perform better when warm. Depending upon the particular solution, this temperature will be in the range of 100 F to 180 F (38 C to 82 C).

The cleaning solution can be reused until it is too weak or too contaminated as determined by pH or concentration analysis. Both decrease as the solution weakens. Experience will establish a contaminant level above which a surface could not be acceptably cleaned.

6.3. Rinsing

The cleaning accomplished is only as good as the rinsing job. All the contaminants may be held in suspension in the cleaning solution. However, if the cleaning solution is not completely flushed from the surface being cleaned, the contaminant in any remaining solution will redeposit on the surface during the drying operation. The surface must never be allowed to dry between the cleaning phase and the rinsing phase because all of the film or residue will not be removed.

Frequently some type of agitation during rinsing is required. This may be by mechanical brushing, fluid impingement, agitation of the parts being cleaned, etc.

The water rinse is often warm to help remove the cleaning solution and aid in the drying process. A method of determining when the rinsing is complete is to monitor the pH of the outlet rinse water. The pH approaches that of the original rinse water as the rinsing progresses.

6.4. Drying

If drying is not completed with the residual heat in the metal, it can be completed with dry oil-free air or nitrogen. If it is desirable that the equipment be maintained in a dry atmosphere before installation or use, the dew point of the contained atmosphere should not be above −30 F (−34 C).

7. ACID CLEANING

This cleaning procedure removes oxides and other contaminants by immersion in a suitable acid solution, usually at room temperature.

7.1. Material

The type of cleaning agent selected will depend, in most cases, on the material to be cleaned. The following general guide can be followed:

(a) Phosphoric acid base cleaning agents can be used for all metals. These agents will remove oxides, light rust, light soils and fluxes.

(b) Hydrochloric acid base cleaning agents are recommended for carbon and low-alloy steels only. These agents will remove rust, scale, and oxide coatings and will strip chromium, zinc and cadmium platings. Certain acidic solutions including hydrochloric or nitric acids should contain an inhibitor to prevent harmful attacks on base metals. Hydrochloric acid should not be used on stainless steel since it may cause stress corrosion.

(c) Chromic acid base cleaning agents and nitric acid base cleaning agents are recommended for aluminum and copper and their alloys. These agents are not true cleaning agents but are used for deoxidizing, brightening and for removing black smut which forms during cleaning with an alkaline solution. Some agents are available as liquids and others as powders and are mixed in concentrations of 5 to 50 percent in water, depending on the cleaning agent and the amount of oxide or scale to be removed.

7.2. Equipment

A storage or immersion tank, acid-resistant recirculation pump and associated piping and valving compatible with the acid solution are required.

7.3. Cleaning Procedure

Common methods of applying acid cleaning agents used for cleaning metals are:

(a) Large areas may be flushed with an appropriate acid solution.
(b) Small parts may be immersed and scrubbed, or agitated in the solution.

Caution: Acid cleaning agents should not be used unless their application and performance are known or are discussed with the cleaning agent manufacturer. The manufacturer's recommendations regarding concentration and temperature should be followed for safe handling of the cleaning agent.

7.4. Rinsing

Rinse thoroughly with cold water. Rinsing must begin as soon as practicable after cleaning to prevent excessive attack on the material being cleaned by the acid cleaning solution.

If there is a chance of any cleaning solution becoming trapped in the equipment being cleaned, a dilute alkaline neutralizing solution can be applied, followed by water rinsing.

7.5. Drying

If drying is not completed with the residual heat in the metal, it can be completed with dry oil-free air or nitrogen. If it is desirable that the equipment be maintained in a dry atmosphere before installation or use, the dew point of the contained atmosphere should not be above −30 F (−34 C).

8. SOLVENT WASHING

8.1. Solvent Washing

Solvent washing may be described as the removal of organic contaminants from the surface to be cleaned by the use of chlorinated hydrocarbons or other suitable solvents.

8.2. Ultrasonic Cleaning

Ultrasonic cleaning may be described as the loosening of oil and grease or other contamination from metal surfaces by the immersion of parts in a solvent or detergent solution in the presence of high-frequency vibrational energy.

8.3. Materials

The solvents frequently used are methylene chloride; refrigerant 11; perchloroethylene; 1, 1, 1,-trichlorethane (methylchloroform); and trichlorethylene. Refer to Section 13.2 for additional precautions. Carbon tetrachloride shall not be used because of its high toxicity, i.e., its low-threshold limit value (TLV). Trichloroethylene should be used only if absolutely necessary because it is 3.5 times as toxic as methylchloroform.

The boiling points, freezing points, toxicity (threshold limit values) and Kauri-Butenol numbers are listed in Table 1.

8.4. Washing Equipment

Washing equipment may consist of a recirculating system for the solvent or a closed container for immersing parts.

Auxiliary control and test equipment might include the following: space heaters, halogen detectors, thermometers, a utility container, funnel and strainer, an Imhoff cone, dry oil-free air or nitrogen and syphon pump.

For ultrasonic cleaning, a high-frequency sound generator and container is substituted for the recirculation system.

Caution: Some plastic tubing including polyvinylchloride (PVC) may have its plasticizer extracted by the solvent and deposited on the surface being cleaned. Rubber and neoprene tubing should not be used with these solvents for the same reason when cleaning oxygen equipment. Nylon and polytetrafluoroethylene

TABLE 1. CHEMICAL AND PHYSICAL PROPERTIES OF CLEANING SOLVENTS

Solvent	Formula	Molecular Weight	Boiling Point (°F)	Freezing Point (°F)	Density at 68 F (lb/ft²)	Latent Heat of Vaporization at Boiling Point (Btu/lb)	Evaporation Rate (Ether = 100)	TLV[1] (ppm)	Residue Weight (%)	Kauri-Butanol Number[2] at 77 F
1,1,1 Trichloroethane (methylchloroform)	$C_2H_3Cl_3$	133.42	161.8	−58	82.1 (77 F)	95.4	37	350	0.001	130
Methylene chloride	CH_2Cl_2	84.94	104.2	−142.1	83.37	141.7	62	100	0.00075	109
Perchloroethylene	C_2Cl_4	165.85	250.2	−8.2	101.5	90.0	12	100	0.001	88
Refrigerant 11 (fluorotrichloromethane)	CCl_3F	137.4	74.8	−168	92.7 (70 F)	78.31	81	1000	0.001	60
Trichloroethylene[3]	C_2HCl_3	131.40	188.6	−122.8	91.42	103.0	30	100	0.0005	130

[1]Threshold limit values (time weighted average) adopted by American Conference of Governmental Industrial Hygienists, 1975.
[2]The higher the Kauri-Butanol number, the greater the solvent action.
[3]Trichloroethylene has a listed flash point of 90 F (32 C) and flammable limits of 12 to 40 percent in air.

(PTFE) tubing are satisfactory with the frequently used solvents.

8.5. Washing Procedure

A sample of new wash solvent should be taken for control purposes when required. Circulate the wash solvent through the equipment for a predetermined period. The desired cleanliness level can be determined by comparing the used solvent with new solvent. A vessel can be considered clean when no distinct color difference exists between the two samples. Additional washings with new solvent may be required to obtain the desired level of cleanliness. Then drain the solvent into a container and ensure that all solvent has been removed from the equipment by using such techniques as temperature and concentration monitoring of the exit purge gas. Immersion wash procedures may be utilized if practical.

If solvent monitoring is desired, measure the used solvent collected, take a representative sample of the solvent and determine its contaminant level, correcting for the amount of contaminant in the original solvent.

The solvent shall be discarded or reclaimed when the cleaning operation does not yield acceptably cleaned surfaces. A useful guide for this determination is when the solvent is dis-colored more than new solvent. Where standards are considered necessary, use ASTM standard D-2108-71 "Color of Halogenated Organic Solvents and Their Admixtures" for guidance.

Caution: Use proper solvent transfer containers (precleaned glass or metal) with no seals that can be dissolved by the solvent.

8.6. Removal of Solvents

After the oil and grease contaminants have been removed or dissolved and the solvent drained, blow down the piping or tubing with dry oil-free air or nitrogen to remove liquid by entrainment. Then circulate the purge gas until the final traces of the solvent have been removed. Purging can be considered complete when the solvent cannot be detected by appropriate methods in the gas venting from the vessel being purged.

A halogen leak detector may be used with chlorinated solvents for determining when a vessel is adequately purged. If the odor of solvent gases is detected in the vicinity of the effluent purge gas, the equipment requires additional purging. The method of test should be agreed upon by the manufacturer and the purchaser.

For equipment being used in oxygen service, it may be desirable to estimate the total quantity of oil or grease removed to justify future extensions of operating periods between washing or omissions of washing operations.

9. VAPOR DEGREASING

Vapor degreasing can be described as the removal of soluble organic materials from the surfaces of equipment by the continuous condensation of solvent vapors and their subsequent washing action.

9.1. Equipment

Commercial degreasers are available for cleaning metals at room temperatures. Vapor degreasing equipment consists essentially of a vaporizer for generating clean vapors from a contaminated solvent and a vessel for holding the parts to be cleaned in the vapor space.

9.2. Materials for Vapor Degreasing

The solvents frequently used for vapor degreasing are methylene chloride; refrigerant 11; perchloroethylene; 1, 1, 1-trichlorethane (methylchloroform); and trichloroethylene. Some of these solvents are flammable in air under certain conditions and have varying degrees of toxicity. Caution should be exercised in their use. Dry oil-free air or nitrogen should be available for purging.

9.3. Vapor Degreasing Procedure

The following procedure is useful for cleaning cold or cryogenic equipment.

The temperature of a vessel must be between the freezing and boiling points of the solvent so that the solvent vapors will condense and wash down by gravity over the equipment surfaces.

This cleaning procedure requires that the solvent be boiled in a vaporizer and the solvent vapors piped into a relatively cold vessel where the vapors condense on the cold surfaces. The equipment should be positioned and connected so that the condensate can be thoroughly drained from the system. Continuous removal of the condensate and its transport back into

the vaporizer will carry the dissolved impurities into the vaporizer where they remain, as fresh pure vapors are released to continue the degreasing operation.

Cleaning can be considered complete when the returning condensate is as clean as the unused solvent.

Note: The vapor degreasing action will stop when the temperature of the vessel reaches the boiling point of the solvent.

9.4. Removing Solvent Vapors

The solvent should be removed by following the procedure in Section 8.6.

10. MECHANICAL CLEANING

This type of cleaning may be accomplished by blast cleaning, wire brushing or grinding.

10.1. Blast Cleaning

Blast cleaning may be described as the use of abrasives propelled through nozzles against the surface of pipe, fittings, or containers to remove mill scale, rust, varnish, paint or other foreign matter. The medium propelling the abrasive shall be oil free unless the oil is to be removed by subsequent cleaning. Specific abrasive materials shall be suitable for performing the cleaning without depositing contaminants that cannot be removed by subsequent cleaning. Care is to be taken when blast cleaning so as not to remove an excessive amount of parent metal. The blasting medium and residue shall be removed to meet the cleanliness levels suggested herein for oxygen service equipment.

10.2. Wire Brushing or Grinding

Accessible surfaces may be wire brushed. Welds may be ground and wire brushed to remove slag, grit or excess weld material. Carbon steel wire brushes shall not be used on aluminum or stainless steel surfaces. Any wire brushes previously used on carbon steel shall not be used on aluminum or stainless steel surfaces.

10.3 Tumbling

Tumbling can be described as a cleaning method that uses a quantity of hard abrasive material placed in a container to clean the internal surfaces. The container and the abrasive are energized so as to impart relative motion between the abrasive material and the container.

10.4. Swabbing and Vacuuming

Equipment, parts or piping may be vacuum cleaned after mechanical cleaning to remove loose particles of dirt and slag.

If vacuum cleaning is not possible the surfaces may be swabbed with a suitable solvent using a clean lint-free cloth to remove loose dirt, slag, etc.

10.5. Blowing and Purging

After the equipment, parts and piping have been mechanically cleaned and any abrasive material removed, the assembled piping should be blown with dry oil-free air or nitrogen to remove small particles and any solvent vapors present.

If drying is not completed with the residual heat in the metal, it can be completed with dry oil-free air or nitrogen. If it is desirable that the equipment be maintained in a dry atmosphere before installation or use, the dew point of the contained atmosphere should not be above −30 F (−34 C).

11. INSPECTION

11.1 Approval of Quality Control—Procedures and Standards

Detailed quality control standards and procedures should be agreed upon between the manufacturer and the purchaser. A source inspection by the purchaser's representative at the manufacturer's location may be desirable.

The purchaser should initially and periodically inspect the manufacturer's facilities and audit cleaning and quality control procedures.

11.1.1. Record Keeping. Records of the following information as applicable should be prepared for the cleaned equipment or assembly, kept on file, and if requested, a copy forwarded to the purchaser.

(a) a descriptive name of the item covered;
(b) its serial number;
(c) its invoice number or other means of identification;
(d) the cleaning specification and method employed;
(e) the dates of inspection for cleanliness;
(f) the method of inspection;
(g) the results of inspection;
(h) the inspector's signature and date signed.

11.2. Inspection Procedures

When specified by the purchaser, any one or combination of the following tests shall be used to assess the cleanliness of a piece of equipment. Failure to pass any of the specified tests requires recleaning and reinspection and may require reevaluation of the cleaning procedures. In-process inspections to assure adequacy of cleaning procedures may be desirable.

11.2.1. Direct Visual Inspection (White Light). This is the most common test used to detect the presence of contaminants such as oils, greases, preservatives, moisture, corrosion products, weld slag, scale, filings and chips and other foreign matter. The item is observed for the absence of contaminants (without magnification 20/20 vision) under strong white light and for the absence of accumulations of lint fibers. This method will detect particulate matter in excess of 50 microns (0.002 in. or 0.00005 m) and moisture oils, greases, etc. in relatively large amounts.

11.2.2. Direct Visual Inspection (Ultraviolet Light). Ultraviolet light causes many common hydrocarbon or organic oils or greases to fluoresce when they may not be detectable by other visual means. The surface is observed in darkness or subdued light using an ultraviolet light radiating at wavelengths between 2500 and 3700 Å units. Ultraviolet (black light) inspection shall indicate that cleaned surfaces are free of any hydrocarbon fluorescence. Accumulations of lint or dust that may be visible under the black light shall be removed by blowing with dry oil-free air or nitrogen, wiping with a clean lint-free cloth or vacuuming. Not all organic oils fluoresce and some materials such as cotton lint that fluoresce are acceptable unless present in excessive amounts. If fluorescence shows up as a blotch, smear, smudge, or film, reclean the fluorescing area.

1.1.2.3. Wipe Test. This test is used to detect contaminants on visually inaccessible areas as an aid in the above visual inspections. The surface is rubbed lightly with a clean white paper or lint-free cloth which is examined under white and ultraviolet light. The area should not be rubbed hard enough to remove any oxide film as this could be confused with normal surface contamination.

11.2.4. Water Break Test. This test may be used to detect oily residues not found by other means. The surface is wetted with a spray of clean water. This should form a thin layer and remain unbroken for at least 5 sec. Beading of the water droplets indicates the presence of oil contaminants. This method is generally limited to horizontal surfaces.

11.2.5. Solvent Extraction Test. This method may be used to supplement visual techniques or to check inaccessible surfaces by using a solvent to extract contaminants for inspection. The surface is flushed, rinsed or immersed in a low-residue solvent. Solvent extraction is limited by the ability of the procedure to reach and dissolve the contaminants present. Components of the equipment tested may also contain material which would be attacked by the solvent and give erroneous results.

The used solvent may be checked to determine the amount of nonvolatile residue. A known quantity of a representative sample of filtered used solvent is evaporated almost to dryness, then transferred to a small weighed beaker for final evaporation, being careful not to overheat the residue. In the same manner, the weight of residue from a similar quantity of clean solvent is determined. The difference in weight of the two residues and the quantity of solvent used should be used to compute the amount of contaminant extracted per square foot of surface area cleaned.

In a similar manner, a 1 liter representative sample of the unfiltered used solvent can be placed in an Imhoff cone and evaporated to dryness. The volume of residue can be measured directly and used to compute the amount of contaminant extracted per square foot of surface area cleaned. Greater sensitivity can be achieved by evaporating successive liters of solvent in the same Imhoff cone.

Another method is to take a sample of known quantity of the used solvent and compare it to a similar sample of new solvent by comparing light transmission through the two samples simultaneously. There should be little, if any, difference in color of the solvents and very few particles.

Hydrocarbon or particulate matter residues determined by the inspection procedure shall not exceed the amount specified by the purchaser. An acceptable contamination level for oxygen service equipment is about 100 mg per square foot but could be more or less depending on the specific application (state of fluid, temperature and pressure).

11.2.6 Particle Count. If the purchaser's requirement includes a particle and fiber count, a representative square foot of surface shall show no particle larger than 1000 microns (0.04 in. or 0.001 m) and no more than 50 particles between 500 (0.02 in. or 0.0005 m) and 1000 microns. Isolated fibers of lint shall be no longer than 2000 microns (0.08 in. or 0.002 m) and there shall be no accumulation of lint fibers.

2. PACKAGING AND LABELING

12.1. Protection from Recontamination

Once a piece of equipment has been cleaned for oxygen service and the cleaning agent completely removed from the equipment, it should be suitably protected as soon as practicable to prevent recontamination during storage and prior to being placed in service.

Following are several ways in which this can

be done. The protection provided will depend on a number of factors such as the type of equipment, length of storage and atmospheric conditions. The type of protection required should be specified by the purchaser.

12.1.1. Protection of Openings. Equipment or parts having small openings may be protected by caps, plugs, cartons, sealing equipment in plastic bags or by other appropriate means.

Openings on large equipment may be sealed, preferably with caps, plugs or blind flanges where appropriate. Taped solid board blanks or other durable covers which cannot introduce contamination into the equipment when removed can also be used to seal such openings.

12.1.2. Pressurization. Equipment with large internal volumes may be filled with dry oil-free air or nitrogen after all openings are sealed and valves closed.

Parts in suitable plastic bags may be inerted or evacuated.

12.2. Labeling

Where purchaser's requirements include labeling to show oxygen cleaning of parts or equipment, a statement, "Cleaned for oxygen service" or other suitable wording should appear on the part, or package as applicable. Additional information which may be included is as follows:

(a) A statement, "This equipment is cleaned in accordance with Oxygen Cleaning Specification No. ————."
(b) Date of inspection and inspector's stamp or marking.
(c) Description of part, including part number if available.
(d) Statement, "Do not open until ready for use."

13. PERSONNEL SAFETY

Cleaning operations for oxygen service equipment shall be carried out in a manner which provides for the safety of personnel performing the work and shall conform to the local ordinances and state and federal regulations.

13.1. Instructions and Supervision

Operators shall be instructed in the safe use of the cleaning agents employed, including any hazards associated with the use of these agents. Written instructions shall be issued whenever special safety considerations are involved. A responsible individual shall direct oxygen cleaning operations.

13.2. Dangerous Chemicals

No highly toxic chemicals shall be used. Carbon tetrachloride shall not be employed in any cleaning operation.

The health hazards associated with the use of any solvent shall be considered in its selection. The user shall ensure that the TLV time weighted average is not exceeded for a specific solvent and consider that some chlorinated solvents are suspected of being carcinogenic. Breathing of solvent fumes and liquid contact with the skin should be avoided.

Caution must be exercised in using solvents commonly referred to as nonflammable that are flammable in air under certain conditions. The concentrations creating a flammable mixture in air are usually well in excess of the concentrations that cause physiological harm. Therefore, on removing solvents to the extent necessary to protect personnel from respiratory harm, it must not be forgotten that purging with air can create a flammable mixture and that failure to purge adequately can leave a flammable mixture which in the presence of heat, flame or sparks may result in a dangerous energy release.

Follow normal industry procedures in the mixing and handling of acids and caustics to eliminate injuries.

13.3. Protective Equipment

Face shields or goggles shall be provided for face or eye protection from cleaning solutions.

Safety glasses with side protection shall be provided for protection from injuries due to flying particles in the air.

Protective clothing shall be provided when required to prevent cleaning solutions from contacting the skin.

A self-contained breathing apparatus (see ANSI Z88.2 Practices for Respiratory Protection) shall be provided wherever there is a possibility of a deficiency of oxygen due to the use of an inert gas purge or if there is any possibility of exceeding allowable TLV values.

13.4. Proper Ventilation

All areas where cleaning compounds and solvents are used should be adequately ventilated. In outdoor operations, locate cleaning operations so that operators can work upwind of solvent vapor accumulations.

13.5. Special Situations

13.5.1 Entering Vessels.

13.5.1.1. Work should not be performed inside a vessel or confined area until the vessel or confined area has been properly prepared and work procedures established that will ensure the safety of the worker.

13.5.1.2. A Hazardous Work Permit (HWP) is an instrument widely used in industry for ensuring safe working conditions and its use is strongly recommended. The HWP should consider at least the following items before anyone enters a vessel or confined space. Other considerations may be required depending on the type of work being performed. For example, a vessel should not be entered until its temperature is at or near the surrounding temperature. All workers involved with any vessel entry should be fully apprised of the total operation prior to any tank entry.

(a) *Isolation.* All lines to a vessel should be suitably isolated to prevent the entry of foreign materials, in particular the atmospheric gases (nitrogen, argon or the rare gases) that cause asphyxiation by oxygen depletion. Oxygen enrichment is also to be avoided because of the associated fire hazard. Acceptable means of isolating vessels are blanking, double block and bleed valves or disconnecting all lines from the vessel.

(b) *Periodic Monitoring.* The need for periodic monitoring of the atmosphere in any vessel or confined space shall be considered before any work is performed.

(c) *Ventilation.* A fresh air supply suitable for breathing is normally supplied to the vessel when personnel are inside.

(d) *Atmospheric Analysis.* The atmosphere in a vessel that has been in service or has been inerted must always be analyzed before entering to determine that the vessel or confined area has been adequately ventilated with fresh air and is safe for personnel.

(e) *Rescue Procedure.* A reliable procedure for removing personnel from any vessel or confined workspace should be available and understood by all workers before any work begins.

(f) *Work Procedure.* When cleaning operations are performed inside oxygen vessels or other such confined spaces a reliable preplanned procedure for quickly removing or protecting personnel in cases of emergency shall be established and understood by all workers before work begins.

(g) *Watcher.* When toxic cleaning agents are used it is recommended that a watcher be stationed immediately outside a vessel or confined space to ensure the safety of those working within. A portable air-breathing supply shall be immediately available.

13.5.1.3. Personnel must not enter any vessel unless its atmosphere has a normal air composition. Normal atmospheric air has 21 percent oxygen by volume. However, it is permissible to work in atmospheres having oxygen concentrations in the range of 19 to 23 percent if the other gases present do not exceed their threshold limit values. In the event that the oxygen concentration deviates from 21 percent, a review of the system is required to assure that oxygen or an asphyxiant is not entering the vessel.

13.5.2. Heating Solvents. Chlorinated solvents upon heating can break down to dangerous compounds. A commonly used solvent, trichlorethylene, decomposes at temperatures not far above the boiling point of water. Ventilation must be adequate to prevent breathing excessive amounts of the solvent vapors or their decomposition products. Air respirators must be used in situations where the concentration of solvent vapors or any other foreign material in the atmosphere exceeds their TLV limit.

13.5.3. Welding Near Solvents. It is important to ensure that parts to be welded shall be free of cleaning solvents. Ultraviolet rays from welding can decompose certain chlorinated solvents to produce phosgene gas. Accordingly, the atmosphere in the vicinity of such operations shall be free from chlorinated solvent vapors.

Individual Compressed Gases and Mixtures: Properties, Manufacture, Uses, Safe Handling, Shipping, Containers

Introduction: Essential
Reference Material

Part II of the *Handbook* provides basic information on the properties, uses and handling of gases that are of current commercial importance. Gases produced and used only in small laboratory quantities are not included because of their limited applications.

Gases are treated individually in separate sections in all but four cases where closely related gases have been grouped together. The sections are arranged in alphabetical order for easy reference. A section on gas mixtures appears at the end of Part II.

Each gas section opens with basic identifying information: the generally accepted chemical name of the gas, its chemical formula, the other names by which it is known, and its DOT classification. The text of each section is divided into the following main subsections:

Physical Constants
Description
Grades Available
Uses
Physiological Effects
Safe Handling and Storage (including handling leaks and emergencies)
Methods of Shipment; Regulations
Containers (specifications, including filling limits)
Methods of Manufacture
References

Physical constants appear in tabular form in the section. Data are given for the following types of physical constants, where appropriate, in conventional United States and equivalent SI metric value:

International symbol
Molecular weight
Vapor pressure
Density of the gas
Specific gravity (compared to air)
Specific volume
Density as a liquid
Boiling point
Melting point
Critical temperature
Critical pressure
Critical density
Triple point
Latent heat of vaporization
Latent heat of fusion
Specific heat
Ratio of specific heats
Solubility in water
Weight of liquid

Where a physical constant listed above is inappropriate for a particular gas, it has either been deleted from the table or noted by a dash (−). Physical constants in addition to those listed above are given when they are important to the safe handling and use of a particular gas. The types of SI metric values used are identified at the end of Chapter 1 of Part I (corresponding conversion factors are also listed there).

QUALIFICATIONS ON PHYSICAL CONSTANTS AND ALL MATERIAL

Data given on the physical constants in the gas sections are based on authoritative scientific and industrial sources, as are all other matters of factual information and recommended practice. In presenting all this material, the publisher and Compressed Gas Association, Inc., assume no legal responsibility whatever for any losses or injuries sustained, or liabilities incurred, by persons or organizations acting in any way on the basis of any part of the material. However, the information and recommendations are believed to be accurate and sound to the best knowledge of the CGA.

The physical constants data that are given generally represent the properties of pure commodities rather than those of commercial grades of the gases. The properties of commercial grades should be expected to differ somewhat from the values for pure grades presented here.

NECESSARY REFERENCE TO SECTIONS IN PART I

Concerning safety with individual gases, readers should refer to certain chapters in Part I of this book for essential detailed information. These chapters are:

Chapter 4, especially Section A, on the general rules for the safe handling of gases, and Section C, on the safe handling of gases used medically. The sections directly and very significantly supplement Part II gas sections.

Chapter 5, on pressure relief device standards for different gases and different types of containers.

Chapter 6, on safety practices with compressed gas cylinders—marking and labeling, testing and inspecting for requalification, repairing and disposition.

Chapter 7, on insuring safety through standard cylinder valve-connection systems in general and for gases used in medicine.

Sections of Part I also directly supplement the gas sections of Part II, as follows: (1) when hydrostatic retesting is identified as required for cylinder requalification in a gas section, the testing meant is that described in Section D of Chapter 6 in Part I; (2) when visual inspection is identified as required for cylinder requalification, the present standards for such visual inspection are those given in Section C of Chapter 6, Part I; and (3) when standard cylinder valve outlet and inlet connections are identified in a gas section, the standard connections cited are those about which further information is given in Chapter 7, Part I.

Persons unfamiliar with the compressed gas field will also find it helpful to refer to Chapter 3 in Part I for descriptions and illustrations of the different types of containers that are merely identified in the following gas sections. Similarly, Chapter 9 of Part I will be helpful for those who need to know how to safely unload bulk shipments of liquefied gas.

SUMMARIES OF SHIPPING REGULATIONS

Summaries of shipping regulations in the Part II gas sections are of course based on the regulations in effect while the *Handbook* was being prepared. For the full and current regulations, the reader is strongly advised to consult current editions of their published forms identified in Chapter 2.

It is also important for readers to understand that the authorized service pressures given in gas sections for cylinders and other containers are only the minimum service pressures cited in DOT or CTC regulations. For example, if cylinders meeting DOT or CTC specifications 3A150 are noted in a gas section as authorized for a given gas, then any other 3A cylinders with higher service pressures are also authorized (such as 3A1000, 3A2000, etc.).

KEY TO ABBREVIATIONS AND TERMS USED

Abbreviations and terms given in the gas sections (and elsewhere in the volume) have these meanings:

atm—atmospheres (or pressure; 1 atm = 14.7 psi)

Btu–British thermal units

C–degrees Celsius

cargo tank–large unitary container fixed to a motor vehicle or motor vehicle trailer or semitrailer

C_p–specific heat at constant pressure

C_v–specific heat at constant volume

cu–cubic

cm–centimeters

F–Farenheit, degrees Farenheit

filling density–the container charging ratio for liquefied gases. It is the maximum limit to which containers are authorized to be charged under DOT and CTC regulations. (See also "SERVICE PRESSURE.") It is given in DOT and CTC regulations as "The percent ratio of the weight of gas in a container to the weight of water that the container will hold at 60 F (1 lb of water = 27.737 cu in. at 60 F)."

ft–feet

g–grams

Gg–gegagram

gal–U.S. gallons only

in.–inches

kg–kilograms

kg/m^3–kilogram per cubic meter (density)

kJ–kilojoule (energy)

kPa–kilopascal (pressure)

lb–pounds avoirdupois

m^3–cubic meter (volume)

m^3/kg–cubic meter per kilogram (specific volume)

portable tank–large transferrable container originally designed to transport compressed gas by motor vehicle

psi–pounds per square inch pressure

psia–pounds per square inch, absolute pressure

psig–pounds per square inch, gage pressure

service pressure–under DOT and CTC regulations it is the authorized pressure marking on the container. For example, for cylinders marked "DOT-3A1800," the service pressure is 1800 psig.

sq–square

TLV–threshold limit value. The concentration of a gas in air to which most workers can be safely exposed on a daily basis.

TMU–ton multi-unit; a TMU container is a large transferrable cylinder originally designed to transport a ton of liquefied chlorine in multiple units on railroad flatcars.

tube trailer–motor vehicle trailer fitted with long, multiple, fixed tubes (usually one built to meet DOT or CTC cylinder specifications 3A, 3AX, 3AA or 3AAX) for transporting nonliquefied gas at high pressure.

Chemical symbols and other standard abbreviations not shown above are also employed.

Acetylene

C$_2$H$_2$
Synonym: Ethine, ethyne
DOT Classification: Flammable gas; flammable gas label

PHYSICAL CONSTANTS

	U.S. Units	Metric Units
International symbol	C$_2$H$_2$	C$_2$H$_2$
Molecular weight	26.04	26.04
Vapor pressure at 70 F (21.1 C)	635 psig	4378 kPa, gage
Density of the gas at 32 F (0 C) and 1 atm	0.07314 lb/ft^3	1.1716 kg/m^3
Specific gravity of the gas at 32 F and 1 atm (air = 1)	0.906	0.906
Specific volume of the gas at 70 F (21.1 C) and 1 atm	14.7* ft^3/lb	0.918* m^3/kg
Specific gravity of liquid at −112 F (−80 C)	0.613	0.613
Density of liquid at 70 F (21.1 C)	24.0 lb/ft^3	384 kg/m^3
Boiling point at 10 psig**	−103 F	−75.0 C
Melting point at 10 psig**	−116 F	−82.2 C
Critical temperature	96.8 F	36.0 C
Critical pressure	907 psia	6250 kPa, abs.
Critical density	14.4 lb/ft^3	231 kg/m^3
Triple point	−116 F at 17.7 psia	−82.2 C at 122 kPa, abs.
Latent heat of varporization at triple point	264 Btu/lb	614 kJ/kg
Latent heat of fusion at −114.7 F (−81.5 C)	41.56 Btu/lb	96.67 kJ/kg

*Based on 1.171 g/liter at 32 F (0 C) and 1 atm.
**Reported at 10 psig instead of at 1 atm because, at 1 atm, acetylene sublimes directly from the solid to the gaseous state without entering the liquid state. Its sublimation point at 1 atm is −118 F (−83.3 C).

	U.S. Units	Metric Units
Specific heat of gas at 60 F (15.6 C) and 1 atm		
C_p	0.383 Btu/(lb) (F)	1.60 kJ/(kg) (C)
C_v	0.304 Btu/(lb) (F)	1.27 kJ/(kg) (C)
Ratio of specific heats	1.26	1.26
Solubility in water, vol/vol at 60 F (15.6 C)	1.1	1.1
Specific volume of the gas at 60 F (15.6 C) and 1 atm	14.5 ft³/lb	0.905 m³/kg
Solubility in water, vol/vol at 32 F (0 C) and 1 atm	1.7	1.7

DESCRIPTION

Acetylene is a compound of carbon and hydrogen in proportions by weight of about 12 parts carbon to 1 part hydrogen (92.3 to 7.7 percent). A colorless flammable gas, it is slightly lighter than air. Acetylene of 100 percent purity is odorless, but gas of ordinary commercial purity has a distinctive, garlic-like odor.

Acetylene burns in air with an intensely hot, luminous and smoky flame. The ignition temperatures of acetylene and of acetylene-air and acetylene-oxygen mixtures vary according to composition, initial pressure, initial temperature and water vapor content. As a typical example, an air mixture containing 30 percent acetylene by volume at atmospheric pressure can be ignited at about 581 F (305 C). The flammable limits of acetylene-air and acetylene-oxygen mixtures similarly depend on initial pressure, temperature and water vapor content. In air at atmospheric pressure the upper flammable limit is about 80 percent acetylene by volume, and the lower limit is 2.5 percent acetylene. If an ignition source is present, 100 percent acetylene will decompose with violence under certain conditions of pressure and container size and shape.

Acetylene can be liquefied and solidified with relative ease. However, in both the liquid and solid states, acetylene explodes with extreme violence when ignited unless special conditions of confinement are employed. A mixture of gaseous acetylene with air or oxygen in certain proportions explodes if ignited. Gaseous acetylene under pressure may also decompose with explosive force under certain conditions, but experience indicates that 15 psig (100 kPa gage) is generally acceptable as a safe upper pressure limit. Generation, distribution through hose or pipe, or utilization of acetylene at pressures in excess of 15 psi gage pressure or 30 psi (200 kPa) absolute pressure for welding and allied purposes should be prohibited.

Pressure exceeding 15 psig can be employed provided specialized equipment is used. Where acetylene is to be utilized for chemical synthesis at pressures in excess of 15 psig, or transported through large-diameter pipelines, means to prevent propagation, should ignition occur, must be employed. Packing large-diameter pipe with small-diameter pipes as a protection against exposure to fires is recommended (see Ref. 1).

Acetylene cylinders avoid the decomposition characteristics of the gas by providing a porous-mass packing material having minute cellular spaces so that no pockets of appreciable size remain where "free" acetylene in gaseous form can collect. This porous mass is saturated with acetone or other suitable solvent in which the acetylene actually dissolves. The combination of these two features—porous filling and solvent—allows acetylene to be contained in such cylinders at moderate pressure without danger of explosive decomposition (the maximum authorized cylinder pressure is 250 psig (1720 kPa

TABLE 1. ACETYLENE GRADES AVAILABLE.

Limiting Characteristics	Grades					
	A	B	C	D	E	F
Acetylene min. assay (v/v)	95%	98%	98%	98%	99.5%	99.5%
Phosphine and arsine (ppm)			500	50	500	50
Hydrogen sulfide (ppm)			500	50	500	50

Note 1. Cylinder acetylene contains variable percentage quantities of a solvent (normally acetone), the amount of solvent present in the expelled gas being dependent upon the vapor pressure of the solvent, the conditions of the cylinder and the conditions of withdrawal. The purities listed in Table 1 are given on a solvent-free basis.

Note 2. Acetylene manufactured from hydrocarbon feedstock is inherently free from phosphine, arsine and hydrogen sulfide.

gage) at 70 F (21.1 C), with a variation of about 2.5 psig (17 kPa gage) rise or fall per degree F of temperature change).

GRADES AVAILABLE

Table 1 presents the assay (purity-minimum) for each grade of gaseous acetylene, plus the component maxima in ppm (v/v). A blank indicates no limiting characteristic (see Ref. 2).

USES

Approximately 80 percent of the acetylene produced annually in the United States is used for chemical synthesis. It is possible to use acetylene for an almost infinite number of organic chemical syntheses, but this use in North America has been less extensive than in Europe owing to the ready availability of petroleum from which competitive syntheses are often possible. Nevertheless, acetylene has come into increasing prominence as the raw material for a whole series of organic compounds, among them acetaldehyde, acetic acid, acetic anhydride, acetone and vinyl chloride. These compounds may be used in turn to produce a diverse group of products including plastics, synthetic rubber, dyestuffs, solvents and pharmaceuticals. Acety-

lene is also utilized to manufacture carbon black.

The remaining 20 percent of the annual United States acetylene production is used principally for oxyacetylene welding, cutting, heat treating, etc. Small amounts are utilized for lighting purposes in buoys, beacons and similar devices.

PHYSIOLOGICAL EFFECTS

Acetylene can be inhaled in rather high concentrations without chronic harmful effects, and it has, in fact, been used as an anesthetic. It it of course a simple asphyxiant if present in concentrations high enough to deprive the lungs of oxygen and produce suffocation. However, the lower flammable limit of acetylene in air would usually be reached long before suffocation could occur as the result of an acetylene leak.

SAFE HANDLING AND STORAGE

Materials Suitable for Containers and Storage

Only steel or wrought iron pipe should be used for acetylene. Joints in piping must be welded or made up of thread or flange fittings. The materials for fittings can be rolled, forged or cast steel, or malleable iron. Cast iron fittings are not permissible.

Under certain conditions acetylene forms readily explosive compounds with copper, silver and mercury. For this reason, contact between acetylene and these metals, or their salts, compounds and high-concentration alloys, is to be avoided. It is generally accepted that brass containing less than 65 percent copper in the alloy, and certain nickel alloys, are suitable for use in acetylene service under normal conditions. In normal service, conditions involving contact with highly caustic salts or solutions, or contact with other materials corrosive to copper or copper alloys, can render the above generally acceptable alloys unsatisfactory for this service. The presence of moisture, certain acids or alkaline materials tends to enhance the formation of copper acetylides. Further information on metallic acetylides is given in the references listed in the concluding "Additional References" section.

For recommendations on acetylene cylinder manifolds and shop piping, users should consult their supplier and recognized safety authorities such as the Underwriters' Laboratories, Inc., the Associated Factory Mutual Fire Insurance Companies, and Compressed Gas Association, Inc.

Regulators and Control Valves

Never use acetylene through blowpipes or other devices equipped with shutoff valves on the acetylene supply connections without reducing the pressure through a suitable regulator attached to the cylinder valve. Acetylene should never be used in equipment outside the cylinder at pressures exceeding 15 psig (100 kPa gage). It is always preferable to use acetylene cylinders in an upright position to avoid loss of solvent and accompanying reduction in flame quality. However, use in a horizontal position, with or without the loss of solvent, does not make the acetylene less stable or less safe.

In preparing to withdraw acetylene from cylinders, never use wrenches or other tools for opening cylinder valves except for those tools that have been provided or approved by the manufacturer of the gas.

After removing the valve-protection cap, slightly open the valve for an instant in order to clear the opening of particles of dust or dirt, being careful to stand so that the valve points away from the body. Avoid blowing dangerous amounts of the gas in confined spaces. Do not "crack" an acetylene cylinder valve near welding work, sparks, open flame or any other possible sources of ignition.

Be sure that all connections are gas tight and remain so, and that the connected hose is in good condition and does not have any leaks.

Always open the acetylene cylinder valve slowly. Never use a hammer or mallet in attempting to open or close a valve.

Do not open an acetylene cylinder valve more than one-and-one-half turns. Do not stand in front of the regulator and gage faces when opening the valve.

Do not pile hose, tools or other objects on top of an acetylene cylinder where they might interfere with quick closing of the valve.

The wrench used for opening the cylinder valve should always be kept on the valve spindle when the cylinder is in use.

Always close the cylinder valve when the work is finished. Be sure the cylinder valve is closed and all gas is released from the regulator before removing the regulator from a cylinder.

Never apply a torch to the side of a cylinder to raise the pressure. Serious accidents have resulted from violation of this rule.

Storage and Handling in Normal Use

Users store acetylene only in cylinders in almost all cases. In storing acetylene cylinders, the user should comply with all local, municipal, and state regulations, and with Standard No. 51 of the National Fire Protection Association (see Ref. 3).

Inside all buildings, acetylene cylinders should not be stored near oxygen cylinders. Unless they are well separated, there should be a noncombustible partition between acetylene cylinders and oxygen cylinders.

Acetylene cylinders stored inside a building

must be limited to a total capacity of 2000 ft³ (56.6 m³) of gas exclusive of cylinders in use or attached for use. Quantities exceeding this total must be stored in a special building or in a separate room as required by NFPA Standard No. 51.

Conspicuous signs must be posted in the storage area forbidding smoking or the carrying of open lights.

While storage in a horizontal position does not make the acetylene in cylinders less stable or less safe, it does increase the likelihood of solvent loss, which will result in a lower flame quality when the cylinder is used. Therefore, it is always preferable to store acetylene cylinders in an upright position.

Handling Acetylene Cylinders. Always call acetylene by its proper name, "acetylene," to promote recognition of its hazards and the taking of proper precautions. Never refer to acetylene merely as "gas."

Never attempt to repair or alter cylinders or valves. This should be done only by the cylinder manufacturer. If a cylinder is leaking, follow the recommendations made in the section on "Handling Leaks and Emergencies."

Never tamper with safety devices in valves or cylinders. Keep sparks and flame away from acetylene cylinders and under no circumstances allow a torch flame to come in contact with safety devices. Should the valve outlet of an acetylene cylinder become clogged by ice, thaw with warm—not boiling—water.

Never, under any circumstances, attempt to transfer the acetylene from one cylinder to another, to refill acetylene cylinders, or to mix any other gas with acetylene in a cylinder.

In welding shops and industrial plants using both oxyacetylene and electric welding apparatus, care must be taken to avoid the handling of this equipment in any manner which may permit the compressed gas cylinders to come in contact with the electric welding apparatus or electrical circuits.

Never use acetylene cylinders as rollers or supports, or for any other purpose than storing acetylene.

Moving Acetylene Cylinders. Cylinders must be protected against dropping when being unloaded from a truck or platform. One method of protection is to use a V-shaped trough as a skid with a wooden or rope bumper at the bottom if the skid is steep. Another method is to use a welded angle-iron cradle or rocker-rack so constructed that the cylinder may be slid horizontally into a steel trough at the top of the cradle. The cradle is then tilted so that the cylinder is upended and can be lowered to the ground. A heavy iron counterweight attached to the cradle will help balance the cylinder during the tilting operation.

Special caution is necessary in transporting acetylene cylinders by crane or derrick. Lifting magnets, slings or rope or chain, or any other device in which the cylinders themselves form a part of the carrier, must never be used for hoisting acetylene cylinders. Instead, when a crane is used, a platform, cage or cradle should be provided to protect the cylinders from being damaged by slamming against obstructions, and to keep them from falling out. A recommended type of cradle to build for this purpose is shown in Fig. 1.

Horizontal movement of cylinders is easily accomplished by the use of a hand truck; how-

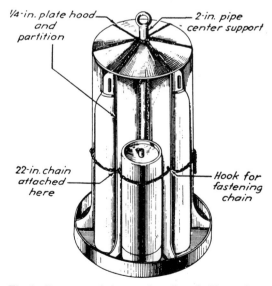

Fig. 1. Recommended type of cradle to hold acetylene cylinders when moved with a crane or derrick.

ever, when a hand truck is used some positive method, such as chaining, should be used to secure a cylinder standing upright in the truck. Cylinders must not be transported lying horizontally on trucks with the valve overhanging in a position to collide with stationary objects. Cylinders should never be dragged from place to place.

Valves should always be closed before cylinders are moved. Unless cylinders are to be moved while secured in an upright position to a suitable truck, pressure regulators should be removed and valve-protecting caps, if provided for in the design, should be attached.

Handling Leaks and Emergencies

Because acetylene and air in certain proportions are explosive, care should be taken to prevent acetylene leakage. Connections must be kept tight and hose maintained in good condition. Points of suspected leakage should be tested by covering them with soapy water. A leak will be indicated by bubbles of escaping acetylene passing through the soap film. *Never test for leaks with an open flame!*

If acetylene leaks around the valve spindle when the valve is open, close the valve and tighten the gland nut. This compresses the packing around the spindle. If this does not stop the leak, close the valve, and attach to the cylinder a tag stating that the valve is unserviceable. Notify the gas supplier and follow his instructions for the cylinder's return.

Acetylene cylinders are equipped with the fusible-metal plugs having a melting point between 208 F and 220 F (212 F nominal; or 97.8 C to 104.4 C, (100 C nominal), which may be located in the top and bottom heads of the cylinders or in a channel in the cylinder valve.

If acetylene leaks from the valve even when the valve is closed, or if rough handling should cause any fusible safety plugs to leak, move the cylinder to an open space well away from any possible source of ignition and plainly tag the cylinder as having an unserviceable valve or fuse plug. Open the valve slightly to let the acetylene escape slowly. Place a sign at the cylinder warn-

ing persons against approaching the cylinder with cigarettes or other open lights. When the cylinder is empty, close the valve. Notify the manufacturer immediately of the serial number of the cylinder and the particulars of its defect, as far as known, and await shipping instructions.

For the safe use of acetylene welding and cutting equipment and to prevent fires, the user should comply with Standard No. 51B of the National Fire Protection Association (see Ref. 4).

METHODS OF SHIPMENT; REGULATIONS

Only cylinders are authorized for shipping acetylene. Acetylene producers using the gas for chemical synthesis store acetylene in low-pressure gas holders for which the recommended material is carbon steel.

Under the appropriate regulations and tariffs, acetylene is authorized for shipment as follows:

By Rail: In cylinders (freight or express).

By Highway: In cylinders on trucks.

By Water: In cylinders via cargo vessels, passenger vessels, passenger or vehicle ferry vessels, and passenger or vehicle railroad car ferry vessels. On barges in cylinders for barges of U.S. Coast Guard classes A and C only.

By Air: Aboard cargo aircraft only, in cylinders as required up to 300 lb (140 kg) maximum net weight per cylinder.

The maximum filling density authorized for acetylene in cylinders that meets the specifications and solvent filling requirements of the DOT is 250 psig at 70 F (1720 kPa gage at 21.1 C), or lower maximum charging pressures at 70 F for cylinders marked with such lower maximum pressures.

Only cylinders that meet DOT specifications 8 or 8-AL, and that also meet requirements for fillings of a porous material and a suitable solvent, are authorized by the DOT for acetylene service.

DOT regulations prohibit shipment of cylinders containing acetylene gas unless they were charged by or with the consent of the owner.

Under present regulations, no periodic hydrostatic retest is required for cylinders 8 and 8-AL which are authorized for acetylene service.

CONTAINERS

Cylinders

Acetylene is most commonly available in cylinders of capacities of 10, 40, 50, 100, 225, 300, 400 and 850 ft³ (0.2 to 24 m³). "Lighthouse" type cylinders—those generally used in acetylene-operated automatic aids to marine navigation—are available in larger sizes, the biggest having a capacity of approximately 1400 ft³ (40 m³).

Do not attempt to charge acetylene into any cylinders except those constructed for acetylene.

Do not charge any other gas but acetylene into an acetylene cylinder.

Do not mix any other gas with acetylene in an acetylene cylinder.

Failure to observe these warnings may result in an explosion.

The following marks are required by the DOT to be plainly stamped on or near the shoulder or top head of all acetylene cylinders as follows: (a) the DOT specification number—DOT-8 or DOT-8AL; (b) a serial number and the user's, purchaser's or maker's identifying symbol (the symbol must be registered with the Bureau of Explosives, 1912 L Street, N.W., Washington, DC 20036; (c) the date of the test to which it was subjected in manufacture; and (d) the tare weight of the cylinder in pounds and ounces.

The markings on cylinders must not be changed except as provided in DOT regulations, which require that serial numbers and ownership marks may be changed only when a detailed report is filed with the Bureau of Explosives. Markings on cylinders must be kept in a readable condition.

Valve Outlet Connections. Standard connection, United States and Canada—No. 510. Alternate standard connection, United States and Canada—No. 300. Additional alternate standard connection, Canada only—No. 410. Small valve series standard connections, United States and Canada—Nos. 200 and 520.

METHOD OF MANUFACTURE

In the United States and Canada calcium carbide is the principal raw material for acetylene manufacture. Calcium carbide and water may be made to react by several methods to produce acetylene, with calcium hydroxide as a by-product. Acetylene is also manufactured by the thermal or arc cracking of hydrocarbons and by a process employing the partial combustion of methane with oxygen.

Acetylene manufactured from carbide made in the United States and Canada normally contains less than 0.4 percent impurities other than water vapor. Apart from water, the chief impurity is air, in concentrations of approximately 0.2 to 0.4 percent. The remainder is mostly phosphine, ammonia, hydrogen sulfide, and in some instances, small amounts of carbon dioxide, hydrogen, methane, carbon monoxide, organic sulfur compounds, silicon hydrides and arsine. Purified cylinder acetylene is substantially free from phosphine, ammonia, hydrogen sulfide, organic sulfur compounds and arsine. The other impurities are nearly the same as in the original gas.

REFERENCES

1. "Acetylene Transmission for Chemical Synthesis" (Pamphlet G-1.3), Compressed Gas Association, Inc.
2. "Commodity Specification for Acetylene" (Pamphlet G-1.1), Compressed Gas Association, Inc.
3. "Standard for the Design and Installation of Oxygen-Fuel Gas Systems for Welding and Cotting" (N.F.P.A. 51), National Fire Protection Association, 470 Atlantic Ave., Boston, MA 02210. (Also available as NBFU Standard No. 51, American Insurance Association, 85 John St., New York, NY 10007.)
4. "Standard for Fire Prevention in Use of Cutting and Welding Processes," Standard No. 51B, National Fire Protection Association.

ADDITIONAL REFERENCES

"Recommendations for Chemical Acetylene Metering" (Pamphlet G-1.2), Compressed Gas Association, Inc.

"Handling Acetylene Cylinders in Fire Situations" (Pamphlet SB-4), Compressed Gas Association, Inc.

"Tentative Recommended Practices for Mobile Acetylene Trailer Systems" (Pamphlet G-1.6T), Compressed Gas Association, Inc.

Metallic Acetylides*

"Copper Acetylides," V. F. Bramfeld, M. T. Clark, and A. P. Seyfang; *J. Soc. Chem. Ind. (London)*, **66**, 346–53 (October 1947).

"The Formation and Properties of Acetylides," paper presented by G. Benson, Shawinigan Chemicals, Ltd., at the Compressed Gas Association Canadian Section, September 17, 1950.

"The Chemistry of Acetylene," ACS Monograph No. 99, Nieuwland and Vogt, New York, Reinhold.

"Conditions of Formation and Properties of Copper Acetylide," unpublished research paper by L'Air Liquide, Paris, France.

"Ueber Bildung and Eigenschaften der Kupferacetylide," von H. Feitnecht and L. Hugi-Carmes', *Schweizer Archiv Angew. Wiss. Tech.*, **10**, 23 (1957).

*Most of the above articles contain additional bibliographies which are also information on this subject.

Air

Synonyms: Compressed air, atmospheric air, the atmosphere
(of the Earth)
DOT Classification: Nonflammable gas; nonflammable gas
label

PHYSICAL CONSTANTS

	U.S. Units	Metric Units
International symbol	Air	Air
Molecular weight	28.975	28.975

Density of the gas
  at 70 F (21.1 C)
  and 1 atm
	0.07493 lb/ft³	1.2000 kg/m³

Specific gravity of the gas
  at 70 F (21.1 C)
  and 1 atm (air = 1)
	1.00	1.00

Specific volume of the gas
  at 70 F (21.1 C)
  and 1 atm
	13.346 ft³/lb	0.8333 m³/kg

Boiling point at 1 atm	-317.8 F	-194.3 C
Freezing point at 1 atm	-357.2 to -312.4 F	-216.2 to -191.3 C
Critical temperature	-221.2 F	-140.7 C
Critical pressure	547 psia	3770 kPa, abs.
Critical density	21.9 lb/ft³	351 kg/m³
Latent heat of vaporization at normal boiling point	88.2 Btu/lb	205 kJ/kg

Specific heat of gas
  at 70 F (21.1 C)
  and 1 atm		

C_p	0.241 Btu/(lb) (F)	1.01 kJ/(kg) (C)
C_v	0.172 Btu/(lb) (F)	0.720 kJ/(kg) (C)
Ratio of specific heats	1.40	1.40
Solubility in water, vol/vol at 32 F (0 C)	0.0292	0.0292
Weight of liquid at normal boiling point	7.29 lb/gal	874 kg/m³

	U.S. Units	Metric Units
Density of liquid at boiling point and 1 atm	54.56 lb/ft^3	874.0 kg/m^3
Liquid/gas ratio (liquid at boiling point, gas at 70 F and atm) vol/vol	1/728.1	1/728.1
Thermal conductivity		
at -148 F (-100 C)	0.0095 Btu/(hr) (ft) (F/ft)	0.0164 W/(m) (C)
at 32 F (0 C)	0.0140 Btu/(hr) (ft) (F/ft)	0.0242 W/(m) (C)
at 212 F (100 C)	0.0183 Btu/(hr) (ft) (F/ft)	0.0317 W/(m) (C)

DESCRIPTION

Air is the natural atmosphere of the Earth— a nonflammable, colorless, odorless gas that consists of a mixture of gaseous elements (with water vapor, a small amount of carbon dioxide and traces of many other constituents). Liquefied air is transparent with a bluish cast and has a milky color when it contains carbon dioxide. Dry air is noncorrosive.

Air is compressed at the point of use for most practical applications. To meet needs for air of special purity or specified composition (as in certain medical, scientific, industrial, fire protection, undersea and aerospace uses), it is purified or compounded synthetically and shipped in cylinders as a nonliquefied gas at high pressures.

The composition of atmospheric air may vary. A typical analysis of dry air at sea level has the following composition:

Component	% by Volume	% by Weight
Nitrogen	78.03	75.5
Oxygen	20.99	23.2
Argon	0.94	1.33
Carbon dioxide	0.03	0.045
Hydrogen	0.005	—
Neon	0.00123	—
Helium	0.0004	—
Krypton	0.00005	—
Xenon	0.000006	—

Atmospheric air also contains varying amounts of water vapor. For most practical purposes, the air composition is taken to be 79 percent nitrogen and 21 percent oxygen by volume, and to be 76.8 percent nitrogen and 23.2 percent oxygen by weight.

GRADES AVAILABLE

Table 1 presents the component maxima in ppm (v/v), unless shown otherwise, for types and grades of air. Gaseous air is denoted as Type 1 and liquefied air as Type II in the table. A blank indicates no maximum limiting characteristic.

USES

Air meeting particular purity specifications has many important applications. Some of these applications are in medical, undersea, aerospace and atomic energy fields. It is also employed in self-contained breathing apparatus used by industrial personnel and firemen, and as a power source for some kinds of pneumatic equipment.

PHYSIOLOGICAL EFFECTS

Air is nontoxic and nonflammable.

SAFE HANDLING AND STORAGE

Materials Suitable for Containers and Storage

Since dry air is noncorrosive, it may be contained in equipment constructed with any common, commercially available metals.

Storage and Handling in Normal Use

Cylinders and other containers charged with air at high pressure must be handled with all

TABLE 1. AIR GRADES AVAILABLE.

Limiting Characteristics	TYPE I (GASEOUS) GRADES									TYPE II (LIQUID)	
	A	B	C	D	E	F	G	H	J	A	B
% O$_2$ (v/v) Balance Predominantly N$_2$ (Note 1)	atm.	atm.	atm. 19.5-23.5	atm/ 19.5-23.5	atm/ 19.5-23.5	atm/ 19.5-23.5	atm. 19.5-23.5	atm/ 19.5-23.5	atm/ 19.5-23.5	atm/ 19.5-23.5	atm/ 19.5-23.5
Water		none condensed (per 5.3.1)	note 2	note 2	note 2	note 2	note 2	note 2	1 —10.4°F		note 2
Hydrocarbons (condensed) in Mg/m^3 of gas at NTP (Note 3)		none (per 5.4.1)	5	5	5						
Carbon Monoxide			50	20	10	5	5	5	1		5
Odor			see 5.1.5	see 5.1.5	see 5.1.5	see 5.1.5	see 5.1.5	see 5.1.5	see 5.1.5		none
Carbon Dioxide				1000	500	500	500	500	0.5		5
Gaseous Hydrocarbons (as methane)						25	15	10	0.5		10
Nitrogen Dioxide							2.5	0.5	0.1		0.5
Nitrous Oxide									0.1		
Sulfur Dioxide							2.5	0.5	0.1		0.5
Halogenated Solvents							10	1	0.1		1
Acetylene									0.05		0.5
Permanent Particulates											—.15

Note 1. The term "atm" (atmospheric) denotes the oxygen content normally present in atmospheric air; the numerical values denote the oxygen limits for synthesized air.

Note 2. The water content of compressed air required for any particular grade may vary with the intended use from saturated to very dry. If a specific water limit is required, it should be specified as a limiting dew point or concentration in ppm (v/v). Dew point is expressed in temperature °F at one atmosphere absolute pressure (760 mmHg). To convert dew point °F to °C, ppm (v/v), or mg/liter.

Note 3. No limits are given for condensed hydrocarbons beyond grade E since the gaseous hydrocarbon limits could not be met if condensed hydrocarbons were present.

Note 4. Specific measurement of odor in Type I air is impractical. A pronounced odor should render the air unsatisfactory for breathing purposes.

the precautions necessary for safety with any nonflammable compressed gas. (See also Part I, Chapter 4, Section A.)

Handling Leaks and Emergencies

See Part 1, Chapter 4, Section A.

METHODS OF SHIPMENT; REGULATIONS

Under the appropriate DOT regulations, air is authorized for shipment as follows:

By Rail: In cylinders as a compressed gas, and as a liquid in special containers for cryogenic gas.

By Highway: In cylinders as a compressed gas, and as a liquid in special containers for liquefied cold compressed gas.

By Water: In cylinders on passenger vessels, cargo vessels, and ferry and railroad car ferry vessels (passenger or vehicle). In cylinders on barges for barges of U.S. Coast Guard classes A, BA, BB, CA and CB.

By Air: In cylinders as a compressed gas, aboard passenger aircraft up to 150 lb (70 kg), or aboard cargo aircraft up to 300 lb (140 kg), maximum net weight per cylinder. In insulated containers meeting specified requirements, liquid air, either low pressure or pressurized, in cargo aircraft only up to 300 lb (140 kg) maximum net weight per container. Nonpressurized liquid air is not accepted for shipment aboard cargo or passenger aircraft.

CONTAINERS

Compressed or liquefied air may be shipped under DOT regulations in qualified containers authorized by the DOT. Liquefied air may also be shipped in special insulated and vented containers for liquefied cold compressed gas at pressures below 25 psig (170 kPa gage).

Filling Limits. The maximum filling limits at 70 F (21.1 C) for compressed air, under DOT regulations, are the authorized service pressures marked on the cylinders. Also under DOT regulations, authorized cylinders meeting special requirements may be filled to a limit of up to 110 percent of their marked service pressures.

Cylinders

Compressed air may be shipped in qualified cylinders authorized by the DOT for nonliquefied compressed gas. (These include cylinders meeting DOT specifications 3A, 3AA, 3AX, 3AAX, 3B, 3C, 3D, 3E, 4, 4A, 4B, 4BA, 4BW and 4C; in addition, continued use of cylinders meeting DOT specifications 3, 25, 26, 33 and 38 is authorized, but new construction is not authorized.

All cylinders authorized for compressed air service must be requalified by hydrostatic retest every 5 or 10 years under present regulations, with the following exceptions: DOT-4 cylinders, every 10 years; and no periodic retest is required for cylinders of types 3C, 3E, 4C and 4L.

Valve Outlet and Inlet Connections. Standard connections, United States and Canada—up to 3000 psi (20,684 kPa), No. 346; 3000 to 10,000 psi (20,684 to 68,948 kPa), see Part I, Chapter 7, Section A, paragraph 2.1.7; cryogenic liquid withdrawal, No. 440. Alternate standard connections, United States and Canada—up to 3000 psi, No. 590; 3000 to 10,000 psi, No. 677; cryogenic liquid withdrawal, No. 295. Limited standard, United States and Canada—up to 3000 psi, No. 1310. Standard yoke connections, United States and Canada (for medical use of the gas)—Nos. 850, 950.

METHOD OF MANUFACTURE

Air of known purity and composition may be compressed from the atmosphere and purified by chemical and mechanical means. It may also be synthetically produced from already purified major components, chiefly nitrogen and oxygen.

REFERENCES

1. "Air, Compressed for Breathing Purposes," Federal Specification BB-A-1034a.
2. "Air, Liquid, for Breathing Purposes," Military Specification MIL-A-27420 (USAF).
3. "Compressed Air for Human Respiration" (Pamphlet G-7), Compressed Gas Association, Inc.
4. "Commodity Specification for Air" (Pamphlet G-7.1), Compressed Gas Association, Inc.

Ammonia (Anhydrous)

NH₃   DOT Classification: Nonflammable gas; nonflammable gas label

PHYSICAL CONSTANTS

	U.S. Units	Metric Units
International symbol	NH_3	NH_3
Molecular weight	17.031	17.031
Vapor pressure		
at 70 F (21.1 C)	114.1 psig	786.7 kPa, gage
at 105 F (40.6 C)	214.2 psig	1476.8 kPa, gage
at 115 F (46.1 C)	251.5 psig	1734.0 kPa, gage
at 130 F (54.4 C)	315.6 psig	2176.0 kPa, gage
Density of the gas at 32 F		
(0 C) and 1 atm	0.0481 lb/ft³	0.771 kg/m³
Specific gravity of the gas at		
32 F and 1 atm (air = 1)	0.5970	0.5970
Specific volume of the gas at		
32 F (0 C) and 1 atm	20.78 ft³/lb	1.297 m³/kg
Density of liquid		
at 70 F (21.1 C)	38.00 lb/ft³	608.7 kg/m³
at 105 F (40.6 C)	36.12 lb/ft³	578.6 kg/m³
at 115 F (46.1 C)	35.55 lb/ft³	569.4 kg/m³
at 130 F (54.4 C)	34.66 lb/ft³	555.2 kg/m³
Boiling point at 1 atm	-28 F	-33 C
Melting point at 1 atm	-107.9 F	-77.72 C
Critical temperature	271.4 F	133.0 C
Critical pressure	1657 psia	11,420 kPa, abs.
Critical density	14.7 lb/ft³	235 kg/m³
Triple point	-107.86 F at	-77.70 C at
	0.88 psia	6.1 kPa, abs.
Latent heat of vaporization		
at boiling point and 1 atm	589.3 Btu/lb	1371 kJ/kg
Latent heat of fusion at		
-107.9 F (-77.72 C)	142.8 Btu/lb	332.2 kJ/kg

	U.S. Units	Metric Units
Specific heat of gas at 59 F (15 C) and 1 atm		
C_p	0.5232 Btu/(lb)(F)	2.191 kJ/(kg)(C)
C_v	0.3995 Btu/(lb)(F)	1.673 kJ/(kg)(C)
Ratio of specific heats	1.3096	1.3096
Solubility in water, vol/vol at 68 F (20 C)	0.848	0.848
Weight of liquid per gallon at 60 F (15.5 C)	5.147 lb/gal	616.7 kg/m^3
Vapor density at -28 F (-33 C) and 1 atm	0.05555 lb/ft^3	0.88983 kg/m^3
Specific gravity of liquid at -28 F (-33 C) compared to water at 39.2 F (4 C)	0.6819	0.6819
Liquid density at -28 F (-33 C) and 1 atm	42.57 lb/ft^3	681.9 kg/m^3
Flammable limits (percent in air, by volume)	16–25 percent	16–25 percent
Ignition temperature	1562 F	850.0 C
Heat of solution at 0 percent concentration by weight	347.4 Btu/lb	808.1 kJ/kg
Heat of solution at 28 percent concentration by weight	214.9 Btu/lb	499.9 kJ/kg

DESCRIPTION

Ammonia is the compound formed by the chemical combination of the two gaseous elements, nitrogen and hydrogen, in the molar proportion of one part nitrogen to three parts hydrogen. This relationship is shown in the chemical symbol for ammonia, NH_3. On a weight basis the ratio is fourteen parts nitrogen to three parts hydrogen or approximately 82 percent nitrogen to 18 percent hydrogen.

The term "ammonia" as used throughout this section is the name of the chemical compound, NH_3, which in its commercial form is commonly called "anhydrous ammonia."

At room temperature and atmospheric pressure ammonia is a pungent, colorless gas. It may be compressed and cooled to a colorless liquid. At atmospheric pressure, ammonia boils at −28 F (−33.3 C) and freezes to a white crystalline solid at −107.9 F (−77.7 C). When heated above its critical temperature of 271.4 F (133.0 C), ammonia exists as a vapor regardless of

the pressure. Between the melting and critical points, liquid ammonia exerts a vapor pressure which increases with rising temperature. When liquid ammonia in a closed container is in equilibrium with ammonia vapor, the pressure within the container bears a definite relationship to the temperature as shown by the curve in Fig. 1.

Liquid ammonia is lighter than water, having a density of 42.57 lb/ft^3 (681.9 kg/m^3) at −28 F (−33.3 C), while as a vapor, ammonia is lighter than air, its relative density being 0.5970 compared to air at a pressure of 1 atm and a temperature of 32 F (0.0 C). Under the latter conditions, 1 lb (0.4536 kg) of ammonia vapor occupies a volume of 20.78 ft^3 (0.5884 m^3).

The relationships of temperature to vapor pressure, density, specific gravity and latent heat for liquid ammonia are shown in Table 1. Vapor pressure-temperature and density-temperature curves are shown in Figs. 1 and 2, respectively.

As a chemical compound ammonia is highly associated and stable. Dissociation begins to occur at 840–930 F (450–500 C) and atmo-

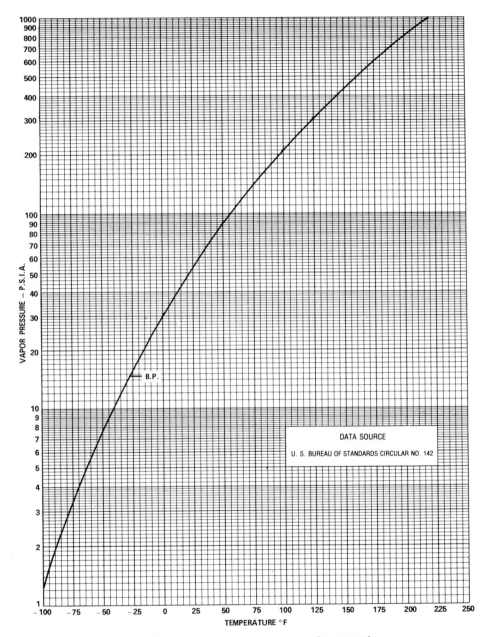

Fig. 1. Vapor pressure–temperature curve for ammonia.

spheric pressure, with the products being nitrogen and hydrogen. Experiments conducted by the Underwriters Laboratories indicate that an ammonia-air mixture in a standard quartz bomb will not ignite at temperatures below 1560 F (850 C). When an iron bomb, having catalytic effect was used, the ignition temperature was 1200 F (651 C). Ammonia gas burns at atmospheric pressure, but only within the limited range of 16–25 percent by volume of ammonia

TABLE 1. PROPERTIES OF LIQUID AMMONIA AT VARIOUS TEMPERATURES.
(Data for Columns 1, 2 and 5 taken from U.S. Bureau of Standards Circular No. 142.
Values for Columns 3 and 4 calculated from Column 2.)

Temperature Degrees F	Vapor Pressure (psig) (1)	Liquid Density		Specific Gravity of Liquid (Compared to Water at 4 C.) (4)	Latent Heat (Btu per pound) (5)
		Pounds Per Cubic Foot (2)	Pounds Per U.S. Gallon (3)		
−28	0.0	42.57	5.69	.682	589.3
−20	3.6	42.22	5.64	.675	583.6
−10	9.0	41.78	5.59	.669	576.4
0	15.7	41.34	5.53	.663	568.9
10	23.8	40.89	5.47	.656	561.1
20	33.5	40.43	5.41	.648	553.1
30	45.0	39.96	5.34	.641	544.8
40	58.6	39.49	5.28	.633	536.2
50	74.5	39.00	5.21	.625	527.3
60	92.9	38.50	5.14	.617	518.1
65	103.1	38.25	5.11	.613	513.4
70	114.1	38.00	5.08	.609	508.6
75	125.8	37.74	5.04	.605	503.7
80	138.3	37.48	5.01	.600	498.7
85	151.7	37.21	4.97	.596	493.6
90	165.9	36.95	4.94	.592	488.5
95	181.1	36.67	4.90	.588	483.2
100	197.2	36.40	4.87	.583	477.8
105	214.2	36.12	4.83	.579	472.3
110	232.3	35.84	4.79	.573	466.7
115	251.5	35.55	4.75	.570	460.9
120	271.7	35.26	4.71	.565	455.0
125	293.1	34.96	4.67	.560	448.9
130	315.6	34.66	4.63	.555	(443)
135	339.4	34.35	4.59	.550	(436)
140	364.4	34.04	4.55	.545	(430)

Note: The figures in parentheses were calculated from empirical equations given in U.S. Bureau of Standards Scientific Papers Nos. 313 and 315 and represent values obtained by extrapolation beyond the range covered in the experimental work.

in air. The products of combustion are nitrogen and water. Due to this low susceptibility to fire, ammonia is classified by DOT as nonflammable.

Ammonia is a highly reactive chemical forming ammonium salts in reactions with inorganic and organic acids; amides in reactions with esters, acid anhydrides, acyl halides, carbon dioxide, or sulfonyl chlorides; amines in reactions with halogen compounds or oxygen-containing compounds such as polyhydric phenols, alcohols, aldehydes and aliphatic ring oxides.

GRADES AVAILABLE

No commodity grade specifications for ammonia have been published as standard for the

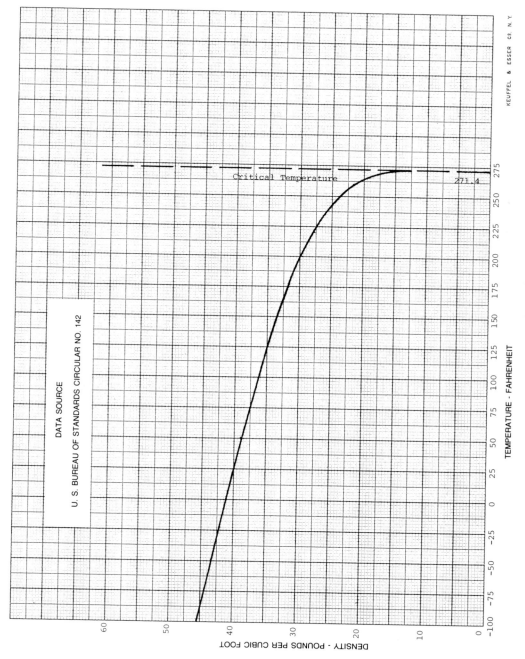

Fig. 2. Density–temperature curve for liquid ammonia.

industry. However, generally accepted grade designations are shown below.

Grade	Ammonia, Percent Weight Minimum
Commercial	99.5
Agricultural	99.7*
Refrigeration	99.95
Metallurgical	99.995

*When 82.0 percent weight minimum nitrogen is guaranteed.

In these grade designations, the assay values are established by difference following the determination of specified impurities, usually water and oil. Federal government specifications for ammonia may be found in Federal Specification O-A-445B.

USES

About 80 percent of all ammonia produced in the United States is used in agriculture as a source of nitrogen which is essential for plant growth. Nitrogen makes up about 16 percent of plant protein. When a fruit, vegetable or grain crop is grown and harvested, nitrogen is removed from the soil. If the fertility of the land is to be maintained, nitrogen and other elements essential to plant growth such as potassium and phosphorus must be restored to the soil by fertilization. Depending upon the particular crop, up to 200 lb of nitrogen may be economically applied per acre.

About four million tons (3.6 teragrams) of ammonia containing 82 percent nitrogen are applied directly to the soil each year in the United States. It can be injected at a depth of several inches below the surface of the soil by specially designed equipment or it can be dissolved in irrigation water.

Ammonia is also used in the production of nitrogen fertilizer solutions which consist of ammonia, ammonium nitrate, urea and water in various combinations. Some are pressure solutions and others are not. Nonpressure and low-pressure solutions are widely used for direct application to the soil. Pressure solutions containing free ammonia are used in the manufacture of high analysis mixed fertilizers.

Ammonia is utilized extensively in the fertilizer industry to produce solid materials such as ammonium salts, nitrate salts and urea. Ammonium sulfate, ammonium nitrate and ammonium phosphate are made directly by neutralizing the corresponding acids—sulfuric acid, nitric acid and phosphoric acid—with ammonia. Urea is an organic compound formed by combining ammonia and carbon dioxide. Ammonium sulfate, ammonium nitrate, sodium nitrate, ammonium phosphate and urea are used for direct application to the soil in dry form and in combination with other phosphate and potassium salts.

In addition to their use as fertilizers, ammonia and urea are used as a source of protein in ruminant livestock feeds. Urea is used in mixed feed supplements to supply the nitrogen needed for the biosynthesis of proteins by the microorganisms in ruminating animals such as cattle, sheep and goats.

Ammonia is oxidized in the production of nitric acid, the principal ammonia derivative used in making explosives. Both industrial and military explosives are divided into two main types: high explosives such as dynamite, nitroglycerine and TNT which detonate rapidly to give a shattering blast for demolition purposes, and low explosives such as nitrocellulose which detonate slowly to give a heaving/pushing effect for propellant or blasting applications. Dynamite, a general term for high explosives used in mining and construction, contains nitroglycerine or other organic nitrogen compounds absorbed in a combustible material. In ammonia dynamites, ammonium nitrate, made by reacting ammonia and nitric acid, replaces all of the nitroglycerine. Blasting gelatin dynamites consist of a colloidal mixture of nitroglycerine and nitrocellulose. The latter is made by treating cellulose with a mixture of nitric and sulfuric acids.

Ammonium nitrate is the principal base material in slurry explosives and lower cost blasting agents. Ammonium nitrate can be converted to an effective blasting agent by properly mixing it

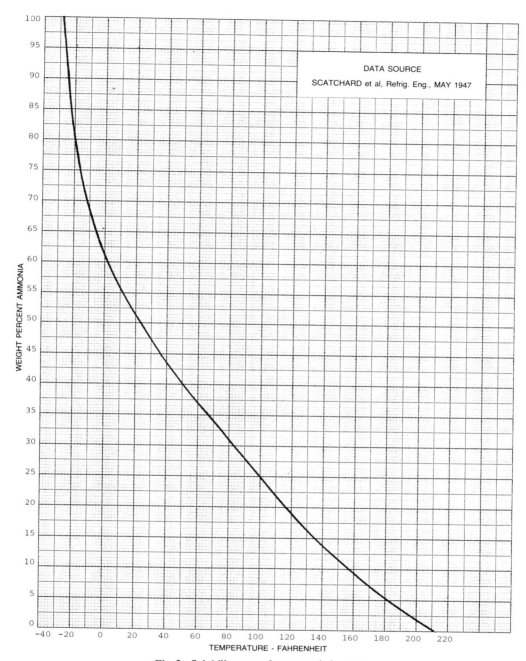

Fig. 3. Solubility curve for ammonia in water.

with a carboniferous material such as fuel oil. Ammonium nitrate/fuel oil mixtures are used extensively in open pit mining and outdoor construction work because of ease of handling, availability, low cost and safety.

The manufacture of ammonium salts and certain alkalies, dyes, pharmaceuticals, synthetic textile fibers and plastics depends upon utilization of ammonia.

Used in both absorption- and compression-type systems, ammonia is the oldest, the most efficient and economical mechanical refrigerant known.

Ammonia or dissociated ammonia is used in such metal-treating operations as nitriding, carbo-nitriding, bright annealing, furnace brazing, sintering and other applications where protective atmospheres are required.

Ammonia is used in pH control, in mineral beneficiation, in neutralizing acidic components during petroleum refining and in the treatment of acidic wastes.

Through the controlled combustion of dissociated ammonia in air, relatively pure nitrogen is produced.

Dissociated ammonia provides a convenient source of hydrogen for hydrogenation and other applications.

Ammonia is used in extracting certain metals such as copper, nickel and molybdenum from their ores.

Ammonia vapor is utilized as the developing agent in the photomechanical process of white-printing and is also employed in the production of diazotype microfilm duplicates.

Ammonia is used in scrubbers to neutralize sulfur oxides in their removal from stack gases in electric power generation and other furnace operations such as in smelting. It is also used to improve the efficiency of electrostatic precipitators in the removal of particulate matter.

Ammonia is highly soluble in water, forming aqueous ammonia (ammonium hydroxide or aqua ammonia) which has many applications. In a very dilute solution (2-5 percent ammonia) it is available as "household ammonia." The solubility of ammonia in water at various temperatures is shown in Fig. 3.

Water solutions of ammonia are used to regenerate weak anion exchange resins.

PHYSIOLOGICAL EFFECTS

Persons having chronic respiratory disease or persons who have shown evidence of undue sensitivity to ammonia should not be employed where they will be exposed to ammonia.

Ammonia is not a cumulative metabolic poison; ammonium ions are actually important constituents of living systems. However, ammonia in the ambient atmosphere has an intense irritating effect on the mucous membranes of the eyes, nose, throat and lungs. High levels of ammonia can produce corrosive effects on tissues and can cause laryngeal and bronchial spasm and edema so as to obstruct breathing. The pungent odor of ammonia affords a protective warning, and conscious persons avoid breathing significantly contaminated air. Unconscious persons, however, are not similarly protected.

Table 2 indicates human physiological response to various concentrations of ammonia in air. Individuals differ in their sensitivity to ammonia, some persons being highly reactive to relatively low concentrations and others showing a significant tolerance to the irritative effects.

TABLE 2. PHYSIOLOGICAL RESPONSE
TO AMMONIA.

Least perceptible odor	5 ppm
Readily detectable odor	20-50 ppm
No discomfort or impairment of health for prolonged exposure	50-100 ppm
General discomfort and eye tearing; no lasting effect on short exposure	150-200 ppm
Severe irritation of eyes, nose and throat; no lasting effect on short exposure	400-700 ppm
Coughing, bronchial spasms	1,700 ppm
Dangerous, less than ½ hour exposure may be fatal	2,000-3,000 ppm
Serious edema, strangulation, asphyxia, rapidly fatal	5,000-10,000 ppm
Immediately fatal	10,000 ppm

Note: Concentrations are for ammonia in air by volume. Exposure levels which are tolerated by normal persons may produce coughing and bronchospasm in others.

In accordance with U.S. Department of Labor Regulations (as set forth in Title 29 Labor, U.S. Code of Federal Regulations, Part 1910, Section 1000), an employee's exposure to ammonia must be limited to a concentration not to exceed 50 ppm of ammonia in air by volume based upon an 8-hour time weighted average. Concentrations in the range of 20–50 ppm are readily detectable and it is therefore unlikely that any individual would become overexposed unknowingly.

Since liquid ammonia vaporizes readily and has a great affinity for water, it may cause severe injury to the skin by freezing the tissue and subjecting it to caustic action. A chemical burn, which may be severe, will result.

SAFE HANDLING AND STORAGE

Materials Suitable for Containers and Storage

Most common metals are not affected by dry ammonia. However, when combined with water vapor, ammonia will attack copper, zinc or alloys containing copper as a major alloying element. Therefore, these materials should not be used in contact with ammonia. Certain high-tensile strength steels have developed stress-corrosion cracking in ammonia service, but such cracking can be prevented by the use of 0.2 percent water by weight in the ammonia as an inhibitor. Ammonia storage tanks and tank cars with their valves and fittings are usually made of steel. DOT regulations covering the construction of ammonia cargo and portable tanks prohibit the use of aluminum, copper, silver, zinc or alloys of these materials. For details regarding materials of construction recommended for piping, tubing, fittings and hose, refer to ANSI Standard K61.1.

Piping and Equipment

Moist ammonia corrodes copper, zinc and their alloys. It is customary, therefore, that iron or steel be used in piping and fittings. Certain aluminum alloys are also used in appurtenances and can be used in other parts of systems for ammonia. Copper, brass or galvanized fittings and pipe should not be used. Unions, valves, gages, pressure regulators and relief valves having copper, brass or bronze parts are not suitable for ammonia service.

Metallic and nonmetallic gasket materials, such as compressed asbestos, graphited asbestos, carbon steel or stainless steel spiral-wound asbestos-filled and aluminum are suitable for ammonia service. Fluorocarbon plastic materials and neoprene have also been found suitable.

All ammonia piping should be extra heavy (Schedule 80) steel when threaded joints are used. Standard weight (Schedule 40) steel may be used when joints are either welded or joined by welding-type flanges. Flow should not be restricted by an excessive number of elbows and bends. All refrigeration system piping should conform to ANSI B31.5, "Refrigeration Piping." Also see ANSI K61.1 requirements.

Welded instead of threaded joints should be used wherever possible, particularly on larger pipe sizes. When threaded joints are used, a suitable pipe thread dope or fluorocarbon plastic tape should be used.

Provision should be made to protect piping against the effects of expansion, contraction, jarring, vibration, settling and external corrosion.

All piping and tubing should be tested after assembly at a pressure not less than the normal operating pressure of the system and proved free from leaks.

Threaded nipples, when used, should be cut from Schedule 80 steel pipe.

All fittings must be extra heavy and nonmalleable metals, such as cast iron, must not be used in their construction. Forged or cast steel valves and fittings should be used in any service subject to significant strain or vibration.

Hose should be manufactured specifically for ammonia service and should have a minimum burst pressure not less than 1750 psig (120.7 bars). Hose couplings should be of steel. Each hose after assembly with couplings should be hydrostatically tested and should be leak free at the maximum rated working pressure of the hose assembly including couplings. Hose assemblies should be inspected visually at regular intervals thereafter depending upon the frequency and severity of service and replaced as necessary.

Storage and Handling in Normal Use

Anhydrous ammonia should be handled with all the precautions necessary for any nonflammable gas. Personnel working with anhydrous ammonia should be thoroughly familiar with safety precautions for handling the gas as well as measures for handling emergencies. (See also Part I, Chapter 4, Section A.)

Handling Leaks and Emergencies

A leak in an ammonia system can be detected by odor. The location of the leak may be determined with moist phenolphthalein or red litmus paper which change color in ammonia vapor. Another means of detection is the use of sulfur dioxide, which forms a white fog in contact with ammonia vapor.

Action if a Leak Occurs. Only personnel trained for and designated to handle emergencies should attempt to stop a leak. Respiratory equipment of a type suitable for ammonia must be worn. All persons not so equipped must leave the affected area until the leak has been stopped.

If ammonia vapor is released, the irritating effect of the vapor will force personnel to leave the area long before they have been exposed to dangerous concentrations. To facilitate their rapid evacuation there should be sufficient well-marked and easily accessible exits. If, despite all precautions, a person should be trapped in an ammonia atmosphere, he should breathe as little as possible and open his eyes only when necessary. Partial protection may be gained by holding a wet cloth over the nose and mouth. Since ammonia vapor in air will rise, a trapped person should remain close to the floor to take advantage of the lower vapor concentrations at that level.

With good ventilation or rapidly moving air currents, ammonia vapor, being lighter than air, can be expected to dissipate readily to the upper atmosphere without further action being necessary. Lacking these conditions, the concentration of ammonia vapor in air can be reduced effectively by the use of adequate volumes of water applied through spray or fog nozzles. Do not put water on a liquid ammonia spill unless sufficient water is available. Sufficient water may be considered as 100 parts of water to one part of ammonia.

Under some circumstances ammonia in a container is colder than the available water supply. At such times water must not be sprayed on the container walls since it would heat the ammonia and aggravate any gas leak. If it is found necessary to dispose of ammonia, as from a leaking container, liquid ammonia may be discharged into a vessel containing water sufficient to absorb it. Sufficient water may be considered as ten parts of water to one part of ammonia. The ammonia must be injected into the water as near the bottom of the vessel as practical.

A leak at a valve stem on an ammonia cylinder in service can usually be stopped by tightening the packing nut which has a left-hand thread. If this fails, the valve should be closed. A cylinder which continues to leak should be removed from the building.

For further information, see "Emergency Services Guide for Selected Hazardous Materials," published by DOT and see CHEM-CARD CC-44 published by the Chemical Manufacturers Association.

Fire Exposure. An ammonia container exposed to a fire should be removed. If for any reason, it cannot be removed, the container should be kept cool with water spray until well after the fire is out. Fire-fighting personnel should be equipped with protective clothing and respiratory equipment. Information on such safety equipment, and on general procedures for security, is given in CGA Pamphlet G-2.

First Aid Suggestions. Call a physician immediately for any person who has been burned or overcome by ammonia.

The patient should be removed to an area free from fumes, preferably a warm room. He should be placed in a reclining position with

head and shoulders elevated and kept warm by
the use of blankets or other cover, if necessary.

Prior to medical aid by the physician, first
aid measures should be taken. Those presented
here are based on what is believed to be common practice in industry. Their adoption in
any specific case should of course be subject
to prior endorsement by a competent medical
advisor.

Eyes. If contacted by ammonia, the eyes
must be flooded immediately with copious
quantities of clean water. Speed is essential.
In isolated areas, water in a squeeze bottle
which can be carried in the pocket is helpful
for emergency irrigation purposes. Eye fountains should be used, but if they are not available, water may be poured over the eyes. In
any case, the eyelids must be held open and
irrigation must continue for at least 15 minutes. The patient must receive prompt attention from a physician, preferably an opthamol-
ogist. Persons subject to ammonia exposure
must not wear contact lenses.

Skin. If liquid ammonia contacts the skin,
the area affected should be immediately flooded
with water. If no safety shower is available, immerse in any available water. Water will have
the effect of thawing out clothing which may
be frozen to the skin. Such clothing should be
removed and flooding with water continued
for at least 15 minutes. Do not apply salves or
ointments or cover burns with dressings, however, the injured area should be protected with
a clean cloth prior to medical care. Do not
attempt to neutralize the ammonia.

Nose and Throat. If ammonia has entered
the nose or throat and the patient can swallow,
have him drink large quantities of water.
*Never give anything by mouth to an uncon-
scious person.*

Inhalation. Any conscious person who has
inhaled ammonia causing irritation should go to
an uncontaminated area and inhale fresh air.

A person overcome by ammonia should immediately be carried to an uncontaminated
area. If breathing has ceased, artificial respiration must be started immediately, preferably
by trained personnel. If breathing is weak or

has been restored by artificial respiration,
oxygen may be administered.

METHODS OF SHIPMENT; REGULATIONS

Under the appropriate regulations and tariffs,
anhydrous ammonia is today shipped primarily
as follows:

By Rail: In insulated and uninsulated tank
cars (shipment by rail in cylinders and in portable tanks is also authorized, as is the currently seldom used method of shipment by rail
in multi-unit tank car tanks).

By Highway: In cargo tank motor vehicles,
including semitrailers and full trailers (shipment by highway in cylinders and portable
tanks is also authorized, as is the currently
seldom used method of shipment by highway
in multi-unit tank car tanks).

By Water: In barges on inland waterways
and tankers on ocean-going routes.

By Air: Aboard cargo aircraft only in appropriate cylinders up to 300 lb (140 kg)
maximum net weight per cylinder.

By Pipeline: In interstate pipelines authorized for liquid service, not gaseous service.

CONTAINERS

Anhydrous ammonia is transported as a
liquefied compressed gas in cylinders, insulated and uninsulated cargo and portable
tanks, insulated and uninsulated tank cars
(and multi-unit tank cars), barges and tankers.
It is stored in bulk in large-capacity containers
installed above or below ground. Normal aboveground storage is in uninsulated, pressure storage tanks. Very large aboveground containers
are often low-pressure, refrigerated and consequently insulated tanks.

Filling Limits. Since the transportation of
ammonia as a vapor is not commercially economical, it is shipped and stored as a liquefied
compressed gas. Because liquid ammonia ex-
pands as its temperature rises, it is possible
with sufficient increase in temperature, for a
container to become liquid-full. The resulting

TABLE 3. MAXIMUM ALLOWABLE FILLING DENSITIES FOR
AMMONIA SHIPPING CONTAINERS.

Type Container	DOT Specification Number		Maximum Allowable Filling Density	
			Percent by Weight	Percent by Volume
Cylinders	(See Table 4)		54	–
Portable tanks	51	(Note 1)	56	82
Cargo tanks	MC-330	(Note 1)	56	82
	MC-331	(Note 1)	56	82
Multi-unit tank car tanks	106A500-X	(Note 2)	50	–
Single-unit tank cars	105A300-W		57	–
	112A340-W	(Notes 3 and 4)	57	–
	112A400-F	(Notes 3 and 4)	57	–
	114A340-W	(Notes 3 and 4)	57	–

Note 1. Uninsulated cargo and portable tanks may be filled to 87.5 percent by volume provided the temperature of the ammonia is not less than 30 F (–1.1 C).

Note 2. Authorized for rail and highway transportation, but current rarely used.

Note 3. Filling density of 58.8 percent permitted November through March inclusive. Storage in transit prohibited.

Note 4. Specification 112 and 114 tank cars built before January 1, 1978 must be equipped with tank head puncture-resistance system (112S and 114S stenciled cars) after December 31, 1981. Specification 112A and 114A tank cars will not be authorized after 12-31-81. New cars built after December 31, 1977 must meet specifications 112S and 114S. Specifications 112T, 112J, 114T and 114J tank cars are also authorized to transport anhydrous ammonia as they are equipped with both thermal protection and tank head puncture-resistance systems.

hydrostatic pressure could cause a container to rupture if it is not protected by a pressure-relief device or to discharge ammonia if the container is equipped with such a device. Therefore, DOT regulations limit the amount of ammonia which may be filled into containers of various types. These regulations provide that the liquid portion of the commodity must not completely fill an insulated tank car, portable tank or cargo tank at 105 F (40.6 C), nor an uninsulated multi-unit tank car or cylinder at 130 F (54.4 C), nor an uninsulated portable tank, cargo tank or single-unit tank car at 115 F (46.1 C). This filling limitation is expressed in terms of maximum filling density for each type of container. The term "filling density" for liquefied gases is defined as "the percent ratio of the weight of gas in a container to the weight of water at 60 F (15.6 C) that the container will hold." For determining

the water capacity of a container in pounds, the weight of a gallon (231 in^3) of water at 60 F shall be 8.32828 lb. (A cubic meter of water at 15.6 C weighs 1000 kg.) Maximum allowable filling densities for shipping containers in common usage are shown in Table 3.

Cylinders

Cylinders meeting the requirements of the following Department of Transportation Specifications are authorized for ammonia service in the United States: DOT-3A480, DOT-3A480X, DOT-3AA480, DOT-3E1800, DOT-4, DOT-4A480 and DOT-4AA480.

The type ammonia cylinder most commonly used is the bottle-type. It is available in seamless (one piece) or welded (two piece) construction and in a number of convenient sizes, the two most widely used having capacities of 100

TABLE 4. APPROXIMATE DIMENSIONS AND WEIGHTS OF TYPICAL AMMONIA CYLINDERS.

DOT Cylinder Spec. No.	Ammonia Capacity (pounds)	Average Tare Weight (pounds)		Overall Length (inches)	Outside Diameter (inches)	Wall Thickness (inches)	Minimum Volume (cubic inches)
		Less Cap	With Cap				
3AA1800	2	5	NA	16	$3\frac{1}{2}$	0.070	108
3A480	100	134	137	59	$12\frac{1}{2}$	0.176	5158
3A480	150	195	198	60	15	0.212	7710
3A480X	100	88	91	57	$12\frac{1}{4}$	0.120	5158
3A480X	150	135	138	58	$14\frac{3}{4}$	0.125	7710
4AA480	25	43	NA	20	$12\frac{1}{4}$	0.153	1344
4AA480	100	114	117	56	$12\frac{1}{4}$	0.153	5158
4AA480	150	158	161	58	$14\frac{3}{4}$	0.185	7710

and 150 lb (45 and 68 kg) of ammonia. Dimensions and weights of typical cylinders are tabulated in Table 4.

The value of each ammonia cylinder having a capacity of 25 lb (11 kg) or more has an internal dip tube connected to it. This makes it possible to withdraw either liquid or vapor by positioning the cylinder so that the end of the dip tube is in the desired phase.

Each cylinder is subject to periodic retesting and reinspection in accordance with 49 CFR 173.34 (e). The periodic retest must include a visual internal and external examination together with a test by interior hydrostatic pressure in a water jacket or other apparatus of suitable form for the determination of the expansion of the cylinder. The internal inspection may be omitted for cylinders of the type and in the service described under Note 1 to Table 5. Requalifying periods and test pressures for cylinders in ammonia service are shown in Table 5.

Withdrawing Ammonia as Liquid. The cylinder should be positioned horizontally with the valve outlet pointing up. In this position the dip tube will point down and be in the liquid phase.

Before disconnecting an empty cylinder, the cylinder valve should be closed and the bleed line valve opened to exhaust the residual am-

TABLE 5. REQUALIFYING PERIODS AND TEST PRESSURES FOR TYPICAL CYLINDERS IN AMMONIA SERVICE.

Specification	Test Period (years)	Min. Retest Pressure (psi)	Note
DOT-3	5	3000	
DOT-4*	10	700	1
DOT-3A480	5 or 10	800	1 2
DOT-3A480X	5 or 10	800	1 2
DOT-3AA480	5	800	1
DOT-3AA1800	5	3000	1
DOT-4A480	5	800	1
DOT-4AA480	5 or 10	960	1 2
DOT-3E1800	retest not required		

*Also see 49 CFR 173.34 (e) (2).

Note 1. Cylinders made in compliance with DOT-4, DOT-3A, DOT-3A480X, DOT-3AA, DOT-4A, DOT-4AA480 and used exclusively for anhydrous ammonia of at least 99.95 percent purity may, in lieu of the periodic hydrostatic retest, be given a complete external visual inspection in accordance with 49 CFR 173.34 (e) (10). When this inspection is used in lieu of hydrostatic retesting, such inspection shall be made at 5-year intervals.

Note 2. A cylinder made in compliance with Specification DOT-3A, DOT-3A480X, or DOT-4AA480, used exclusively for anhydrous ammonia, commercially free from corroding components, and protected externally by suitable corrosion-resistant coatings (such as paint, etc.) may be retested every 10 years instead of every 5 years.

monia from the connecting hose. If frost forms on the hose the presence of liquid ammonia is indicated and the connector should not be removed until the frost has melted.

The practice of manifolding to withdraw liquid ammonia from two or more cylinders simultaneously is hazardous because under certain temperature conditions, it is possible for liquid to flow into one cylinder until it is completely filled. If the valve of this completely filled cylinder were then closed, any rise in temperature would result in the development of hydrostatic pressure due to liquid ammonia expanding and could result in the rupture of the cylinder. The practice of manifolding cylinders to withdraw liquid ammonia should be avoided, but if necessary, backflow check valves should be installed in the piping from each cylinder.

Withdrawing Ammonia as Vapor. When a cylinder is vertical, the dip tube is normally in the vapor phase. When placed horizontally for vapor withdrawal, a cylinder should have its valve outlet pointing downward which points the dip tube upward into the vapor phase.

A cylinder in a vertical position may discharge liquid ammonia if it had been stored in a location warm enough to cause the liquid level to rise above the lower end of the dip tube.

The flow rate of ammonia vapor from a cylinder is dependent on the rate of vaporization of the liquid ammonia. Since heat transfer from the surrounding atmosphere through the cylinder wall is greatest through the portion of the wall that is wet with liquid internally, the rate of vaporization will diminish as the liquid level in the cylinder decreases.

A cylinder of 150 lb (68 kg) size at room temperature standing vertically may be expected to deliver up to 30 ft^3 (0.8 m^3) of ammonia vapor per hour continuously until empty.

Frost forming on a cylinder indicates that the refrigeration effect of vaporization has cooled the liquid below 32 F (0 C) and as a result, the flow rate and pressure may be inadequate. An additional cylinder can be added

to the manifold to increase the heat transfer surface. Shutting off the valve and allowing a frosted cylinder to warm up will restore pressure and flow rate. The procedure can be repeated until the cylinder is empty.

It should be emphasized that a cylinder must never be warmed directly by water bath, steam or flame. This is a very dangerous practice which might create pressure sufficient to rupture the cylinder.

Determining When Cylinders Are Empty. The best way to determine if an ammonia cylinder is empty is to weigh it, without cap, and compare the weight with the tare weight stamped on the cylinder.

Valve Outlet and Inlet Connections. Standard connections, United States and Canada— Nos. 240, 705. Standard yoke connections, United States and Canada—Nos. 800, 845.

Portable Tanks

Portable tanks used for interstate transportation of ammonia must comply with DOT Specification 51. They are designed to be temporarily attached to a motor vehicle, other vehicle, railroad car other than tank car, or vessel, and equipped with skids, mountings or accessories to facilitate handling of the tank by mechanical means. If a portable tank is used as a cargo tank it must comply with all the requirements prescribed for cargo tanks.

Portable tanks complying with DOT Specification 51 must have a capacity in excess of 1000 lb (454 kg) of water. If shipped on vessels under the jurisdiction of the United States Coast Guard, the gross weight must not exceed 55,000 lb (25,000 kg).

DOT Specification 51 prescribes fabrication in accordance with the ASME Boiler and Pressure Vessel Code, Section VIII and a minimum design pressure of 265 psig (18.3 bars) is required. Steel of a thickness less than $\frac{3}{16}$ in. (4.763 mm) shall not be used for the shell, heads or protective housings.

In most states, portable tanks used for intrastate transportation must be fabricated in accordance with the ASME Boiler and Pressure

Vessel Code, Section VIII or DOT Specification 51 with a minimum design pressure of 250 psig (17.2 bars). Some states, however, require design pressure to be 265 psig (18.3 bars) minimum. Portable tanks built in accordance with the ASME Code do not have to comply with paragraphs UG-125 through UG-128 inclusive, paragraph UG-132 and paragraph UG-133. Steel of a thickness less than $\frac{3}{16}$ in. (4.763 mm) must not be used for the shell, heads or protective housings.

Systems Mounted on Farm Wagons

Special farm vehicles have been developed for intrastate handling of ammonia used as fertilizer. Design of the containers should be in accordance with requirements for portable tanks used for intrastate transportation. Containers called "applicator tanks" are mounted on farm equipment and used as mobile storage supplying ammonia as it is applied to the soil. "Nurse tanks" are containers mounted on farm wagons used to store and transport ammonia to replenish applicator tanks and may also be used as applicator tanks or to feed ammonia into irrigation streams. Portable skid tanks are sometimes used instead of "nurse tanks." Other vehicles are used to transport ammonia in bulk to replenish storage tanks at various locations on farms. Such vehicles usually operate locally and therefore, state, provincial and local safety and design regulations should be consulted.

A suitable "stop" or "stops" must be installed on the farm wagon or on the container in such a way that the container shall not be dislodged from its mounting due to the farm wagon coming to a sudden stop. Back slippage shall also be prevented by proper methods.

A suitable "hold-down" device must be provided which will anchor the container to the farm wagon at one or more places on each side of the container.

When containers are mounted on four-wheel farm wagons, care must be taken to insure that the weight is distributed evenly over both axles.

When the cradle and the container are not welded together, suitable material must be used between them to eliminate metal-to-metal friction.

Further information on appurtenances and marking for these tanks, and on other requirements for nurse tanks and applicator tanks, is given in CGA Pamphlet G-2.

Multi-Unit Tank Car Tanks

The DOT authorizes the transportation of ammonia in containers of DOT specification 106A500X. The containers are uninsulated, have a test pressure of 500 psig (3450 kPa) and a water capacity of 1500–2600 lb (680–1179 kg) each.

Cargo Tanks

Cargo tanks are containers designed to be permanently attached to a motor vehicle such as tank trucks and trailers, and must comply with DOT Specifications MC-330 (before 1967) or MC-331 (new construction). Straight truck-mounted units are referred to as "tank trucks" or "bob-tails" whereas larger trailer-mounted units are referred to as "trailer transports" or "tank wagons."

The DOT does not limit the size of cargo tanks, except if shipped on vessels under the jurisdiction of the United States Coast Guard, where the gross weight must not exceed 55,000 lb (25,000 kg). State regulations limit the gross weight of the vehicle.

DOT Specification MC-331 prescribes fabrication under the ASME Boiler and Pressure Vessel Code, Section VIII and a minimum design pressure of 265 psig (1830 kPa) is required. Steel of a thickness less than $\frac{3}{16}$ in. (4.763 mm) shall not be used for the shell, heads or protective housings.

All containers, except those filled by weight, must be equipped with gaging devices to indicate the maximum permitted liquid level. Permitted types for ammonia are the rotary tube, adjustable slip tube, and fixed length dip tube. If other gaging devices are installed, they must not be used as primary controls for filling cargo tanks.

Tank Cars

Ammonia is authorized for shipment in special tank cars, either insulated or uninsulated. Insulated cars must be constructed to DOT Specification 105A300-W. Each has a pressure-relief valve set to start-to-discharge at 225 psig or 247.5 psig (1550 kPa). Uninsulated tank cars must be constructed to DOT Specification 112S340W or 114S340W. Class 105A, 112S, and 114S cars having higher pressures may also be used. DOT 106A500-X tank car tanks are authorized for ammonia, but are seldom used.

Tank cars with a nominal water capacity of 11,000 gal (42 m^3), hold approximately 26 tons (24,000 kg) of ammonia and may be referred to as "standard" cars. Tank cars with a nominal water capacity of 33,500 gal (127 m^3), hold about 80 tons (73,000 kg) of ammonia and are referred to as "jumbo" cars.

Ammonia tank cars must be unloaded on private tracks. If they are equipped with check valves they can also be unloaded on carrier tracks provided they are unloaded into permanent storage tanks of sufficient capacity to receive the entire contents of the car. See 49 CFR 174.560 (b) (2).

Tank cars used in seasonal service must be suitable for the commodities transported and they should be purged completely of the previous commodity when service is changed.

Further information on tank car valves, marking and placarding, shipping and unloading is given in CGA Pamphlet G-2.

Barges and Tankers

Barges and tankers for the transportation of ammonia are especially designed and fabricated for this service. Barges are widely used on inland waterways while tankers are limited to ocean-going routes. Both require suitable terminals for loading and unloading.

Most ammonia transported by water is refrigerated to -28 F (-33.3 C). Older barges were not refrigerated and required pressure storage at the receiving terminal. After unloading into high-pressure storage, the ammonia was refrigerated and transferred to low-pressure storage.

Most barges for transporting ammonia have two refrigerated tanks with a total capacity of 2500 tons (2.3 gigagrams). There is refrigeration equipment on board to maintain the ammonia at -28 F (-33.3 C). Ocean-going tankers have capacities of up to 25,000 tons (22.7 Gg), but most are 9000 to 14,000 tons (8.2-12.7 Gg).

The design and fabrication of barges and tankers for ammonia service must be approved by the authorities in the country where the vessel is fabricated. Any vessel operating in United States waters must be approved by the Coast Guard. Any modification after fabrication must also be approved.

Pipelines

Pipeline companies are common carriers regulated by the Office of Pipeline Safety Operations of the Department of Transportation. For the purpose of the regulations, pipelines conveying ammonia are for liquid service, not gaseous service.

Pipelines typically are constructed of 6 to 8 in. (152-203 mm) pipe having maximum operating pressures of 500-1200 psig (3400-8300 kPa). With adequate pumping horsepower they can transport 3000-5000 tons (2.7-4.5 Gg) per day of liquid ammonia.

Pumping stations for ammonia pipelines commonly are equipped with suction and discharge pressure controllers and with shutdown devices for such malfunctions as low suction pressure, high discharge pressure and mechanical failure. Pipelines commonly have, at regular intervals, shutdown valves which may be remotely operated either automatically or manually.

Ammonia transported by pipeline is at ambient temperature. Relatively small terminals using pressure storage can take ammonia directly from the pipeline. However, most storage operated in conjunction with pipelines is large-scale refrigerated storage.

Pipelines can connect multiple producing

plants with multiple storage locations. The storage location may be hundreds of miles away from the producing plant.

Receipt of ammonia by pipeline requires careful study and planning among producer, pipeline operator and storage operator.

Steels for pipelines transporting ammonia are specified by ANSI Standard B31.4 entitled, "Liquid Petroleum Transportation Piping Systems," 1974 edition, which also requires that the ammonia be inhibited with at least 0.2 percent water by weight.

Stationary Containers for Bulk Storage

Pressure containers mounted on foundation piers are used primarily as ammonia storage tanks. Such tanks may be insulated although most are not. For refrigerated containers see below.

Size limitations may be imposed by regulations or local conditions. The most frequently constructed sizes range from 500–45,000 gal (2–170 m^3).

Ammonia storage containers must be designed for at least 250 psig (1720 kPa), constructed in accordance with the ASME Boiler and Pressure Vessel Code, Section VIII and should be stress relieved after fabrication. Since it is possible that the design of the container, its capacity and its location, may be influenced by state, provincial, or municipal regulations and insurance restrictions, a thorough investigation of pertinent requirements should be made prior to fabrication. All containers must be inspected by a National Board Inspector who shall certify the tank as being in compliance with applicable requirements of the ASME Code and registered with the National Board of Boiler and Pressure Vessel Inspectors.

Refrigerated containers are used for storage of relatively large quantities of ammonia. These systems should be properly engineered and the advice of manufacturers or experienced engineering firms should be obtained before proceeding with such installations. See also American National Standard K61.1.

METHOD OF MANUFACTURE

In the United States, ammonia was first produced on a commercial basis as a by-product of the destructive distillation of coal in the manufacture of coke and coal gas beginning around 1890.

The first commercially successful synthetic ammonia plant in the United States was put into operation in the year 1921. It utilized the Haber–Bosch process in which a preheated mixture of nitrogen and hydrogen was subjected to pressure in the presence of a contact catalyst.

Most ammonia produced commercially today is manufactured by processes which are modifications of the Haber-Bosch process. Several sources of hydrogen are used including natural gas, refinery gas or coke-oven gas. Nitrogen may be supplied by introducing compressed air into the process stream or by introducing nitrogen from an air separation unit. In modern plants, natural gas, air and steam are reacted at high temperatures in the presence of catalysts to yield a mixture of hydrogen, nitrogen and oxides of carbon. Following conversion or removal of the oxides, the remaining hydrogen-nitrogen mixture is compressed and passed over catalysts where ammonia synthesis occurs at elevated temperatures.

Production of ammonia in the United States in 1976 is estimated at 16 million tons (15 teragrams). Worldwide production is estimated at 67 million tons (61 Tg).

REFERENCES

1. "American National Standard Metric Practice Guide," Z210.1 (ASTM E 380).
2. "American National Standard Power-Operated Pumps for Anhydrous Ammonia and L-P Gas," B166.1 (UL-51).
3. "American National Standard Refrigeration Piping," B31.5.
4. "American National Standard Safety Code for Mechanical Refrigeration," B9.1 (ASHRAE 15-70).
5. "American National Standard Safety Requirements for the Storage and Handling

of Anhydrous Ammonia," K61.1 (CGA–G-2.1, TF1-M-1).

6. "American National Standard Slow-Moving Vehicle Identification Emblem," B114.1 (ASAE S276.2, SAE J943).

7. Canadian Transport Commission, "Regulations for the Transportation of Dangerous Commodities by Rail." Available from Information Canada, Publications Division, 171 Slater Street, Ottawa, Canada.

8. Code of Federal Regulations, Title 29—Labor, Chapter XVII—Occupational Safety and Health Administration, Department of Labor (Parts 1900–1999). Available from Superintendent of Documents, U.S. Government Printing Office, Washington, DC 20402.

9. Code of Federal Regulations, Title 46—Shipping, Chapter 1—Coast Guard, Department of Transportation (Parts 0–199). Available from Superintendent of Documents, U.S. Government Printing Office, Washington, DC 20402.

10. Code of Federal Regulations, Title 49—Transportation, Chapter 1—Materials Transportation Bureau, Department of Transportation (Parts 100–199). Available from Superintendent of Documents, U.S. Government Printing Office, Washington, DC 20402.

11. Code of Federal Regulations, Title 49—Transportation, Chapter 11—Federal Railroad Administration, Department of Transportation (Parts 200–299). Available from Superintendent of Documents, U.S. Government Printing Office, Washington, DC 20402.

12. Code of Federal Regulations, Title 49—Transportation, Chapter 11—Federal Highway Administration, Department of Transportation (Parts 300–399). Available from Superintendent of Documents, U.S. Government Printing Office, Washington, DC 20402.

13. Code of Federal Regulations, Title 49—Transportation, Chapter IV—Coast Guard, Department of Transportation (Parts 400–499). Available from Superintendent of Documents, U.S. Government Printing Office, Washington, DC 20402.

14. Compressed Gas Association, Inc., "Anhydrous Ammonia" (CGA Pamphlet G-2). Available from the Compressed Gas Association, 500 Fifth Ave., New York, NY 10110.

15. Federal Specification, Ammonia Technical, O-A-445B, February 25, 1975. Single copies available without charge from Business Service Centers at the General Services Administration Regional Office at Washington, DC and other cities.

16. Henderson, Y. and Haggard, H. W. "Noxious Gases," Chemical Catalog Company, New York, 1927.

17. MCA Chem-Card—Transportation Emergency Guide CC-44, "Anhydrous Ammonia." Available from the Chemical Manufacturers Association, Inc., 1825 Connecticut Avenue, N.W., Washington, DC 20009.

18. "Marine Regulations for the Transport of Dangerous Goods." Available from Canadian Coast Guard, Transport Canada, Ottawa, K1A ON5, Canada.

19. Scatchard, G. et al. "Thermodynamic Properties—Saturated Liquid and Vapor of Ammonia—Water Mixtures," Refrigerating Engineering, May 1974, p. 413.

20. Underwriters' Laboratories, "The Comparative Life, Fire and Explosion Hazards of Common Refrigerants," November 1933.

21. U.S. Bureau of Standards Circular No. 142, "Tables of Thermodynamic Properties of Ammonia," April 1923.

22. U.S. Bureau of Standards Scientific Paper No. 501, "Specific Heat of Superheated Ammonia Vapor," March 1925.

23. U.S. Department of Transportation, "Emergency Services Guide for Selected Hazardous Materials." Available from Superintendent of Documents, U.S. Government Printing Office, Washington, DC 20402.

Argon

Ar   DOT Classification: Nonflammable gas; nonflammable gas label

PHYSICAL CONSTANTS

	U.S. Units	Metric Units
International symbol	Ar	Ar
Molecular weight	39.944	39.944
Density of the gas at 70 F (21.1 C) and 1 atm	0.1034 lb/ft^3	1.656 kg/m^3
Specific gravity of the gas at 70 F and 1 atm (air = 1)	1.38	1.38
Specific volume of the gas at 70 F (21.1 C) and 1 atm	9.67 ft^3/1b	0.604 m^3/kg
Boiling point at 1 atm	-302.6 F	-185.9 C
Freezing point at 1 atm	-308.7 F	-189.3 C
Critical temperature	-188.1 F	-122.3 C
Critical pressure	1710.4 psia	4898 kPa, abs.
Critical density	33.44 lb/ft^3	535.6 kg/m^3
Triple point	-308.8 F at 9.99 psia	-189.3 C at 68.9 kPa, abs.
Latent heat of vaporization at normal boiling point	69.7 Btu/lb	162.0 kJ/kg
Latent heat of fusion at triple point	12.7 Btu/lb	29.5 kJ/kg
Specific heat of gas at 70 F (21.1 C) and 1 atm		
C_p	0.125 Btu/(lb) (F)	0.523 kJ/(kg)(C)
C_v	0.075 Btu/(lb) (F)	0.314 kJ/(kg)(C)
Ratio of specific heats	1.67	1.67
Solubility in water, vol/vol at 32 F (0 C)	0.056	0.056
Weight of liquid at -302.6 F (-185.9 C)	11.63 lb/gal	1394 kg/m^3
Density of gas at boiling point	0.356 lb/ft^3	5.70 kg/m^3
Density of liquid at boiling point	86.98 lb/ft^3	1393 kg/m^3

DESCRIPTION

Argon belongs to the family of inert, rare gases of the atmosphere. It is plentiful compared to the other inert atmospheric gases, 1 million ft^3 (28,317 m^3) of dry air containing 9300 ft^3 (263 m^3) of argon. Argon is colorless, odorless, tasteless and nontoxic. It is extremely inert, and forms no known chemical compound. It is slightly soluble in water.

Argon Mixtures. Mixtures of argon and helium, oxygen, hydrogen, nitrogen, or carbon dioxide are used for lamp-bulb filling and other purposes, and are available in any desired combinations. Small users may obtain these in high-pressure cylinders, mixed according to specifications. Large users purchase the individual gases in proportionate amounts and mix them at the point of use.

GRADES AVAILABLE

Table 1 presents the component maxima in ppm (v/v), unless shown otherwise, for types and grades of argon. Gaseous argon is denoted as Type I and liquefied argon as Type II in the table. A blank indicates no maximum limiting characteristic (see Ref. 1).

USES

Argon is extensively used in filling incandescent and fluorescent lamps, and electronic tubes; as an inert gas shield for arc welding and cutting; as a blanket in the production of titanium, zirconium and other reactive metals; to flush molten metals to eliminate porosity in castings; and to provide a protective shield for growing silicon and germanium crystals.

PHYSIOLOGICAL EFFECTS

Argon is nontoxic. Due to its ability to displace air, it is a simple asphyxiant.

SAFE HANDLING AND STORAGE

Materials Suitable for Containers and Storage

Argon gas may be contained in any commonly available metals since it is inert and noncorrosive. Among materials suitable for use with argon liquefied at low temperatures are 18-8 stainless steel and other austenitic nickel-chromium alloys, copper, monel, brass and aluminum. Ordinary carbon steels and most alloy steels lose their ductility at the tempera-

TABLE 1. ARGON GRADES AVAILABLE.

Limiting Characteristics	TYPE I GASEOUS			TYPE II LIQUEFIED	
	GRADES				
	A	B	C	D	E
Argon Min. % (v/v)	99.985	99.996	99.997	99.998	99.999
Water	23.0	14.3	10.5	3.5	1.5
Dewpoint °F	−65	−72	−76	−90	−100
Oxygen	50	7	5	2	1
Nitrogen	50	15	20	10	5
Hydrogen	50	1	1	1	1
Total Hydrocarbons (as methane)		5*	3*	0.5	0.5
Carbon Dioxide				0.5	0.5
Permanent Particulates	Type II may require filtering. **				

**In order to reduce the amount of permanent particles in liquefied argon, the liquid may be filtered just prior to entering the transport container using a 10-40 (10 microns normal, 40 microns absolute) filter constructed of a metal screen assembly installed in the transfer line.

tures of liquid argon, and hence are usually considered unsafe for use with it.

Storage and Handling in Normal Use

Liquid argon must be handled with all the precautions required for safety with a gas at extremely low temperatures. In particular, severe burn-like injuries will result if liquid argon remains in contact with the skin for more than a few seconds. Delicate tissues, such as those of the eyes, can be damaged by a contact with liquid argon too brief to affect the skin of the hands or face. (See also Part I, Chapter 4, Section A.)

Handling Leaks and Emergencies

See Part I, Chapter 4, Section A.

METHODS OF SHIPMENT; REGULATIONS

Under the appropriate regulations and tariffs, argon is authorized for shipment as follows:

By Rail: Gaseous and liquid argon in cylinders (by freight and express), and in tank cars.

By Highway: Gaseous and liquid argon in cylinders on trucks, and liquid argon in tank trucks (by DOT exemption for trucks operating at pressures at or above 25 psig (170 kPa gage); otherwise, as a commodity not regulated by the DOT).

By Water: Gaseous and liquid argon in cylinders in cargo vessels, passenger vessels, passenger ferries and railroad car ferries. Gaseous argon also in tank cars via cargo vessels and railroad car ferries. Gaseous argon in cylinders on barges of U.S. Coast Guard classes A, BA, BB, CA and CB.

By Air: Gaseous argon in cylinders up to 150 lb (70 kg) in passenger aircraft and in cylinders up to 300 lb (140 kg) in cargo aircraft. Low-pressure and pressurized liquid argon in cylinders up to 300 lb (140 kg) in cargo aircraft only. Nonpressurized liquid argon in containers as specified up to 13.2 gal (50 liters) in passenger and cargo aircraft. (Quantities are maximum net weight per container.)

CONTAINERS

While argon is available in gaseous form, it is increasingly being shipped as a liquid for reasons of economy. Argon gas is authorized for shipment in cylinders and tank cars. Liquid argon is shipped in cylinders, and in tank trucks and tank trailers. Most trucks used for shipping liquid argon maintain it at pressures below 25 psig (170 kPa gage); argon at such pressures is not classified or regulated as a "hazardous material" by the DOT, and hence is not subject to any DOT regulations. Trucks for shipping liquid argon at pressures at or above 25 psig (170 kPa gage) are authorized by exemption of the DOT. Some liquid argon has also been shipped in tank cars at pressures below 25 psig (170 kPa gage) or by special DOT permit or exemption.

Filling Limits. The maximum filling limits authorized for argon in shipping containers are as follows:

Gaseous argon in cylinders—up to their marked service pressure at 70 F (21.1 C) and, in the case of cylinders meeting DOT 3A, 3AA, 3AX or 3AAX specifications and special requirements, up to 10 percent in excess of their marked service pressure.

Pressurized liquid argon in DOT-4L200 cylinders—115 percent (percent water capacity by weight).

Argon is uninsulated DOT-107A tank cars— up to $\frac{7}{10}$ of the marked test pressure at 130 F (54.4 C).

Cylinders

Gaseous argon may be shipped in any cylinders authorized by the DOT for nonliquefied compressed gas. (These include cylinders meeting DOT specifications 3A, 3AA, 3AX, 3AAX, 3B, 3C, 3D, 3E, 4, 4A, 4B, 4BA, 4BW and 4C; in addition, continued use of cylinders meeting DOT specifications 3, 25, 26, 33 and 38 is authorized, but new construction is not authorized.)

Pressurized liquid argon may be shipped in cylinders meeting DOT specifications 4L200 (and in cylinders of the 4L-type with a higher marked service pressure).

All cylinders authorized for gaseous argon service must be requalified by hydrostatic retest every 5 or 10 years under present regulations, with the following exceptions: DOT-4 cylinders, every 10 years; and no periodic retest is required for cylinders of types 3C, 3E, 4C and 4L.

Valve Outlet and Inlet Connections. Standard connections, United States and Canada—up to 3000 psi (20,684 kPa), No. 580; 3000–10,000 psi (20,684–68,948 kPa), (see Part I, Chapter 7, Section A, paragraph 2.1.7), cryogenic liquid withdrawal, No. 245. Alternate standard connections, United States and Canada—3000–10,000 psi, No. 677; cryogenic liquid withdrawal, No. 440.

Cargo Tanks

Bulk shipments of liquid argon are made in tank trucks and tank trailers operated not subject to DOT regulations (pressure below 25 psig or 170 kPa gage) or by special DOT permit, as previously noted.

Tank Cars

Argon is authorized for shipment in uninsulated multitube tank cars meeting DOT specification 107A. No DOT regulations apply to liquid argon shipped in insulated single-unit tank cars at pressures below 25 psig (170 kPa gage).

Bulk Storage Systems

For large-scale argon users, bulk storage systems for low-temperature liquid argon are available through suppliers. The systems are equipped with units for converting the liquid to gas at ordinary temperatures unless the liquid argon is drawn through insulated piping for special use at low temperatures.

METHOD OF MANUFACTURE

Argon is manufactured in oxygen-nitrogen plants by means of fractional distillation after the liquefaction of air. In the distillation columns, liquid nitrogen is the product from the bottom of the high-pressure column, followed by liquid oxygen containing argon and some krypton and xenon, gaseous oxygen, gaseous nitrogen and crude neon gas. Crude gas is drawn off in the middle of the column for further processing to obtain high purity.

REFERENCE

1. "Commodity Specification For Argon" (Pamphlet G-11.1), Compressed Gas Association, Inc.

ADDITIONAL REFERENCE

"Saturated Liquid Densities of Oxygen, Nitrogen, Argon, and Parahydrogen," NBS Tech. Note 361, National Bureau of Standards.

Boron Trifluoride

BF$_3$
Synonym: Boron fluoride
DOT Classification: Nonflammable gas; nonflammable gas and
   poison label

PHYSICAL CONSTANTS

	U.S. Units	Metric Units
International symbol	BF$_3$	BF$_3$
Molecular weight	67.82	67.82
Density of the gas		
at 32 F (0 C) and 1 atm	0.192 lb/ft^3	3.08 kg/m^3
Specific gravity of the gas		
at 70 F (21.1 C) and 1 atm		
(air = 1)	2.38	2.38
Specific gravity of liquid		
at boiling point (water = 1)	1.57	1.57
Specific volume of the gas		
at 70 F (21.1 C) and 1 atm	5.6 ft^3/lb	35 m^3/kg
Density of Liquid		
at −196.8 F (−127.1 C)	99.2 lb/ft^3	1589.0 kg/m^3
Boiling point at 1 atm	−148.7 F	−100.4 C
Melting point at 1 atm	−196.8 F	−127.1 C
Critical temperature	10.5 F	−11.9 C
Critical pressure	723 psia	4985 kPa, abs.
Critical density	36.9 lb/ft^3	591.1 kg/m^3
Triple point (approx.)		
at 1 atm	−196.8 F	−127.1 C
Latent heat of vaporization		
at −149.6 F (−100.9 C)	122.6 Btu/lb	285.2 kJ/kg
Latent heat of fusion		
at −199.7 F (−128.7 C)	26.4 Btu/lb	61.4 kJ/kg
Specific heat of gas		
at 78 F (25.6 C) and 1 atm		
C_p	0.178 Btu/(lb) (F)	0.745 kJ/(kg) (C)

	U.S. Units	Metric Units
Solubility in water, weight % at 32 F (0 C) and 1 atm	322 percent	322 percent
Weight of liquid at melting point and 1 atm	13.6 lb/gal	1629.6 kg/m^3

DESCRIPTION

Boron trifluoride is a colorless gas which has a persistent, irritating, acidic odor and which hydrolyzes in moist air to form dense white fumes. It is shipped as a nonliquified compressed gas at varying pressures in the neighborhood of 2000 psig (13,790 kPa gage). Boron trifluoride reacts readily with water with the evolution of heat to form the hydrates $BF_3 \cdot H_2O$ and $BF_3 \cdot 2H_2O$—which are relatively strong acids. Inhalation of the gas irritates the respiratory system, and high concentrations in contact with the skin can cause burns like hydrogen fluoride but less severe. Boron trifluoride readily dissolves in water and in organic compounds containing oxygen or nitrogen. The reaction is so rapid that if boron trifluoride from a cylinder is fed beneath the surface of these liquids, there is danger of suck-back into the cylinder unless a vacuum break or trap is provided in the feed line. Boron trifluoride catalyzes a variety of reactions and forms a great many addition compounds.

GRADES AVAILABLE

Boron trifluoride is available for commercial and industrial use in technical grades having much the same component proportions from one producer to another. The specification for a typical technical grade is as follows:

Component	Specification (Percent by Weight)
Boron trifluoride	99.0 minimum
Sulfur dioxide	0.04 maximum
Sulfate	0.05 maximum
Silicon tetrafluoride	0.1 maximum
Noncondensables (air)	0.6 maximum

USES

Boron trifluoride is used as a catalyst for polymerizations, alkylations and condensation reactions; as a gas flux for internal soldering or brazing; as an extinguisher for magnesium fires; and as a source of B^{10} isotope.

PHYSIOLOGICAL EFFECTS

Boron trifluoride irritates the nose, mucous membranes and other parts of the respiratory system, and concentrations as low as 1 ppm in air can be detected by the sense of smell.

A TLV of 1 ppm (3 mg/m^3) for an 8-hour work day for boron trifluoride has been adopted by the ACGIH.

The irritating sensation and white fumes produced by boron trifluoride in air give easily noticed warning of the escape of even small amounts of the gas; moreover, personnel do not get used to its odor, and tend to seek fresh air if traces of it are inhaled. High concentrations are not only injurious if inhaled but in contact with the skin can cause dehydrating burns similar to those inflicted by acids. Contact of the vapor or liquid with the eyes should also be avoided.

In case of burns or other serious exposures, call a physician immediately, and administer first aid measures that have been previously provided for in consultation with medical authorities.

Protective clothing and equipment required for personnel working with boron trifluoride includes at least rubber gauntlets, goggles and face shields worn as minimum protection. It is advisable as well that long-sleeved shirts buttoned at the wrists be worn.

SAFE HANDLING AND STORAGE

Materials Suitable for Containers and Storage

Dry boron trifluoride does not react with the common metals of construction, but if moisture is present, the hydrate acids identified above (in the "Description" section) can corrode all common metals rapidly. Consequently, lines and pressure-reducing valves in boron trifluoride service must be well protected from the entrance of moist air between periods of use. Cast iron must not be used because active fluorides attack its structure. If steel piping is used for boron trifluoride, forged steel fittings must be used with it instead of cast iron fittings. Among materials suitable for gaskets are teflons and other appropriate fluorocarbon or chlorofluorocarbon plastics. Most plastics become embrittled in boron trifluoride service, but tubing of neoprene or butyl rubber can be used temporarily where pressure is not involved.

Storage and Handling in Normal Use

Boron trifluoride users rarely if ever have storage facilities for the gas, and instead draw supplies from cylinders or tube trailers as needed. Users of smaller quantities usually transfer the gas to process through seamless tubing, with a needle valve for control. Materials may be steel or stainless steel.

The high-pressure system should be designed for a working pressure equal to or greater than the assigned working pressure of the service cylinder.

Large-quantity users usually employ systems of extra-heavy-duty steel pipe with forged steel fittings and large valves of special alloy bronze for control. In the systems, pressure gages must have all-steel internal parts, pressure regulators must have only steel or nickel alloy internal parts, and any brass interior parts used must be silver plated. Cast iron lines or fittings must not be used, since active fluorides cause deterioration of the structure of cast iron.

Any boron trifluoride transfer system must include vacuum breaks or effective check valves to prevent backflow of process materials into cylinders or tubes supplying the gas.

Handling Leaks and Emergencies

Leaks can very readily be detected by the dense white fumes formed by boron trifluoride in moist air, or by its sharp and irritating odor. Personnel who are not using complete breathing, eye and skin protective equipment should leave the vicinity of any leaks at once until the leak or leaks can be ended and the area thoroughly purged of vapors that have leaked. Emergency procedures for dealing with leaks should be fully established and practiced in advance.

Full-protection rubber or plastic garments and breathing apparatus with self-contained air supplies must be available for emergencies that may make it necessary for personnel to enter an area containing a high concentration of boron trifluoride.

METHODS OF SHIPMENT: REGULATIONS

Under the appropriate regulations and tariffs, boron trifluoride is authorized for shipment as follows:

By Rail: In cylinders.

By Highway: In cylinders on trucks, and in tube trailers with tubes meeting cylinder specifications.

By Water: In cylinders aboard cargo vessels only. In cylinders on barges of U.S. Coast Guard classes A, BA, BB, CA and CB only.

By Air: In appropriate cylinders aboard cargo aircraft only up to 300 lb (140 kg) maximum net weight per cylinder.

CONTAINERS

Boron trifluoride is authorized for shipment in cylinders, and is also shipped in motor vehicle tube trailers with tubes built to comply with cylinder specifications.

Filling Limits. The maximum filling limit at 70 F (21.1 C) permitted for boron trifluoride is the service pressure of the container.

Cylinders

Boron trifluoride is frequently shipped and used in cylinders meeting DOT specifications 3A and 3AA and having service pressures ranging from 1800–2400 psig (12,410–16,550 kPa gage). Common commercial cylinder sizes have capacities from 6–62 lb or about 345 ft³ (2.7–28 kg or about 9764 liters) of gas. Under DOT regulations, boron trifluoride is also authorized for shipment in any other cylinders specified as appropriate for nonliquified compressed gas (which include cylinders meeting DOT specifications 3B, 3C, 3D, 3E, 4, 4A, 4BA and 4C; cylinders meeting specifications 3, 7, 25, 26, 33 and 38 may also be continued in boron trifluoride service, but new construction is not authorized).

Cylinders of types 3A and 3AA used in boron trifluoride service must be requalified by hydrostatic retest every 5 years under present regulations. (All other cylinders authorized for boron trifluoride must similarly be requalified by hydrostatic retest every 5 years, with the following exceptions: DOT-4 cylinders, every 10 years; and no periodic retest is required for cylinders of types 3C, 3E, 4C and 7.)

Valve Outlet Connections. Standard Connection, United States and Canada—No. 330.

Tube Trailers

Bulk shipment of boron trifluoride is made in high-pressure tube trailers, for which the tubes have been built to comply with DOT cylinder specifications 3A or 3AA and to have service pressures of around 2000 psig (13,790 kPa gage). Common capacities of these trailers vary from 9000–15,000 lb (4081–6802 kg) of gas.

METHOD OF MANUFACTURE

Basically all methods of preparation of boron trifluoride entail the reaction of a boron compound with a fluorine-containing compound in the presence of an acid. In one major method of commercial preparation, boron trifluoride is produced from boric oxide and hydrofluoric acid (in the reaction $B_2O_3 + 6HF \rightarrow 2BF_3 + 3H_2O$), the product being purified and compressed before packaging in cylinders. Among other methods are ones which employ reactions between boron trichloride or borax and hydrofluoric acid.

REFERENCES

1. Booth, H. S., and Martin, D. R., "Boron Trifluoride and Its Derivatives," New York, John Wiley & Sons, 1949.
2. Booth, H. S., and Martin, D. R., *J. Am. Chem. Soc.,* **64,** 2198–2205 (1942).
3. Eucken, A., and Schröder, E., *Z. Physik, Chem.,* **B41,** 307–319 (1938).
4. Fischer, W., and Weidemann, W., *Z. Anorg. Allgem. Chem.,* **213,** 106–114 (1933).
5. LeBoucher, L., Fischer, W., and Blitz, W., *Z. Anorg. Allgem. Chem.,* **207,** 61–72 (1932).
6. Topchiev, A. V., Zavgorodnii, S. V., and Paushkin, Y. M. (Greaves, J. T., trans.), "Boron Fluoride and Its Compounds as Catalysts in Organic Chemistry," New York, Perachemon Press, 1959.

Butadiene (1,3-Butadiene)

$H_2C:CHCH:CH_2$ (or $CH_2:CHCH:CH_2$)
Synonyms: Vinylethylene, biethylene, erythene, bivynl, divynl B
DOT Classification (butadiene, inhibited): Flammable gas; flammable gas label

PHYSICAL CONSTANTS

	U.S. Units	Metric Units
International symbol	—	—
Molecular weight	54.088	54.088
Vapor pressure		
at 70 F (21.1 C)	21.35 psig	147.2 kPa, gage
at 90 F (32.2 C)	35.8 psig	247 kPa, gage
at 110 F (43.3 C)	54.8 psig	378 kPa, gage
at 130 F (54.4 C)	77 psig	530 kPa, gage
Density of the gas at 70 F		
(21.1 C)	0.37 lb/ft³	5.9 kg/m³
Density of liquid		
at 60 F (15.6 C)	39.05 lb/ft³	625.5 kg/m³
at 70 F (21.1 C)	38.69 lb/ft³	619.8 kg/m³
at 105 F (40.6 C)	37.0 lb/ft³	593 kg/m³
at 115 F (46.1 C)	36.57 lb/ft³	585.8 kg/m³
at 130 F (54.4 C)	35.69 lb/ft³	571.7 kg/m³
Specific gravity of the gas		
at 60 F (15.6 C) and		
1 atm (air = 1)	1.9153	1.9153
Specific gravity of liquid,		
60 F (15.6 C), at satura-		
tion pressure (absolute		
value from weights in		
vacuum for the air-satu-		
rated liquid)	0.6272	0.6272
Specific volume of the gas		
at 70 F (21.1 C) and		
1 atm	6.9 ft³/lb	0.43 m³/kg

	U.S. Units	Metric Units
Specific heat of liquid at 1 atm	0.5079 Btu/(lb) (F)	2.126 kJ/(kg) (C)
Specific heat of ideal gas at 60 F (15.6 C) and 1 atm		
C_p	0.3412 Btu/(lb) (F)	1.429 kJ/(kg) (C)
C_v	0.3045 Btu/(lb) (F)	1.275 kJ/(kg) (C)
Ratio of specific heats	1.121	1.121
Boiling point at 1 atm	24.06 F	-4.411 C
Melting point at 1 atm	-164.05 F	-108.92 C
Critical temperature	306 F	152 C
Critical pressure	628 psia	4330 kPa, abs.
Critical density	15.3 lb/ft^3	245 kg/m^3
Triple point	-164.05 F at 0.010 psia	-108.92 C at 0.069 kPa, abs.
Solubility in water, weight %, at 74 F (23 C) and 1 atm	0.0501 percent	0.0501 percent
Weight of liquid at 70 F (21.1 C)	5.172 lb/gal	619.7 kg/m3
Latent heat of vaporization at boiling point	179.6 Btu/lb	417.7 kJ/kg
Latent heat of fusion at melting point	63.5 Btu/lb	148 kJ/kg
Gross heat of combustion		
Real gas at 60 F (15.6 C) and 1 atm	2954.8 Btu/ft^3	110,090 kJ/m^3
Liquid at 60 F and saturation pressure	20,095 Btu/lb	46,741 kJ/kg
Liquid at 60 F and saturation pressure	104,898 Btu/gal	29.237 × 10^6 kJ/m^3
Net heat of combustion		
Real gas at 60 F and 1 atm	2800 Btu/ft^3	104,300 kJ/m^3
Liquid at 60 F and saturation pressure	19,035 Btu/lb	44,275 kJ/kg
Liquid at 60 F and saturation pressure	99,364 Btu/gal	27,695 × 10^6 kJ/m^3

DESCRIPTION

Butadiene (1,3-butadiene) is a flammable, colorless gas with a mild aromatic odor. It is highly reactive and readily polymerizes, and is authorized for shipment only if inhibited. (Among inhibitors often used are tertiary butyl-catechol, di-n-butylamine, and phenyl-beta-naphthylamine.) Inhibited butadiene is shipped as a liquefied compressed gas under its own low vapor pressure of about 21 psig at 70 F (145 kPa gage at 21.1 C). (Only 1,3-butadiene is treated in this section; the information given here should not be assumed to apply to 1,2-butadiene nor to other butadienes.)

GRADES AVAILABLE

Butadiene (1,3-butadiene) is available for commercial and industrial use in various grades having much the same component proportions from one producer to another.

A polymerization inhibitor contained in all grades is 115 ppm tertiary-butylcatechol. Distillation or washing with dilute caustic solution is employed for removing the inhibitor when removal is desired.

USES

One major use of butadiene has been in the making of synthetic rubber (styrene-butadiene and nitrile-butadiene rubbers, to a large extent; *cis*-polybutadiene is also an extender and substitute for rubber, and *trans*-polybutadiene is a type of rubber with unusual properties). Butadiene is also used extensively for various polymerizations in manufacturing plastics. Copolymers with high proportions of styrene have found applications as stiffening resins for rubber, in water-base and other paints, and in high-impact plastics. Butadiene also serves as a starting material for nylon 66 (adiponitrile) and an ingredient in rocket fuel (butadiene-acrylonitrile polymer).

PHYSIOLOGICAL EFFECTS

If inhaled in high concentrations, butadiene has an anesthetic or mild narcotic action which appears to vary with individuals. Inhalation of a 1 percent concentration in air has been reported to have had no effect on the respiration or blood pressure of individuals, but such exposures may cause the pulse rate to quicken and give a sensation of prickling and dryness in the nose and mouth. Inhalation in higher concentrations has brought on blurring of vision and nausea in some persons. Inhalation in excessive amounts leads to progressive anesthesia, and exposure to a 25 percent concentration for 23 minutes proved fatal in one instance. No cumulative action on the blood, lungs, liver or kidneys has been evidenced.

The maximum allowable exposure in an 8-hour period shall not exceed the 8-hour time-weighted average of 1000 ppm as specified by OSHA (Occupational Safety and Health Act) (*Federal Register*, Vol. 37, No. 202, pp. 22139-22140).

Contact with excessive concentrations of butadiene vapors also irritates the eyes, lungs and nasal passages. Prolonged contact between liquid butadiene and the skin causes freezing of the tissues. Butadiene liquid evaporates rapidly, and delayed skin burns may result if liquid butadiene is allowed to remain trapped in clothing or in shoes.

SAFE HANDLING AND STORAGE

Materials Suitable for Containers and Storage

Butadiene is noncorrosive, and may be used with any common metals. Steel is recommended for tanks and piping in butadiene service by some authorities. Welded rather than threaded connections are similarly recommended because butadiene tends to leak through even extremely small openings. Before being exposed to butadiene that is not inhibited, iron surfaces should be treated with an appropriate reducing agent like sodium nitrite because polymerization is accelerated by oxygen (even if present in ferrous oxide) as well as by heat.

Storage and Handling in Normal Use

All the precautions necessary for the safe handling of a flammable compressed gas must also be observed with butadiene, and special precautions must be taken against its possible polymerization.

Butadiene must be kept inhibited in storage to prevent polymerization and the formation of spontaneously flammable peroxides. The inhibitor content of butadiene stored for any appreciable period should be regularly measured and maintained at safe levels. Ignition within a storage tank can be prevented by diluting the vapor phase with a sufficient proportion of inert gas.

Because of its high volatility, butadiene is

usually stored under pressure, or in insulated tanks at reduced temperatures, preferably below 35 F (1.6 C).

All precautions necessary for storage tanks containing flammable compressed gas must be taken with butadiene storage installations. Tanks should be located outdoors, isolated from boilerhouses and other possible sources of ignition, and provided with adequate diking or drainage to confine or discard the content should the tank rupture. Processes employing butadiene should be designed so that personnel are not exposed to butadiene vapor or liquid. Installations must comply with all local regulations, and should be designed with the help of authorities thoroughly familiar with butadiene.

Handling Leaks and Emergencies

Detailed suggestions concerning safety precautions, protective equipment and first aid are given in Safety Data Sheet SD-55 of the Chemical Manufacturers Association (see Ref. 1).

METHODS OF SHIPMENT; REGULATIONS

Under the appropriate regulations and tariffs, inhibited butadiene is authorized for shipment as follows:

By Rail: In cylinders, and in single-unit tank cars and TMU (ton multi-unit) tank cars.

By Highway: In cylinders on trucks, and in cargo tanks and portable tanks.

By Water: In cylinders and portable tanks (tanks meeting DOT-51 Specifications and not over 20,000 lb or 9072 kg gross weight) aboard cargo vessels only. In authorized tank cars aboard trainships only, and in authorized motor vehicle tank trucks aboard trailerships and trainships only. In cylinders on barges of U.S. Coast Guard classes A, CA and CB only. In cargo tanks aboard tankships and tank barges (to maximum filling densities by specific gravity as stated in Coast Guard regulations).

By Air: In cylinders aboard cargo aircraft only up to 300 lb (136 kg) maximum net weight per cylinder.

CONTAINERS

Inhibited butadiene is authorized for shipment in cylinders, single-unit tank cars, TMU tank cars, and motor vehicle cargo tanks and portable tanks.

Filling Limits. The maximum filling limits authorized for inhibited butadiene are as follows:

In cylinders—Not in excess of the cylinder service pressure at 70 F (21.1 C) nor in excess of $\frac{5}{4}$ of the service pressure at 130 F (54.4 C). In addition, the liquid portion of the content at 130 F (54.4 C) must not completely fill the container.

In other authorized containers—filling limits as with liquefied petroleum gas; these maximum filling densities are prescribed according to the specific gravity of the liquid material at 60 F (15.6 C) in detailed tables that are part of the DOT regulations. Producers and suppliers who charge inhibited butadiene containers other than cylinders should consult these tables in the current regulations. The lower and upper limits of the maximum filling densities authorized in the present regulations for such containers are as follows (percent water capacity by weight):

In single-unit tank cars and TMU tank cars—from 45.500 percent (insulated tanks, April through October) for 0.500 specific gravity, to 61.57 percent (uninsulated tanks, November through March with no storage in transit) for 0.635 specific gravity. (In addition, filling must not exceed various specified limits of pressure and liquid content at temperatures of 105 F (40.6 C), 115 F (46.1 C), 130 F (54.4 C), as given in DOT regulations.)

In cargo tanks and portable tanks—from 38 percent (tanks of 1200 gal or 4.542 m³ capacity or less) for 0.473 to 0.480 specific gravity, to 60 percent (tanks of over 1200 gal or 4.542 m³ capacity) for 0.627 specific gravity and over (except when using fixed length dip tubes or other fixed maximum liquid level indicators). Moreover, the tank must be not liquid full at 105 F (40.6 C) if insulated nor at 115 F (46.1 C) if uninsulated, and the gage vapor pressure at 115 F (46.1 C) must not exceed the tank's design pressure.

CYLINDERS

Inhibited butadiene is authorized by the DOT for shipment in any cylinders specified by liquefied compressed gas; such cylinders include those that meet the following DOT specifications: 3A, 3AA, 3B, 3BN, 3D, 3E, 4, 4A, 4B, 4BA, 4B-ET, 4BW, 9 and 39 (cylinders complying with DOT specifications 3, 25, 26, 38, 40 and 41 may also be continued in butadiene service, but new construction is not authorized).

Cylinders of all types authorized for inhibited butadiene service must be requalified by hydrostatic retest every 5 years under present regulations with the following exceptions:

(1) no periodic retest is required for 3E cylinders;

(2) 10 years is the required retest interval for specification 4 cylinders; and

(3) external visual inspection may be used in lieu of hydrostatic retest for cylinders that are used exclusively for inhibited butadiene which is commercially free from corroding components and that are of the following types (including cylinders of these types with higher service pressure): DOT 3A, 3AA, 3A480X, 3B, 4B, 4BA, 4BW, and ICC-26-240 and ICC-26-300. (Continued use of existing cylinders of the 26-240 and 26-300 types is authorized but no new construction is authorized.)

Valve Outlet Connections. Standard connection, United States and Canada—No. 510.

Cargo Tanks and Portable Tanks

Inhibited butadiene may be shipped by motor vehicle under DOT regulations in cargo tanks meeting DOT specifications MC-330 or MC-331, and in portable tanks complying with DOT-51 specifications. The minimum design pressure required for these tanks is 100 psig (689 kPa gage).

Single-Unit Tank Cars and TMU Tank Cars

Single-unit tank cars and TMU tank cars are authorized by the DOT for the shipment of inhibited butadiene as follows:

At pressures not exceeding 75 psig (517 kPa gage) at 105 F (40.6 C)—in single-unit tank cars which comply with DOT specifications 105A100-W or 111A100-W-4 (provided that they have excess flow valves as required; shipment may also be continued in DOT 105A100 cars, but new construction is not authorized); also, in TMU tank cars which meet DOT specifications 106A500-X.

At pressures not exceeding 255 psig (1758 kPa gage) at 115 F (46.1 C)—in single-unit tank cars which comply with DOT specifications 112A340-W and 114A340-W.

At pressures not exceeding 300 psi inhibited at 115 F—DOT 112A400-W and 114A400-W.

METHOD OF MANUFACTURE

Butadiene is made commercially by dehydrogenating butanes or butenes in the presence of a catalyst, by reacting ethanol and acetaldehyde, and by the cracking of naphtha and light oil. It is also derived as a by-product in ethylene production.

REFERENCE

1. "Butadiene" (Safety Data Sheet SD-55), Chemical Manufacturers Association, 1825 Connecticut Avenue, N.W., Washington, DC 20009.

Carbon Dioxide

CO_2 DOT Classification: Nonflammable gas; nonflammable
gas label

PHYSICAL CONSTANTS

	U.S. Units	Metric Units
International symbol	CO_2	CO_2
Molecular weight	44.01	44.01
Vapor pressure		
at 70 F (21.1 C)	838 psig	5778 kPa, gage
at 32 F (0 C)	491 psig	3385 kPa, gage
at 2 F (−16.7 C)	302 psig	2082 kPa, gage
at −20 F (−28.9 C)	200 psig	1379 kPa, gage
at −69.9 F (−56.6 C)	60.4 psig	416 kPa, gage
at −109.3 F (−78.5 C)	0 psig	0 kPa, gage
Density of the gas		
at 70 F (21.1 C) and 1 atm	0.1144 lb/ft^3	1.833 kg/m^3
at 32 F (0 C) and 1 atm	0.1234 lb/ft^3	1.977 kg/m^3
Specific gravity of the gas		
at 70 F (21.1 C) and		
1 atm (air = 1)	1.522	1.522
at 32 F (0 C) and		
1 atm (air = 1)	1.524	1.524
Specific volume of the gas		
at 70 F (21.1 C) and 1 atm	8.741 ft^3/lb	0.5457 m^3/kg
at 32 F (0 C) and 1 atm	8.104 ft^3/lb	0.5059 m^3/kg
Density of liquid, saturated		
at 70 F (21.1 C)	47.6 lb/ft^3	762 kg/m^3
at 32 F (0 C)	58.0 lb/ft^3	929 kg/m^3
at 2 F (−16.7 C)	63.3 lb/ft^3	1014 kg/m^3
at −20 F (−28.9 C)	66.8 lb/ft^3	1070 kg/m^3
at −69.9 F (−56.6 C)	73.5 lb/ft^3	1177 kg/m^3
Sublimation temperature (1 atm)	−109.3 F	−78.5 C
Critical temperature	87.9 F	31.1 C
Critical pressure	1070.6 psia	7382 kPa, abs.

	U.S. Units	Metric Units
Critical density	29.2 lb/ft³	468 kg/m³
Triple point	−69.9 F at	−56.6 C at
	60.4 psig	416 kPa, gage
Latent heat of vaporization		
at 32 F (0 C)	100.8 Btu/lb	234.5 kJ/kg
at 2 F (−16.7 C)	119.0 Btu/lb	276.8 kJ/kg
at −20 F (−28.9 C)	129.6 Btu/lb	301.4 kJ/kg
Latent heat of fusion		
at −69.9 F (−56.6 C)	85.6 Btu/lb	199 kJ/kg
Specific heat of gas		
at 77 F (25 C) and 1 atm		
C_p	0.203 Btu/(lb) (F)	0.850 kJ/(kg) (C)
C_v	0.157 Btu/(lb) (F)	0.657 kJ/(kg) (C)
Ratio of specific heats		
at 59 F (15 C)	1.304	1.304
Solubility in water, vol/vol		
at 68 F (20 C)	0.90	0.90
Weight of liquid		
at 2 F (16.7 C)	8.46 lb/gal	1014 kg/m³
Cylinder pressure at 68 percent filling density (42.5 lb/ft; 681 kg/m³)		
at 70 F (21.1 C)	838 psig	5778 kPa, gage
at 100 F (37.8 C)	1450 psig	10,000 kPa, gage
at 130 F (54.4 C)	2250 psig	15,500 kPa, gage
Latent heat of sublimation		
at −109.3 F (−78.5 C)	245.5 Btu/lb	571.0 kJ/kg
Viscosity of saturated liquid		
at 2 F (−16.7 C)	0.287 lb/(ft) (hr)	0.119 (g/(cm) (sec) × 10⁻² (centi-poise)

DESCRIPTION

Carbon dioxide is a compound of carbon and oxygen in proportions by weight of about 27.3 percent carbon to 72.7 percent oxygen. A gas at normal atmospheric temperatures and pressures, carbon dioxide is colorless, odorless and about 1.5 times as heavy as air. A slightly acid gas, it is felt by some persons to have a slight pungent odor and biting taste.

Carbon dioxide gas is relatively nonreactive and nontoxic. It will not burn, and it will not support combustion or life. When dissolved in water, carbonic acid (H_2CO_3) is formed. The pH of carbonic acid varies from 3.7 at atmospheric pressure to 3.2 at 23.4 atm.

Carbon dioxide may exist simultaneously as a solid, liquid and gas at a temperature of −69.9 F (−56.6 C) and a pressure of 60.4 psig (416 kPa gage), its triple point. Figure 1 shows the triple point and full equilibrium curve for carbon dioxide.

At temperatures and pressures below the triple point, carbon dioxide may be either a solid ("dry ice") or a gas, depending upon conditions. Solid carbon dioxide at a temperature of −109 F (−78.5 C) and atmospheric pressure transforms directly to a gas (sublimes) without pass-

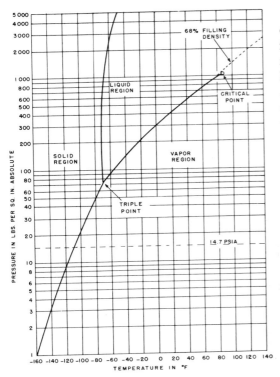

Fig. 1. Equilibrium curve for carbon dioxide.

The component maxima for Type I carbon dioxide which are analyzed in the vapor phase include carbon monoxide, hydrogen sulfide, nitric oxide and water; those analyzed in the liquid phase are nitrogen dioxide and ammonia. All volatile component maxima for Type II carbon dioxide are in the liquid phase. The component maxima for Type III solid carbon dioxide are in ppm for a solid sample. A blank indicates no maximum limiting characteristic (see Ref. 1).

USES

Solid carbon dioxide is used quite extensively to refrigerate dairy products, meat products, frozen foods and other perishable foods while in transit. It is also used as a cooling agent in many industrial processes, such as grinding heat-sensitive materials, rubber tumbling, cold-treating metals, shrink fitting of machinery parts, vacuum cold traps, etc.

Gaseous carbon dioxide is used to carbonate soft drinks; for pH control in water treatment; in chemical processing; as a food preservative; as an inert "blanket" in chemical and food processing and metal welding; as a growth stimulant for plant life; for hardening molds and cores in foundries; and in pneumatic devices.

Liquid carbon dioxide is used as an expendable refrigerant for freezing and chilling food products; for low-temperature testing of aviation, missile and electronic components; stimulation of oil and gas wells; for rubber tumbling; and for controlling chemical reactions. Liquid carbon dioxide is also used as a fire extinguishing agent in portable and built-in fire extinguishing systems.

ing through the liquid phase. Lower temperatures will result if solid carbon dioxide sublimes at pressures less than atmospheric.

At temperatures and pressures above the triple point and below 87.9 F (31.1 C), carbon dioxide liquid and gas may exist in equilibrium in a closed container. Within this temperature range the pressure in a closed container holding carbon dioxide liquid and gas in equilibrium bears a definite relationship to the temperature. Above the critical temperature, which is 87.9 F (31.1 C), carbon dioxide cannot exist as a liquid regardless of the pressure.

GRADES AVAILABLE

Table 1 presents the component maxima in ppm (v/v), unless otherwise stated, for grades available for three types of carbon dioxide: Type I, carbon dioxide in cylinders at ambient temperatures; Type II, liquid carbon dioxide at subambient temperatures in bulk containers; and Type III, solid carbon dioxide (dry ice).

PHYSIOLOGICAL EFFECTS

Carbon dioxide is present in the atmosphere at a concentration of 0.03 percent by volume and is a normal product of human and animal metabolism. Carbon dioxide acts upon vital functions in a number of ways, including respiratory stimulation, regulation of blood circulation and acidity of body fluids, and the concentration of carbon dioxide in the air af-

TABLE 1. CARBON DIOXIDE GRADES AVAILABLE.

COMPONENT	MAXIMA FOR TYPES I AND II				MAXIMA FOR TYPE III
	A	B	C	D	
Carbon Dioxide Min. % (v/v)	99.5	99.5	99.5	99.5	99.5
Water Dewpoint °F.	120 −40	32 −60	120 −40	56.3 −52	
Volatile Hydrocarbons as Methane	20	20	20		
Non-volatile Residues PPM (wt/wt)	10	10	10		500
Oxygen			30		
Carbon Monoxide				10	
Hydrogen Sulfide			1	1	
Nitrogen Oxides			5	5	
Phosphine			0.3	0.3	
Sulfur Dioxide			5	5	
Carbonyl Sulfide			0.5		
Meets USP				YES	
Color					White Opaque
Odor	FREE FROM ODOR				

fects all of these. However, high concentrations may become dangerous upon extended exposure, due to increased breathing and heart rates and a change in body acidity. Extremely high concentrations of CO_2 displace oxygen and suffocation occurs, requiring artificial respiration.

The current OSHA (Occupational Safety and Health Act) standard for maximum allowable concentration of carbon dioxide in air is 0.5 percent for eight continuous hours of exposure. However, the National Institute for Occupational Safety and Health has recommended that the maximum time weighted average exposure to carbon dioxide in air be set at 1.0 percent for a 10-hour shift in a 40-hour week. They also recommend a ceiling concentration of 3.0 percent over a 10-minute period. (Further details may be found in "Criteria for a Recommended Standard, Occupational Exposure to Carbon Dioxide," HEW Publication No. (NIOSH) 76-194, obtainable from Division of Criteria Documentation and Standards Development, National Institute for Occupational Safety and Health, 5600 Fishers Lane, Rockville, MD 20852.

Contact between the skin and dry ice, which is formed when liquid carbon dioxide is discharged directly into the atmosphere, can result in frostbite, and must be avoided.

SAFE HANDLING AND STORAGE

Materials Suitable for Containers and Storage

The common commercially available metals can be used for carbon dioxide installations (those not handling carbon dioxide in aqueous solutions). Any carbon dioxide system at the user's site must be designed to contain safely the pressures involved, and must conform with all state and local regulations. For low-pressure carbon dioxide systems (up to 400 psig or 2758 kPa gage), containers and related equipment should have design pressures rated at least 10 percent above the normal maximum operating pressure. For such systems, schedule 80 threaded steel pipe with forged steel fittings rated at 2000 psi (13,790 kPa) or seamless schedule 40 steel pipe with welded joints are recommended; alternate recommendations include stainless steel, copper or brass pipe, stainless steel or copper tubing. Special materials and construction are required for containers operating at temperatures below −20 F (−28.9 C).

Since carbon dioxide forms carbonic acid when dissolved in water, systems handling carbon dioxide in aqueous solutions must be fabricated from such acid-resistant materials as certain stainless steels, "Hastelloy" metals or "Monel" metal.

Storage and Handling in Normal Use

Carbon dioxide is contained, shipped and stored in either liquefied or solid form. Applications using gaseous carbon dioxide are supplied by gas converted from liquid or solid carbon dioxide.

Being denser than air, carbon dioxide gas may accumulate in low or confined areas under certain conditions of use or storage. Precautions with regard to ventilation are required.

Appropriate warning signs should be affixed outside of those areas where high concentrations of carbon dioxide gas may accumulate. Suggested wording for such a sign is:

CAUTION—CARBON DIOXIDE GAS
Ventilate before entering.
A high CO_2 gas concentration
may occur in this area and
may cause suffocation.

When entering low or confined areas where a high concentration of carbon dioxide gas is present, do not use air-breathing or filter-type gas masks. Gas masks of the self-contained type, or the type which feeds clean outside air to the breathing mask, are required.

Liquefied Carbon Dioxide. Users of liquefied carbon dioxide must comply with all state, municipal and other local regulations.

Storage containers of liquefied carbon dioxide are noninsulated and nonrefrigerated (except for bulk low-pressure containers, discussed below). The contained carbon dioxide is, therefore, at ambient temperature and relatively high pressure.

Cylinders and high-pressure tubes charged with liquid carbon dioxide must never be allowed to reach a temperature exceeding 130 F (54.4 C). Storage should never be near furnaces, radiators or any other source of heat.

Bulk containers for storing liquid carbon dioxide at low pressures are well insulated and equipped with a means, usually mechanical refrigeration, to control and limit internal temperatures and pressures. Storage temperatures are maintained well below ambient, usually in the range of −20 to −10 F (−28.9 to −12.2 C) with corresponding carbon dioxide pressures of 200–345 psig (1379–2379 kPa gage).

Further information on safe handling and storage of liquefied carbon dioxide is given in the Appendix to the Carbon Dioxide Section.

Solid Carbon Dioxide. Solid carbon dioxide (dry ice) has a temperature of −109.3 F (−78.5 C) and must be protected during storage with thermal insulation in order to minimize loss through sublimation.

Dry ice should be stored in well-insulated storage containers, preferably in a cool, non-confined, or ventilated area.

Do not handle dry ice with bare hands. Use heavy gloves or dry ice tongs.

Handle blocks of dry ice carefully as injuries can occur if one is accidentally dropped on the feet.

A suggested wording for a caution label for dry ice follows:

SOLID CARBON DIOXIDE (Dry Ice)
WARNING—EXTREMELY COLD (−109 F) Avoid contact with skin and eyes; use gloves. Do not taste. Keep out of children's reach. Liberates gas which may cause suffoca-

tion. Do not put in stoppered or closed containers.

Solid carbon dioxide may be transformed into a liquid in dry ice converters. These converters should be located in areas where they will never be subjected to temperatures of more than 130 F (54.4 C). Converter locations must be protected so that unauthorized persons cannot tamper with fittings and valves. Adequate protection should be provided to prevent heavy objects from shearing off piping, valves or safety relief devices.

Safety relief devices on converters located in confined areas where the carbon dioxide discharged cannot be dissipated shall be vented outdoors remote from personnel. Such piping must not be capped on the end or equipped with valves or other means of stopping or restricting the flow of the gas.

Further information on safe handling and storage of solid carbon dioxide is given in the Appendix to the Carbon Dioxide Section.

METHODS OF SHIPMENT; REGULATIONS

Under the appropriate regulations and tariffs, liquid or gaseous carbon dioxide is authorized for shipment as follows:

By Rail: In cylinders (by freight, express or baggage), and in insulated tank cars.

By Highway: In cylinders, in insulated tank trucks and in insulated portable tanks.

By Water: In cylinders aboard passenger vessels, cargo vessels and all types of ferry vessels. In authorized tank cars, tank trucks and portable tanks (maximum 20,000 lb or 9000 kg gross weight) aboard cargo vessels and railroad car ferry vessels (passenger or vehicle); and in tank trucks aboard passenger or vehicle ferry vessels. In appropriate cylinders on barges of U.S. Coast Guard classes A, BA, BB, CA and CB.

By Air: In cylinders aboard passenger aircraft up to 150 lb (68 kg) maximum net weight per cylinder; and in cylinders aboard cargo aircraft up to 300 lb (136 kg) maximum net weight per cylinder.

Shipment of solid carbon dioxide by rail or highway is not subject to DOT regulations (under which it is not designated as a dangerous article), while shipment by water and air must meet only certain labeling and packaging requirements.

CONTAINERS

Liquefied carbon dioxide is shipped in cylinders, insulated portable tanks, insulated tank trucks and in insulated tank cars. In high-pressure supply systems, it may be stored and used from single or manifolded cylinders. In low-pressure systems, it is stored in insulated pressure vessels with controlled refrigeration systems and auxiliary heating when required.

Normally, solid blocks of carbon dioxide weighing 50–60 lb (22.7–27.2 kg) are wrapped in heavy kraft paper bags. They are stored and shipped in insulated containers and storage boxes of varying size. Extrusions are normally stored and shipped in bulk or bags in insulated containers.

Filling Limits. The maximum allowable filling densities authorized for carbon dioxide are:

In cylinders—the weight of CO_2 in the cylinder must not exceed 68 percent of the weight of water the cylinder will hold at 60 F (15.6 C).

In tank cars—so that the liquid portion of the gas does not completely fill the tank at 0 F (–17.8 C).

In cargo tanks and portable tanks on trucks—95 percent by volume.

Cylinders

There are two kinds of liquefied carbon dioxide cylinders in commercial use: the standard type and the siphon type. The standard cylinder, in an upright position, discharges gas; inverted, it discharges liquid. The valve on the siphon cylinder is equipped with a dip tube, extending to the bottom of the cylinder so as to discharge liquid when the cylinder is in the upright position. With the exception of fire extinguisher cylinders, all siphon-type cylinders are clearly identified by the word "siphon," "dip tube" or other descriptive phrase. Under no circumstances shall a pressure regulator be attached to a siphon cylinder.

Cylinders containing liquid carbon dioxide under balanced thermal conditions will have a pressure of 733 psig (5054 kPa gage) at an average room temperature of 60 F (15.6 C).

Cylinders that meet the following DOT specifications are authorized for liquefied carbon dioxide service: 3A1800, 3HT2000, 3AA1800, 3E1800, 3T1800, 3AX1800, 3AAX1800 and 39; other cylinders may be used by special permit; DOT-3 cylinders may also be continued in carbon dioxide service, but new construction is not authorized.

Under present regulations, cylinders authorized for liquefied carbon dioxide service must be requalified by hydrostatic retest every 5 years, with three exceptions: 3HT cylinders must be requalified by retest every 3 years, and no periodic retest is required for 3E cylinders, and 39 cylinders are nonreusable and nonrefillable.

Valve Outlet and Inlet Connections. Standard connection, United States and Canada—No. 320. Standard yoke connection, United States and Canada (for medical use of the gas)—No. 940.

Cargo Tanks

Liquefied carbon dioxide is authorized for shipment in cargo tanks on trucks complying with DOT specifications MC-330 and MC-331. The minimum design pressure for these tanks must be 200 psig (1379 kPa gage) (or, if built to requirements in "Low Temperature Operation of the ASME Boiler and Pressure Vessel Code, Section VIII, Unfired Pressure Vessels," the design pressure may be reduced to 100 psig (689 kPa gage) or the controlled pressure, whichever is greater).

Portable Tanks

Liquefied carbon dioxide is also authorized for shipment in portable tanks conforming to specification DOT-51. The minimum design pressure is the same as for cargo tanks.

Tank Cars

Tank cars may be used if they meet DOT specification 105A500 or 105A500-W and are fitted as required with insulation and pressure-regulating valves.

METHOD OF MANUFACTURE

Unrefined carbon dioxide gas is obtained from the combustion of coal, coke, natural gas, oil or other carbonaceous fuels; from by-product gases from steam-hydrocarbon reformers, lime kilns, etc.; from fermentation processes; and from gases found in certain wells and natural springs. The gas obtained from these sources is liquefied and purified by several different processes to a purity of about 99.9 percent or better.

In general, the process involved in producing solid carbon dioxide is as follows: First, cold liquid carbon dioxide is piped into either a special hydraulic press or an extruder. As the liquid boils and evaporates, the vapors are vented or pumped off. The remaining liquid cools until it finally freezes into solid carbon dioxide particles (dry ice snow).

After the vapor pressure has been reduced to atmospheric pressure by pumping or bleeding the vapors away, the solid carbon dioxide particles are pressed into a block or extruded through dies.

REFERENCES

1. "Commodity Specification for Carbon Dioxide" (Pamphlet G-6.2), Compressed Gas Association, Inc.

ADDITIONAL REFERENCES

"Carbon Dioxide" (Pamphlet G-6), Compressed Gas Association, Inc.

"Standard for Low Pressure Carbon Dioxide Systems at Consumer Sites" (Pamphlet G-6.1), Compressed Gas Association, Inc.

Ashrae Handbook of Fundamentals, American Society of Heating, Refrigerating and Air Conditioning Engineers, 1972.

Thermophysical Properties of Carbon Dioxide, M. P. Vukalovich and V. V. Altunin, 1965 (Trans. by D. S. Gaunt 1968), Collet's Ltd., London.

Thermodynamic Functions of Gases, vol. 1, Newitt, Pai, Kuloor, and Huggill, edited by F. Din, Butterworth's, London, 1956.

American Institute of Physics Handbook, 3rd ed., McGraw-Hill, New York, 1972.

Handbook of Chemistry and Physics, Chemical Rubber Publishing Co., 48th ed., 1967.

Chemical Engineer's Handbook, 4th ed.

"Carbon Oxides: CO and CO_2," C. L. Yaws, K. Y. Li, and C. H. Kuo, *Chemical Engineering*, September 30, 1974, p. 115.

APPENDIX TO CARBON DIOXIDE SECTION: FURTHER INFORMATION ON SAFE HANDLING AND STORAGE

Carbon Dioxide Cylinder Manifolding Systems

Piping from cylinders to point of use in carbon dioxide cylinder manifolding systems must be of correct high-pressure design (of at least Schedule 80 high-pressure steel pipe or type "K" copper tubing, with provisions made in the piping for adequate safety relief devices). Piping should be adequately braced.

Transfer of liquid carbon dioxide from one high-pressure carbon dioxide cylinder to another may be accomplished by several means, including direct transfer by pressure differential, or, more usually, by means of a pump. For refilling fire extinguishers, or any other carbon dioxide cylinder, consult the supplier and follow his recommendations.

When a depleted carbon dioxide cylinder is removed from a manifold supply line, close the valve first and leave it closed to prevent air from entering the so-called "empty" cylinder.

Outlets from safety relief valves should be piped to the outside to prevent accumulation of heavy carbon dioxide vapors. Such piping must not be capped on the end or equipped with valves or other means of stopping the flow of gas.

No attempt should be made to use carbon dioxide vapor without a pressure-reducing regulator of suitable design and in good condition.

Liquid Carbon Dioxide in Bulk Containers at Low Pressures

Bulk containers used for storing liquid carbon dioxide at low pressures should be protected from tampering by unauthorized individuals. If a small enclosed location is used, the outlet from the safety relief valves must be piped to the outside or other point where a discharge of carbon dioxide vapor will not result in a high concentration of carbon dioxide. Such piping must not be equipped with valves or other means of stopping the flow of gas.

The storage container should preferably be located in an area that is not subject to unduly high temperatures. If the ambient temperature is above 110 F (43.3 C) for long periods of time, it may be necessary to provide additional refrigeration capacity. If the ambient temperature falls below 0 F (−17.8 C) for a prolonged period, no harm will result, but the carbon dioxide pressure may fall below the desired range.

Dusty, oily locations should be avoided because of the tendency of dust and oil to collect on the refrigerator condenser and thus reduce its efficiency. A dry, well-ventilated location is preferable.

The storage container should not be located in an area where it might be struck by heavy moving or falling objects. A break or tear in the outer shell or covering of the insulation will destroy the vapor seal and allow water vapor to enter the insulation with eventual losses of insulation efficiency.

Whenever liquid carbon dioxide is discharged directly to the atmosphere, as in those cases where sudden cooling is desired, extreme caution should be exercised to guard against and counteract the heavy recoil inherent with the discharge of a dense liquid (weighing more than water) under high pressure. Lines should be anchored firmly against this recoil by means of positive mechanical devices installed prior to use.

All lines from a bulk carbon dioxide container should be anchored in such a fashion that shrinkage of the piping or tubing due to the passage of the cold liquid carbon dioxide through them will not tear them loose. Such piping must be protected with adequate safety relief devices to prevent undue pressure buildup from entrapped liquid carbon dioxide. Valves used in such lines should have a design pressure not less than 350 psig (2413 kPa gage).

The rapid discharge of liquid carbon dioxide through a line which is not grounded will result in a buildup of static electricity potential, which may be dangerous to operating personnel. Such lines should, therefore, be grounded before use.

Flexible hoses used with liquid carbon dioxide should have a minimum working pressure of 500 psig (3447 kPa gage) for low-temperature operation. They should be reinforced with wire braid and attached to couplings of sufficient strength to resist separation from the hose at pressures below the minimum bursting pressure of the hose. Hose lines must be equipped with pressure-relief valves if liquid carbon dioxide can be trapped between two valves.

After use of such flexible hose, it becomes quite rigid and may contain loose dry ice snow. Do not fold, bend or distort the hose, or point it in any direction where pressure buildup within the hose will eject the dry ice snow so as to endanger personnel. The hoses should be examined periodically for wear and damage to both the hose body and couplings. Any hose showing evidence of weakening should be removed from service immediately.

Whenever liquid carbon dioxide is discharged into confined spaces to reduce temperature rapidly, large volumes of carbon dioxide gas will be evolved, amounting to some 8.5 ft^3 of gas per pound (0.53 m^3/kg). These spaces must therefore have some provision incorporated to vent this vapor to the atmosphere to prevent pressure rise within the container.

Solid Carbon Dioxide and Converters

The same precautions for handling high-pressure gas or liquid carbon dioxide apply equally to solid carbon dioxide converters. The only differences between liquid and solid storage are the method of filling and the amount of liquefied carbon dioxide contained. Dry ice converters are charged with dry ice and the amount charged must not exceed the rated capacity of the converters. *"Do not overfill."*

Carbon Monoxide

CO   DOT Classification: Flammable gas; flammable gas label

PHYSICAL CONSTANTS

	U.S. Units	Metric Units
International symbol	CO	CO
Molecular weight	28.01	28.01
Density of the gas at 70 F (21.1 C) and 1 atm	0.0725 lb/ft^3	1.161 kg/m^3
Specific gravity of the gas at 70 F (21.1 C) and 1 atm (air = 1)	0.9676	0.9676
Specific volume of the gas at 70 F (21.1 C) and 1 atm	13.8 ft^3/lb	0.862 m^3/kg
Density of liquid at -312.95 F (-191.5 C)	49.37 lb/ft^3	790.8 kg/m^3
Boiling point at 1 atm	-312.7 F	-191.5 C
Melting point at 1 atm	-340.6 F	-207.0 C
Critical temperature	-220.4 F	-140.2 C
Critical pressure	507.5 psig	3499 kPa, abs.
Critical density	18.79 lb/ft^3	301 kg/m^3
Triple point	-337.1 F at 2.2 psia	-205.1 C at 15.2 kPa, abs.
Latent heat of vaporization at -312.7 F (-191.5 C)	92.79 Btu/lb	215.8 kJ/kg
Latent heat of fusion at -340.6 F (-207.0 C)	12.85 Btu/lb	29.89 kJ/kg
Specific heat of gas at 60 F (15.6 C) and 1 atm		
C_p	0.2478 Btu/(lb)(F)	1.037 kJ/(kg)(C)
C_v	0.1766 Btu/(lb)(F)	0.7394 kJ/(kg)(C)
Ratio of specific heats	1.403	1.403
Solubility in water, vol/vol of water at 32 F (0 C)	0.035	0.035
Weight of liquid at -317.2 F (-194 C)	6.78 lb/gal	812 kg/m^3
Net heat of combustion at 77 F (25 C)	4343.6 Btu/lb	10,103 kJ/kg
Flammable limits in air by volume	12.5-74 percent	12.5-74 percent

DESCRIPTION

Carbon monoxide is a toxic, flammable gas with no color and no odor. If inhaled, concentrations of 0.4 percent prove fatal in less than an hour, while inhalation of high concentrations can cause sudden collapse with little or no warning. Pure carbon monoxide has a negligible corrosive effect on metals at atmospheric pressures. Impure carbon monoxide, containing water vapor, sulfur compounds and/or other reactive impurities, causes stress corrosion to many metals at elevated pressures. Chemically, carbon monoxide is stable with respect to decomposition. At temperatures of 570–2700 F (299–1482 C) it reduces many metal oxides to lower metal oxides, metals or metal carbides. Hydrogenation of carbon monoxide yields products varying according to catalysts and conditions, which include methane, benzene, olefins, paraffin waxes, hydrocarbon high polymers, methanol, higher alcohols, ethylene glycol, glycerol and other oxygenated products. Carbon monoxide also combines with the alkali and alkaline earth metals, reacts with chlorine, bromine, sulfur, Grignard reagents, and sodium alkyls, adds to alcohols and enters into many other reactions.

GRADES AVAILABLE

Carbon monoxide is available for commercial and industrial use in various grades having much the same component proportions from one producer to another.

USES

Carbon monoxide is used in the chemical industry to produce such commodities as methanol and phosgene, and in organic synthesis; and in metallurgy, to recover high-purity nickel from crude ore in the Mond process, and for special steels and reducing oxides. It is also used to obtain powdered metals of high purity, such as zinc white pigments; to form certain metal catalysts applied in synthesizing hydrocarbons or organic oxygenating compounds. It is used as well in the manufacture of acids, esters and

hydroxy acids, such as acetic and propionic acids and their methyl esters and glycolic acid.

PHYSIOLOGICAL EFFECTS

A chemical asphyxiant, carbon monoxide acts toxically by combining with the hemoglobin of the red blood cells to form the stable compound, carbon monoxide-hemoglobin. It thus prevents the hemoglobin from taking up oxygen and cuts off needed oxygen from the body. The affinity of carbon monoxide for hemoglobin is about 300 times the affinity of oxygen for hemoglobin. The inhalation of concentrations as low as 0.04 percent will result in headache and discomfort within 2 to 3 hours, and, inhalation of a 0.4 percent concentration proves fatal in less than 1 hour. Lacking odor and color, carbon monoxide gives no warning of its presence, and inhalation of heavy concentrations can cause sudden, unexpected collapse.

The maximum concentration allowable for a daily 8-hour exposure, according to the American Conference of Governmental Industrial Hygienists, is 50 ppm.

SAFE HANDLING AND STORAGE

Materials Suitable for Containers and Storage

Steels and other common metals are satisfactory for use with dry, sulfur-free carbon monoxide at pressures below 2000 psig (13,790 kPa gage). Iron, nickel and other metals can react with carbon monoxide at elevated pressures to form carbonyl liquids or vapors in small quantities, and the presence of moisture and sulfur-containing impurities in carbon monoxide appreciably increases its corrosive action on steel at any pressure. High-pressure plant equipment is often lined with copper for increased resistance to carbon monoxide attack, and very highly alloyed chrome steels are sufficiently resistant to corrosion by carbon monoxide with small amounts of sulfur-bearing impurities. Users are strongly urged to make stress-corrosion tests of samples of proposed construction materials in order to select ones which will withstand high-pressure use of carbon monoxide under actual conditions.

Systems for storing and handling carbon monoxide should be designed to meet the appropriate ASME standards, and must conform to all applicable state and local regulations.

Storage and Handling in Normal Use

All precautions necessary for the safe handling of any flammable gas must be observed with carbon monoxide. Among these, special care should be taken to avoid storing carbon monoxide cylinders with cylinders containing oxygen or other highly oxidizing or flammable materials. It is recommended that carbon monoxide cylinders in use be grounded, and protected by check valves to prevent suck-back of reaction contents into the cylinders. Areas in which cylinders are being used must be free of all ignition sources and hot surfaces.

Handling Leaks and Emergencies

As in the case of any flammable gas, never use a flame in trying to detect carbon monoxide leaks. Soapy water painted over the suspected area will indicate leaks by the formation of bubbles. Carbon monoxide alarm detectors must be installed in all indoor areas in which the gas is regularly used in more than small laboratory amounts. Alarm detector units based on infrared absorption or measurement of the heat of reaction occurring in the catalytic conversion of CO to CO_2 are available.

First-Aid Suggestions

Personnel accidentally overcome by carbon monoxide should be given first aid prior to a physician's arrival. The first aid measures presented here are based upon what is believed to be common practice in industry, but they should be reviewed and amplified into a complete first-aid program by a competent medical advisor before adoption in any specific case.

It is extremely important to hasten the elimination of carbon monoxide from the bloodstream, should poisoning occur. Such elimination is best effected by inhalation of a mixture of oxygen containing 7 to 10 percent carbon dioxide by volume. The inhalation of pure oxygen will eliminate carbon monoxide much more rapidly than inhalation of fresh air, and oxygen alone could be used instead of the oxygen-carbon dioxide mixture. The mixture, however, acts as a powerful respiratory and cardiac stimulant, inducing deep breathing and rapid ventilation of the lungs. Administering oxygen alone after a prolonged and severe asphyxia does not prove very effective since the victim generally breathes poorly, although spontaneously.

Any person showing symptoms of carbon monoxide poisoning must be moved immediately to fresh, but not cold, air. Keep him warm and place him on his stomach with his face turned to one side. If he is breathing, he should be given oxygen containing 7 to 10 percent carbon dioxide to inhale. If he is not breathing, start manual artificial respiration at once with simultaneous administration of the oxygen-carbon dioxide mixture. Inhalation of the mixture should continue for 15 to 30 minutes after spontaneous respiration returns. The aftertreatment consists of general measures to prevent pneumonia from developing.

Drugs are of little value in the treatment of carbon monoxide poisoning. Coffee may be given if the patient is able to hold a cup. Alcohol should not be given.

Detailed information on the treatment of carbon monoxide asphyxia is presented in the volume, "Noxious Gases and the Principles of Respiration Influencing Their Action."

METHODS OF SHIPMENT; REGULATIONS

Under the appropriate regulations and tariffs, carbon monoxide is authorized for shipment as follows:

By Rail: In cylinders (freight, or express in one outside container to a maximum of 150 lb (70 kg)).

By Highway: In cylinders on trucks.

By Water: In cylinders via cargo vessels only.

In cylinders on barges of U.S. Coast Guard classes A and C only.

By Air: Aboard cargo aircraft only in appropriate cylinders up to 150 lb (70 kg) maximum net weight per cylinder.

CONTAINERS

Carbon monoxide is presently being shipped in cylinders and bulk cryogenic containers.

Filling Limits. The maximum pressure authorized for carbon monoxide cylinders is 1000 psig at 70 F (6894 kPa gage at 21.1 C) except, if the gas is dry and sulfur-free, the cylinders can be charged to $\frac{5}{6}$ the service pressure but never more than 2000 psig at 70 F (13,788 kPa gage at 21.1 C). Tariff BOE-6000 173 : 302-F.

Cylinders

Only cylinders that meet the following DOT specifications and that have service pressures of 1800 psi (12,409 kPa) or higher are authorized for carbon monoxide service: 3A, 3AX, 3AA, 3AAX, 3E, 3T and 3.

Under present regulations, the cylinders authorized for carbon monoxide service must be requalified by hydrostatic test every 5 years with the exception of type 3E (for which periodic hydrostatic retest is not required).

Valve Outlet Connections. Standard connection, United States and Canada—No. 350.

METHOD OF MANUFACTURE

Pure carbon monoxide is made commercially from synthesis gas, blast-furnace gas or coke-oven gas by two methods: (1) absorption of carbon monoxide by an ammoniacal cuprous salt solution at elevated pressure, followed by pressure release; (2) low-temperature condensation and fractionation.

Chlorine

Cl₂   DOT Classification. Nonflammable gas; nonflammable gas and poison label

PHYSICAL CONSTANTS

	U.S. Units	Metric Units
International symbol	Cl_2	Cl_2
Molecular weight	70.906	70.906
Vapor Pressure		
at 32 F (0 C)	38.459 psig	265.17 kPa, gage
at 70 F (21.1 C)	85.46 psig	589.2 kPa, gage
at 105 F (40.6 C)	151.12 psig	1041.9 kPa, gage
at 115 F (46.1 C)	174.69 psig	1204.5 kPa, gage
at 130 F (54.4 C)	213.88 psig	1474.7 kPa, gage
Density of the gas at 32 F		
(21.1 C) and 1 atm	0.2003 lb/ft³	3.2085 kg/m³
Specific gravity of the gas at		
32 F and 1 atm (air = 1)	2.482	2.482
Specific volume of the gas at		
70 F (21.1 C) and 1 atm	4.992 ft³/lb	0.3116 m³/kg
Density, saturated liquid		
at 32 F (0 C)	91.67 lb/ft³	1468 kg/m³
at 70 F (21.1 C)	87.72 lb/ft³	1405 kg/m³
at 105 F (40.6 C)	83.82 lb/ft³	1343 kg/m³
at 115 F (46.1 C)	82.65 lb/ft³	1324 kg/m³
at 130 F (54.4 C)	80.89 lb/ft³	1296 kg/m³
Boiling point at 1 atm	-29.9 F	-34.05 C
Melting point at 1 atm	-149.76 F	-100.98 C
Critical temperature	291.2 F	144 C
Critical pressure	1118.4 psia	7711.1 kPa, abs.
Critical density	35.765 lb/ft³	573.07 kg/m³
Triple point	-149.76 F at 0.1991 psia	-100.98 at 1.373 kPa, abs.
Latent heat of vaporization		
at boiling point	123.7 Btu/lb	287.8 kJ/kg

	U.S. Units	Metric Units
Latent heat of fusion at melting point	38.86 Btu/lb	90.39 kJ/kg
Specific heat of dry gas at 30–80 F (–1.1–27 C) at or below 100 psia		
C_p	0.113 Btu/(lb) (F)	0.473 kJ/(kg) (C)
C_v	0.0832 Btu/(lb) (F)	0.348 kJ/(kg) (C)
Ratio of specific heats	1.355	1.355
Viscosity at 68 F (20 C)		
liquid	0.3518 centipoise	0.3518 mPa · s
gas	0.0132 centipoise	0.0132 mPa · s
Weight of liquid at 70 F (21.1 C)	11.73 lb/gal	1405 kg/m³
Specific gravity of liquid at 30 F and 53.155 psia (0 C and 366.49 kPa, abs. or 3.617 atm)	1.468	1.468

DESCRIPTION

Chlorine is a greenish-yellow, nonflammable gas with a distinctive, pungent odor. It is almost 2.5 times as heavy as air. The gas acts as a severe irritant if inhaled. Chlorine liquid has the color of clear amber and is about half again as heavy as water. It is shipped as a compressed liquefied gas with a vapor pressure of 85.5 psig at 70 F (590 kPa gage at 21.1 C). Chlorine is nonflammable and nonexplosive in both gaseous and liquid states. However, like oxygen, it is capable of supporting the combustion of certain substances. Many organic chemicals react readily with clorine, in some cases with explosive violence. Chlorine usually forms univalent compounds, but it can combine with a valence of 3, 4, 5 or 7.

Chlorine is only slightly soluble in water. When it reacts with pure water, weak solutions of hydrochloric and hypochlorous acids are formed. Chlorine hydrate ($Cl_2 \cdot 8H_2O$), may crystallize below 49.3 F (9.6 C).

Chlorine unites, under specific conditions, with most of the elements; these reactions may be extremely rapid. At its boiling point it reacts with sulfur. It does not react directly with oxy-gen or nitrogen. The oxides and nitrogen compounds are well known, but can be prepared only by indirect methods. Mixtures of chlorine and hydrogen composed of more than 5 percent of either component can react with explosive violence, forming hydrogen chloride.

The preparation of soda and lime bleaches (sodium and calcium hypochlorite) are typical reactions of chlorine with the alkalies and alkaline earth metal hydroxides. The hypochlorites formed are powerful oxidizing agents. Because of its great affinity for hydrogen, chlorine removes hydrogen from some of its compounds, such as the reaction with hydrogen sulfide to form hydrochloric acid and sulfur.

Chlorine reacts with organic compounds much the same as with inorganics to form chlorinated derivatives and hydrogen chloride. Some of these reactions can be explosive, including those with hydrocarbons, alcohols and ethers, and proper methods must be applied in operations in which they are involved.

At ordinary temperatures dry chlorine reacts with aluminum, arsenic, gold, mercury, selenium, tellurium, tin and titanium. At certain temperatures sodium and potassium burn in chlorine gas. Carbon steel ignites in it at 483 F (251 C).

GRADES AVAILABLE

Chlorine for commercial and industrial use has much the same quality from all producers. High-purity grades (99.9 percent) are available from specialty gas suppliers.

USES

The largest quantities of chlorine are used in manufacturing chemicals—among them: such solvents as carbon tetrachloride, trichloroethylene, 1,1,1-trichloroethane, perchloroethylene and methylene chloride; pesticides and herbicides; such plastics and fibers as vinyl chloride and vinylidene chloride; and refrigerants and propellants like the halocarbons and methyl chloride. It is also an ingredient in the widely used bleach, deodorizer and disinfectant, sodium hypochlorite. Chlorine is also widely employed in bleaching pulp, paper and textiles; for drinking and swimming water purification; in the sanitation of industrial and sewage wastes; and in the degassing of aluminum melts.

PHYSIOLOGICAL EFFECTS

Chlorine gas is primarily a respiratory irritant. It is so intensely irritating that low concentrations in the air are readily detectable by the normal person. In higher concentrations the severely irritating effect of the gas makes it unlikely that any person will remain in a chlorine-contaminated atmosphere unless he is unconscious or trapped.

Liquid chlorine will cause skin and eye burns upon contact.

A TLV of 1 ppm (3 mg/m^3) for an 8-hour work day for chlorine has been adopted by the ACGIH. According to American National Standards Institute Z37.25-1974, the acceptable 8-hour time weighted average concentration is 1 ppm (3 mg/m^3), the acceptable ceiling concentration is 2 ppm (6 mg/m^3) and the accept-able maximum for peaks above the acceptable ceiling concentration is 3 ppm (9 mg/m^3) for periods of 5 minutes.

Acute Toxicity. When a sufficient concentration of chlorine gas is present, it will irritate the mucous membranes, the respiratory system and the skin. Large amounts cause irritation of eyes, coughing and labored breathing. If the duration of exposure or the concentration of chlorine is excessive, general excitement of the person affected, accompanied by restlessness, throat irritation, sneezing and copious salivation results. The symptoms of exposure to high concentrations are retching and vomiting, followed by difficult breathing. In extreme cases, the difficulty of breathing may increase to the point where death can occur from suffocation. Liquid chlorine in contact with the eyes or skin will cause local irritation and/or burns.

Systemic and Chronic Effects. Chlorine produces no known systemic effect. All symptoms and signs result directly or indirectly from the local irritant action. Low concentrations of chlorine gas in the air may have a minor irritating effect or may produce slight symptoms after several hours' exposure, but careful examination of persons repeatedly exposed to such conditions reportedly have shown no chronic effect.

SAFE HANDLING AND STORAGE

Materials Suitable for Containers and Storage

Carbon steel is a very satisfactory material for use in dry chlorine (gas and liquid) service at temperatures below 300 F (150 C). Iron, copper, nickel and lead are also resistant to dry chlorine at moderate temperature. Nickel and certain nickel alloys resist corrosion by dry chlorine at temperatures of 600–1000 F (300–500 C). Even small amounts of moisture mixed with chlorine form hypochlorous and hydrochloric acids that are very corrosive to most metals. Tantalum and titanium resist attack by moist chlorine and chlorinated water of any concentration. Tantalum is not affected by wet or dry chlorine at temperatures up to 250 F (120 C) but titanium must not be used in dry chlorine. Silver and platinum resist the action of wet chlorine gas fairly well, and have been used in special pieces of chlorine equipment. Certain nonmetals such as glass, ceramics and various plastics withstand corrosion by wet or dry chlorine. Plastics must be used with great care; careful attention must be given to operating pressures and temperatures.

Storage and Handling in Normal Use

All precautions necessary for the safe handling of any nonflammable toxic gas must be observed with chlorine (see also Part I, Chapter 4, Section A).

Employee Protection and Training. Persons afflicted with asthma, bronchitis and other chronic lung conditions or irritations of the upper respiratory tract should not be employed in areas where chlorine is handled. All employees working with or around chlorine should be given preemployment and periodic physical examinations, including X-rays.

All employees handling or working around chlorine should be trained to handle it properly and safely with special emphasis placed on actions to be taken in case of emergencies such as leaks. Each employee should be trained in the proper use of the several types of respiratory equipment and be familiar with the conditions under which each type must be used. Each employee should also be trained in first-aid procedures, particularly in administering artificial respiration. Quiz sessions on actions to be taken in emergencies, the proper use of respiratory equipment, and first-aid measures should be held at regular intervals.

Personal Protective Equipment. Suitable gas masks should be available for handling emergencies. Gas masks should be located outside the probable location of any leak. They should be routinely inspected and maintained in good condition. They should be cleaned after each use and at regular intervals. Equipment used by more than one person should be sanitized after each use. Respiratory equipment must be approved by the National Institute of Occupational Safety and Health.

Protective clothing is not required for performing routine plant operations. Resistant plastic or rubber gloves should be worn by personnel who may come in contact with ferric chloride during maintainance operations. Rubber boots, as well as protective equipment described above, should be worn by persons entering tanks.

While not specific for chlorine, safety glasses or goggles, hard hats and safety shoes should be worn or available as dictated by the special hazards of the area or by plant practices.

Safety harnesses and attached life lines are required for persons entering chlorine tanks. All such equipment should be used and maintained in accord with the manufacturer's instructions.

Oxygen administration equipment should be available and personnel trained in its use.

Shower baths and bubble-type fountains or the equivalent should be installed where contact of the skin, eyes or clothing with liquid chlorine or chlorinated water is a possibility.

Handling Leaks and Emergencies

Immediate steps should be taken to find and stop chlorine leaks as soon as there is any indication of chlorine in the air. Chlorine leaks never get better; they always get worse, unless promptly corrected. Authorized, trained personnel equipped with suitable gas masks should investigate whenever a chlorine leak occurs. All other persons must be kept away from the affected area until the cause of the leak has been found and remedied. If the leak is extensive, all persons in the path of the fumes must be warned. If outdoors, keep all persons upwind from the leak. Also, if possible, keep all persons in locations higher than the leak. At sites involving the handling of chlorine outdoors, it is advisable to have a wind sock or weather vane installed in a prominent location. Unless there is a fire-caused updraft, gaseous chlorine tends to lie close to the ground or floor because it is approximately $2\frac{1}{2}$ times as heavy as air.

Finding Leaks. To find a leak tie a cloth to the end of a stick, soak the cloth with ammonia water and hold close to the suspected area. Alternatively, vapor from a squeeze bottle containing strong ammonia can be used. (Avoid contact of ammonia water with brass.) A white cloud of ammonium chloride will result if there is any chlorine leakage. A supply of strong ammonia water (commercial 26 Be') always should be available (household ammonia is not strong enough). Containers, piping and equipment should be checked for leaks daily.

Emergency Assistance. If a chlorine leak cannot be handled promptly by the user's personnel, the nearest office or plant of the supplier should be called for assistance. If the supplier cannot be immediately reached, CHEMTREC (TEAP in Canada) should be called (as noted in the opening pages of this book). The Chlorine Institute's CHLOREP system is in operation throughout the United States and Canada to assist anyone who needs help with an emergency involving chlorine. This assistance can be summoned day and night by calling CHEMTREC (toll free) in the United States and TEAP in Canada. The telephone numbers of the supplier and CHEMTREC or TEAP should be posted in suitable places so that they will be quickly available, these should be checked periodically to be sure that the numbers are correct.

When phoning for assistance the following should be given:

(1) Your company name, address, telephone number, and the person or persons to contact for further information.

(2) Type (and serial number if possible) of container or other equipment which is leaking.

(3) Nature, location and extent of the leak.

(4) Corrective measures that are being applied.

(5) Chlorine supplier's name, address and telephone number.

In Case of Fire. In case of fire, chlorine containers should be removed from the fire zone immediately. Tank cars or barges should be disconnected and pulled out of the danger area. If no chlorine is escaping, water should be applied to cool any containers that cannot be moved. All unauthorized persons should be kept at a safe distance.

Do Not Use Water with Leaks. Never use water on a chlorine leak. Chlorine is only slightly soluble in water, also, the corrosive action of chlorine and water always will make a leak worse. In addition, the heat supplied by even the coldest water applied to a leaking container will cause liquid chlorine to evaporate faster. Never immerse or throw a leaking container into a body of water because the leak will be aggravated.

Equipment and Piping Leaks. If a leak occurs in equipment in which chlorine is being used, the supply of chlorine should be shut off and the chlorine which is under pressure at the leak should be disposed of.

Valve Leaks. Leaks around valve stems usually can be stopped by tightening the packing nut or gland by turning clockwise. If this does not stop the leak, the container valve should be closed, and the chlorine which is under pressure in the outlet piping should be disposed of. If a container valve does not shut off tight, the outlet cap or plug should be applied. Ton containers have two valves; in case of a valve leak, the container should be rolled so the valves are in a ver-vertical plane with the leaky valve on top.

Container Body Leaks. If confronted with container leaks other than at the valves, one or more of the following steps should be taken:

(1) If possible turn the container so that gas instead of liquid escapes. The quantity of chlorine that escapes from a gas leak is about $\frac{1}{15}$ the amount that escapes from a liquid leak through the same size hole.

(2) Apply appropriate emergency kit device, if available.

(3) Call for emergency assistance.

(4) If practical, reduce pressure in the container by removing the chlorine as a gas (not as a liquid) to process or a disposal system.

(5) Move the container to an isolated spot where it will do the least harm.

Leaks in Transit. If a chlorine leak develops in transit through a populated area, it is generally advisable to keep the vehicle or tank car moving until open country is reached in order to minimize the hazards of escaping gas. Appropriate emergency measures should then be taken as quickly as possible.

If a motor vehicle transporting chlorine containers is wrecked, any leaking containers should be positioned, if possible, so that only gas escapes, and if necessary, moved to an isolated area before attempts are made to stop the leaks. If a tank car or tank truck is wrecked and chlorine is leaking, a danger area should be evacuated and emergency clearing operations should not be started until safe working conditions have been restored.

Preparations for Handling Emergencies. List at least several physicians who could be summoned in the event of an emergency. Request all physicians listed to familiarize themselves with the treatment of persons exposed to chlorine. Provide phone numbers of these physicians along with supplier and CHEMTREC numbers to the plant telephone operator. Post a similar list in the area where chlorine is handled. Also, list and post the phone numbers of the nearest hospital, fire and police departments. Arrange for a telephone extension in areas where chlorine is handled and keep these extensions open at night, on weekends and on holidays.

Alkali Absorption. At regular points in areas of chlorine storage and use, provisions should be made for emergency disposal of chlorine from leaking containers. Chlorine may be absorbed in solutions of caustic soda or soda ash, or in agitated hydrated lime slurries. Caustic soda is recommended as it absorbes chlorine most readily. The proportions of alkali and water recommended for this purpose, in the amounts needed to absorb indicated quantities of chlorine, are given in Table 1. A suitable

tank to hold the solution should be provided in a convenient location. Never immerse any chlorine container in the solution or a body of water. Chlorine should be passed into the solution through plastic pipe or rubber hose properly weighted to hold it under the surface.

Emergency Kits. Most chlorine suppliers have emergency kits and skilled technicians to use them. These kits are designed to control most leaks in chlorine shipping containers. Kit A is for chlorine cylinders; Kit B for ton containers and Kit C for tank cars and cargo tanks. Many consumers have found it advisable to purchase kits and train employees in their use. Additional information is available from the Chlorine Institute (see Ref. 1).

First Aid

Prompt treatment of anyone overcome or seriously exposed to chlorine is of the utmost importance. The patient should be removed from the contaminated area. Obtain medical assistance as soon as possible.

Contact with Skin or Mucous Membranes. (If the patient has also inhaled chlorine, first aid for inhalation should be given first.)

If chlorine has contaminated the skin or clothing, the emergency shower or any other means of washing with copious amounts of water should be used immediately. Contaminated clothing should be removed under the shower and the chlorine should be washed off with very large quantities of water. Skin areas should be washed with large quantities of soap and

TABLE 1. RECOMMENDED ALKALINE SOLUTIONS
FOR ABSORBING CHLORINE.

Chlorine Container Capacity (lb net)	Caustic Soda (100%)		Soda Ash		Hydrated Lime*	
	Alkali (lb)	Water (gal)	Alkali (lb)	Water (gal)	Alkali (lb)	Water (gal)
100	125	40	300	100	125	125
150	188	60	450	150	188	188
2000	2500	800	6000	2000	2500	2500

*Hydrated lime solution must be continuously and vigorously agitated while chlorine is to be absorbed.

water. Never attempt to neutralize the chlorine with chemicals. Salves or ointments should not be applied unless directed by a physician.

Contact With the Eyes. If even minute quantities of liquid chlorine enter the eyes, or if the eyes have been exposed to strong concentrations of chlorine gas, they should be flushed immediately with copious quantities of running water for at least 15 minutes. Never attempt to neutralize with chemicals. The eyelids should be held apart during this period to ensure contact of water with all accessible tissues of the eyes and lids.

Call a physician, preferably an eye specialist, at once. If a physican is not immediately available, the eye irrigations should be continued for a second period of 15 minutes. No oils or oily ointments, or any medications, should be installed unless ordered by the physican.

Inhalation. If breathing has not ceased, the patient should be placed in a comfortable position. He should be kept warm and remain at rest until medical help arrives. Call a physician immediately. *Caution:* Never give anything by mouth to an unconscious patient.

If breathing has apparently ceased, artificial respiration should be started immediately. If available, oxygen should be administered.

METHODS OF SHIPMENT; REGULATIONS

Under the appropriate regulations and tariffs, chlorine is authorized for shipment as follows:

By Rail: In cylinders (by freight, express or baggage), single-unit tank cars and TMU tank cars.

By Highway: In cylinders and ton containers (by truck or on special trailers in lots of 15) and cargo tank trucks.

By Water: In cylinders, single-unit tank cars and ton containers aboard cargo vessels only. In bulk in steel tank barges. (In an emergency involving life or health, and on application made to the commandment of the Coast Guard, limited shipments of chlorine may be made under conditions authorized by the commandment aboard passenger vessels and ferry vessels.)

In cylinders on barges of U.S. Coast Guard classes A, BA, CA and CB.

By Air: In cylinders aboard cargo aircraft only up to 150 lb (70 kg) maximum net weight per cylinder.

CONTAINERS

Chlorine is stored and shipped as liquefied gas under pressure in cylinders, tank trucks, TMU tanks, tank cars and tank barges.

Filling Limits. The maximum filling density authorized for chlorine containers is 125 percent water capacity by weight.

Cylinders

Chlorine is authorized for shipment in cylinders with not over 150 lb (68 kg) capacity which comply with DOT specifications 3A480, 3AA480, 25, 3, 3BN480, 3E1800 and B. E. 25. Class 3A and 3AA cylinders having higher service pressures may also be used.

Storing Cylinders. In addition to the precautions required for the safe handling and storage of any compressed gas cylinders, the following practices must be observed for chlorine cylinders. Never store chlorine cylinders next to cylinders containing other compressed gases. Similarly, never store chlorine containers near turpentine, ether, anhydrous ammonia, finely divided metals, hydrocarbons such as oil, grease and gasoline, or any flammable materials. The storage area must be well ventilated, and storage below ground should be avoided.

Valve Outlet Connections. Standard connection, United States and Canada—No. 820. Alternate standard connections, United States and Canada—No. 660 and No. 840.

Ton Containers

Chlorine is authorized for shipment in ton containers with 200 lb (970 kg) capacity which comply with DOT specifications 106A500X, 106A500, DOT 27 and BE 27. Valve outlet connections are the same as for cylinders.

Ton containers are authorized for motor vehicle transport as well as rail transport as a part of a TMU tank car. Standard TMU cars are designed to carry 15 containers.

Cargo Tanks

Chlorine is authorized for shipment in motor vehicle cargo tanks complying with DOT specifications MC331 or MC330. These are generally 15–20 tons (14–18 metric tons) in capacity but range as high as 32 tons (29 metric tons).

Tank Cars

Chlorine is authorized for shipment in tank cars meeting DOT specifications 105A300W, 105A500W, 104A300, 105A500 or 105. Class 105A cars having higher marked test pressures also may be used. Chlorine tank cars must comply with the special requirements in DOT 173.314 (C).

Bulk Storage Facilities

Consumers of chlorine in bulk quantities who receive shipment in single-unit tank cars often withdraw the chlorine direct to process, without transferring it to bulk storage facilities of their own. This practice avoids both the expense of storage facilities and such requirements of safe storage as having specialized, experienced personnel available at all times for possible emergencies.

Bulk chlorine consumers supplied by tank barges, however, usually unload into their own storage facilities. Very few chlorine barges are built with capacities of less than 600 tons (544 metric tons) and capacities this large make it impractical to hold the barge while unloading direct to process. Piping and unloading systems for use with tank barges must be approved by the U.S. Coast Guard. Whether supplied by tank car or tank barge, storage facilities must be designed by experienced engineers and must conform to all applicable state and local regulations.

METHOD OF MANUFACTURE

Chlorine is produced largely by the electrolysis of salt. The two leading devices for electrolysis are diaphragm cells and mercury (amalgam) cells. Together, these account for about 95 percent of world chlorine production.

REFERENCE

1. "Chlorine Manual" (4th ed., 1969), The Chlorine Institute, 342 Madison Ave, New York, NY 10017.

ADDITIONAL REFERENCES

"Thermodynamic Properties of Chlorine," R. M. Kapoor, and J. J. Martin, University of Michigan Press (available from the Chlorine Institute).

"First Aid and Medical Management of Chlorine Exposures" (2nd ed. 1975), The Chlorine Institute, 342 Madison Avenue, New York, NY 10017.

Cyclopropane

C_3H_6 [or $(CH_2)_3$]
Synonym: Trimethylene
DOT Classification: Flammable gas; flammable gas label

PHYSICAL CONSTANTS

	U.S. Units	Metric Units
International symbol	C_3H_6	C_3H_6
Molecular weight	42.078	42.078
Vapor pressure		
at 70 F (21.1 C)	80.0 psig	552 kPa, gage
at 105 F (40.6 C)	141.5 psig	975.6 kPa, gage
at 115 F (46.1 C)	164.0 psig	1131 kPa, gage
at 130 F (54.4 C)	200.3 psig	1381 kPa, gage
at 68 F (20 C)	79.2 psig	546 kPa, gage
Density of the gas at 70 F (21.1C)		
and 1 atm	0.109 lb/ft^3	1.75 kg/m^3
Specific gravity of the gas at 70 F		
(21.1 C) and 1 atm (air = 1)	1.48	1.48
Specific volume of the gas at 70 F		
(21.1 C) and 1 atm	9.2 ft^3/lb	0.5743 m^3/kg
Density (or specific gravity) of liquid		
at 70 F (21.1 C)	38.59 lb/ft^3	618.2 kg/m^3
at 105 F (40.6 C)	36.73 lb/ft^3	588.4 kg/m^3
at 115 F (46.1 C)	36.16 lb/ft^3	579.2 kg/m^3
at 130 F (54.4 C)	35.27 lb/ft^3	565.0 kg/m^3
Boiling point at 1 atm	27.15 F	-2.694 C
Melting point at 1 atm	-197.7 F	-127.6 C
Critical temperature	255.9 F	124.4 C
Critical pressure	797 psia	5495 kPa, abs.
Critical density	16.32 lb/ft^3	261.4 kg/m^3
Latent heat of vaporization at		
boiling point	205 Btu/lb	477 kJ/kg
Latent heat of fusion at melting		
point	55.9 Btu/lb	130 kJ/kg

	U.S. Units	Metric Units
Specific heat of gas at -27.15 F (-32.86 C) and 1 atm C_p	0.4599 Btu/(lb)(F)	1.926 kJ/(kg)(C)
Solubility in water, vol/vol at 68 F (20 C)	0.296	0.296
Weight of liquid at -110 F (-78.9 C)	6.0 lb/gal	720 kg/m^3
Flammable limits, by volume		
In air	2.40–10.3 percent	2.40–10.3 percent
In oxygen	2.48–60.0 percent	2.48–60.0 percent
Autoignition temperature		
In air	928 F	498 C
In oxygen	849 F	454 C

DESCRIPTION

Cyclopropane is a colorless, flammable gas with a sweet, distinctive odor resembling that of petroleum naphtha. It is shipped and used as a liquefied compressed gas in cylinders, with a vapor pressure of about 68 psig at 68 F (469 kPa gage at 20 C). Cyclopropane is an anesthetic and is used as a medical gas. Chemically, it reacts with hydrogen iodide or bromide and with bromine to give the corresponding halopropanes or dihalopropanes, and with chlorine to form chlorocyclopropanes.

GRADES AVAILABLE

For its major use as an anesthetic medical gas, cyclopropane is (and must be) supplied according to purity standards of the U.S.P. (*Pharmacopeia of the United States or National Formulary*).

USES

Cyclopropane is used chiefly as an inhalant anesthetic in medicine. It is also employed for organic synthesis in the chemical industry.

PHYSIOLOGICAL EFFECTS

A general anesthetic, cyclopropane can produce all levels of inhalation anesthesia and max-

imum muscle relaxation in patients of all ages and body types. It is not altered or combined in the body; the major part is exhaled within 10 minutes, while full desaturation takes several hours. It tends to irritate the circulatory system; laryngeal spasm, emergence delirium and nausea after anesthesia with it are common. Concentrations (by volume) or 6 to 8 percent result in unconsciousness; of 7 to 14 percent, in moderate anesthesia; and of 14 to 23 percent, in deep anesthesia. Concentrations ranging from 23 to 40 percent are lethal through respiratory failure.

SAFE HANDLING AND STORAGE

Materials Suitable for Containers and Storage

A noncorrosive gas, cyclopropane may be contained by any commercially available metals.

Storage and Handling in Normal Use

The principal hazard met in handling cyclopropane stems from its high flammability, and all the precautions necessary for the safe handling of any flammable gas must be observed in its use (see Part I, Chapter 4, Section A).

All the precautions necessary for the safe handling and storage of a medical gas must also be observed with cyclopropane when it is used medicinally (see Part I, Chapter 4, Section C).

Handling Leaks and Emergencies

See Part I, Chapter 4, Sections A and C.

METHODS OF SHIPMENT; REGULATIONS

Under the appropriate regulations and tariffs, cyclopropane is authorized for shipment as follows:

By Rail: In cylinders (freight, express or baggage).

By Highway: In cylinders on trucks.

By Water: In cylinders on cargo or passenger vessels, and on ferry or railroad ferry vessels (either passenger or vehicle). In cylinders on barges of U.S. Coast Guard classes A and C only.

By Air: Aboard cargo aircraft only in appropriate cylinders up to 300 lb (140 kg) maximum net weight per cylinder.

CONTAINERS

Cyclopropane is shipped in cylinders as a liquefied compressed gas at gage pressures which range from about 53 psi at 50 F (365 kPa at 10 C) to 199 psi at 30 F (1372 kPa at -1.1 C).

Filling Limits. The maximum filling density authorized by the DOT for cyclopropane cylinders is 55 percent (percent of water capacity by weight).

Cylinders that meet the following DOT specifications are authorized for cyclopropane service: 3A225, 3A480X, 3AA225, 3B225, 4A225, 4AA480, 4B225, 4BA225, 4BW225, 4B240ET, 3E1800 and 39 (cylinders manufactured under the specifications 3 and 7-300 may be continued in service but new construction is not authorized).

Requalification by hydrostatic retest is required every 10 years for 3A and 3AA cylinders and every 5 years for all other types of cylinders authorized for cyclopropane under present regulations.

However, the following types of cylinders can be requalified by visual inspection every 5 years under present DOT regulations if cyclopropane is shipped in them as a liquefied hydrocarbon gas and if the cylinders are used exclusively for liquefied hydrocarbon gas which is commercially free from corroding components: 3A480X, 3B225, 4B225 and 4BA225. Cylinders of the same types but with higher service pressures that are used as specified can also be requalified by visual inspection.

Valve Outlet and Inlet Connections. Standard connection, United States and Canada—No. 510. Standard yoke connection, United States and Canada (for medical use of the gas)—No. 920.

Medical Gas Cylinders

For use as an anesthetic, cyclopropane is supplied in medical gas cylinders of standard styles. See Section C of Chapter 4 in Part I, on gases used medicinally, for data on cyclopropane in these standard cylinder styles and an explanation of their safe handling.

METHOD OF MANUFACTURE

One major method used to produce cyclopropane commercially is the reduction of a water solution of trimethylene-chlorobromide in the presence of zinc at 200 F (93.3 C). It is also made by the progressive thermal chlorination of propane.

Dimethyl Ether

CH_3OCH_3 [or $(CH_3)_2O$]
Synonyms: Methyl ether, methyl oxide, wood ether
DOT Classification: Flammable gas; flammable gas label

PHYSICAL CONSTANTS

	U.S. Units	Metric Units
International symbol	C_2H_6O	C_2H_6O
Molecular weight	46.07	46.07
Vapor pressure		
at 70 F (21.1 C)	62 psig	427.47 kPa, gage
at 90 F (32.2 C)	90 psig	620.52 kPa, gage
at 110 F (43.3 C)	125 psig	861.84 kPa, gage
at 130 F (54.4 C)	170 psig	1172.11 kPa, gage
Density of the gas at 70 F		
(21.1 C) and 1 atm	0.119 lb/ft^3	1.906 kg/m^3
Specific gravity of the gas at 70 F		
and 1 atm (air = 1)	1.59	1.59
Specific volume of the gas at 70 F		
(21.1 C) and 1 atm	8.4 ft^3/lb	0.5244 m^3/kg
Density of liquid		
at 70 F (21.1 C)	41.2 lb/ft^3	659.96 kg/m^3
at 105 F (40.6 C)	39.2 lb/ft^3	627.92 kg/m^3
at 115 F (46.1 C)	38.6 lb/ft^3	618.31 kg/m^3
at 130 F (54.4 C)	37.7 lb/ft^3	603.90 kg/m^3
Boiling point at 1 atm	-12.8 F	-24.9 C
Melting point at 1 atm	-223 F	-141.7 C
Critical temperature	264 F	128.9 C
Critical pressure	772 psia	5322.75 kPa, abs.
Critical density	16.94 lb/ft^3	271.4 kg/m^3
Triple point at 1 atm	-222.9 F	-141.6 C
Latent heat of vaporization at		
-12.6 F (-24.7 C)	201 Btu/lb	467.53 kJ/kg
Latent heat of fusion at melting		
point	46.13 Btu/lb	107.30 kJ/kg

	U.S. Units	Metric Units
Specific heat of gas at -18 F (-27.8 C) and 1 atm	0.96 Btu/(lb)(F)	4.02 kJ/(kg)(C)
Ratio of specific heats, gas at 6-30 C and 1 atm		
C_p/C_v	1.11	1.11
Solubility in water, weight percent at 75 F (23.9 C) and 5 atm	35 percent	35 percent
Weight of liquid at 70 F (21.1 C)	5.51 lb/gal	660.24 kg/m^3
Flammable limits in air, by volume	3.4-26.7 percent	3.4-26.7 percent
Flash point (closed cup)	-42 F	-41.1 C
Autoignition temperature	662 F	350 C

DESCRIPTION

Dimethyl ether is a colorless, flammable gas easily compressed to a colorless liquid. It has a faint sweetish odor and leads to anesthesia when inhaled in fairly large concentrations. It readily forms complexes with inorganic compounds and acts as a methylating agent.

GRADES AVAILABLE

Dimethyl ether is available for commercial and industrial use in various grades having much the same component proportions from one producer to another.

USES

Dimethyl ether is used as a propellant in aerosol sprays and as a refrigerant (mixed with a fluorocarbon to reduce flammability). It is also used as a methylating agent in the dye industry, and as a chemical reaction medium, a solvent, and a catalyst and stabilizer in polymerization.

PHYSIOLOGICAL EFFECTS

Studies of dimethyl ether have found that inhalation of a 7.5 percent concentration for 12 minutes resulted in a feeling of intoxication and some lack of attention; of a 10 percent concentration for 64 minutes, in nauseous sickness; and of a 20 percent concentration for 17 minutes, in unconsciousness. Early experiments investigating dimethyl ether as an anesthetic with animals and humans found no permanent residual effects (see Ref. 1).

Prolonged contact of liquid dimethyl ether with the skin causes freezing or frostbite of the skin.

SAFE HANDLING AND STORAGE

Materials Suitable for Containers and Storage

Any commercially available metals may be used with dimethyl ether, as it is noncorrosive.

Storage and Handling in Normal Use

Precautions necessary for the safe handling of any flammable gas must be observed with dimethyl ether (see part I, Chapter 4, Section A).

Storage and handling equipment for dimethyl ether must be designed to keep it from contact with the air, as it may form peroxides when exposed to atmospheric oxygen. Unloading and storage systems must be purged of all air before dimethyl ether is introduced into them. Compressed nitrogen is among substances recommended for such purging; carbon dioxide must not be used because it is highly soluble in dimethyl ether. Indoor storage areas must be located only in fire-resistive buildings and fitted with sprinkler systems to keep the storage container cool, should fire occur, because dimethyl ether exerts extreme pressure when heated.

Ventilation must be provided at the floor level, since dimethyl ether vapors are heavier than air and collect in low spots.

Handling Leaks and Emergencies.

Should dimethyl ether ignite, recommended extinguishing agents for fire-fighting equipment include dry chemical, carbon dioxide and carbon tetrachloride (see also Part I, Chapter 4, Section A).

METHODS OF SHIPMENT; REGULATIONS

Under the appropriate regulations and tariffs, dimethyl ether is authorized for shipment as follows:

By Rail: In cylinders (freight or express), and in single-unit tank cars and TMU tank cars.

By Highway: In cylinders on trucks, and in TMU (ton multi-unit) tanks on trucks.

By Water: In cylinders via cargo vessels only, and in authorized tank cars via trainships only. In cylinders on barges of U.S. Coast Guard classes A and C only.

By Air: Aboard cargo aircraft only in appropriate cylinders up to 300 lb (140 kg) maximum net weight per cylinder.

CONTAINERS

Dimethyl ether is authorized for shipment by the DOT in cylinders, in single-unit tank cars, in TMU tank cars, and in TMU tanks on trucks.

Filling Limit. The maximum filling density authorized for dimethyl ether in cylinders is the maximum cylinder service pressure at 70 F (21.1 C). The maximum filling densities authorized for dimethyl ether in other containers are (percent water capacity by weight): for single-unit tank cars, 62 percent; for TMU tanks, 59 percent.

Cylinders

Dimethyl ether is authorized for shipment in cylinders of any type currently approved by the DOT for liquefied compressed gases (these are cylinders that meet DOT specifications 3A, 3AA, 3B, 3BN, 3D, 3E, 4, 4A, 4B, 4BA, 4B-ET, 9, 40 and 41; cylinders meeting DOT specifications 3, 25, 26 and 38 may be continued in dimethyl ether service, but new construction of them is not authorized).

All cylinders authorized for dimethyl ether service must be requalified by hydrostatic retest every 5 years with the exceptions of: type 4, for which the retest period is 10 years; type 3E, for which periodic retest is not required; and types 40 and 41, which are small inside containers that it is illegal to refill.

Valve Outlet Connections. Standard connection, United States and Canada—No. 510.

TMU Tanks

Shipment of TMU tanks of DOT specifications 106A500-X and 110A500-W on trucks or by rail is authorized.

Tank Cars

Dimethyl ether is authorized for shipment in single-unit tank cars meeting DOT specifications 105A300-W (provided that they have properly fitted loading and unloading valves).

METHOD OF MANUFACTURE

Dimethyl ether is produced by the dehydration of methanol, either with sulfuric acid or over alumina at high temperatures and pressures.

REFERENCE

1. Brown, W. E., *J. Pharmacol. Exp. Therapeutics,* **23**, 485–496 (1924); Davidson, B. M., **26**, 43–48 (1925).

Ethane

C_2H_6 (or CH_3CH_3)
Synonyms: Bimethyl, dimethyl, ethyl hydride, methyl-methane
DOT Classification: Flammable gas; flammable gas label

PHYSICAL CONSTANTS

	U.S. Units	Metric Units
International symbol	C_2H_6	C_2H_6
Molecular weight	30.068	30.068
Vapor pressure		
at 70 F (21.1 C)	544 psig	3753 kPa, gage
Density of the gas		
at 70 F (21.1 C) and 1 atm	0.07990 lb/ft³	1.2799 kg/m³
Specific gravity of the gas		
at 60 F and 1 atm (air = 1)	1.0469	1.0469
Specific volume of the gas		
at 60 F (15.6 C) and 1 atm	12.5151 ft³/lb	0.7813 m³/kg
Density of liquid at saturation pressure		
at 60 F (15.6 C)	23.52 lb/ft³	376.7 kg/m³
at 70 F (21.1 C)	22.40 lb/ft³	358.8 kg/m³
Boiling point at 1 atm	−127.53 F	−88.630 C
Melting point at 1 atm	−297.76 F	−183.2 C
Critical temperature	86.96 F	32.20 C
Critical pressure	708.35 psia	4883.9 kPa, abs.
Critical density	12.67 lb/ft³	203.0 kg/m³
Triple point at 1 atm	−297.89 F	−183.27
Latent heat of vaporization		
at boiling point and 1 atm	210.41 Btu/lb	489.41 kJ/kg
Latent heat of fusion		
at triple point	40.9 Btu/lb	95.1 kJ/kg
Specific heat of gas		
at 60 F (15.6 C) and 1 atm		
C_p	0.4097 Btu/(lb) (F)	1.7153 kJ/(kg) (C)
C_v	0.3436 Btu/(lb) (F)	1.4386 kJ/(kg) (C)

	U.S. Units	Metric Units
Ratio of specific heats	1.192	1.192
Solubility in water, vol/vol of water, at 68 F (20 C) and 1 atm	0.047/liter	0.047/liter
Pressure in typical full cylinder (approx.) at 130 F (54.4 C)	2474 psig	17,044 kPa, gage
Critical volume	0.0788 ft^3/lb	0.00492 m^3/kg
Specific gravity of liquid, 60 F/ 60 F (15.6 C/15.6 C), at saturation pressure (absolute value from weights in vacuum for the air-saturated liquid)	0.3771	0.3771
Gross heat of combustion Gas at 60 F (15.6 C) and 1 atm (of the real gas)	1783.7 Btu/ft^3	66,453 kJ/m^3
Liquid at 77 F (25 C) and saturation pressure	22,169 Btu/lb	51,565 kJ/kg
Net heat of combustion Gas at 60 F (15.6 C) and 1 atm, (of the real gas)	1631.5 Btu/ft^3	60,783 kJ/m^3
Liquid at 77 F (25 C) and saturation pressure	20,281 Btu/lb	47,173 kJ/kg
Air required for combustion, volume of air per volume of the real gas, at 60 F (15.6 C) and 1 atm	16.845	16.845
Air required for combustion, weight of air per weight of gas	16.090	16.090
Flammable limits in air, by volume	3.0–12.4 percent	3.0–12.4 percent
Flash point	−211 F	−135 C

DESCRIPTION

Ethane is a colorless, odorless, flammable gas that is relatively inactive chemically and is considered nontoxic. It is shipped as a liquefied compressed gas under its vapor pressure of 544 psig at 70 F (3751 kPa gage at 21.1 C).

GRADES AVAILABLE

Ethane is typically available for commercial and industrial purposes in a C.P. grade (minimum purity of 99.0 mole percent), or a technical grade (minimum purity of 95.0 mole per-

cent). A typical analysis of a technical grade of ethane is as follows:

Ethane	97.41 mole percent
Methane	0.03 mole percent
Ethylene	0.4 mole percent
Hydrogen	2.15 mole percent
n-Butane	0.01 mole percent
Acetylene	not detectable

USES

Major uses of ethane include its employment as a fuel, in organic synthesis (it can be chlorinated to give ethyl chloride, for example, and can yield ethylene with a greater heat input than is required for obtaining ethylene by propane cracking), and as a refrigerant.

PHYSIOLOGICAL EFFECTS

Inhalation of ethane in concentrations in air up to 5 percent produces no definite symptoms, but inhalation of higher concentrations has an anesthetic effect. It can act as a simple asphyxiant by displacing the oxygen in the air. Contact between liquid ethane and skin can cause freezing of the tissues, and should be avoided.

SAFE HANDLING AND STORAGE

Materials Suitable for Containers and Storage

Ethane is noncorrosive and may be contained in installations constructed of any common metals to withstand the pressures involved.

Storage and Handling in Normal Use

All the precautions required for the safe handling of any flammable compressed gas must be observed with ethane (see Part I, Chapter 4, Section A).

Handling Leaks and Emergencies

See Part I, Chapter 4, Section A.

METHODS OF SHIPMENT; REGULATIONS

Under the appropriate regulations and tariffs, ethane is authorized for shipment as follows:

By Rail: In cylinders (via freight or express).

By Highway: In cylinders on trucks and in tube trailers.

By Water: In cylinders on cargo vessels only. In cylinders on barges of U.S. Coast Guard classes A, CA and CB only.

By Air: In cylinders aboard cargo aircraft only up to 300 lb (140 kg) maximum net weight per cylinder.

CONTAINERS

Ethane is authorized by the DOT for shipment in cylinders.

Filling Limits. The maximum filling densities authorized for ethane in cylinders are: 35.8 percent (percent water capacity by weight) in cylinders meeting DOT specifications 3A1800, 3AA1800, 3, 3E1800 and 39, or 36.8 percent in cylinders meeting DOT specifications 3A2000, 3AA2000 and 39.

Cylinders

Cylinders that comply with the following DOT specifications are authorized for ethane service: 3A1800, 3AA1800, 3E1800 and 39. (Containers of the same types with higher service pressures may also be used.) DOT-3 cylinders may also be continued in service, but new construction is not authorized.

All types of cylinders authorized for ethane service must be requalified by periodic hydrostatic retest every 5 years under present regulations, except that no periodic retest is required for 3E cylinders.

Valve Outlet Connections. Standard connection, United States and Canada— No. 350.

METHOD OF MANUFACTURE

Ethane is produced commercially from the cracking of light petroleum fractions, and also by fractionation from natural gas.

Ethylene

$CH_2 : CH_2$ (or $H_2C : CH_2$)
Synonym: Ethene (also olefiant gas, bicarbuttetted hydrogen, elayl, or etherin)
DOT Classification: Flammable gas; flammable gas label

PHYSICAL CONSTANTS

	U.S. Units	Metric Units
International symbol	C_2H_4	C_2H_4
Molecular weight	28.05	28.05
Density of the gas at 32 F (0 C) and 1 atm	0.0787 lb/ft^3	1.261 kg/m^3
Specific gravity of the gas at 32 F and 1 atm (air = 1)	0.978	0.978
Specific volume of the gas at 70 F (21.1 C) and 1 atm	12.7 ft^3/lb	0.793 m^3/kg
Density of liquid at boiling point	35.42 lb/ft^3	567.37 kg/m^3
Boiling point at 1 atm	154.8 F	-103.8 C
Melting point at 1 atm	-272.9 F	-169.4 C
Critical temperature	49.82 F	9.900 C
Critical pressure	742.15 psia	5117.0 kPa, abs.
Critical density	14.2 lb/ft^3	228 kg/m^3
Triple point	-272.47 F at 0.0147 psia	169.15 C at 0.1014 kPa, abs.
Latent heat of vaporization at boiling point	208 Btu/lb	484 kJ/kg
Latent heat of fusion at melting point	51.2 Btu/lb	119 kJ/kg
Specific heat of gas at 59 F (15 C) and 1 atm		
C_p	0.3622 Btu/(lb) (F)	1.516 kJ/(kg) (C)
C_v	0.2914 Btu/(lb) (F)	1.220 kJ/(kg) (C)
Ratio of specific heats	1.243	1.243
Solubility in water, vol/vol of water at 32 F (0 C)	0.26/liter	0.26/liter

	U.S. Units	Metric Units
Weight of liquid at boiling point	4.735 lb/gal	567.4 kg/m^3
Gross heat of combustion	21,625 Btu/lb	50,300 kJ/kg
Flammable limits in air, by volume		
Lower	2.7–3.1 percent*	2.7–3.1 percent
Upper	16–36 percent*	16–36 percent
Flammable limits in oxygen, by volume		
Lower	2.9 percent	2.9 percent
Upper	79.9 percent	79.9 percent
Autoignition temperature		
In air	914 F**	490 C
In oxygen	905 F	485 C

*Other source gives 3.1–32.0 percent.
**Other source gives 1009 F (542.8 C).

DESCRIPTION

Ethylene is a colorless, highly flammable gas with a faint odor that is sweet and musty. It is nontoxic, being used as an anesthetic, and is hazardous only as a flammable substance or as a simple asphyxiant. Chemically, it reacts chiefly by addition to give saturated paraffins, or derivatives of paraffin hydrocarbons, and is widely used as a raw material in the synthetic organic chemical industry. It is shipped as a gas at about 1250 psig at 70 F (8618 kPa gage at 21.1 C). Below 50 F (10 C) at such charging pressure, it is a liquefied gas in the cylinder.

GRADES AVAILABLE

Ethylene is typically available for commercial and industrial purposes in a C.P. grade (minimum purity of 99.5 mole percent), and a technical grade (minimum purity of 98.0 mole percent). A typical lot analysis of technical grade is as follows:

Ethylene	98.5 mole percent
Methane	0.4 mole percent
Ethane	1.0 mole percent
Propane	0.1 mole percent
Oil and foreign material	none
Dew point	–50 F (–45.6 C)

USES

Roughly half of the ethylene produced in the United States has been used to make ethyl alcohol, with another substantial portion going into the production of ethylene glycol. Other chemical raw materials made with ethylene include ethyl chloride, dichloroethane and vinyl chloride, ethyl ether, methyl acrylate and styrene.

Ethylene is also employed as an anesthetic (as noted), a refrigerant, and a fuel for metal cutting and welding. It is also used to accelerate plant growth and fruit ripening.

PHYSIOLOGICAL EFFECTS

As an anesthetic drug, ethylene is a nontoxic gas found pleasant and nonirritating by patients. Prolonged inhalation of substantial concentrations results in unconsciousness, light and moderate anesthesia is attained, and deep anesthesia seldom occurs. Inhalation is fatal only if the gas acts as a simple asphyxiant, depriving the lungs of necessary oxygen.

No deleterious action by ethylene on circulatory, respiratory or other systems or organs has been observed, and the gas is not altered or combined in the body with any tissue. Exhalation eliminates the major portion of ethylene within minutes, although complete desaturation from body fat takes several hours. Minute traces can be detected in the blood a number of hours after anesthesia has ended.

SAFE HANDLING AND STORAGE

Materials Suitable for Containers and Storage

Any common commercially available metals may be used with ethylene because it is noncorrosive.

Storage and Handling in Normal Use

Ethylene poses hazards to personnel through its high flammability, and the precautions necessary for the safe handling of any flammable gas must be observed in its use (see also Part I, Chapter 4, Sections A and C).

Handling Leaks and Emergencies

See Part I, Chapter 4, Sections A and C.

METHODS OF SHIPMENT; REGULATIONS

Under the appropriate regulations and tariffs, ethylene is authorized for shipment as follows:

By Rail: In cylinders (freight, express or baggage), and by DOT exemption in 113A tank cars in liquid, low-temperature form.

By Highway: In cylinders on trucks, and in tube trailers; by DOT exemption, in insulated truck cargo tanks liquefied at low temperatures.

By Water: In cylinders via only cargo vessels. In cylinders on barges of U.S. Coast Guard classes A and C only.

By Air: Aboard cargo aircraft only in appropriate cylinders up to 300 lb (140 kg) maximum net weight per cylinder.

CONTAINERS

Ethylene is authorized for shipment in cylinders under DOT regulations. It is also shipped in bulk quantities under DOT exemption.

Filling Limits: The maximum filling limits prescribed for ethylene in cylinders are as follows (percent water capacity by weight): for cylinders of 1800 psig (12,409 kPa gage) maximum service pressure, 31 percent; for cylinders of 2000 psig (13,788 kPa gage) maximum service pressure, 32.5 percent; and for cylinders of 2400 psig (16,546 kPa gage) maximum service pressure, 35.5 percent. It is also shipped by highway in tube trailers. Under DOT exemption, ethylene is also shipped liquefied at low temperatures in insulated cargo tanks on trucks and truck trailers, and in insulated, low-temperature tank cars of the DOT 113A type. Bulk shipment of ethylene is also made over relatively short distances by pipeline.

Cylinders

Cylinders that meet the following DOT specifications are authorized for ethylene service: 3A1800, 3AA1800, 3E1800, 3A2000, 3AA2000, 3AA2265, 3A2400, 3AA2400 and 39 (cylinders manufactured under the now obsolete specification DOT-3 may be continued in service but new construction is not authorized).

Cylinders of types 3A and 3AA (as well as 3) used in ethylene service must be requalified by hydrostatic retest every 10 years under present regulations. For cylinders of type 3E, no periodic retest is required.

Valve Outlet and Inlet Connections. Standard connection, United States and Canada—No. 350. Standard yoke connection, United States and Canada (for medical use of the gas)—No. 900.

Medical Gas Cylinders

For use as an anesthetic, ethylene is shipped and stored in special medical gas cylinders. See Section C of Chapter 4 in Part I for descriptions of these cylinders and an explanation of their safe handling.

METHOD OF MANUFACTURE

The most commonly used of a number of methods for producing ethylene commercially is high-temperature coil cracking of propane or of ethane and propane. Recovery of ethylene from the cracked gases is often accomplished by low-temperature high-pressure straight fractionation. Another customary manufacturing method is by catalytic decomposition of ethyl alcohol.

Fluorine

F$_2$ DOT Classification: Nonflammable gas; poison and oxidizer label

PHYSICAL CONSTANTS

	U.S. Units	Metric Units
International symbol	F$_2$	F$_2$
Molecular weight	37.996	37.996
Density of the gas		
at 32 F (0.0 C) and 1 atm	0.106 lb/ft^3	1.70 kg/m^3
at 70 F (21.1 C) and 1 atm	0.098 lb/ft^3	1.57 kg/m^3
Specific gravity of the gas at 70 F		
and 1 atm (air = 1)	1.31	1.31
Specific volume of the gas at		
70 F (21.1 C) and 1 atm	10.2 ft^3/lb	0.637 m^3/kg
Density of liquid		
at −306.0 F (−188.1 C)	94.0 lb/ft^3	1.507 kg/m^3
at −320.4 F (−195.8 C)	97.9 lb/ft^3	1.568 kg/m^3
Boiling point at 1 atm	−306.6 F	−188.1 C
Melting point at 1 atm	−363.3 F	−219.6 C
Critical temperature	−200.4 F	−129.1 C
Critical pressure	808.5 psia	5574 kPa, abs.
Critical density	380 lb/ft^3	6087 kg/m^3
Triple point	−363.3 F at	−219.6 C at
	0.0324 psia	0.223 kPa, abs.
Latent heat of vaporization at		
−306.6 F (−188.1 C)	74.6 Btu/lb	173 kJ/kg
Latent heat of fusion at		
−363.3 F (−219.6 C)	5.76 Btu/lb	13.40 kJ/kg
Weight of liquid per gallon		
at −306.6 F (−188.1 C)	12.6 lb/gal	1509.8 kg/m^3
at −320.4 F (−195.8 C)	13.1 lb/gal	1569.7 kg/m^3
Molar heat capacity, Cp at	7.5183 Btu/lb	
32.0 F (0.0 C)	mole-R	31.478 J/mole · K
Thermal conductivity of gas at		
32.0 F (21.1 C) and 1 atm	0.172 Btu · in/hr F	0.248 W/m · K

	U.S. Units	Metric Units
Viscosity		
of vapor at 32.0 F (21.1 C) and		
1 atm	0.0527 lb/ft hr	0.0784 kg/m hr
of liquid at -306.3 F (-18.79 C)	0.621 lb mass/ft hr	0.924 kg mass/m hr
of liquid at -340.5 F (-206.9 C)	1.002 lb mass/ft hr	1.491 kg mass/m hr
Surface tension		
of liquid at -306.3 F (-187.9 C)	0.0010 lb-force/ft	0.00146 N/m
of liquid at -340.5 F (-206.9 C)	0.0012 lb-force/ft	0.00179 N/m

DESCRIPTION

Fluorine is a highly toxic, pale yellow gas about 1.7 times as heavy as air at atmospheric temperature and pressure. When cooled below its low boiling point (-306.6 F or -188.1 C), it is a liquid about 1.1 times as heavy as water.

Fluorine is the most powerful oxidizing agent known, reacting with practically all organic and inorganic substances. Exceptions are metal fluorides and a few completely fluorinated organic compounds in pure form. However, the latter may also react with fluorine if they are contaminated with a combustible material.

Heats of reaction with fluorine are always high and most reactions take place with ignition.

Fluorine at low pressures and concentrations reacts slowly with many metals at room temperatures, however, and the reaction often results in formation of a metal fluoride film on the metal's surface; in the case of some metals, this film retards further action.

Fluorine readily displaces the other halogens from their compounds, but such reactions are not always feasible for preparing fluorides. It reacts with water to form a mixture containing principally oxygen and hydrogen fluoride plus small amounts of ozone, hydrogen peroxide and oxygen fluoride.

GRADES AVAILABLE

Fluorine is available for commercial and industrial use in various grades having much the same component proportions from one producer to another.

USES

Fluorine is used in producing uranium hexafluoride, sulfur hexafluorides, the halogen fluorides and other fluorine compounds that require the high reactivity of elemental fluorine for their preparation.

PHYSIOLOGICAL EFFECTS

Fluorine gas is a powerful caustic irritant and is highly toxic (see Ref. 1). In one series of animal experiments, inhalation of acute exposures of 10,000 ppm for 5 minutes, 1000 ppm for 30 minutes, and 500 ppm for 1 hour produced 100 percent mortality in rats, mice, guinea pigs and rabbits. Inhalation of 100 ppm for 7 hours produced wide variation in species mortaility, ranging from 0 percent in guinea pigs to 96 percent in mice. Daily subacute exposures to a concentration of 2 ppm for periods of time varying from totals of 30 to 176 hours resulted in high mortality, rabbits appearing to be the most susceptible species and guinea pigs the least. Pulmonary irritation varying from severe at 16 ppm in some species to mild at 2 ppm represented the major pathological change, while similar subacute exposure at 0.5 ppm resulted in no significant pathology but some retention of fluorine in osseous tissues.

Contact between the skin and high concentrations of fluorine gas under pressure will produce burns comparable to thermal burns; contact with lower concentrations results in a chemical type of burn resembling that caused by hydrofluoric acid.

A TLV of 1 ppm (2 mg/m^3) for an 8-hour day for fluorine has been adopted by the American Conference of Governmental Industrial Hygienists (ACGIH).

SAFE HANDLING AND STORAGE

Materials Suitable for Containers and Storage

Nickel, iron, aluminum, magnesium, copper and certain of their alloys are quite satisfactory for handling fluorine at room temperature, for these are among the metals with which forma-tion of a surface fluoride film retards further action. Listed in Table 1 are various materials that have been used with satisfactory results in gaseous fluorine service at normal temperatures and liquid service at low temperatures.

Nickel and "Monel" are generally considered to be by far the best materials for fluorine service at high temperatures and pressures, but selection of suitable materials for service at elevated temperatures and pressures must be based on the conditions of the specific application.

TABLE 1. MATERIALS GIVING SATISFACTORY RESULTS IN FLUORINE SERVICE.*

Type of Equipment	Gaseous Service, Normal Temp.	Liquid Service, Low Temp.
Storage tanks	Stainless steel 304L Aluminum 6061 Mild steel (low pressure)	"Monel" Stainless steel 304L Aluminum 6061
Lines and fittings	Nickel "Monel" Copper Brass Stainless steel 304L Aluminum 2017, 2024, 5052, 6061 Mild steel (low pressure)	"Monel" Stainless steel 304L Copper Aluminum 2017, 2024, 2050
Valve bodies	Stainless steel 304 Bronze Brass	"Monel" Stainless steel 304 Bronze
Valve seats	Copper Aluminum 1100 Stainless steel 303 Brass "Monel"	Copper Aluminum 1100 "Monel"
Valve plugs	Stainless steel 304 "Monel"	Stainless steel 304 "Monel"
Valve packing	Tetrafluoroethylene polymer	Tetrafluoroethylene polymer
Valve bellows	Stainless steel 300 series "Monel" Bronze	Stainless steel 300 series "Monel" Bronze
Gaskets	Aluminum 1100 Lead Copper Tin Tetrafluoroethylene polymer Red rubber (5 psig) Neoprene (5 psig)	Aluminum 1100 Copper

*It is not necessarily implied that other materials not listed would not give adequate service.

Storage and Handling in Normal Use

All precautions necessary for the safe handling of any flammable gas must be observed with fluorine, in addition to the precautions outlined below. Fluorine fires accidentally breaking out may be most simply extinguished by cutting off the fluorine supply at a primary point, then employing conventional fire-fighting methods. The dry types of extinguishers are recommended.

Before introducing any application of fluorine, users should fully work out all details of first aid and treatment with the medical personnel who would be called to administer aid in case of accident.

Only trained and competent personnel should be permitted to handle fluorine. It is recommended that they work in pairs and within sight and sound of each other, but not in the same working area. Supervisory personnel should make frequent checks of the operation.

Essential additional precautions in the handling of fluorine cylinders are outlined in the subsequent section on cylinders. (See also Part I, Chapter 4, Section A.)

Personal Protective Equipment. Clean neoprene gloves must be worn when handling equipment which contains or has recently contained fluorine. This precaution affords not only limited protection against fluorine contact but protection against contact with possible films of hydrofluoric acid that are formed by escaping fluorine and air moisture and that collect on valve handles and other surfaces. Neoprene coats and boots afford overall body protection for short intervals of contact with low-pressure fluorine. All protective clothing must be designed and used so that it can be shed easily and quickly. Safety glasses must be worn at all times.

Face shields made preferably of transparent, highly fluorinated polymers like "Aclar" (registered trademark) should be worn whenever operators must approach equipment containing fluorine under pressure. Face shields made of any conventional materials afford limited, though valuable, protection against air-diluted blasts of fluorine.

Leak Detection

All areas containing fluorine under pressure should be inspected for leaks at suitable intervals, and any leaks discovered should be repaired at once after fluorine has been removed from the system. Ammonia vapor expelled from a squeeze bottle of ammonium hydroxide at suspected points of leakage may be used to detect leaks. Filter paper moistened with potassium iodide provides a very sensitive means for detecting fluorine (effective down to about 25 ppm); in using it, hold the paper with metal tongs or forceps about 18 to 24 in. (45 to 60 cm) long.

Equipment Preparation and Decontaminating

Equipment to be used for fluorine service should first be thoroughly cleaned, degreased and dried, then treated with increasing concentrations of fluorine gas so that any impurities will be burned out without the simultaneous ignition of the equipment. The passive metal fluoride film thus formed will inhibit further corrosion by fluorine.

Before opening or refilling equipment that has contained fluorine, thoroughly purge it with a dry inert gas (such as nitrogen) and evacuate it if possible. Minor quantities of fluorine to be vented and purged can be converted to harmless carbon fluoride gases by passage through a lump-charcoal-packed column. Large quantities to be purged require a purge system with a fluorine-hydrocarbon-air burner, scrubber and stack to prevent any undue exit hazards. Should a purged fluorine system require evacuation, a soda-lime tower followed by a drier should be included in the vacuum system to pick up trace amounts of fluorine in order to protect the vacuum pump.

Liquid Fluorine Spills

In the event of a large spillage of liquid fluorine, the contaminated area can be neutralized with sodium carbonate. The dry powder can be sprayed on the spill area from a fluidized system similar in principle to that of dry-chemical fire extinguishers. If major spillages occur in areas where the formation of hydrofluoric acid liquid and vapor pose no undue danger, water in the form of a fine mist or fog is recommended. The major portion of the fluorine will be converted to hot, light, gaseous products which rise vertically and diffuse quickly into the atmosphere.

First-Aid Suggestions

It is unlikely that persons not injured or trapped would continue to inhale highly toxic concentrations of fluorine because of its strong odor and its irritation of eyes, nose and mucous membranes. Should exposure by inhalation occur, immediately remove the patient to fresh air and call a physician. Administer oxygen as necessary to help prevent pulmonary irritation.

Liquid fluorine will severely burn the skin and eyes. For recommended first aid and medical treatment in the event of hydrofluoric acid exposure, see Safety Data Sheet SD-25 of the Chemical Manufacturers Association (see Ref. 2).

METHODS OF SHIPMENT; REGULATIONS

Under the appropriate regulations and tariffs, fluorine is authorized for shipment as follows:

By Rail: In cylinders (via freight, and express to a maximum quantity of 6 lb in one outside container).

By Highway: In cylinders on trucks, and, under special DOT permit, in trailer-mounted tank transports.

By Water: In cylinders on cargo vessels only. On barges of U.S. Coast Guard classes A, CA and CB only.

By Air: Not acceptable for shipment.

CONTAINERS

Fluorine is authorized for shipment in cylinders as a compressed gas under DOT regulations, and as a liquefied, low-temperature gas in liquid-nitrogen refrigerated tanks mounted on truck trailers by special permit of the DOT.

Filling Limits. The maximum filling density authorized for fluorine in cylinders is 400 psig at 70 F (3000 kPa gage at 21.1 C), and cylinders must not contain over 6 lb. (2.7 kg) of fluorine gas.

Cylinders

Cylinders that meet DOT specifications 3A1000, 3AA1000 and 3BN400 are authorized for fluorine service. The cylinders must not be equipped with safety relief devices, and must be fitted with valve-protection caps. Commonly available sizes of cylinders are 0.5 lb, 4.9 lb and 6 lb (0.2 kg, 2.2 kg, and 2.7 kg) net weight.

All cylinders authorized for fluorine must be requalified by hydrostatic retest every 5 years under current regulations.

Valve Outlet Connections. Standard connection, United States and Canada—No. 670.

Safe Handling and Storage of Cylinders. Personnel working with fluorine cylinders must be protected by use of a cylinder enclosure or barricade and remote-control valves, preferably ones operated by manual extension handles passing through the barricade. The main function of a barricade is to dissipate and prevent the breakthrough of any flame or flow of molten metal which, in case of equipment failure, could issue from any part of a system containing fluorine under pressure. Barricades of $\frac{1}{4}$-in. steel plate, brick or concrete provide satisfactory protection for fluorine in cylinder quantities. Adequate ventilation of enclosed working spaces is essential. Installation in a fume hood is recommended for laboratory use of fluorine cylinders.

Fluorine cylinders should be securely supported while in use to prevent movement or straining of connections.

Store full or empty fluorine cylinders in a well-ventilated area, making sure that they are protected from excessive heat, located away from organic or flammable materials, and chained in place to prevent falling. Valve-protection caps and valve-outlet caps must be securely attached to cylinders not in use.

Always protect fluorine cylinders from mechanical shock or abuse and never heat them with a torch or heat lamp.

Additional precautions required for the safe handling of fluorine have been explained in a preceding section.

Trailer-Mounted Tanks

Fluorine is authorized for shipment as a liquid at low temperature and atmospheric pressure in tanks mounted on motor vehicle trailers under exemption of the DOT. These tanks commonly have a 5000 lb (2000 kg) capacity.

Such trailer-mounted units consist of three concentric tanks. The liquid fluorine is contained in an inner baffled tank made of "Monel" metal or stainless steel. The inner tank is enclosed by a stainless steel tank filled with liquid nitrogen. The third, outer tank is made of carbon steel and the annular space between it and the liquid-nitrogen cooling jacket is filled with insulation and evacuated.

Vaporization of the liquid nitrogen (at -320.4 F or -195.8 C and 1 atm) in the specially constructed units keeps the liquid fluorine below its boiling point (of -306.6 F or 188.1 C at 1 atm), and radiation heat loss is minimized by the outer insulation. The ullage or vacant space in the liquid fluorine tank is brought to atmospheric pressure with helium to prevent subsequent in-leakage of moist air in case of valve or piping failure.

The liquid fluorine tank has two connections, one to a vapor-space line and the other to a dip line in the liquid. Both lines are double-valved. There is no rupture disk or safety valve for relieving excess pressure, but a pressure gage and high-pressure alarm are installed in the vapor line.

More than 300 gal (1.14 m^3) of liquid nitrogen are held by the inner cooling jacket, which has fill, vent and drain lines, each protected as required with a safety valve and rupture disk. Gages showing the level and pressure of the liquid nitrogen are on the control panel. The outer jacket of insulation is protected from excess pressure and has an electronic vacuum gage attached.

Safety equipment on the trailer body includes a water fire extinguisher, a dry-chemical fire extinguisher and a special tool chest (containing hand tools, safety clothing, tubing and breathing apparatus for use by drivers only, in case of emergency on the road, and not for unloading). Full information on unloading the units is available from fluorine producers operating them.

Storage and Piping Equipment

Stationary installations for storing and piping fluorine must be designed with due regard for the reactivity of fluorine and must be made of materials proven in the type of fluorine service planned, such as high-temperature, high-pressure or liquid cold compressed gas. Extreme care must be taken to keep all lines, fittings, tanks and other equipment clean. Pipe and fittings for service in lines not to be dismantled should be welded and, in general, the number of non-welded lines should be kept to a minimum. Valves should have seatings of dissimilar metals in order to prevent galling, and packless stem seals should be used if possible.

METHOD OF MANUFACTURE

Fluorine is produced commercially by electrolytic decomposition of an anhydrous hydrofluoric acid, potassium bifluoride (KF.HF) solution. The melt formed (approximately KF.2HF) is a solid at normal ambient temperatures and liquefies at approximately 160 F (71.1 C). The commercial electrolytic cell uses carbon anodes and a metal cathode with a method for separate collection of the gas released at each electrode. (These cells are heated when not operating or operating at low current rates to prevent their freezing, and are cooled when operated at normal rates to prevent overheating.) When voltage is applied to the cell, current passing through the melt decomposes the hydrofluoric acid with the release of fluorine at the anode and hydrogen at the cathode. Continuous addition of hydrofluoric acid replaces the acid decomposed.

The fluorine thus produced is purified and then either used directly in a production process, compressed for cylinder filling, or condensed to a liquid for charging into refrigerated containers.

REFERENCES

1. Voegtlin, C., and Hodge, H. C., *Pharmacology and Toxicology of Uranium Compounds*, National Nuclear Energy Series, Division VI, vol. 1, pp. 1021–1042.

2. "Hydrofluoric Acid" (Safety Data Sheet SD-25), Chemical Manufacturers Association, 1825 Connecticut Ave., N.W., Washington, DC 20009.

Fluorocarbons

12, Dichlorodifluoromethane: CCl_2F_2
DOT Classification: Nonflammable gas; nonflammable gas label

13, Chlorotrifluoromethane: $CClF_3$
DOT Classification: Nonflammable gas; nonflammable gas label
DOT Shipping Name: Monochlorotrifluoromethane

13B1, Bromotrifluoromethane (also Halon, 1301, fire extinguishant): $CBrF_3$
DOT Classification: Nonflammable gas; nonflammable gas label
DOT Shipping Name: Monobromotrifluoromethane

22, Chlorodifluoromethane: $CHClF_2$
DOT Classification: Nonflammable gas; nonflammable gas label
DOT Shipping Name: Monochlorodifluoromethane

115, Chloropentafluoroethane: $CClF_2CF_3$
DOT Classification: Nonflammable gas; nonflammable gas label
DOT Shipping Name: Monochloropentafluoroethane

142b, Chlorodifluoroethane $CClF_2CH_3$
DOT Classification: Flammable gas; flammable gas label
DOT Shipping Name: Difluoromonochloroethane

152a, Difluoroethane: CHF_2CH_3
DOT Classification: Flammable gas; flammable gas label

500, Dichlorodifluoromethane(12)/Difluoroethane(152a):
  CCl_2F_2/CHF_2CH_3 (73.8: 26.2 percent by weight (azeotropic composition))
DOT Classification: Nonflammable gas; nonflammable gas label
DOT Shipping Name: Dichlorodifluoromethane and Difluoroethane mixture

502, Chlorodifluoromethane(22)/Chloropentafluoroethane(115):
  $CHClF_2/CClF_2CF_3$ (48.8: 51.2 percent by weight (azeotropic composition))
DOT Classification: Nonflammable gas; nonflammable gas label

503, Trifluoromethane(23)/Chlorotrifluoromethane(13): $CHF_3/CClF_3$ (40.1:
  59.9 percent by weight (azeotropic composition))
DOT Classification: Nonflammable gas; nonflammable gas label

504, Difluoromethane(32)/Chloropentafluoroethane(115): $CH_2F_2/CClF_2CF_3$
  (48.2: 51.8 percent by weight (azeotropic composition))
DOT Classification: Nonflammable gas; nonflammable gas label

505, Dichlorodifluoromethane(12)/Chlorofluoromethane(31): CCl_2F_2/CH_2ClF
(78:22 percent by weight (azeotropic composition))
DOT Classification: Nonflammable gas; nonflammable gas label

506, Chlorofluoromethane(31)/Dichlorotetrafluoroethane(114):
$CH_2ClF/C_2Cl_2F_4$ (55.1:44.9 percent by weight (azeotropic composition))
DOT Classification: Nonflammable gas, nonflammable gas label

Note. This section includes only fluorocarbons which are defined as compressed gases and are regulated in interstate commerce by the Department of Transportation. There are other fluorocarbons which are: not defined as compressed gases; and still more which are shipped under DOT exemption but are not in wide commercial use. These other fluorocarbons are not included in this section.

The number which precedes each chemical name listed above is the standard designation of the gas in the system developed for identifying refrigerant gases by the American Society of Heating, Refrigerating and Air Conditioning Engineers (ASHRAE). The system is American National Standard B79.1-1967.

Within the fluorocarbon industry, a system of letter prefixes has been employed widely to indicate the intended use of the material: "P" for Aerosol Propellent or "R" for Refrigerant. Thus, for instance, P-12 (Propellent 12) and R-12 (Refrigerant 12), are the aerosol propellent and refrigerant designation for dichlorodifluoromethane. Chemically, the two are identical.

The chemical names above conform to generally accepted industry practice in the United States:

(1) The prefix "mono" is used when only one atom of a particular halogen appears in a given compound.
(2) Halogens are listed in alphabetical order; e.g.,

   bromo-
   chloro-
   fluoro-
   iodo-

(3) The chemical name and/or ASHRAE designation of the higher-pressure material is listed first in the case of mixtures. (*Exceptions.* In certain aerosol propellent mixtures, nonfluorine-containing halocarbons or hydrocarbons are always listed last, regardless of their pressure characteristics. These mixtures may also contain fluorocarbons which are not classified as compressed gases; in this case, fluorocarbons are listed in order of decreasing pressure, and nonfluorocarbons last.)

PHYSICAL CONSTANTS

(12, Dichlorodifluoromethane)

	U.S. Units	Metric Units
International symbol	CCl_2F_2	CCl_2F_2
Molecular weight	120.93	120.93
Vapor pressure		
at 0 F (−17.8 C)	23.85 psia	164.44 kPa, abs.
at 70 F (21.1 C)	84.90 psia	584.56 kPa, abs.
at 105 F (40.6 C)	141.30 psia	974.23 kPa, abs.
at 115 F (46.1 C)	161.50 psia	1113.50 kPa, abs.
at 130 F (54.4 C)	195.70 psia	1349.30 kPa, abs.
Density of the gas		
at 70 F (21.1 C) and 1 atm	0.319 lb/ft³	5.110 kg/m³
Specific gravity of the gas		
at 70 F and 1 atm (air = 1)	4.26	4.26
Specific volume of the gas		
at 70 F (21.1 C) and 1 atm	3.14 ft³/lb	0.1960 m³/kg
Density of liquid		
at 0 F (−17.8 C)	90.66 lb/ft³	1452.2 kg/m³
at 70 F (21.1 C)	82.72 lb/ft³	1325.0 kg/m³
at 105 F (40.6 C)	78.09 lb/ft³	1250.9 kg/m³
at 115 F (46.1 C)	76.65 lb/ft³	1227.8 kg/m³
at 130 F (54.4 C)	74.37 lb/ft³	1191.3 kg/m³
Boiling point at 1 atm	−21.6 F	−29.80 C
Freezing point at 1 atm	−252 F	−157.80 C
Critical temperature	233.6 F	112.0 C
Critical pressure	597 psia	4116.2 kPa, abs.
Critical density	34.84 lb/ft³	558.1 kg/m³
Latent heat of vaporization		
at boiling point	71.04 Btu/lb	165.2 kJ/kg
Specific heat of gas		
at 86 F (30 C) and 1 atm		
C_p	0.148 Btu/(lb)(F)	0.620 kJ/(kg)(C)
C_v	0.130 Btu/(lb)(F)	0.544 kJ/(kg)(C)
Ratio of specific heats	1.139	1.139
Solubility in water, weight percent,		
at 77 F (25 C) and 1 atm	2.8 percent	2.8 percent
Density of liquid		
at 70 F (21.1 C)	11.06 lb/gal	1325.3 kg/m³
Density of saturated vapor		
at 70 F (21.1 C)	2.09 lb/ft³	33.48 kg/m³
Critical volume	0.0287 ft³/lb	0.001792 m³/kg
Specific heat of liquid		
at 86 F (30 C)	0.235 Btu/(lb)(F)	0.984 kJ/(kg)(C)

PHYSICAL CONSTANTS

(13, Chlorotrifluoromethane)

	U.S. Units	Metric Units
International symbol	$CClF_3$	$CClF_3$
Molecular weight	104.47	104.47
Vapor pressure		
at 0 F (−17.8 C)	176.80 psia	1218.99 kPa, abs.
at 70 F (21.1 C)	473.40 psia	3263.98 kPa, abs.
Density of the gas		
at 70 F (21.1 C) and 1 atm	0.277 lb/ft^3	4.437 kg/m^3
Specific gravity of the gas		
at 70 F and 1 atm (air = 1)	3.70	3.70
Specific volume of the gas		
at 70 F (21.1 C) and 1 atm	3.61 ft^3/lb	0.2254 m^3/kg
Density of Liquid		
at 0 F (−17.8 C)	76.98 lb/ft^3	1233.1 kg/m^3
at 70 F (21.1 C)	56.46 lb/ft^3	904.4 kg/m^3
Boiling point at 1 atm	−114.6 F	−81.44 C
Freezing point at 1 atm	−294 F	−181.11 C
Critical temperature	83.9 F	28.83 C
Critical pressure	561 psia	3867.0 kPa, abs.
Critical density	36.1 lb/ft^3	578.3 kg/m^3
Latent heat of vaporization		
at boiling point	63.85 Btu/lb	148.5 kJ/kg
Specific heat of gas		
at −30 F (−34.4 C) and 1 atm		
C_p	0.14 Btu/(lb)(F)	0.59 kJ/(kg)(C)
C_v	0.12 Btu/(lb)(F)	0.50 kJ/(kg)(C)
Ratio of specific heats	1.17	1.17
Solubility in water, weight percent,		
at 77 F (25 C) and 1 atm	0.9 percent	0.9 percent
Density of liquid		
at 70 F (21.1 C)	7.55 lb/gal	904.7 kg/m^3
Density of saturated vapor		
at 70 F (21.1 C)	17.34 lb/ft^3	277.76 kg/m^3
Critical volume	0.0277 ft^3/lb	0.001729 m^3/kg
Specific heat of liquid		
at −30 F (−34.4 C)	0.240 Btu/(lb)(F)	1.005 kJ/(kg)(C)

PHYSICAL CONSTANTS

(13B1, Bromotrifluoromethane)

	U.S. Units	Metric Units
International symbol	$CBrF_3$	$CBrF_3$
Molecular weight	148.93	148.93
Vapor pressure		
at 0 F (−17.8 C)	71.16 psia	490.63 kPa, abs.
at 70 F (21.1 C)	213.70 psia	1473.41 kPa, abs.
at 105 F (40.6 C)	335.10 psia	2310.43 kPa, abs.
at 115 F (46.1 C)	377.70 psia	2605.53 kPa, abs.
at 130 F (54.4 C)	448.90 psia	3095.06 kPa, abs.
Density of the gas		
at 70 F (21.1 C) and 1 atm	0.397 lb/ft^3	6.36 kg/m^3
Specific gravity of the gas		
at 70 F and 1 atm (air = 1)	5.30	5.30
Specific volume of the gas		
at 70 F (21.1 C) and 1 atm	2.52 ft^3/lb	0.157 m^3/kg
Density of liquid		
at 0 F (−17.8 C)	112.48 lb/ft^3	1801.76 kg/m^3
at 70 F (21.1 C)	97.79 lb/ft^3	1566.45 kg/m^3
at 105 F (40.6 C)	87.79 lb/ft^3	1406.26 kg/m^3
at 115 F (46.1 C)	84.21 lb/ft^3	1348.91 kg/m^3
at 130 F (54.4 C)	77.61 lb/ft^3	1243.19 kg/m^3
Boiling point at 1 atm	−71.95 F	−57.75 C
Freezing point at 1 atm	−270 F	−167.77 C
Critical temperature	152.6 F	67.00 C
Critical pressure	575 psia	3964.49 kPa, abs.
Critical density	46.5 lb/ft^3	744.86 kg/m^3
Latent heat of vaporization		
at boiling point	51.08 Btu/lb	118.81 kJ/kg
Specific heat of gas		
at 86 F (30 C) and 1 atm		
C_p	0.113 Btu/(lb)(F)	0.473 kJ/(kg)(C)
C_v	0.099 Btu/(lb)(F)	0.414 kJ/(kg)(C)
Ratio of specific heats	1.143	1.143
Solubility in water, weight percent,		
at 77 F (25 C) and 1 atm	3 percent	3 percent
Density of liquid		
at 70 F (21.1 C)	13.07 lb/gal	1566.13 kg/m^3
Density of saturated vapor		
at 70 F (21.1 C)	7.44 lb/ft^3	119.18 kg/m^3
Critical volume	0.0215 ft^3/lb	0.00134 m^3/kg
Specific heat of liquid		
at 86 F (30 C)	0.21 Btu/(lb)(F)	0.88 kJ/(kg)(C)

PHYSICAL CONSTANTS

(22, Chlorodifluoromethane)

	U.S. Units	Metric Units
International symbol	$CHClF_2$	$CHClF_2$
Molecular weight	86.48	86.48
at 0 F (-17.8 C)	38.70 psia	266.83 kPa, abs.
at 70 F (21.1 C)	136.10 psia	938.38 kPa, abs.
at 105 F (40.6 C)	225.40 psia	1554.01 kPa, abs.
at 115 F (46.1 C)	257.50 psia	1775.40 kPa, abs.
at 130 F (54.4 C)	311.50 psia	2147.72 kPa, abs.
Density of the gas		
at 70 F (21.1 C) and 1 atm	0.231 lb/ft^3	3.700 kg/m^3
Specific gravity of the gas		
at 70 F and 1 atm (air = 1)	3.08	3.08
Specific volume of the gas		
at 70 F (21.1 C) and 1 atm	4.33 ft^3/lb	0.270 m^3/kg
Density of liquid		
at 0 F (-17.8 C)	83.83 lb/ft^3	1342.83 kg/m^3
at 70 F (21.1 C)	75.47 lb/ft^3	1208.91 kg/m^3
at 105 F (40.6 C)	70.47 lb/ft^3	1128.82 kg/m^3
at 115 F (46.1 C)	68.88 lb/ft^3	1103.35 kg/m^3
at 130 F (54.4 C)	66.31 lb/ft^3	1062.18 kg/m^3
Boiling point at 1 atm	-41.4 F	-40.77 C
Freezing point at 1 atm	-256 F	-160.00 C
Critical temperature	204.8 F	96.00 C
Critical pressure	721.9 psia	4977.3 kPa, abs.
Critical density	32.8 lb/ft^3	525.4 kg/m^3
Latent heat of vaporization		
at boiling point	100.45 Btu/lb	233.65 kJ/kg
Specific heat of gas		
at 86 F (30 C) and 1 atm		
C_p	0.158 Btu/(lb)(F)	0.662 kJ/(kg)(C)
C_v	0.134 Btu/(lb)(f)	0.561 kJ/(kg)(C)
Ratio of specific heats	1.18	1.18
Solubility in water, weight percent,		
at 77 F (25 C) and 1 atm	30 percent	30 percent
Density of liquid		
at 70 F (21.1 C)	10.09 lb/gal	1209.05 kg/m^3
Density of saturated vapor		
at 70 F (21.1 C)	2.477 lb/ft^3	39.68 kg/m^3
Critical volume	0.0305 ft^3/lb	0.00190 m^3/kg
Specific heat of liquid		
at 86 F (30 C)	0.306 Btu/(lb)(F)	1.278 kJ/(kg)(C)

PHYSICAL CONSTANTS

(115, Chloropentafluoroethane)

	U.S. Units	Metric Units
International symbol	$CClF_2CF_3$	$CClF_2CF_3$
Molecular weight	154.48	154.48
Vapor pressure		
at 0 F (-17.8 C)	36.94 psia	254.69 kPa, abs.
at 70 F (21.1 C)	119.10 psia	821.17 kPa, abs.
at 105 F (40.6 C)	195.40 psia	1347.24 kPa, abs.
at 115 F (46.1 C)	222.60 psia	1534.77 kPa, abs.
at 130 F (54.4 C)	268.80 psia	1853.31 kPa, abs.
Density of the gas		
at 70 F (21.1 C) and 1 atm	0.415 lb/ft^3	6.648 kg/m^3
Specific gravity of the gas		
at 70 F and 1 atm (air = 1)	5.54	5.54
Specific volume of the gas		
at 70 F (21.1 C) and 1 atm	2.41 ft^3/lb	0.150 m^3/kg
Density of liquid		
at 0 F (-17.8 C)	91.78 lb/ft^3	1470.17 kg/m^3
at 70 F (21.1 C)	81.38 lb/ft^3	1303.58 kg/m^3
at 105 F (40.6 C)	74.97 lb/ft^3	1200.90 kg/m^3
at 115 F (46.1 C)	72.89 lb/ft^3	1167.59 kg/m^3
at 130 F (54.4 C)	69.44 lb/ft^3	1112.32 kg/m^3
Boiling point at 1 atm	-38.4 F	-39.1 C
Freezing point at 1 atm	-159 F	-106.1 C
Critical temperature	175.9 F	79.94 C
Critical pressure	458 psia	3157.80 kPa, abs.
Critical density	38.3 lb/ft^3	613.5 kg/m^3
Latent heat of vaporization		
at boiling point	54.20 Btu/lb	126.07 kJ/kg
Specific heat of gas		
at 86 F (30 C) and 1 atm		
C_p	0.176 Btu/(lb)(F)	0.737 kJ/(kg)(C)
C_v	0.163 Btu/(lb)(F)	0.682 kJ/(kg)(C)
Ratio of specific heats	1.08	1.08
Solubility in water, weight percent,		
at 77 F (25 C) and 1 atm	0.6 percent	0.6 percent
Density of liquid		
at 70 F (21.1 C)	10.88 lb/gal	1303.71 kg/m^3
Density of saturated vapor		
at 70 F (21.1 C)	4.107 lb/ft^3	65.788 kg/m^3
Critical volume	0.0261 ft^3/lb	0.00163 m^3/kg
Specific heat of liquid		
at 86 F (30 C)	0.263 Btu/(lb)(F)	1.10 kJ/(kg)(C)

PHYSICAL CONSTANTS

(142b, Chlorodifluoroethane)

	U.S. Units	Metric Units
International symbol	$CClF_2CH_3$	$CClF_2CH_3$
Molecular weight	100.5	100.5
Vapor pressure		
at 0 F (-17.8 C)	10.62 psia	73.22 kPa, abs.
at 70 F (21.1 C)	42.50 psia	293.03 kPa, abs.
at 105 F (40.6 C)	77.30 psia	532.96 kPa, abs.
at 115 F (46.1 C)	89.90 psia	619.84 kPa, abs.
at 130 F (54.4 C)	111.80 psia	770.83 kPa, abs.
Density of the gas		
at 70 F (21.1 C) and 1 atm	0.272 lb/ft^3	4.36 kg/m^3
Specific gravity of the gas		
at 70 F and 1 atm (air = 1)	3.63	3.63
Specific volume of the gas		
at 70 F (21.1 C) and 1 atm	3.68 ft^3/lb	0.230 m^3/kg
Density of liquid		
at 0 F (-17.8 C)	75.72 lb/ft^3	1212.92 kg/m^3
at 70 F (21.1 C)	69.93 lb/ft^3	1120.17 kg/m^3
at 105 F (40.6 C)	66.68 lb/ft^3	1068.11 kg/m^3
at 115 F (46.1 C)	65.68 lb/ft^3	1052.10 kg/m^3
at 130 F (54.4 C)	64.11 lb/ft^3	1026.94 kg/m^3
Boiling point at 1 atm	14.4 F	-9.78 C
Freezing point at 1 atm	-204 F	-131.11 C
Critical temperature	278.8 F	137.11 C
Critical pressure	598 psia	4123.06 kPa, abs.
Critical density	27.2 lb/ft^3	435.70 kg/m^3
Latent heat of vaporization		
at boiling point	92.51 Btu/lb	215.18 kJ/kg
Specific heat of gas		
at 86 F (30 C) and 1 atm		
C_p	0.201 Btu/(lb)(F)	0.842 kJ/(kg)(C)
C_v	0.181 Btu/(lb)(F)	0.758 kJ/(kg)(C)
Ratio of specific heats	1.11	1.11
Solubility in water, weight percent,		
at 77 F (25 C) and 1 atm	slight	slight
Density of liquid		
at 70 F (21.1 C)	9.35 lb/gal	1120.38 kg/m^3
Density of saturated vapor		
at 70 F (21.1 C)	0.850 lb/ft^3	13.62 kg/m^3
Critical volume	0.0368 ft^3/lb	0.00230 m^3/kg
Specific heat of liquid		
at 86 F (30 C)	0.286 Btu/(lb)(F)	1.20 kJ/(kg)(C)

PHYSICAL CONSTANTS

(152a, Difluoroethane)

	U.S. Units	Metric Units
International symbol	CHF_2CH_3	CHF_2CH_3
Molecular weight	66.05	66.05
Vapor pressure		
at 0 F (-17.8 C)	19.80 psia	136.52 kPa, abs.
at 70 F (21.1 C)	70.20 psia	484.01 kPa, abs.
at 105 F (40.6 C)	134.30 psia	928.03 kPa, abs.
at 115 F (46.1 C)	155.30 psia	1070.76 kPa, abs.
at 130 F (54.4 C)	191.50 psia	1320.35 kPa, abs.
Density of the gas		
at 70 F (21.1 C) and 1 atm	0.177 lb/ft³	2.84 kg/m³
Specific gravity of the gas		
at 70 F and 1 atm (air = 1)	2.36	2.36
Specific volume of the gas		
at 70 F (21.1 C) and 1 atm	5.65 ft³/lb	0.353 m³/kg
Density of liquid		
at 0 F (-17.8 C)	62.20 lb/ft³	996.35 kg/m³
at 70 F (21.1 C)	56.75 lb/ft³	909.05 kg/m³
at 105 F (40.6 C)	53.54 lb/ft³	857.63 kg/m³
at 115 F (46.1 C)	52.54 lb/ft³	841.61 kg/m³
at 130 F (54.4 C)	50.95 lb/ft³	816.14 kg/m³
Boiling point at 1 atm	-13.0 F	-25.00 C
Freezing point at 1 atm	-178.6 F	-117.00 C
Critical temperature	236.3 F	113.50 C
Critical pressure	652 psia	4495.38 kPa, abs.
Critical density	22.79 lb/ft³	365.06 kg/m³
Latent heat of vaporization		
at boiling point	136.96 Btu/lb	318.57 kJ/kg
Specific heat of gas		
at 86 F (30 C) and 1 atm		
C_p	0.255 Btu/(lb)(F)	1.07 kJ/(kg)(C)
C_v	0.225 Btu/(lb)(F)	0.942 kJ/(kg)(C)
Ratio of specific heats	1.133	1.133
Solubility in water, weight percent,		
at 77 F (25 C) and 1 atm	slight	slight
Density of liquid		
at 70 F (21.1 C)	7.59 lb/gal	909.48 kg/m³
Density of saturated vapor		
at 70 F (21.1 C)	1.014 lb/ft³	16.24 kg/m³
Critical volume	0.0439 ft³/lb	0.00274 m³/kg
Specific heat of liquid		
at 86 F (30 C)	0.418 Btu/(lb)(F)	1.75 kJ/(kg)(C)

PHYSICAL CONSTANTS

(500, (12) Dichlorodifluoromethane/(152a) Difluoroethane)

	U.S. Units	Metric Units
International symbol	CCl_2F_2/CHF_2CH_3 (73.8 : 26.2)	
Molecular weight	99.31	99.31
Vapor pressure		
at 0 F (-17.8 C)	27.98 psia	192.92 kPa, abs.
at 70 F (21.1 C)	101.70 psia	701.20 kPa, abs.
at 105 F (40.6 C)	167.10 psia	1152.11 kPa, abs.
at 115 F (46.1 C)	191.20 psia	1318.28 kPa, abs.
at 130 F (54.4 C)	231.90 psia	1598.89 kPa, abs.
Density of the gas		
at 70 F (21.1 C) and 1 atm	0.262 lb/ft^3	4.20 kg/m^3
Specific gravity of the gas		
at 70 F and 1 atm (air = 1)	3.50	3.50
Specific volume of the gas		
at 70 F (21.1 C) and 1 atm	3.82 ft^3/lb	0.238 m^3/kg
Density of liquid		
at 0 F (-17.8 C)	80.46 lb/ft^3	1288.84 kg/m^3
at 70 F (21.1 C)	72.98 lb/ft^3	1169.03 kg/m^3
at 105 F (40.6 C)	68.63 lb/ft^3	1099.35 kg/m^3
at 115 F (46.1 C)	67.26 lb/ft^3	1077.40 kg/m^3
at 130 F (54.4 C)	65.08 lb/ft^3	1042.48 kg/m^3
Boiling point at 1 atm	-28.3 F	-33.50
Freezing point at 1 atm	-254 F	-158.89 C
Critical temperature	221.9 F	105.50 C
Critical pressure	641.9 psia	4425.74 kPa, abs.
Critical density	31.0 lb/ft^3	496.57 kg/m^3
Latent heat of vaporization		
at boiling point	86.47 Btu/lb	201.13 kJ/kg
Specific heat of gas		
at 86 F (30 C) and 1 atm		
C_p	0.176 Btu/(lb)(F)	0.737 kJ/(kg)(C)
C_v	0.154 Btu/(lb)(F)	0.645 kJ/(kg)(C)
Ratio of specific heats	1.14	1.14
Solubility in water, weight percent,		
at 77 F (25 C) and 1 atm	slight	slight
Density of liquid		
at 70 F (21.1 C)	9.76 lb/gal	1169.50 kg/m^3
Density of saturated vapor		
at 70 F (21.1 C)	2.059 lb/ft^3	32.98 kg/m^3
Critical volume	0.0323 ft^3/lb	0.00202 m^3/kg
Specific heat of liquid		
at 86 F (30 C)	0.290 Btu/(lb)(F)	1.214 kJ/(kg)(C)

PHYSICAL CONSTANTS

(502, (22)Chlorodifluoromethane/(115)Chloropentafluoroethane)

	U.S. Units	Metric Units
International symbol	$CHClF_2/CClF_2CF_3$(48.8:51.2)	
Molecular weight	111.63	111.63
Vapor pressure		
at 0 F (-17.8 C)	45.90 psia	316.47 kPa, abs.
at 70 F (21.1 C)	151.30 psia	1043.18 kPa, abs.
at 105 F (40.6 C)	244.40 psia	1685.08 kPa, abs.
at 115 F (46.1 C)	277.30 psia	1911.92 kPa, abs.
at 130 F (54.4 C)	332.70 psia	2293.89 kPa, abs.
Density of the gas		
at 70 F (21.1 C) and 1 atm	0.290 lb/ft³	4.65 kg/m³
Specific gravity of the gas		
at 70 F and 1 atm (air = 1)	3.87	3.87
Specific volume of the gas		
at 70 F (21.1 C) and 1 atm	3.45 ft³/lb	0.215 m³/kg
Density of liquid		
at 0 F (-17.8 C)	86.68 lb/ft³	1388.48 kg/m³
at 70 F (21.1 C)	77.06 lb/ft³	1234.38 kg/m³
at 105 F (40.6 C)	71.02 lb/ft³	1137.63 kg/m³
at 115 F (46.1 C)	69.02 lb/ft³	1105.59 kg/m³
at 130 F (54.4 C)	65.66 lb/ft³	1051.77 kg/m³
Boiling point at 1 atm	-49.8 F	-45.44 C
Critical temperature	179.9 F	82.17 C
Critical pressure	591.0 psia	4074.80 kPa, abs.
Critical density	34.97 lb/ft³	560.17 kg/m³
Latent heat of vaporization		
at boiling point	74.18 Btu/lb	172.54 kJ/kg
Specific heat of gas		
at 86 F (30 C) and 1 atm		
C_p	0.168 Btu/(lb)(F)	0.703 kJ/(kg)(C)
C_v	0.147 Btu/(lb)(F)	0.615 kJ/(kg)(C)
Ratio of specific heats	1.14	1.14
Solubility in water, weight percent,		
at 77 F (25 C) and 1 atm	slight	slight
Density of liquid		
at 70 F (21.1 C)	10.30 lb/gal	1234.21 kg/m³
Density of saturated vapor		
at 70 F (21.1 C)	3.736 lb/ft³	59.845 kg/m³
Critical volume	0.0286 ft³/lb	0.00179 m³/kg
Specific heat of liquid		
at 86 F (30 C)	0.295 Btu/(lb)(F)	1.235 kJ/(kg)(C)

PHYSICAL CONSTANTS

(503, (23)Trifluoromethane/(13)Chlorotrifluoromethane)

	U.S. Units	Metric Units
International symbol	$CHF_3/CClF_3(40.1:59.9)$	
Molecular weight	87.5	87.5
Vapor pressure		
at 0 F (−17.8 C)	245.30 psia	1691.28 kPa, abs.
Density of the gas		
at 70 F (21.1 C) and 1 atm	0.227 lb/ft³	3.636 kg/m³
Specific gravity of the gas		
at 70 F and 1 atm (air = 1)	3.03	3.03
Specific volume of the gas		
at 70 F (21.1 C) and 1 atm	4.40 ft³/lb	0.275 m³/kg
Density of liquid		
at 0 F (−17.8 C)	72.33 lb/ft³	1158.62 kg/m³
Boiling point at 1 atm	−126.1 F	−87.83 C
Critical temperature	67.1 F	19.50 C
Critical pressure	632.2 psia	4358.87 kPa, abs.
Critical density	30.67 lb/ft³	491.29 kg/m³
Latent heat of vaporization		
at boiling point	77.12 Btu/lb	179.38 kJ/kg
Specific heat of gas		
at −30 F (−34.4 C) and 1 atm		
C_p	0.43 Btu/(lb)(F)	1.800 kJ/(kg)(C)
C_v	0.36 Btu/(lb)(F)	1.507 kJ/(kg)(C)
Ratio of specific heats	1.19	1.19
Solubility in water, weight percent,		
at 77 F (25 C) and 1 atm	slight	slight
Density of liquid		
at 60 F (15.5 C)	6.65 lb/gal	796.85 kg/m³
Critical volume	0.0326 ft³/lb	0.00204 m³/kg
Specific heat of liquid		
at −30 F (−34.4 C)	0.383 Btu/(lb)(F)	1.604 kJ/(kg)(C)

PHYSICAL CONSTANTS

(504, (32)Difluoromethane/(115)Chloropentafluoroethane)

	U.S. Units	Metric Units
International symbol	$CH_2F_2/CClF_2CF_3$ (48.2 : 51.8)	
Molecular weight	79.2	79.2
Vapor pressure		
at 0 F (–17.8 C)	77.40 psia	533.65 kPa, abs.
at 70 F (21.1 C)	250.80 psia	1729.20 kPa, abs.
at 105 F (40.6 C)	405.30 psia	2794.44 kPa, abs.
at 115 F (46.1 C)	459.10 psia	3165.38 kPa, abs.
at 130 F (54.4 C)	551.00 psia	3799.01 kPa, abs.
Density of the gas		
at 70 F (21.1 C) and 1 atm	0.207 lb/ft^3	3.316 kg/m^3
Specific gravity of the gas		
at 70 F and 1 atm (air = 1)	2.76	2.76
Specific volume of the gas		
at 70 F (21.1 C) and 1 atm	4.83 ft^3/lb	0.302 m^3/kg
Density of liquid		
at 0 F (–17.8 C)	78.0 lb/ft^3	1249.44 kg/m^3
at 70 F (21.1 C)	67.5 lb/ft^3	1081.25 kg/m^3
at 105 F (40.6 C)	60.3 lb/ft^3	965.91 kg/m^3
at 115 F (46.1 C)	57.6 lb/ft^3	922.66 kg/m^3
at 130 F (54.4 C)	52.7 lb/ft^3	844.17 kg/m^3
Boiling point at 1 atm	–71.0 F	–57.22 C
Critical temperature	151.5 F	66.39 C
Critical pressure	609.5 psia	4202.35 kPa, abs.
Critical density	30.86 lb/ft^3	494.33 kg/m^3
Latent heat of vaporization		
at boiling point	101.63 Btu/lb	236.39 kJ/kg
Specific heat of gas		
at 86 F (30 C) and 1 atm		
C_p	0.185 Btu/(lb)(F)	0.775 kJ/(kg)(C)
C_v	0.156 Btu/(lb)(F)	0.653 kJ/(kg)(C)
Ratio of specific heats	1.19	1.19
Solubility in water, vol/vol		
at 77 F (25 C)	slight	slight
Density of liquid		
at 70 F (21.1 C)	9.02 lb/gal	1080.83 kg/m^3
Density of saturated vapor		
at 70 F (21.1 C)	4.791 lb/ft^3	76.74 kg/m^3
Critical volume	0.0324 ft^3/lb	0.0020 m^3/kg
Specific heat of liquid		
at 86 F (30 C)	0.400 Btu/(lb)(F)	1.675 kJ/(kg)(C)

PHYSICAL CONSTANTS

(505, (12)Dichlorodifluoromethane/(31)Chlorofluoromethane)

	U.S. Units	Metric Units
International symbol	CCl_2F_2/CH_2ClF (78:22)	
Molecular weight	103.5 (average)	103.5 (average)
Vapor pressure		
at 0 F (-17.8 C)	23.67 psia	163.20 kPa, abs.
at 70 F (21.1 C)	85.70 psia	590.88 kPa, abs.
at 105 F (40.6 C)	143.40 psia	988.71 kPa, abs.
at 115 F (46.1 C)	164.20 psia	1132.12 kPa, abs.
at 130 F (54.4 C)	199.40 psia	1374.81 kPa, abs.
Density of the gas		
at 70 F (21.1 C) and 1 atm	0.273 lb/ft^3	4.373 kg/m^3
Specific gravity of the gas		
at 70 F and 1 atm (air = 1)	3.64	3.64
Specific volume of the gas		
at 70 F (21.1 C) and 1 atm	3.66 ft^3/lb	0.228 m^3/kg
Density of liquid		
at 0 F (-17.8 C)	88.05 lb/ft^3	1410.43 kg/m^3
at 70 F (21.1 C)	80.52 lb/ft^3	1289.81 kg/m^3
at 105 F (40.6 C)	76.28 lb/ft^3	1221.88 kg/m^3
at 115 F (46.1 C)	74.97 lb/ft^3	1200.90 kg/m^3
at 130 F (54.4 C)	72.92 lb/ft^3	1168.07 kg/m^3
Boiling point at 1 atm	-21.3 F	-29.6 C
Critical temperature	243.97 F	117.76 C
Critical pressure	685.59 psia	4726.98 kPa, abs.
Critical density	33.51 lb/ft^3	536.78 kg/m^3
Critical volume	0.0298 ft^3/lb	0.00186 m^3/kg
Latent heat of vaporization		
at boiling point	84.40 Btu/lb	196.31 kJ/kg
Specific heat of gas		
at 86 F (30 C) and 1 atm		
C_p	0.152 Btu/(lb)(F)	0.636 kJ/(kg)(C)
C_v	0.131 Btu/(lb)(F)	0.548 kJ/(kg)(C)
Ratio of specific heats	1.16	1.16
Specific heat of liquid		
at 86 F (30 C)	0.250 Btu/(lb)(F)	1.0467 kJ/(kg)(C)
Solubility in water, weight percent,		
at 77 F (25 C) and 1 atm	slight	slight
Weight of liquid		
at 70 F (21.1 C)	10.77 lb/gal	1290.53 kg/m^3
Density of saturated vapor		
at 70 F (21.1 C)	1.790 lb/ft^3	28.67 kg/m^3

PHYSICAL CONSTANTS

(506, (31)Chlorofluoromethane/(114)Dichlorotetrafluoroethane)

	U.S. Units	Metric Units
International symbol	$CH_2ClF/C_2Cl_2F_4$ (55.1 : 44.9)	
Molecular weight	93.69 (average)	93.69 (average)
Vapor pressure		
at 0 F (−17.8 C)	11.70 psia	80.67 kPa, abs.
at 70 F (21.1 C)	49.50 psia	341.29 kPa, abs.
at 105 F (40.6 C)	143.40 psia	988.71 kPa, abs.
at 115 F (46.1 C)	164.20 psia	1132.12 kPa, abs.
at 130 F (54.4 C)	199.40 psia	1374.81 kPa, abs.
Density of the gas		
at 70 F (21.1 C) and 1 atm	0.248 lb/ft^3	3.973 kg/m^3
Specific gravity of the gas		
at 70 F and 1 atm (air = 1)	3.31	3.31
Specific volume of the gas		
at 70 F (21.1 C) and 1 atm	4.03 ft^3/lb	0.252 m^3/kg
Density of liquid		
at 0 F (−17.8 C)	86.38 lb/ft^3	1383.67 kg/m^3
at 70 F (21.1 C)	80.01 lb/ft^3	1281.64 kg/m^3
at 105 F (40.6 C)	76.50 lb/ft^3	1225.41 kg/m^3
at 115 F (46.1 C)	75.44 lb/ft^3	1208.43 kg/m^3
at 130 F (54.4 C)	73.80 lb/ft^3	1182.16 kg/m^3
Boiling point at 1 atm	9.60 F	12.44 C
Critical temperature	287.7 F	142.06 C
Critical pressure	749.4 psia	5166.93 kPa, abs.
Critical density	33.7 lb/ft^3	539.82 kg/m^3
Critical volume	0.0297 ft^3/lb	0.00185 m^3/kg
Latent heat of vaporization		
at boiling point	102.96 Btu/lb	239.48 kJ/kg
Specific heat of gas		
at 86 F (30 C) and 1 atm		
C_p	0.169 Btu/(lb)(F)	0.708 kJ/(kg)(C)
C_v	0.145 Btu/(lb)(F)	0.607 kJ/(kg)(C)
Ratio of specific heats	1.166	1.166
Specific heat of liquid		
at 86 F (30 C)	0.270 Btu/(lb)(F)	1.130 kJ/(kg)(C)
Solubility in water, weight percent,		
at 77 F (25 C) and 1 atm	slight	slight
Weight of liquid		
at 70 F (21.1 C)	10.70 lb/gal	1282.14 kg/m^3
Density of saturated vapor		
at 70 F (21.1 C)	0.8879 lb/ft^3	14.22 kg/m^3

DESCRIPTION

Strictly speaking, fluorocarbon compounds contain only the elements carbon, fluorine and sometimes hydrogen. However, in industrial applications such as refrigerants and aerosol propellents, the term fluorocarbon is used to include compounds containing chlorine and/or bromine atoms as well. The industrial products have somewhat similar chemical and physical properties. Their relatively inert character and wide range of vapor pressures and boiling points make then especially well suited for refrigerants and aerosol propellents in a variety of applications.

Some of the fluorocarbon compounds are listed by name in the regulations of the Department of Transportation and others are shipped under DOT exemption. The prefix "mono" is used in DOT listings to show the presence of one atom of the indicated element but is omitted in modern terminology.

The fluorocarbons are relatively inert, in general are nonflammable (in all concentrations in air under ordinary conditions) and are low in toxicity. Shipped as liquefied compressed gases under their own vapor pressures, they are colorless as liquids and freeze to white solids. The fluorocarbons are odorless in concentrations of less than 20 percent by volume in air but some have a faint and ethereal odor in higher concentrations. Chemically, they are analogs of hydrocarbons in which all or nearly all of the hydrogen has been replaced by fluorine and/or bromine or chlorine. The presence of fluorine atoms in the molecule accounts for their pronounced stability. They are more dense than corresponding hydrocarbons, have lower refractive indices, lower solubilities and lower surface tension. The viscosity is comparable to that of hydrocarbons. They also have relatively high dielectric strength.

The fluorocarbons are unusually stable for organic compounds. Resistance toward thermal decomposition, in general, is high but varies with each product. When decomposition does occur, toxic products may be formed, especially in intimate contact with flames. The toxic products are very irritating and ususally give adequate warning of their presence even in very low concentrations in air. Resistance toward hydrolysis is also high especially in neutral or acid solutions. However, some products containing hydrogen, such as fluorocarbon 22, are quite susceptible to alkaline hydrolysis.

GRADES AVAILABLE

Fluorocarbons are available for commercial and industrial use in various grades with the same composition from one producer to another.

USES

The fluorocarbons are most widely used as refrigerants, as blowing agents in the manufacture of polymerized foams used in insulation and comfort cushioning, and as polymer intermediates in the chemical industry. They are also used as aerosol propellants for products applied in foam or spray form, solvents, cleaning liquids and fire extinguishing agents. Other applications include dielectric fluid, the vapor in wind-tunnel experiments in aerodynamics, power-transmitting fluids and food freezing agents.

Special mixtures of two or more fluorocarbons, or fluorocarbons and hydrocarbons, are often used to provide desired special properties in particular refrigeration or aerosol propellent applications.

PHYSIOLOGICAL EFFECTS

In general, the fluorocarbons have low levels of toxicity. This is illustrated by the following toxicity group classifications of the Underwriters' Laboratory:

ASHRAE Designation of the Gas	Underwriters' Laboratory Toxicity Group Classification
12	6
13	6
13B1	6
22	5a
115	6

ASHRAE Designation of the Gas	Underwriters' Laboratory Toxicity Group Classification
142b	5a
152a	6
500	5a
502	5a
503	6
504	6

The Underwriters' Laboratory classification is for acute exposures, and lists chemicals in a series of six groups, with Group 6 being the least toxic (20 percent concentration for 2 hours showed no injury), while Group 4 compounds produce serious or lethal effects at 2–2.5 percent during expsosures of more than 1 hour and within about 2 hours. Group 5a is defined as much less toxic than Group 4, but more toxic than Group 6.

A threshold limit value (TLV) represents the concentration under which it is believed that nearly all workers may be repeatedly exposed day after day without adverse effect, i.e., for 8 hours/day, 40 hours/week. Apart from carbon dioxide (threshold limit value 5000 ppm) and simple asphyxiants such as nitrogen, the highest threshold limit value (least toxic) is 1000 ppm. A threshold limit value of 1000 ppm has been assigned for some fluorocarbons, and comparative data suggest similar values are reasonable for the other fluorocarbons discussed in this section. The threshold limit values as established by the American Conference of Governmental Industrial Hygienists (ACGIH), have recently been adopted by the Occupational Safety and Health Administration (OSHA) as industrial standards.

Due to their low boiling points, the fluorocarbons evaporate very quickly at ambient temperature, minimizing dermal, eye and ingestion toxicity. The rapid evaporation and resultant chilling can cause tissue freezing and frostbite when contact is made with the boiling liquid. Should liquid splash into the eyes, wash them thoroughly with water for at least 15 minutes and call an eye specialist at once.

The critical mode of entry into the body is by inhalation. These products generally show inhalation effects similar to anesthetics, causing central nervous system (CNS) depression, and some increase in activity with an initial feeling of intoxication and euphoria (psychological effects). Human experiments have shown that the onset of psychomotor effects becomes statistically detectable with fluorocarbon 12 at 1 percent (10,000 ppm) in a 2.5 hour exposure, i.e., at ten times the threshold limit value. Under conditions of progressively greater exposure, there occurs loss of coordination, loss of consciousness and eventually death. No adverse effects have been observed during or after long exposures at the threshold limit value.

SAFE HANDLING AND STORAGE

Materials Suitable for Containers and Storage

The fluorocarbons are generally compatible with most of the common metals except at high temperatures. At elevated temperatures, the following metals resist fluorocarbon corrosion (and are named in decreasing order of their corrosive resistance): "Inconel," stainless steel, nickel, steel and bronze. Water or water vapor in fluorocarbon systems will corrode magnesium alloys or aluminum containing over 2 percent magnesium. These metals are not recommended for use with fluorocarbon systems in which water may be present.

Systems using fluorocarbons as refrigerants should be dry in order to prevent the possibility of malfunctioning of components such as regulating valves, bellows, diaphragms, hermetically sealed coils and so forth. Most fluorocarbon compounds can be used with elastomeric and plastic packing and gasketing materials, but there are some exceptions. Consult manufacturers for specific information or refer to ASHRAE Handbook and Product Directory, *Fundamentals* (1977) volume.

Storage and Handling in Normal Use

All the precautions necessary for the safe handling of any nonflammable or flammable gas must be observed with the fluorocarbons (see also Part I, Chapter 4, Section A).

Handling Leaks and Emergencies

Large liquid leaks in fluorocarbon systems may be detected visually. As the material escapes, moisture in the air surrounding the leak condenses and then freezes around the leak due to the refrigerating effect of the vaporizing fluorocarbons. The frost thus formed is readily apparent. Smaller leaks may be located through the use of:

(1) A solution of liquid detergent in water, applied directly to the area being tested. The formation of bubbles indicates a leak.

(2) A halide torch which is equipped with a hose through which air is sucked to the flame. The flame will change color when it comes in contact with a leak.

(3) Electronic leak detectors, capable of sensitivities far greater than the other methods—often in terms of fractions of an ounce of fluorocarbon per year. When the probe of the instrument is placed near a leak, positive identification of the leak is indicated by a flashing light, meter deflection or by audible means.

It should be noted that the vapors of these fluorocarbons are all much heavier than air and in the absence of good ventilation will tend to collect in low areas.

The vapors will undergo decomposition when drawn through a flame or if in contact with very hot surfaces (see the "Description" section). The products of decomposition include hydrogen fluoride and hydrogen chloride, and perhaps small quantities of carbonyl compounds such as phosgene. The halogen acids are both toxic and intensely irritating to the nose and throat. The irritating action of these decomposition products is readily noticeable before hazardous levels are reached. If such a situation develops, the affected areas should be vacated, the heat source and leak eliminated and the area well ventilated before resuming work.

Monitoring Concentrations in Air

Instruments and analytical methods are available for monitoring the concentration of fluorocarbon gases in air. Consult your supplier for specific information.

First Aid

Accidental exposures to concentrations higher than the TLV should be treated by prompt removal to fresh air. Severe exposures requiring medical attention should not be treated with stimulants or adrenalin, since high concentrations of these fluorocarbons may result in a sensitization of the heart to adrenalin (a relatively common effect of many volatile organic compounds).

Liquid fluorocarbons in contact with the skin can cause severe freezing or frostbite because of their low temperatures. Affected parts should be warmed gradually using body heat or warm (not hot) water. Should liquid splash into the eyes, wash them thoroughly with water for at least 15 minutes and call an eye specialist at once.

METHODS OF SHIPMENT; REGULATIONS

Fluorocarbons are authorized for shipment by rail, highway, water and air in interstate commerce in the *Code of Federal Regulations (Title 49, Parts 100–199,* in the regulations in effect as of January 1, 1977). Due to frequency of change in these regulations, it is recommended that current editions of them be consulted for information at the time needed. Tables 1 and 2 in the following section on "Containers" list the types of containers and maximum filling densities for fluorocarbons that are authorized in these regulations as of January 1977.

CONTAINERS

Fluorocarbons are stored and shipped as liquefied compressed gases under their own vapor pressures in cylinders, portable tanks, truck cargo tanks, TMU (ton multi-unit) tanks and single-unit tank cars as shown in Table 1.

The commodity names are given in Table 1 as they appear in current regulations; where applicable, nomenclature used in current industry practice is shown parenthetically. Pre-

TABLE 1. AUTHORIZED SHIPPING CONTAINERS FOR FLUOROCARBONS.

Gas Number and DOT Shipping Name	Cylinders	Railcars (TMU, Single-Unit)	Motor Vehicle Tanks (Portable, Cargo)
12 Dichlorodifluoro- methane	3A225, 3AA225, 3B225, 4A225, 4B225, 4BA225, 4BW225, 4B240ET, 4E225, 9, 39, 41, 3E1800	106A500-X 110A500-W 112A340-W 114A340-W 105A300-W	DOT-51 MC-330, MC-331
13 Monochlorotrifluoro- methane	3A1800, 3AA1800, 3, 3E1800, 39	None specified	None specified
13B1 Monobromotrifluoro- methane (bromotri- fluoromethane)	4DA500, 3HT900, 4DS500, 3E1800, 3A, 3AA, 4B, 4BA, 4BW, 4D, 39	110A800-W	None specified
22 Monochlorodifluoro- methane (chlorodi- fluoromethane)	3A240, 3AA240, 3B240, 4B240, 4BA240, 4BW240, 4B240ET, 4E240, 39, 41, 3E1800	106A500-X 110A500-W 105A300-W	DOT-51 MC-330, MC-331
115 Monochloropenta- fluoroethane (chloro- pentafluoroethane)	3A225, 3AA225, 3B225, 4A225, 4B225, 4BA225, 4BW225, 39, 3E1800	None specified as of 1/3/77	None specified as of 1/3/77
142b Difluoromonochloro- ethane: 1,1 difluoro 1-chloroethane (chlorodifluoro ethane: 1-chloro 1,1-difluoroethane)	3A150, 3AA150, 3B150, 4B150, 4BA225, 4BW225, 39, 3E1800	106A500-X 110A500-W 105A100-W	MC-330, MC-331
152a Difluoroethane	3A150, 3AA150, 3B150, 4B150, 4BA225, 4BW225, 3E1800	106A500-X 110A500-W 105A300-W	MC-330, MC-331,
500 Dichlorodifluoro- methane and difluoro- ethane mixture: con- stant boiling mixture (dichlorodifluoro- methane and 1,1 difluoroethane mixture)	3A240, 3AA240, 3B240, 4A240, 4B240, 4BA240, 4BW240, 9, 39, 3E1800	See railcars authorized for R-12	MC-330, MC-331

Note. Other fluorocarbons are shipped under exemptions granted by the DOT. This table is based on DOT regulations at the time the *Handbook* went to press; consult the regulations for current information.

TABLE 2. FLUOROCARBONS—MAXIMUM FILLING DENSITIES PERMITTED FOR AUTHORIZED SHIPPING CONTAINERS (PERCENT WATER CAPACITY BY WEIGHT).

Gas Number	Cylinders Filling Density (percent)	Railcar (TMU, Single-Unit)		Motor Vehicle Tanks (Portable, Cargo)	
		Specification	Filling Density (percent)	Filling Density (percent)	Minimum Design Pressure (psig)
12 (Note 1)	119	106A500-X (Note 2)	119	119	150
		110A500-W (Note 2)	119		
		105A300-W	125		
		112A340-W	123		
		114A340-W	123		
13	100	None specified		–	–
13B1	not liquid full at 130 F	110A800-W (Note 2)	124	–	–
22 (Note 1)	105	106A500-X (Note 2)	105	105	250
		110A500-W (Note 2)	105		
		105A300-W	110		
115	110	–	–	–	–
142b (Note 1)	100	106A500-X (Note 2)	100	100	100
		110A500-W (Note 2)			
		105A100-W (Note 2)			
152a	79	106A500-X (Note 2)	79	79	150
		110A500-W (Note 2)	79		
		105A300-W	84		
500 (Note 1)	not liquid full at 130 F	112A340-W (Note 4)		(Note 5)	250
		114A340-W (Note 4)			
		106A500-X (Note 2)			
		110A500-W (Note 2)			
		105A300-W			

Note 1. This gas may be transported in authorized tank cargo tanks or portable tanks marked "Disperant Gas" or "Refrigerant Gas."

Note 2. Specification 106A or 110A tanks authorized only for transportation by rail freight and by highway.

Note 3. For tank cars other than DOT 106 class used for the transportation of liquefied flammable gases, interior pipes of loading and unloading valves must be equipped with excess flow-check valves of approved design.

Note 4. The gas pressure at 105 F (40.6 C) in any insulated tank car tank of the DOT-105A and 109A-W class or specification DOT-111A100-W-4; at 115 F (46.1 C) in any uninsulated tank car tank of the DOT-112A-W and 114A-W class; or at 130 F (54.4 C) in any uninsulated tank car tank of the DOT-106A and 110A-W class must not exceed $\frac{3}{4}$ times the prescribed retest pressure of the tank. The gas pressure at 130 F in any uninsulated tank car tank of the DOT-107A series must not exceed $\frac{7}{10}$ of the marked test pressure of the tank.

Note 5. The liquid portion of the gas must not fill the tank at 105 F if the tank is insulated nor at 115 F if uninsulated. The vapor pressure (psig) at 115 F must not exceed the design pressure of the portable tank or cargo tank container.

Note 6. When transporting flammable gases in tank cars of the 112A/114A classes, the car must be equipped with head shield of a design approved by the DOT, *or* shipping papers must bear the following notation: DOT 112A (or 114A): Must be handled in accordance with F.R.A. E.O. #5. *All* tank cars of the DOT 112A/114A classes must have been fitted with approved head shields by December 31, 1977.

ceding each commodity name is the ASHRAE developed number designating that gas.

Filling Limits. Table 2 lists maximum permitted filling densities (as given in current regulations) for each of the fluorocarbons. Filling density is expressed as percentage of the water capacity of the container, in pounds. For these calculations, the weight of water at 60 F (15.6 C) in air is 8.32828 lb/gal (997.94781 kg/m^3), 1 gal (0.0038 m^3) occupying 231 in^3 or 0.1336 ft^3.

Valve Outlet Connections. Standard connection, United States and Canada—No. 668. As an alternate, standards 660, 165 or 182 may be used on large cylinders. For small cylinders, standard connection for United States and Canada—No. 180. As an alternate, standard 110 and 170 may be used.

Bulk Storage and Handling. Shipments of fluorocarbons in single-unit tank cars and cargo or portable tanks can range from approximately 30,000–200,000 lb (10,000–90,000 kg) net weight subject to weight limit regulations at local, regional and federal levels. Filling and unloading of bulk quantities of fluorocarbons is accomplished as outlined in Part I, Chapter 3 of this *Handbook*.

Fluorocarbon storage vessels are commonly fabricated of steel and are built according to the ASME Code for Unfired Pressure Vessels. All tanks and installations must also comply with the requirements of appropriate insurance and regulatory agencies having jurisdiction within the locality of the installations. Recommended design pressures (or working pressures) are based upon the material to be stored; these all should be at least as great as the design pressures of the bulk container in which the commodities are normally transported, or as required in appropriate codes or regulations, whichever is the greater.

Piping may be of copper, preferably type K, with silver-soldered fittings. ASTM schedule 40 steel pipe with welded joints may also be used. Screwed fittings, also widely used, exhibit a greater tendency to work loose, resulting in leaks, particularly in areas subject to vibration due to operation of machinery. Polytetrafluoroethylene luting and gaskets are often recommended for fluorocarbon service.

A recommended safe practice is to install hydrostatic safety relief valves in all pipelines wherever the possibility exists that liquid fluorocarbons may be trapped (such as between two valves) to prevent piping rupture through thermal expansion of the fluorocarbon.

Installations for the storage and handling of fluorocarbons should be designed, fabricated and installed in consultation with experts who are thoroughly familiar with the fluorocarbons and their handling.

METHOD OF MANUFACTURE

In general, fluorocarbons are produced commercially by the reaction of hydrofluoric acid with chlorocarbons, or by the disproportionation of other fluorocarbons. Fluorocarbon 12, for example, is made by the reaction of carbon tetrachloride and hydrofluoric acid in the presence of antimony chloride as a catalyst. Fluorocarbon 13 is produced by the disproportionation of fluorocarbon 12 in the vapor phase in the presence of aluminum chloride or bromide. Fluorocarbon 22 is made by treating chloroform with hydrofluoric acid in the presence of a small amount of antimony chloride at elevated temperatures and pressures. Fluorocarbon 152a is a component of fluorocarbon 500 and is manufactured by the addition of hydrofluoric acid to acetylene with boron trifluoride as a catalyst.

REFERENCES

1. Air Conditioning and Refrigeration Institute, *Properties of Commonly Used Refrigerants* (1967 ed.), The Institute, 1815 N. Fort Myer Drive, Arlington, VA 22209.
2. American Society of Heating, Refrigerating and Air Conditioning Engineers (ASHRAE), *Thermodynamic Properties of Refrigerants* (1969 ed.), ASHRAE, 345 E. 47th Street, New York, NY 10017.
3. ASHRAE, *Thermophysical Properties of Refrigerants* (1969 ed.), ASHRAE.
4. ASHRAE, *ASHRAE Handbook and Product Directory* (4, vol.: *Fundamentals*, 1977; *Systems*, 1976; *Equipment*, 1975; *Applications*, 1974), ASHRAE.
5. DOT, *Code of Federal Regulations, Title 49, Parts 100–199*, U.S. Government Printing Office, Washington, DC.

Helium

He   DOT Classification: Nonflammable gas; nonflammable gas label

PHYSICAL CONSTANTS

	U.S. Units	Metric Units
International Symbol	He	He
Molecular weight	4.003	4.003
Density of the gas		
at 70 F (21.1 C) and 1 atm	0.01034 lb/ft^3	0.1656 kg/m^3
Specific gravity of the gas		
at 70 F and 1 atm (air = 1)	0.138	0.138
Specific volume of the gas		
at 70 F (21.1 C) and 1 atm	96.71 ft^3/lb	6.039 m^3/kg
Boiling point at 1 atm	−452.0 F	−268.9 C
Freezing point at 1 atm	none*	none*
Critical temperature	−450.2 F	−267.9 C
Critical pressure	33.0 psia	227 kPa, abs.
Critical density	4.347 lb/ft^3	69.64 kg/m^3
Latent heat of vaporization		
at boiling point	8.77 Btu/lb	20.4 kJ/kg
Specific heat of gas		
at 70 F (21.1 C) and 1 atm		
C_p	1.24 Btu/(lb) (F)	5.19 kJ/(kg) (C)
C_v	0.744 Btu/(lb) (F)	3.11 kJ/(kg) (C)
Ratio of specific heats	1.667	1.667
Solubility in water, vol/vol		
at 68 F (21 C)	0.0086	0.0086
Weight of Liquid		
at normal boiling point	1.043 lb/gal	125.0 kg/m^3
Density of gas		
at boiling point	1.06 lb/ft^3	17.0 kg/m^3

*Helium will not solidify at 1 atm. The approximate pressure required for solidification near absolute zero (which is −459.69 F (−273.16 C) is calculated to be 25 atm or 367.7 psia (2535 kPa, abs.) at −458 F (−272 C).

	U.S. Units	Metric Units
Density of liquid at boiling point	7.798 lb/ft^3	124.9 kg/m^3
Liquid/gas ratio (liquid at boiling point, gas at 70 F and 1 atm, vol/vol	1/754.2	1/754.2

DESCRIPTION

Helium is the second lightest element; only hydrogen is lighter. It is one-seventh as heavy as air. Helium is one of the rare gases of the atmosphere, in which it is present in a concentration of only 5 ppm. Natural gas containing up to 2 percent helium has been found in the American Southwest. Other helium-bearing natural gas fields have been discovered in Saskatchewan, Canada, and near the Black Sea.

Helium is chemically inert. It has no color, odor or taste. Liquid helium is extremely important in cyrogenic research since it is the only known substance to remain fluid at temperatures near absolute zero, and hence has a unique use as a refrigerant in cryogenics. It is also the only known nuclear reactor coolant that does not become radioactive. Helium is nonflammable and is only slightly soluble in water. It is shipped at high pressures—at or above 2200 psig at 70 F (15,000 kPa gage at 21.1 C) in cylinders and in bulk units. It is also shipped as a liquid cold compressed gas.

GRADES AVAILABLE

Table 1 presents the assay (purity-minimum) for each grade, plus the contaminant maxima in ppm (v/v). A blank indicates no specified limiting characteristic (see Ref. 1).

USES

Helium is used as an inert gas shield in arc welding; as a lifting gas for lighter-than-air aircraft; as a gaseous cooling medium in nuclear reactors; to provide a protective atmosphere for growing germanium and silicon crystals for transistors; to provide a protective atmosphere in the production of such reactive metals as titanium and zirconium; to fill cold-weather fluorescent lamps; to trace leaks in refrigeration and other closed systems; and to fill neutron and gas thermometers. It is used in cryogenic research, and, in mixtures with oxygen, it has medical applications. Radioactive mixtures of helium with krypton are available to users licensed by the Atomic Energy Commission.

PHYSIOLOGICAL EFFECTS

Helium is nontoxic, and its gas poses hazards only as a simple asphyxiant that can deprive the lungs of needed oxygen. Liquid helium and cold helium gas given off by the liquid can damage living tissues with injuries like high-temperature burns on contact.

SAFE HANDLING AND STORAGE

Materials Suitable for Containers and Storage

Because helium gas is inert, any commercially available metals may be used to contain it in installations designed to meet requirements for the pressures involved. Special materials, such as certain stainless steels, and insulation are required for equipment handling liquid helium.

Storage and Handling in Normal Use

Helium gas must be handled with all the precautions required for handling any nonflammable, nontoxic compressed gas.

High-pressure tube trailers, tube banks or cylinder banks often serve as the storage supply for gaseous helium.

For minimizing helium losses, the shipping

TABLE 1. HELIUM GRADES AVAILABLE.

Limiting Characteristics	GRADES (note 1)						
	A	B	C	D	E	F	G
Assay v/v	99%	99.8%	99.99%	99.995%	99.997%	99.999%	99.9999%
Water (note 2)	none condensed	none condensed	← see footnote →				Sum of all impurities less than 1 ppm v/v
Hydrocarbons (condensed)	none	none	none				
Hydrocarbons CH₄				5	1	1	
Oxygen			5	3	3	1	
Nitrogen			20	14	5	5	
Neon					23		
Hydrogen					1		
CO	10						

Note 1. Grades apply equally to Type I (gas) and Type II (liquid) helium since the liquid must be vaporized before it can be analyzed. Note that Type II helium when passed through a 10 micron nominal, 40 micron absolute filter at less than 5.2 K may be considered pure helium.

Note 2. The water content of helium required for any particular grade may vary with intended use. If a specific limit is required it should be specified as a limiting concentration expressed in ppm (v/v) or the dew point equivalent.

container for liquid helium is normally used for storage.

Special Handling Precautions for Liquid Helium. Users of liquid helium must also take special precautions in addition to those necessary for the safe handling of such inert liquefied gases as nitrogen and argon. Two properties of liquid helium make these special precautions imperative: its extreme cold solidifies all other gases; and its cold causes oxygen to condense on any uninsulated or inadequately insluated pipe through which it passes.

In consequence, liquid helium must not be allowed to come in contact with air, and must be equipped with pressure-relief devices that prevent back leakage of air in order to prevent backflow of air into liquid helium equipment. Air backflow and plugging by solidified air constitute a serious safety hazard.

Similarly, if air enters and plugs the vent of a helium container, a serious hazard is created; therefore, the vents of liquid helium containers must be tested on delivery and periodically checked to make sure their vents remain clear. The use of open-neck Dewar flasks for liquid helium also increases the possibility of neck-tube plugging from transfer or gaging of contents. Users of liquid helium should obtain full information on safe handling precautions and equipment from their suppliers (see also Part I, Chapter 4, Section A, and Part I, Chapter 9).

Handling Leaks and Emergencies

See Part I, Chapter 4, Section A.

METHODS OF SHIPMENT; REGULATIONS

Under the appropriate regulations and tariffs, helium is authorized for shipment as follows (in gaseous form, unless liquid is noted; liquid helium at pressures lower than 25 psig or 170 kPa gage is not classified by the DOT as a "hazardous article" and no DOT regulations apply to its shipment at such pressures):

By Rail: In cylinders (by freight or express) and in tank cars.

By Highway: In cylinders on trucks, and in tube trailers. Liquid helium, in special containers on trucks and in insulated tank trucks or tank trailers either at pressures below 25 psig (170 kPa gage) or by DOT exemption.

By Water: In cylinders on passenger and cargo vessels and on ferry vessels of all types; in authorized tank cars on cargo vessels and railroad car ferries (passenger or vehicle). In cylinders on barges of U.S. Coast Guard classes A, BA, BB, CA and CB.

By Air: In cylinders on passenger aircraft up to 150 lb (70 kg) and on cargo aircraft up to 300 lb (140 kg). In containers meeting special requirements, pressurized liquid helium on cargo aircraft only up to 300 lb (140 kg) and low-pressure liquid helium on cargo aircraft only up to 660 lb (300 kg). (Quantities are maximum net weight per container).

CONTAINERS

Helium gas is shipped in cylinders, tank cars, special tanks on trucks and tank semitrailers. Liquid helium is shipped in special insulated cylinders and tank trucks, as a commodity not regulated by the DOT (when kept at pressures below 25 psig or 170 kPa gage), or by DOT

exemption. Atmospheric helium gas is shipped in cylinders and in liter flasks of pyrex or soft glass.

Filling Limits: The maximum filling limits authorized for gaseous helium in shipping containers are as follows:

In cylinders and tube trailers—the authorized service pressure marked on the cylinders or tube assemblies at 70 F (21.1 C).

In cylinders and vehicle-mounted tubes that meet DOT specifications 3A, 3AA, 3AX or 3AAX and special requirements—up to 10 percent in excess of their marked service pressure.

In uninsulated rail tank cars of the DOT 107A type—the authorized service pressure marked on the tank at 70 F (21.1 C) or, at 130 F (54.4 C), 10 percent in excess of the marked service pressure.

Liquid Containers

A large part of the liquid helium used in the United States is shipped in special containers. One type, insulated by a liquid nitrogen shield between vacuum jackets, is usually made in 25-liter (6.6-gal) and 50-liter (13.2-gal) capacities. Containers of this type stand some 4 ft (1 m) high and are about 20–25 in. (50–63 cm) in diameter. Those of a second type, built in 50-liter (13.2-gal) and 100-liter (24.6-gal) capacities, use a high-efficiency insulation. These containers stand over 5 ft (2 m) high and are about 22 in. (55 cm) in diameter. Both types are vented through relief valves set at about 0.5 psig (3 kPa gage), and have helium evaporation rates of 1 to 2 percent capacity per day. Helium shipped at such low pressures is not regulated by the DOT.

Some use has also been made of specially insulated and equipped containers which meet DOT cylinder specifications 4L for shipping liquid helium (either at pressures below 25 psig, 170 kPa gage, or by DOT exemption).

Cylinders

Cylinders authorized by the DOT for the shipment of any nonliquefied compressed gas may be used to ship helium gas. (These are cylinders meeting DOT specifications 3A, 3AA, 3AX, 3AAX, 3B, 3C, 3D, 3E, 4, 4B, 4BA, 4BW, and 4C. In addition, cylinders meeting DOT specifications 3, 25, 26, 33 and 38 may be continued in helium service, but new construction is not authorized.)

All cylinders authorized for helium gas service must be requalified by hydrostatic retest every 5 or 10 years, except as follows: DOT-4 cylinders, every 10 years; and no retest is required for cylinders of types 3C, 3E and 4C.

Valve Outlet and Inlet Connections. Standard connections, United States and Canada—up to 3000 psi (20,684 kPa), No. 580; 3000–10,000 psi (20,684–68,948 kPa), No. 677 (see Part I, Chapter 7, Section A, Paragraph 2.1.7); cryogenic liquid withdrawal, No. 792. Alternate standard connection, United States and Canada—3000–10,000 psi, No. 677. Standard yoke connection, United States and Canada (for medical use of the gas)—No. 930.

Cargo Tanks

Bulk shipment of helium gas at high pressure is also made in tube trailers at pressures in excess of 2200 psig at 70 F (15,000 kPa gage at 21.1 C). Liquid helium is shipped at pressures above 25 psig (170 kPa gage) in motor vehicle tank semitrailers under DOT exemption.

Tank Cars

Tank cars meeting DOT specification 107A are authorized for the shipment of helium gas.

METHOD OF MANUFACTURE

The principal source of helium is natural gas, from which it is recovered in essentially a stripping operation involving liquefaction and purification. Helium may also be recovered from the atmosphere by fractionation.

REFERENCE

1. "Commodity Specification for Helium" (Pamphlet G-9.1), Compressed Gas Association Inc.

ADDITIONAL REFERENCE

"Standard Density Data, Atmospheric Gases and Hydrogen" (Pamphlet P-6), Compressed Gas Association, Inc.

Hydrogen

H$_2$ DOT Classification: Flammable gas; flammable gas label

PHYSICAL CONSTANTS

	U.S. Units	Metric Units
International symbol	H$_2$	H$_2$
Molecular weight	2.016	2.016
Density of the gas at 70 F (21.1 C) and 1 atm	0.005209 lb/ft^3	0.0834401 kg/m^3
Specific gravity of the gas at 32 F and 1 atm (air = 1)	0.06950	0.06950
Specific volume of the gas at 70 F (21.1 C) and 1 atm	192.0 ft^3/lb	11.99 m^3/kg
Boiling point at 1 atm	-423.0 F	-252.7 C
Melting point at 1 atm	-434.6 F	-259.2 C
Critical temperature	-399.91 F	-239.95 C
Critical pressure	190.8 psia	1316 kPa, abs.
Critical density	1.96 lb/ft^3	31.396 kg/m^3
Triple point	-434.56 F at 1.0414 psia	-259.2 C at 7.1802 kPa, abs.
Latent heat of vaporization at boiling point	192.7 Btu/lb	448.2 kJ/kg
Latent heat of fusion at triple point	25.08 Btu/lb	58.336 kJ/kg
Specific heat of gas at 70 F (21.1 C) and 1 atm		
C_p	3.416 Btu/(lb) (F)	14.30 kJ/(kg) (C)
C_v	2.430 Btu/(lb) (F)	10.17 kJ/(kg) (C)
Ratio of specific heats	1.41	1.41
Solubility in water, vol/vol at 60 F (15.6 C)	0.019/1	0.019/1
Weight of liquid at boiling point	0.5920 lb/gal	70.94 kg/m^3
Density of gas at boiling point and 1 atm	0.084 lb/ft^3	1.35 kg/m^3

	U.S. Units	Metric Units
Density of liquid at boiling point and 1 atm	4.28 lb/ft^3	68.6 kg/m^3
Liquid/gas ratio (liquid at boiling point, gas at 70 F and 1 atm), vol/vol	1/850.1	1/850.1
Heat of combustion		
gross	325 Btu/ft^3	12,100 kJ/m^3
net	275 Btu/ft^3	10,200 kJ/m^3

DESCRIPTION

Hydrogen is colorless, odorless, tasteless, flammable and nontoxic. It exists as a gas at atmospheric temperatures and pressures. It is the lightest gas known, being only some seven-hundredths as heavy as air. Hydrogen is present in the atmosphere, occurring in concentrations of only about 0.00005 percent by volume at lower altitudes.

Hydrogen burns in air with a pale blue, almost invisible flame. Its ignition temperature will not vary greatly from the range 1050–1074 F (565.5–578.9 C) in mixtures with either air or oxygen at atmospheric pressure. Its flammable limits in dry air at atmospheric pressure are 4.1–74.2 percent hydrogen by volume. In dry oxygen at atmospheric pressure, the flammable limits are 4.7–93.9 percent hydrogen by volume. Its flammable limits in air or oxygen vary somewhat with initial pressure, temperature and water vapor content.

When cooled to its boiling point of –423 F (–252.8 C), hydrogen becomes a transparent liquid only one-fourteenth as heavy as water. All gases except helium become solids at the temperature of liquid hydrogen. Because of its extremely low temperature, it can make ductile or pliable materials with which it comes in contact brittle and easily broken (an effect that must be considered whenever liquid hydrogen is handled). Liquid hydrogen has a relatively high thermal coefficient of expansion compared with other cryogenic liquids.

The hydrogen molecule exists in two forms: *ortho* and *para*, named according to their types of nuclear spin. (*Ortho*-hydrogen molecules have a parallel spin; *para*-hydrogen molecules, an anti-parallel spin.) There is no difference in the chemical properties of these forms, but there is a difference in physical properties. *Para*-hydrogen is the form preferred for rocket fuels. Hydrogen consists of about three parts *ortho* and one part *para* as a gas at room temperature. The equilibrium concentration of *para* increases with decreasing temperature until, as a liquid, the *para* concentration is nearly 100 percent. If hydrogen should be cooled and liquefied rapidly, the relative three-to-one concentration of *ortho* to *para* would not immediately change. Conversion to the *para* form takes place at a relatively slow rate and is accompanied by the release of heat. For each pound (or kilogram) of rapidly cooled liquid hydrogen that changes to the *para* form, enough heat is liberated to vaporize 1.5 lb (0.68 kg) of liquid hydrogen. However, if a catalyst is used in the liquefaction cycle, *para*-hydrogen can be produced directly without loss from self-generated heat.

Throttled expansion from high to low pressure at ordinary temperatures cools most common gases (such as oxygen, nitrogen and carbon dioxide). Hydrogen, though, is an exception, becoming heated to a slight extent under these conditions (increasing about 10 F (–12.2 C) in temperature, for example, when throttled from 2000 psig (13,790 kPa gage) to atmospheric pressure).

Hydrogen diffuses rapidly through porous materials and through some metals at red heat. It may leak out of a system which is gas-tight for air or common gases at equivalent pressures.

In its chemical properties, hydrogen is funda-

TABLE 1. HYDROGEN GRADES AVAILABLE (units in ppm (v/v) unless shown otherwise)

Limiting Characteristic	Type I (Gaseous) GRADES						Type II (Liquefied) GRADES	
	A	B	C	D	E	F	A	B
Min. % H_2 (v/v)	99.8	99.95	99.95	99.99	99.995	99.995*	99.995	99.9997*
Water	None Condensed	32.0	7.8	3.5	3.5	1.5		
Dew Point °F		-60	-80	-90	-90	-100		
Oxygen		10	10	5.0	5.0	1.0		
Argon							1.0	1.0
Nitrogen		400	400	25	20	2.0	9.0**	2.0
Total Hydrocarbons	10	10	10	5.0	1.0	0.5		
Helium							39	
Carbon Dioxide	10	10	10	0.5			1.0	
Carbon Monoxide	10	10	10	1.0				
Hg Vapor (ppb)			4					
Para Content Min. %							95	95
Permanent Particulate							Filtering req.***	

*May include up to 50 ppm neon + helium.
**Includes water.
***In order to reduce the amount of permanent particulates in liquefied hydrogen, the liquid may be filtered just prior to its entering the transport container. The equipment used is a 10-40 (10 microns nominal-40 microns absolute) filter constructed of a metal screen assembly which is installed in the transfer line.

mentally a reducing agent and is frequently applied as such in organic chemical technology.

GRADES AVAILABLE

Table 1 presents the component maxima in ppm (v/v), unless shown otherwise, for the types and grades of hydrogen. Gaseous hydrogen is denoted as Type I and liquefied hydrogen as Type II in the table. A blank indicates no maximum limiting characteristic (see Ref. 1).

USES

Large quantities of hydrogen are consumed in chemical syntheses, primarily of ammonia and methanol.

In the organic chemical field, hydrogenation of edible oils in soy beans, fish, cotton seed and corn yields solids used as shortenings and other foods. Various alcohols are produced by the hydrogenation of the corresponding acids or aldehydes.

Hydrogen is used with oxygen in oxyhydrogen welding and cutting, being employed largely in certain brazing operations, welding aluminum and magnesium (especially in thin sections), and welding lead. The oxyhydrogen flame has a temperature of about 4000 F (2204 C), and is well suited for such comparatively low-temperature welding and brazing. It is also used to some extent in cutting metals, particularly in underwater cutting because hydrogen can be safely compressed to the pressures necessary to overcome water pressures at the depths involved in salvage operations. The oxyhydrogen flame is also applied in the working and fabrication of quartz and glass.

Atomic hydrogen welding, another important application of hydrogen, is particularly suitable for thin stock and can be used with practically all nonferrous metals and alloys as well as with ferrous alloys; jobs performed with this welding process range from the fabrication of nickel and monel tanks to the welding of aluminum aircraft parts and propellers and the repair of steel molds and dies. In the process, an arc with a temperature of about 11,000 F (6093 C) is maintained between two nonconsumable metal electrodes. Molecular hydrogen fed into the arc is transformed into atomic hydrogen, which transmits heat from the arc to the weld zone. At the relatively colder surface of the weld area, the atomic hydrogen recombines to molecular hydrogen with the release of heat. The hydrogen also shields the weld area from the air.

Hydrogen also serves as a nonoxidizing shield, alone or with other gases, in annealing, furnace brazing, producing parts from sintered carbides and other metals, and in the refining and heat treatment of such metals as tungsten and molybdenum.

Large electrical generators are sometimes run in a hydrogen atmosphere to reduce windage losses and remove heat.

Liquid hydrogen has assumed importance as a fuel for powering missiles and rockets. It is also employed in laboratory research on the properties of materials at cryogenic temperatures, among them the superconductivity of metals (a state in which they have extremely low electrical resistance).

PHYSIOLOGICAL EFFECTS

Hydrogen is nontoxic, but it can act as a simple asphyxiant by displacing the oxygen in the air. Unconsciousness from inhaling air which contains a sufficiently large amount of hydrogen can occur without any warning symptoms such as dizziness. Still lower concentrations than those which could lead to unconsciousness would be flammable, for the lower flammable limit of hydrogen in air is only some 4 percent by volume. All the precautions necessary for the safe handling of any flammable gas must of course be observed with hydrogen.

Liquid hydrogen and also the cold gas evolving form the liquid can produce severe burns similar to thermal burns upon contact with the skin and other tissues, and eyes can be injured by exposure to the cold gas or splashed liquid too brief to affect the skin of the hands or face. Contact between unprotected parts of the body with uninsulated piping or vessels containing liquid hydrogen can cause the flesh to stick and tear when an attempt is made to withdraw.

SAFE HANDLING AND STORAGE

Materials Suitable for Containers and Storage

Hydrogen gas is noncorrosive, and may be contained at normal temperatures by any common metals used in installations designed to have sufficient strength for the working pressures involved. Metals used for liquid hydrogen equipment must have satisfactory properties at very low operating temperatures. Ordinary carbon steels lose their ductility at liquid hydrogen temperatures and are considered too brittle for this service. Suitable materials include austenitic chromium-nickel steels, copper, copper-silicon alloys, aluminum, "Monel" and some brasses and bronzes.

Equipment and piping built to utilize hydrogen should be selected with consideration of the possibility of embrittlement particularly at elevated pressures and temperatures above 450 F (232 C). Compressed gas containers of present materials designated as suitable are not embrittled by hydrogen service.

Regulators and Control Valves. Contrary to general practice with other gas cylinders, it is inadvisable to "crack" hydrogen cylinder valves before connecting them to a regulator or manifold since "self-ignition" of the issuing hydrogen may occur.

Storage and Handling in Normal Use

Hydrogen gas should be treated with the same care given to all flammable compressed gases. It is most often stored at the user's location in cylinders or tube trailers. Recommended standards for gaseous hydrogen systems are found in NFPA Standard 50-A (see Ref. 2). Standards for the storage of hydrogen cylinders used with oxygen for welding and cutting are presented in NFPA Pamphlet No. 51 (see Ref. 3).

Users of liquid hydrogen should obtain full information on safe handling, protective equipment and first aid from their suppliers. Measures to insure safe handling are also described in Pamphlet G-5 of Compressed Gas Association, Inc. (see Ref. 4).

Liquefied hydrogen is stored in systems that range in capacity to two million ft^3 of gas or more and that frequently include a vaporizer to convert the liquid to gas at ambient temperatures.

Systems for handling liquefied hydrogen, as well as piping and manifold systems for gaseous hydrogen, should be constructed only under the supervision of competent engineers who are thoroughly familiar with the problems incident to the piping of cryogenic fluids and flammable gases. Care should be taken to comply with the requirements of all state, municipal and insurance authorities.

Recommended standards for liquefied hydrogen systems are given in NFPA Standard 50-B (see Ref. 5).

Extensive information on equipment for handling liquid hydrogen appears in a handbook issued by the U.S. Department of Commerce (see Ref. 6).

Electrical equipment in areas for the storage of large quantities of liquid hydrogen should be installed in accordance with Article 500 of NFA Pamphlet No. 70 (see Ref. 7). Separate rooms or buildings devoted to the handling of liquid hydrogen in substantial quantities should conform with the recommendations of NFPA Pamphlet No. 68 (see Ref. 8).

Hydrogen cylinders used with oxygen for welding and cutting and stored inside a building should be limited to a total capacity of 2000 ft^3 (56.63 m^3) exclusive of cylinders in use or attached for use. Quantities exceeding this total should be stored in a special building or separate room as recommended in Pamphlet No. 51 of the National Fire Protection Association (see Ref. 3), or should be stored out-of-doors.

Storage of hydrogen gas not used in conjunction with oxygen should conform with the recommendations of NFPA Standard 50-A (see Ref. 2). Storage of liquefied hydrogen in liquid cylinders should conform to NFPA Standard 50-B (see Ref. 5).

The potential hazards of liquid hydrogen stem mainly from three important properties: (1) its extremely low temperature; (2) its very large liquid-to-gas expansion ratio (with 1 volume of liquid giving rise to 850 volumes of gas at room temperature); and (3) its wide range of flammable limits after it has vaporized to a gas.

Only persons who have been thoroughly instructed in the hazards of liquid hydrogen and corresponding protective measures should be allowed to handle it. All precautions necessary for the safe handling of any gas liquefied at extremely low temperatures must be observed with liquid hydrogen. In addition, liquid hydrogen must be handled in closed systems so that air does not enter vents and lines and plug them by solidifying. If air, oxygen or another gas condenses and solidifies in the openings of a liquid hydrogen container, pressure may build up to a point of damaging or bursting the container.

Quantities of liquid hydrogen up to 5 liters may be handled for laboratory or test purposes in open-mouth Dewar vessels, but only if the vessels are stoppered down to the smallest opening needed for the work and if they have vents which permit the release of hydrogen vapor but exclude the entry of air. Glass is not recommended as a material for Dewar flasks or other vessels that are to be filled with liquid hydrogen because of the possibility that the liquid may burst a vessel into which it is being poured with explosive violence unless the vessel is completely prepared for the liquid.

Handling Leaks and Emergencies

Leaking hydrogen cylinders should be guarded with special care. If hydrogen leaks from the cylinder valve even when the valve is closed, or if a leak occurs at the safety device, carefully remove the cylinder to an open space that is outdoors and well away from any possible source of ignition. Plainly tag the cylinder as having an unserviceable valve or safety device, and immediately notify the cylinder's supplier and ask for his instructions. Extreme care is recommended in protecting access to the defective cylinder because the leaking hydrogen may ignite in the absence of any normally apparent source of ignition and if so, will burn with an almost invisible flame that can instantly injure anyone coming into contact with it.

METHODS OF SHIPMENT; REGULATIONS

Under the appropriate regulations and tariffs, hydrogen is authorized for shipment as follows (hydrogen gas, except where liquid hydrogen is noted):

By Rail: In cylinders (via freight, express or baggage) and in tank cars. Liquid hydrogen, in tank cars and in portable containers under DOT exemption.

By Highway: In cylinders on trucks; in tube trailers. Liquid hydrogen, in special portable containers on trucks and in tank trucks under DOT exemption.

By Water: In cylinders on cargo vessels only, and in DOT authorized tank cars aboard trainships only. In cylinders on barges of U.S. Coast Guard classes A, BA, BB, CA and CB only. Liquid hydrogen is not authorized for shipment by water under present Coast Guard regulations, but it is shipped under special permit in tank barges.

By Air: In cylinders aboard cargo aircraft only up to 300 lb (140 kg) maximum net weight per cylinder. Liquid hydrogen is not accepted for air shipment under present regulations.

CONTAINERS

Hydrogen gas is authorized for shipment in cylinders (including trailer DOT tubes) and tank cars.

Liquid hydrogen is authorized for shipment in specially designed, insulated tank cars and cylinders. Liquid hydrogen is also shipped in insulated tank trucks under DOT exemption.

Filling Limits. The maximum filling limits authorized for hydrogen are as follows (hydrogen gas, unless liquid hydrogen is noted):

In cylinders—the service pressure marked on the cylinder at 70 F (21.1 C) and not in excess of $\frac{5}{4}$ of the marked service pressure at 130 F (54.4 C) (hydrogen gas is usually shipped in cylinders at pressures of around 2200–2400 psig (15,170–16,550 kPa gage)). Cylinders may be shipped under DOT exemption at pressures (70 F) up to 110 percent of the marked service

pressure. Liquid hydrogen may be shipped in
DOT-4L cylinders. The pressure in liquefied
hydrogen cylinders must be limited by a pressure–controlling valve, as specified in DOT regulations set to limit pressure to not more than
17 psig (120 kPa gage).

In tank cars—not more than seven-tenths of
the marked test pressure at 130 F (54.4 C) in
uninsulated cars of the DOT 107A type. Liquid
hydrogen in tank cars (meeting DOT specifications 113A60-W-2), to a maximum filling density of 6.6 percent (percent water capacity by
weight), and each car must have a pressure-
controlling valve set at a maximum of 17 psig
(120 kPa gage).

Cylinders

Cylinders built to comply with DOT specifications 3A, 3AA, 3AX or 3AAX are the types
chiefly used to ship hydrogen gas, but it is also
authorized for shipment in any cylinders approved for nonliquefied gases (these include
cylinders meeting DOT specifications 3B, 3C,
3D, 3E, 4, 4A, 4B, 4BA and 4C; in addition,
cylinders meeting DOT specifications 3, 25,
26, 33 and 38 may be continued in gaseous
hydrogen service, but new construction is not
authorized).

The pressure in a cylinder charged with a nonliquefied gas such as hydrogen is related to both
the amount of gas in the cylinder and the temperature of the gas. When the temperature of
hydrogen in a cylinder is close to room or atmospheric temperature, the volume of hydrogen may be estimated by use of the chart
shown in Fig. 1. For example, if the room temperature is 70 F (21.1 C) and the cylinder pressure is 1200 psig (8200 kPa gage), follow the
diagonal line 70 F until it crosses the vertical
line representing 1200 psig. From the point of
intersection of these two lines trace horizontally across the chart to read the figures at the
left. This gives the volume of hydrogen in a
cylinder with 2640 in^3 (0.0430 m^3) capacity
of about 120 ft^3 (3.40 m^3) at 70 F and 1200
psig. To use the chart for hydrogen cylinders
with volumes differing from 2640 in^3, multiply the hydrogen volume figure read off the

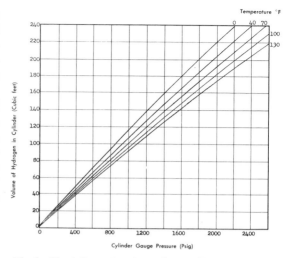

Fig. 1. Chart for estimating volume of hydrogen in a
cylinder having a volumetric capacity of 2640 in^3.
(Based on data in Bureau of Standards Miscellaneous
Publication M191 of November 17, 1948.)

left side of the chart by $V/2640$, where V is
the volume in cubic inches of the given cylinder.

All cylinders authorized for gaseous hydrogen
service must be requalified by hydrostatic retest
every 5 or 10 years under present regulations,
with the following exceptions: DOT-4 cylinders, every 10 years; and no periodic retest is
required for cylinders of types 3C, 3E and 4C.

Valve Outlet and Inlet Connections. Standard connections, United States and Canada—
up to 3000 psi (20,684 kPa), No. 350; 3000–
10,000 psi (20,684–68,948 kPa), see Part I,
Chapter 7, Section A, paragraph 2.1.7; cryogenic liquid withdrawal, No. 795. Alternate
standard connections, United States and Canada—3000–10,000 psi, No. 677; cryogenic liquid withdrawal, No. 792.

Cargo Tanks

Liquefied hydrogen is shipped in insulated
cargo tanks on trucks or truck trailers under
DOT exemption. Approximately 12,000 gal
(45 m^3) is a common capacity for the tanks,
which consist of an inner shell surrounded by
very efficient vacuum insulation and enclosed
in an outer case of $\frac{3}{16}$ or $\frac{1}{2}$-in. (4.8 or 12.7

mm) steel. Though often equipped with an initial pressure-relief device set at 17 psig (120 kPa gage), the tanks do not normally vent any hydrogen in transit, since they are ordinarily filled at a pressure of only some 2 psig (10 kPa gage) at the plant and will rise in pressure to only about 6–8 psig (40–56 kPa gage) during the journeys of up to 3000 miles (4830 km). Should the insulation deteriorate in transit, drivers are usually required to leave the highway and go to a safe area in which they reduce the pressure with a manual blowdown valve.

Tube Trailers

Hydrogen gas is authorized for bulk highway shipment at high pressure (around 2400 psig (16,550 kPa gage)) in tube trailers. These trailers are made up of a number of large tubes, constructed to DOT specifications 3A, 3AA, 3AX or 3AAX and manifolded together to a common header. Valves must be installed on each tube and closed while in transit. Such trailers are usually towed to the point of utilization and replace an empty unit which is then towed to the supplier's plant for refilling. Capacities of the tube trailers range up to 125,000 ft³ (3540 m³).

Tank Cars

Gaseous hydrogen is authorized for rail shipment in tank cars that comply with DOT specifications 107A. These cars consist of a number of seamless containers permanently mounted on a special car frame. DOT regulations require that on such cars the safety relief device header outlet must be equipped with an approved ignition device which will instantly ignite any hydrogen discharged through the safety devices. Other requirements are also set forth in the regulations.

Liquefied hydrogen containing a minimum of 95 percent *para*-hydrogen is authorized for shipment in tank cars meeting DOT specifications 113A60-W-2 which, as previously noted, must be outfitted with pressure-controlling valves set at a pressure not exceeding 17 psig (120 kPa

gage). A common size for these cars is 68-ft (20.7 m) length with 28,300-gal capacity. The cars are so well insulated for internal cooling that they usually do not vent any hydrogen in being shipped distances up to 3000 miles (4830 km); should the pressure approach the 17 psig relief setting, excess hydrogen is mixed with air in a special device to a concentration of less than half its lower flammable limit before being vented.

Portable Containers

Portable containers which comply in some cases with DOT-4L specifications are used for the shipment of liquid hydrogen under exemption of the DOT. These containers employ a "super" insulating material to keep the liquid at its boiling point for weeks at a time, and also rely on venting a small portion of the vapor in the container to remove what heat is absorbed. The containers maintain the liquid under low positive pressure and are outfitted with complex venting, safety relief device, and liquid withdrawal equipment. One commonly used size stands about 5 ft (1.52 m) high and is some 20 in. (50.8 cm) in diameter; it holds about 40 gal (151.4 liters) of liquid hydrogen. One of the largest such containers has a capacity of some 270 gal (1.02 m³).

METHOD OF MANUFACTURE

Hydrogen is produced industrially by several methods. A large part of the hydrogen produced in the United States is consumed at or within pipeline distance of the production site.

Hydrogen is frequently obtained as a by-product of cracking operations using petroleum liquids or vapors as feedstock.

Electrolysis of water, aqueous acid, or alkali solutions, is also used to produce hydrogen. Some plants have been built to make hydrogen as the primary desired product from the electrolysis of water. Considerable amounts of hydrogen are also obtained as by-products of other types of electrolytic operations designed to yield different chemicals as main products.

Passing steam over heated, spongy iron reduces the steam to hydrogen with accompanying formation of iron oxide. Several varieties of this "steam-iron" process are employed to make hydrogen commercially.

The reaction of steam with incandescent coke or coal (called the "water gas reaction") is also used as a source of hydrogen, with carbon monoxide as an additional product. In a catalytic version of the water gas process, excess steam breaks down to form more hydrogen while oxidizing the carbon monoxide to carbon dioxide.

Natural gas or light hydrocarbons may also be used as a raw material for a method of hydrogen production in which they react with steam in the presence of a catalyst. Hydrogen is also produced through the dissociation of ammonia.

REFERENCES

1. "Commodity Specification for Hydrogen" (Pamphlet G-5.3), Compressed Gas Association, Inc.
2. "Standard for Gaseous Hydrogen Systems at Consumer Sites" (Pamphlet 50-A), National Fire Protection Association, 470 Atlantic Ave., Boston, MA 02210.
3. "Standard for the Design and Installation of Oxygen-Fuel Gas Systems for Welding and Cutting" (Pamphlet No. 51), National Fire Protection Association.
4. "Hydrogen" (Pamphlet G-5), Compressed Gas Association, Inc.
5. "Standard for Liquefied Hydrogen Systems at Consumer Sites" (Pamphlet 50-B), National Fire Protection Association.
6. "Handbook for Hydrogen Handling Equipment" (PB161835), U.S. Department of Commerce, Office of Technical Services, Washington, DC 20025. (Deals with liquid hydrogen only.)
7. "National Electric Code" (Pamphlet No. 70), National Fire Protection Association.
8. "Guide for Explosion Venting" (Pamphlet No. 68), National Fire Protection Association.

ADDITIONAL REFERENCE

"Standard Density Data, Atmospheric Gases and Hydrogen" (Pamphlet P-6), Compressed Gas Association, Inc.

Hydrogen Chloride, Anhydrous

HCl
Synonym: Hydrochloric acid, anhydrous
DOT Classification: Nonflammable gas; nonflammable gas label

PHYSICAL CONSTANTS

	U.S. Units	Metric Units
International symbol	HCl	HCl
Molecular weight	36.465	36.465
Vapor pressure		
at 70 F (21.1 C)	613 psig	4227 kPa, gage
at 77 F (25 C)	676 psig	4661 kPa, gage
at 105 F (40.6 C)	950 psig	6550 kPa, gage
at 115 F (46.1 C)	1075 psig	7412 kPa, gage
at 124.5 F (51.4 C)	1185 psig	8170 kPa, gage
Density of gas		
at −50 F (−45.6 C)	0.7370 lb/ft^3	11.81 kg/m^3
at 2 F (−16.7 C)	1.735 lb/ft^3	27.79 kg/m^3
at 32 F (0 C)	0.102 lb/ft^3	1.634 kg/m^3
at 70 F (21.1 C) and 1 atm	0.0950 lb/ft^3	1.522 kg/m^3
Density of liquid		
at −121 F (−85 C)	74.3 lb/ft^3	1190 kg/m^3
at −50 F (−45.6 C)	67.80 lb/ft^3	1086 kg/m^3
at 2 F (−16.7 C)	62.9 lb/ft^3	1005 kg/m^3
at 70 F (21.1 C)	52.7 lb/ft^3	842 kg/m^3
at 105 F (40.6 C)	43.8 lb/ft^3	700 kg/m^3
at 115 F (46.1 C)	39.7 lb/ft^3	634 kg/m^3
at 124.5 F (51.4 C)	26.3 lb/ft^3	420 kg/m^3
Specific gravity of the gas		
at 32 F and 1 atm (air = 1)	1.268	1.268
Specific volume of the gas		
at 70 F (21.1 C) and 1 atm	10.6 ft^3/lb	0.6617 m^3/kg

	U.S. Units	Metric Units
Boiling point at 1 atm	−121 F	−85 C
Melting point at 1 atm	−168 F	−111.1 C
Critical temperature	124.5 F	51.4 C
Critical pressure	1198 psia	8260 kPa, abs.
Critical density	26.2 lb/ft³	420 kg/m³
Triple point	−167.8 F at	−111 C at
	2.61 psia	18 kPa, abs.
Latent heat of vaporization		
at boiling point	190.5 Btu/lb	443.1 kJ/kg
Latent heat of fusion		
at melting point	23.49 Btu/lb	54.64 kJ/kg
Specific heat of gas		
at 312.8 F (156 C) and 1 atm		
C_p	0.1939 Btu/(lb) (F)	0.8118 kJ/(kg) (C)
C_v	0.1375 Btu/(lb) (F)	0.5757 kJ/(kg) (C)
Ratio of specific heats	1.41	1.41
Solubility in water, wt/wt		
of water, at 32 F (0 C)	0.823/1	0.823/1
Weight of liquid		
at 2 F (−16.7 C)	8.346 lb/gal	1000 kg/m³

DESCRIPTION

Anhydrous hydrogen chloride is a colorless gas which fumes strongly in moist air and has a highly irritating effect on body tissues. It has a sharp, suffocating odor, and is shipped in cylinders as a liquefied compressed gas under its vapor pressure of about 613 psig at 70 F (4226 kPa gage at 21.1 C).

Chemically, hydrogen chloride is relatively inactive and noncorrosive in the anhydrous state. However, it is readily absorbed by water to yield the highly corrosive hydrochloric (muriatic) acid. It also dissolves readily in alcohol and ether and reacts rapidly (violently, in some cases) with many organic substances. At high temperatures (3240 F or 1782 C and above), hydrogen chloride tends to dissociate into its constituent elements.

GRADES AVAILABLE

Anhydrous hydrogen chloride is typically available for commercial and industrial purposes in a technical grade (minimum purity of 99.0 percent). A typical technical grade analysis is as follows:

Hydrogen chloride (liquid phase)	99.5 weight percent
Hydrocarbons (ethylene, 1, 1- dichloroethane, and ethyl chloride)	0.04 weight percent
Water	0.01 weight percent
Carbon dioxide	0.01 weight percent
Inert materials	0.1 weight percent

For use as an etchant for semiconductor crystals in the electronics industry, anhydrous hydrogen chloride is available in an electronic grade with a minimum purity of 99.99 percent plus the following specifications for impurities:

Nitrogen	50 max. ppm
Oxygen	10 max. ppm
Hydrocarbons (as methane)	10 max. ppm
Moisture	10 max. ppm
Carbon dioxide	10 max. ppm

USES

An old and important industrial chemical, hydrogen chloride is widely used in the manufacture of rubber, pharmaceuticals and both inorganic and organic chemicals, as well as in gasoline refining, metals processing and wool reclaiming. Rubber hydrochloride, which results from the treatment of natural rubber with hydrogen chloride, can be cast in film from solutions; such rubber hydrochloride films provide a strong, water-resistant packaging material for meats and other foods, paper products and textiles.

The chemicals industry uses hydrogen chloride to produce a large variety of chlorinated derivatives through both addition and substitution reactions with organic compounds—as in the manufacture of ethyl chloride from ethylene in making vinyl plastics, of chloromethanes and monochlorobenzene, of alkyl chlorides, and of methyl chloride from methyl alcohol. Hydrogen chloride is utilized in the gasoline industry as a promoter for the aluminum chloride catalyst for converting n-butane to isobutane.

In the metals industry, uses of gaseous hydrogen chloride include application as a flux in bonding Babbitt metal to steel strip in manufacturing insert bearing blanks, and for treating steel strip at elevated temperatures to improve the bond in later hot galvanizing. The textiles industry uses hydrogen chloride to decompose vegetable fibers with which wool has been woven in reclaiming wool for fabrics. Cotton seeds are also delinted and disinfected by being tumbled in a current of gaseous hydrogen chloride.

In the electronic industry, a high-purity grade of hydrogen chloride is used as an etchant for semicondutor crystals.

PHYSIOLOGICAL EFFECTS

Hydrogen chloride is toxic, causing severe irritation of the eyes and the skin on contact, and severe irritation of the upper respiratory tract on inhalation. Should contact occur, the eyes and eyelids must be washed immediately and thoroughly with large quantities of flowing water in order to avoid impairment of vision or even loss of sight. Prompt washing of the skin is also required after contact to prevent severe burns. Repeated exposure of the skin to concentrated anhydrous hydrogen chloride vapor may also result in burns or dermatitis.

Inhalation of excessive quantities produces coughing, burning of the throat and a choking sensation. Occasionally, ulceration of the nose, throat and larynx, or edema of the lungs, has resulted. Prolonged inhalation of high concentrations may cause death. The irritating character of the vapors provides warning of dangerous concentrations well before injury can result, unless personnel are trapped or disabled.

A concentration of 50 ppm in air cannot be tolerated for more than one hour, and concentrations of some 1500–2000 ppm are fatal for human beings in a few minutes.

A TLV of 5 ppm (7 mg/m^3) for an 8-hour work day for hydrogen chloride has been adopted by the ACGIH.

SAFE HANDLING AND STORAGE

Materials Suitable for Containers and Storage

Piping, valves and other equipment used in direct contact with anhydrous hydrogen chloride should be of stainless steel or of cast or mild steel. Carbon steel may be employed in some components, but only if their temperature is controlled to remain below about 265 F (129 C). Hydrogen chloride corrodes zinc, copper and copper alloys, and galvanized pipe or brass or bronze fittings must not be used with it. In the presence of moisture, however, hydrogen chloride will corrode most metals other than silver, platinum or tantalum.

Storage and Handling in Normal Use

Safety precautions include adequate ventilation of working areas, readily accessible safety shower and eye bath facilities, rubber or plastic aprons and gloves as protective clothing with wool or other acid-resistant outer garments, chemical safety goggles, the availability of proper full-face gas masks and thorough familiarity with safety equipment and measures on the part of personnel. Should eye contact, skin contact, or inhalation occur, immediately give first aid and call a physician.

Detailed information on safety precautions and first aid in the event of hydrogen chloride exposure is given in Safety Data Sheet SD-39 of the Chemical Manufacturers Association (see Ref. 1).

Hydrogen chloride is most often unloaded from special tank cars and tank trucks directly to process, without being stored by users. Unloading piping and valves must be of stainless steel or of cast or mild steel (Charpy tested to −50 F or −10 C and capable of withstanding the high pressures of the unloading process). Galvanized or copper-alloyed metals such as brass or bronze are corroded by hydrogen chloride and must not be used in pipe or fittings. Design and specification of components for unloading systems should be made with the help of suppliers or engineers thoroughly familiar with anhydrous hydrogen chloride (see also Part I, Chapter 4, Section A).

Handling Leaks and Emergencies

Leaks in anhydrous hydrogen chloride system connections may be detected by the white fumes which form when the gas comes into contact with the moisture of the atmosphere. Small leaks may be found with an open bottle of concentrated ammonium hydroxide solution (which forms dense white fumes in the presence of hydrogen chloride) or with wet blue litmus paper (which is turned pink by hydrogen chloride). (See Part I, Chapter 4, Section A.)

METHODS OF SHIPMENT; REGULATIONS

Under the appropriate regulations and tariffs, anhydrous hydrogen chloride is authorized for shipment as follows:

By Rail: In cylinders (freight or express), and in insulated tank cars (by DOT exemption).

By Highway: In cylinders on trucks; also in insulated tank trucks and tube trailers.

By Water: In cylinders via cargo vessels, passenger vessels, and ferry and railroad car ferry vessels (passenger or vehicle). In cylinders on barges of U.S. Coast Guard classes A, BA, BB, CA and CB.

By Air: Aboard cargo aircraft only in appropriate cylinders up to 300 lb (140 kg) maximum net weight per cylinder.

CONTAINERS

Anhydrous hydrogen chloride is shipped as a compressed liquefied gas under its own vapor pressure in cylinders and in tube trailers. It is also shipped in liquefied form in insulated tank cars and tank trucks by DOT exemption or by special permit of the Canadian Transport Commission.

Filling Limit. The maximum filling density authorized for hydrogen chloride in cylinders by the DOT is 65 percent (percent water capacity by weight).

Cylinders

Cylinders that meet the following DOT specifications are authorized for hydrogen chloride service: 3A1800, 3AA1800, 3AX1800, 3AAX-1800 and 3E1800 (cylinders manufactured under the now obsolete specification DOT-3 may be continued in service but new construction is not authorized). DOT-3A2015 cylinders are also often used in anhydrous hydrogen chloride service. DOT-3T tubes may also be used.

Cylinders of types 3A and 3AA (as well as 3) used in hydrogen chloride service must be requalified by hydrostatic retest every 5 years under present regulations. For 3E cylinders, no periodic retest is required.

Cylinder Manifolding. Hydrogen chloride cylinders are manifolded by users, but care must be taken to have any special manifolding equipment meet all safety requirements.

Valve Outlet Connections. Standard connection, United States and Canada—No. 330.

Tank Trucks

Under exemption of the DOT, hydrogen chloride is authorized for shipment in insulated tank trucks of special design. Constructed of stainless steel with urethane foam insulation, the tanks conform to DOT specification MC-330 and are loaded with liquid hydrogen chloride at −70 F and 50 psig (−56.6 C and 345 kPa gage). Unloading of the truck must be started within 3 days, unless special methods are used to keep truck pressure between 50–80 psig (345–552 kPa gage).

Tube Trailers

Shipment of hydrogen chloride is also made in tube trailers.

Special Single-Unit Insulated Tank Cars

Exemption of the DOT is also required for shipment of hydrogen chloride in single-unit tank cars of special design (in conformance with DOT specification 105A600-W, made of carbon steel with 10-in. (254-mm) insulation, and cooled by external coils). The cars are loaded with liquid hydrogen chloride at −50 F (−45.6 C), and are designed for unloading without additional refrigeration within 25 to 35 days after filling.

METHOD OF MANUFACTURE

Hydrogen chloride is produced as a by-product from the chlorination of benzene and other hydrocarbons, and by burning hydrogen, methane or water gas in a chlorine atmosphere.

REFERENCE

1. "Hydrogen Chloride" (Safety Data Sheet SD-39), Chemical Manufacturers Association, 1825 Connecticut Ave., NW., Washington, DC 20009.

Hydrogen Cyanide

HCN

Synonyms: Hydrocyanic acid, prussic acid, formonitrile

DOT Classification (as hydrocyanic acid liquefied): Poison A; flammable gas; flammable gas and poison label

PHYSICAL CONSTANTS

	U.S. Units	Metric Units
International symbol	HCN	HCN
Molecular weight	27.03	27.03
Vapor pressure		
at 70 F (21.1 C)	12.3 psia	84.8 kPa, abs.
at 105 F (40.6 C)	25.6 psia	177 kPa, abs.
at 115 F (46.1 C)	29.8 psia	205 kPa, abs.
at 130 F (54.4 C)	42.0 psia	290 kPa, abs.
Density of the gas at 70 F		
(21.1 C) and 1 atm	71.5 lb/ft^3	1145 kg/m^3
Specific gravity of the gas at 87.8 F		
(31 C) and 1 atm (air = 1)	0.947	0.947
Specific volume of the gas at 70 F		
(21.1 C) and 1 atm.	0.014 ft^3/lb	0.00087 m^3/kg
Density of liquid		
at 32 F (0 C)	44.64 lb/ft^3	715.1 kg/m^3
at 70 F (21.1 C)	42.85 lb/ft^3	686.4 kg/m^3
at 105 F (40.6 C)	41.23 lb/ft^3	660.4 kg/m^3
at 115 F (46.1 C)	40.75 lb/ft^3	652.8 kg/m^3
at 130 F (54.4 C)	40.05 lb/ft^3	641.5 kg/m^3
Boiling point at 1 atm	78.3 F	25.7 C
Melting point at 1 atm	8.2 F	-13.3 C
Critical temperature	362.3 F	183.5 C
Critical pressure	782 psia	5390 kPa, abs.
Critical density	12.17 lb/ft^3	194.9 kg/m^3
Triple point	8.03 F at	-13.3 C at
	2.702 psia	18.63 kPa, abs.
Latent heat of vaporization at		
77 F (25 C)	444.2 Btu/lb	1033 kJ/kg

	U.S. Units	Metric Units
Latent heat of fusion at 9.3 F (-12.6 C)	133.8 Btu/lb	311.2 kJ/kg
Specific heat of gas at 80.6 F (27 C) and 1 atm		
C_p	0.3185 Btu/(lb)(F)	1.333 kJ/(kg)(C)
C_v	0.241 Btu/(lb)(F)	1.01 kJ/(kg)(C)
Ratio of specific heats C_p/C_v	1.321	1.321
Solubility in water, vol/vol, at 70 F (21.1 C)	all proportions	
Weight of liquid		
at 32 F (0 C)	5.97 lb/gal	715 kg/m^3
at 68 F (20 C)	5.74 lb/gal	68 kg/m^3
Specific heat		
at -27.5 F (-33.1 C)	0.516 Btu/(lb)(F)	2.16 kJ/(kg)(C)
liquid, at 69.6 F (20.9 C)	0.6270 Btu/(lb)(F)	2.625 kJ/(kg)(C)
Heat of combustion	10,555 Btu/lb	24,551 kJ/kg
Flammable limits in air, by volume	6-41 percent	6-41 percent
Autoignition temperature	1000.4 F	538 C
Flash point (closed cup)	-0.04 F	-17.8 C

DESCRIPTION

Hydrogen cyanide (hydrocyanic acid) is a colorless liquid with an atmospheric boiling point of 78.3 F (25.7 C). The vapor phase is also colorless and has a faint odor of bitter almonds. Hydrogen cyanide vapors are highly toxic, and hydrogen cyanide is designated a poison gas of the "extremely dangerous" class A variety by the DOT. Its gas is slightly lighter than air, having a specific gravity of about 0.93 compared to air, and its liquid is lighter than water (specific gravity about 0.7). Hydrogen cyanide polymerizes spontaneously with explosive violence when not completely pure or stabilized; it is commonly stabilized by adding acids, such as 0.05 percent phosphoric or sulfuric acid. Unstabilized hydrogen cyanide is not accepted for shipment under DOT regulations. Hydrogen cyanide is miscible in all proportions with water or alcohol, and miscible in ether, benzene and most other organic solvents. Alkali cyanides react with water, steam, acids or acid fumes to form very toxic fumes of cyanides.

GRADES AVAILABLE

Commercial hydrogen cyanide typically has the following specifications: 99.5 mole percent minimum hydrogen cyanide; 0.5 mole percent maximum water; 0.06-0.10 percent acidity (calculated as H_2SO_4; color not darker than APHA 20).

USES

Hydrogen cyanide is used as an intermediate in producing acrylic plastics, including "Plexiglas," and in the manufacture of dyes, fumigants, rubber, cyanide salts, acrylonitrile, acrylates, chelates and adiponitrile (for nylon).

PHYSIOLOGICAL EFFECTS

Acute cyanide poisoning can render a person unconscious with extreme rapidity but the victim can be saved with prompt first aid. Less acute cases result in headache, dizziness, unsteadiness and a sense of suffocation and nausea.

Hydrogen cyanide can enter the body through absorption by the skin as well as by inhalation. Concentrations of from 100–200 ppm in air can be lethal if inhaled for 30–60 minutes. Among persons who recover, disability rarely occurs. Hydrogen cyanide and other cyanides act as protoplasmic poisons, combining in the tissues with enzymes necessary for cellular oxidation. Suspension of tissue oxidation lasts only while the cyanide is present, with normal functions returning upon its removal if death has not occurred.

A TLV of 10 ppm (11 mg/m^3) for an 8-hour work day for hydrogen cyanide has been adopted by the American Conference of Governmental Hygienists (ACGIH).

SAFE HANDLING AND STORAGE

Materials Suitable for Containers and Storage

Moderately low-carbon steels are among materials ordinarily used for hydrogen cyanide equipment. Certain stainless steels are employed in applications or equipment that must avoid contamination, or operate at high temperatures, or resist corrosion by other chemicals being handled with hydrogen cyanide.

Storage and Handling in Normal Use

All the precautions necessary for the safe handling of any flammable gas, as well as all precautions required with a highly toxic gas, must be observed with hydrogen cyanide (see also Part I, Chapter 4, Section A).

Handling Leaks and Emergencies

See Part I, Chapter 4, Section A. Detailed information on safety precautions and first aid with hydrogen cyanide is given in Safety Data Sheet SD-67 of the Chemical Manufacturers Association (see Ref. 1).

First Aid. Should a person be exposed to hydrogen cyanide, carry him to fresh air, have him lie down, and immediately start first aid and then call a physician. Also, if clothing has been splashed by the liquid, remove contaminated garments but keep the patient warm.

If hydrogen cyanide gas has been inhaled, break an amyl nitrite pearl in a cloth and hold the cloth lightly under the patient's nose for 15 seconds, repeating this five times at intervals of about 15 seconds. If breathing has stopped use a resuscitator such as a Pneolator resuscitator or give artificial respiration.

If liquid hydrogen cyanide has been splashed on the skin or eyes, immediately flush the skin for at least 15 minutes. Get medical attention if cyanide has entered the eyes.

METHODS OF SHIPMENT; REGULATIONS

Under the appropriate regulations and tariffs, hydrogen cyanide is authorized for shipment as follows (only if stabilized; unstabilized hydrogen cyanide is not accepted for shipment under DOT regulations):

By Rail: In cylinders (via freight only), and in single-unit tank cars.

By Highway: In cylinders on trucks.

By Water: Not accepted.

By Air: Not accepted. (Accepted only if in solution not exceeding 5 percent, aboard cargo aircraft only in special containers to a maximum of 26.4 lb (12 kg) net weight per container.)

CONTAINERS

Hydrogen cyanide is authorized for shipment in cylinders and tank cars. In addition, liquid hydrogen cyanide completely absorbed in inert material is authorized for shipment in wooden or fiberboard boxes which meet detailed DOT requirements.

Filling Limits. The maximum filling densities permitted for hydrogen cyanide are:

In cylinders—Not more than 0.6 lb (0.27 kg) of liquid for each 1-lb (0.45-kg) water capacity of the cylinder.

In tank cars—Not more than 0.63 lb (0.29 kg) of liquid for each 1-lb (0.45 kg) water capacity of the tank.

Cylinders

Cylinders that meet the following DOT specifications are authorized for hydrogen cyanide service: 3A480, 3AA480 and 3A480X. In addition, these cylinders must be of not over 278-lb (126 kg) water capacity (nominal); valve-protection caps must also be used, and be at least $\frac{3}{16}$-in. (4.76 mm) thick, gas-tight, with $\frac{3}{16}$-in. faced seat for gasket and with United States standard form thread. These caps must be capable of preventing injury or distortion of the valve when it is subjected to an impact caused by allowing the cylinder, prepared as if for shipment, to fall from an upright position with the side of the cap striking a solid steel object projecting not more than 6 in. (152.4 mm) above floor level. Additional requirements are that each filled cylinder must be tested for leakage before shipment and must show absolutely no leakage. This test must consist in passing over the closure of the cylinder, without the protection cap attached, a piece of Guignard's sodium picrate paper to detect any escape of hydrocyanic acid from the cylinder. Other equally efficient test methods may also be used instead of picrate paper.

The 3A and 3AA cylinders authorized for hydrogen cyanide service must be requalified by hydrostatic retest every 5 years under present regulations.

Valve Outlet Connections. Standard connection, United States Canada—No. 750. Alternate standard connection, United States and Canada—No. 160.

Tank Cars

Bulk shipment of hydrogen cyanide is authorized by the DOT in single-unit tank cars meeting specifications 105A500W or 105A600W, with these further requirements: (1) the tank car must be equipped with safety valves of the type and size used on the specification 105A300W tank car and restenciled 105A300W; (2) the tank must be equipped with approved dome fittings and safety devices, and with cork insulation at least 4 in. (102 mm) thick; (3) the tank must be stenciled on both sides, in letters at least 2 in. (51 mm) high, "HYDROCYANIC ACID ONLY"; and (4) a written statement of procedure covering details of tank car appurtenances, dome fittings and safety devices, and marking, loading handling, inspection and testing practices, must be filed with and approved by the Bureau of Explosives (1920 "L" Street N.W., Washington, DC 20036) before any car may be offered for transportation of hydrogen cyanide.

METHOD OF MANUFACTURE

Hydrogen cyanide is produced on a large scale by reacting ammonia and air with methane in the presence of a catalyst. Other methods by which it is made include treatment of a cyanide with dilute sulfuric acid; recovery from coke oven gas; decomposition of formamide; and synthesis from ammonia and hydrocarbons in an electrofluid reactor.

REFERENCE

1. "Hydrocyanic Acid" (Safety Data Sheet SD-67), Chemical Manufacturers Association, 1825 Connecticut Ave., N.W., Washington, DC 20009.

Hydrogen Sulfide

H₂S

H$_2$S
Synonym: Sulfuretted hydrogen
DOT Classification: Flammable gas; flammable gas and poison label

PHYSICAL CONSTANTS

	U.S. Units	Metric Units
International symbol	H$_2$S	H$_2$S
Molecular weight	34.08	34.08
Vapor pressure of saturated gas at		
59 F (15 C)	211.2 psig	1456.2 kPa gage
70 F (21.1 C)	249.8 psig	1722.3 kPa gage
105 F (40.6 C)	409.0 psig	2820.0 kPa gage
115 F (46.1 C)	465.0 psig	3206.1 kPa gage
130 F (54.4 C)	557.0 psig	3840.4 kPa gage
Density of saturated gas at		
59 F (15 C)	1.658 lb/ft^3	26.55 kg/m^3
70 F (21.1 C)	1.938 lb/ft^3	31.04 kg/m^3
Specific gravity of gas at		
59 F (15 C) and 1 atm		
(101.325 kPa abs.)	1.189	1.189
Specific volume of gas at		
1 atm (101.325 kPa abs.)		
59 F (15 C)	10.99 ft^3/lb	0.686 m^3/kg
70 F (21.1 C)	11.24 ft^3/lb	0.701 m^3/kg
Density of saturated liquid at		
-76.6 F (-60.3 C)	57.1 lb/ft^3	915.3 kg/m^3
59 F (15 C)	49.2	787.8
70 F (21.1 C)	48.3	774.2
105 F (40.6 C)	45.5	728.1
115 F (46.1 C)	44.5	712.5
130 F (54.4 C)	43.0	688.7
Boiling point at 1 atm		
(101.325 kPa abs.)	-76.6 F	-60.3 C

375

	U.S. Units	Metric Units
Melting point at 1 atm (101.325 kPa abs.)	-117.2 F	-82.9 C
Critical temperature	212.7 F	100.4 C
Critical pressure	1306.5 psia	9008 kPa abs.
Critical density	21.8 lb/ft^3	349 kg/m^3
Latent heat of vaporization at boiling point, 1 atm (101.325 kPa abs.)	236.3 Btu/lb	549.3 kJ/kg
Latent heat of fusion at freezing point	29.99 Btu/lb	69.7 kJ/kg
Specific heat of gas at 70 F (21.1 C) and 1 atm (101.325 kPa abs.)		
C_p	0.2532 Btu/(lb) (F)	1.059 kJ/(kg) (K)
C_v	0.1918	0.803
Ratio of specific heats	1.32	1.32
Solubility in water at 80 F (26.67 C) wt/wt	32 percent	32 percent
Mass of saturated liquid at 70 F (21.1 C)	6.46 lb/US. gal.	774.1 kg/m^3
Heat of combustion (H_2S to liquid water and SO_2 gas	8340 Btu/lb	19385.9 kJ/kg
Flammability limits in air	4.3–46 percent	4.3–46 percent

DESCRIPTION

Hydrogen sulfide is a colorless, flammable, toxic gas with an offensive odor often referred to as smelling like "rotten eggs." It is instantly fatal if inhaled in very high concentrations, and irritating to the eyes and upper respiratory tract in low concentrations. Its burning in air forms sulfur dioxide and water. A reducing agent, hydrogen sulfide reacts readily with all the metals in the electromotive series down to and including silver. It is soluble in water, alcohol, petroleum solvents and crude petroleum.

GRADES AVAILABLE

Hydrogen sulfide is available in a technical or commercial grade which is 98 percent minimum hydrogen sulfide. It is also available in a high purity or C. P. Grade, 99.5 percent minimum hydrogen sulfide.

USES

Hydrogen sulfide is used commercially to purify hydrochloric and sulfuric acid, to precipitate sulfides of metals and to manufacture elementary sulfur. Chemical production processes using hydrogen sulfide include manufacture of mercaptans, ethylene, nylon, soda ash, sodium hydrosulfide, heavy water and others. It is used as a reducing agent in cresylic acid recovery. It is also employed as a reagent in analytical chemistry.

PHYSIOLOGICAL EFFECTS

A toxic and asphyxiant gas, hydrogen sulfide in concentrations of from 20-150 ppm irritates the eyes, often causing severe pain and incapacitating workers for 10 days or more. Inhalation of slightly higher concentrations irritates the upper respiratory tract and, if prolonged, may result in pulmonary edema. Hydrogen sulfide acts on the nervous system. Inhalation of 500 ppm for 30 minutes produces headache, dizziness, excitement, staggering and disorders of the digestive tract, followed in some cases by bronchitis or bronchial pneumonia. Exposure to concentrations above 800–1000 ppm can be fatal within 30 minutes through respiratory paralysis. As noted previously, exposure to very high concentrations is almost immediately fatal.

Relying on the strong, rotten egg-like odor of hydrogen sulfide to warn of its presence can prove very dangerous under certain conditions. High concentrations have a very sweetish odor and rapidly paralyze the sense of smell. Even low concentrations will exhaust the sense of smell, after prolonged exposure, to the point where exposed personnel may fail to detect the presence of the gas.

A TLV of 10 ppm (15 mg/m^3) for an 8-hour work day for hydrogen sulfide has been adopted by the American Conference of Governmental Industrial Hygienists (ACGIH).

SAFE HANDLING AND STORAGE

Materials Suitable for Containers and Storage

Dry hydrogen sulfide is satisfactorily handled under pressure in steel or black iron piping. Aluminum and certain stainless steels are recommended for use with either wet or dry hydrogen sulfide. Brass valves, though tarnished by dry hydrogen sulfide, have been found to withstand years of service without appreciable malfunction.

Storage and Handling in Normal Use

All the precautions required for the safe handling of any flammable toxic gas must be observed with hydrogen sulfide (see also Part I, Chapter 4, Section A).

Handling Leaks and Emergencies

Leaks can be investigated by coating suspected areas with soapsuds and watching for bubbles, or by the use of a moist lead acetate paper.

First-Aid Suggestions. First-aid measures and treatment of personnel affected by hydrogen sulfide must be conducted under the direction of competent medical authorities. A detailed description of safety precautions, protective equipment and first aid for hydrogen sulfide hazards is given in Safety Data Sheet SD-36 of the Chemical Manufacturers Association (see Ref. 1). Among suggested steps to take if personnel should be overcome by inhaling hydrogen sulfide, move the person to fresh air and cover him with a blanket to keep him warm. Recovery should be quick if he is breathing and conscious. If he is not breathing, call a physician and start artificial respiration at once. Continue artificial respiration until breathing has been restored and give the patient oxygen until completely recovered (see also Part I, Chapter 4, Section A).

METHODS OF SHIPMENT; REGULATIONS

Under the appropriate regulations and tariffs, hydrogen sulfide is authorized for shipment as follows:

By Rail: In cylinders (freight or express), in TMU tank cars; and in 105A600-W tank cars under DOT exemption.

By Highway: In cylinders on trucks, in TMU tanks on trucks, and in MC-331 tank trucks under DOT exemption.

By Water: In cylinders via cargo vessels only, and in authorized tank cars aboard trainships only. In cylinders on barges of U.S. Coast Guard classes A and C only.

By Air: Aboard cargo aircraft only, in appropriate cylinders up to 300 lb (140 kg) maximum net weight per cylinder.

CONTAINERS

Hydrogen sulfide is shipped as a flammable gas in cylinders, TMU (ton multi-unit) tank cars, TMU tanks on trucks, tank cars and trucks.

Filling Limits. The maximum filling density

authorized for hydrogen sulfide are (percent water capacity by weight): in cylinders—62.5 percent, in TMU tanks—68 percent.

Cylinders

Cylinders that meet the following DOT specifications are authorized for hydrogen sulfide service: 3A480, 3AA480, 3B480, 4A480, 4BA480, 26-480 and 3E1800.

Under present regulations, cylinders of all types authorized for hydrogen sulfide service must be requalified by hydrostatic test every 5 years with the exception of type 3E (for which periodic hydrostatic retest is not required).

Valve Outlet Connections. Standard connection, United States and Canada—No. 330.

TMU Tanks

Hydrogen sulfide is authorized for shipment in TMU tanks meeting DOT specifications 106A800-X (with the further requirements that each tank must be equipped with adequate safety relief devices of the fusible plug type having a yield temperature not over 170 F (76.7 C), not less than 157 F (69.5 C). Each device must be resistant to extrusion of the fusible alloy and leak-tight at 130 F (54.4 C).) Each valve outlet must be sealed by threaded cap or a threaded solid plug. In addition, the valves must be protected by a metal cover. Such tanks are authorized for shipment by rail freight car and by highway.

METHOD OF MANUFACTURE

Most hydrogen sulfide produced today is made as a by-product of other processes. The largest tonnage currently made originates as a by-product in the natural gas industry, which mainly converts the hydrogen sulfide derived to commercial sulfur. Most of the hydrogen sulfide sold commercially in high-pressure cylinders is a purified, liquefied by-product of the chemical industry.

Many methods have been used in the past to produce hydrogen sulfide, with the classical method for many years consisting of iron sulfide plus an acid (usually hydrochloric). It can be made by combining sulfur with hydrogen in a noncatalytic reactor at elevated pressure and temperature.

REFERENCE

1. "Hydrogen Sulfide" (Safety Data Sheet SD-36), Chemical Manufacturers Association, 1825 Connecticut Ave., N.W., Washington, DC 20009.

Liquefied Petroleum Gases

Butane (Synonyms—Normal butane, *n*-butane, butyl hydride): C_4H_{10}
Butylenes (Synonym—butenes): C_4H_8
   1-butene (Synonyms—Ethylethylene, alpha-butene): $CH_2:CHCH_2CH_3$
   cis-2-butene (Synonyms—Dimethylethylene, beta-butylene, "high-boiling" butene-2): $CH_3CH:CHCH_3$
   trans-2-Butene (Synonyms—Dimethylethylene, beta-butylene, "low-boiling" butene-2): $CH_3CH:CHCH_3$
   Isobutene (Synonyms—2-methylpropene, isobutylene): $(CH_3)_2C:CH_2$
Isobutane (Synonyms—2-methylpropane, trimethylmethane): C_4H_{10} [or $(CH_3)_2CHCH_3$]
Propane (Synonym—Dimethylmethane): C_3H_8
Propylene (Synonym—Propene): C_3H_6 (or $CH_3CH:CH_2$)
DOT Classification: Flammable gas; flammable gas label

PHYSICAL CONSTANTS
(for Research Grade Products)

Butane

	U.S. Units	Metric Units
International symbol	LP-gas or LPG	LP-gas or LPG
Molecular weight	58.124	58.124
Vapor pressure		
at 70 F (21.1 C)	16.54 psig	114.04 kPa, gage
at 100 F (37.8 C)	36.92 psig	254.55 kPa, gage
at 115 F (46.1 C)	50.26 psig	346.53 kPa, gage
at 130 F (54.4 C)	66.03 psig	455.26 kPa, gage
Density of the gas at		
70 F (21.1 C) and 1 atm	0.15537 lb/ft³	2.489 kg/m³
Specific gravity of the gas		
at 70 F and 1 atm (air = 1)	2.0064	2.0064

	U.S. Units	Metric Units
Specific volume of the gas		
at 60 F (15.56 C) and 1 atm	6.3356 ft³/lb	0.3955 m³/kg
Density of liquid at saturation		
pressure*		
at 60 F (15.6 C)	36.39 lb/ft³	582.91 kg/m³
at 70 F (21.1 C)	35.95 lb/ft³	575.86 kg/m³
at 105 F (40.6 C)	34.38 lb/ft³	550.71 kg/m³
at 115 F (46.1 C)	34.01 lb/ft³	544.79 kg/m³
at 130 F (54.4 C)	33.38 lb/ft³	534.70 kg/m³
Boiling point at 1 atm	31.10 F	-0.51 C
Freezing point at 1 atm	-217.05 F	-138.36 C
Critical temperature	305.65 F	152.03 C
Critical pressure	550.7 psia	3796.94 kPa, abs.
Critical density	14.2 lb/ft³	227.46 kg/m³
Latent heat of vaporization		
at 31.10 F (0.51 C)	153.59 Btu/lb	357.25 kJ/kg
Latent heat of fusion at		
217.05 F (138.36 C)	10.64 Btu/lb	24.75 kJ/kg
Specific heat of gas		
at 60 F (15.6 C) and 1 atm		
C_p	0.3991 Btu/(lb)(F)	1.671 kJ/(kg)(C)
C_v	0.3649 Btu/(lb)(F)	1.528 kJ/(kg)(C)
Ratio of specific heats	1.094	1.094
Solubility in water, vol/vol		
at 100 F (37.78 C)	0.000061	0.000061
Weight of liquid at saturation		
pressure and 60 F	4.865 lb/gal	582.955 kg/m³
Specific heat of liquid		
at 1 atm	0.5636 Btu/(lb)(F)	2.3597 kJ/kg (C)
Gross heat of combustion		
Ideal gas at 60 F (15.6 C)		
and 1 atm	3262.1 Btu/ft³	121,542 kJ/m³
Liquid at 60 F (15.6 C)		
and saturation pressure	21139. Btu/lb	49,169.3 kJ/kg
Liquid at 60 F (15.6 C)		
and saturation pressure	102989. Btu/gal	28,704,713 kJ/m³
Net heat of combustion		
Ideal gas at 60 F (15.6 C)		
and 1 atm	3010.4 Btu/ft³	112,164.3 kJ/m³
Liquid at 77 F (25.0 C)		
and saturation pressure	19494. Btu/lb	45,343.0 kJ/kg
Net heat of combustion		
Liquid at 77 F (25.0 C)		
and saturation pressure	93201. Btu/gal	25,976,637 kJ/m³

*Apparent value from weight in air of the air-saturated liquid.

	U.S. Units	Metric Units
Air required for combustion		
Vol of air per 1-unit volume		
of the ideal gas	31.02 ft^3, air	0.8784 m^3, air
Weight of air per 1-unit		
weight of the ideal gas	15.459 lb, air	7.0121 kg, air
Flammable limits in air,		
percent by volume		
Lower	1.8 percent	1.8 percent
Upper	8.4 percent	8.4 percent
Flash point	−101 F	−73.9 C

PHYSICAL CONSTANTS
1-Butene

	U.S. Units	Metric Units
International symbol		
Molecular weight	56.108	56.108
Vapor pressure		
at 70 F (21.1 C)	23.45 psig	161.68 kPa, gage
at 100 F (37.8 C)	47.59 psig	328.12 kPa, gage
at 115 F (46.1 C)	63.27 psig	436.23 kPa, gage
at 130 F (54.4 C)	81.72 psig	563.44 kPa, gage
Density of the gas		
at 70 F (21.1 C) and 1 atm	0.14949 lb/ft^3	2.395 kg/m^3
Specific gravity of the gas		
at 70 F and 1 atm (air = 1)	1.9368	1.9368
Specific volume of the gas		
at 60 F (15.6 C) and 1 atm	6.551 ft^3/lb	0.4090 m^3/kg
Density of liquid at saturation pressure		
at 60 F (15.6 C)	37.43 lb/ft^3	599.57 kg/m^3
at 70 F (21.1 C)	36.82 lb/ft^3	589.80 kg/m^3
at 105 F (40.6 C)	35.26 lb/ft^3	564.81 kg/m^3
at 115 F (46.1 C)	34.69 lb/ft^3	555.68 kg/m^3
at 130 F (54.4 C)	33.90 lb/ft^3	543.03 kg/m^3
Boiling point at 1 atm	20.75 F	−6.25 C
Freezing point at 1 atm	−301.63 F	−185.35 C
Critical temperature	295.5 F	146.39 C
Critical pressure	583.0 psia	4019.64 kPa, abs.
Critical density	14.6 lb/ft^3	233.87 kg/m^3
Latent heat of vaporization		
at boiling point	167.94 Btu/lb	390.63 kJ/kg
Latent heat of fusion	0.06501 Btu/lb	0.15121 kJ/kg
Specific heat of gas		
at 60 F (15.6 C) and 1 atm		
C_p	0.3543 Btu/(lb)(F)	1.483 kJ/(kg)(C)
C_v	0.3189 Btu/(lb)(F)	1.335 kJ/(kg)(C)
Ratio of specific heats	1.11	1.11
Solubility in water, vol/vol		
at 100 F (37.8 C)	0.00022	0.00022
Weight of liquid at saturation pressure at 60 F (15.6 C)	5.004 lb/gal	599.61 kg/m^3
Heat of combustion		
gross	3177.9 Btu/ft^3	118405.2 kJ/m^3
net	2970.3 Btu/ft^3	110670.2 kJ/m^3

PHYSICAL CONSTANTS
cis-2-Butene

	U.S. Units	Metric Units
International symbol		
Molecular weight	56.108	56.108
Vapor pressure		
at 70 F (21.1 C)	12.67 psig	87.36 kPa, gage
at 100 F (37.8 C)	31.37 psig	216.29 kPa, gage
at 115 F (46.1 C)	43.73 psig	301.51 kPa, gage
at 130 F (54.4 C)	58.44 psig	402.93 kPa, gage
Density of the gas		
at 70 F (21.1 C) and 1 atm	0.15004 lb/ft^3	2.4034 kg/m^3
Specific gravity of the gas		
at 70 F and 1 atm (air = 1)	1.9368	1.9368
Specific volume of the gas		
at 60 F (15.6 C) and 1 atm	6.523 ft^3/lb	0.4072 m^3/kg
Density of liquid at saturation pressure		
at 60 F (15.6 C)	39.04 lb/ft^3	625.36 kg/m^3
at 70 F (21.1 C)	38.50 lb/ft^3	616.71 kg/m^3
at 105 F (40.6 C)	37.01 lb/ft^3	592.84 kg/m^3
at 115 F (46.1 C)	36.55 lb/ft^3	585.47 kg/m^3
at 130 F (54.4 C)	35.86 lb/ft^3	574.42 kg/m^3
Boiling point at 1 atm	38.69 F	3.72 C
Freezing point at 1 atm	−218.06 F	−138.92 C
Critical temperature	324.32 F	162.4 C
Critical pressure	610 psia	4206 kPa, abs.
Critical density	15.0 lb/ft^3	240.28 kg/m^3
Latent heat of vaporization		
at boiling point	178.91 Btu/lb	416.14 kJ/kg
Latent heat of fusion		
at −218.06 F (−138.92 C)	0.12347 Btu/lb	0.28719 kJ/kg
Specific heat of gas		
at 60 F (15.6 C) and 1 atm		
C_p	0.3222 Btu/(lb)(F)	1.349 kJ/(kg)(C)
C_v	0.2868 Btu/(lb)(F)	1.201 kJ/(kg)(C)
Ratio of specific heats	1.123	1.123
Weight of liquid at saturation pressure at 60 F (15.6 C)	5.219 lb/gal	625.374 kg/m^3
Heat of combustion		
gross	3168 Btu/ft^3	118036 kJ/m^3
net	2960.5 Btu/ft^3	110305 kJ/m^3

PHYSICAL CONSTANTS

trans-2-Butene

	U.S. Units	Metric Units
International symbol		
Molecular weight	56.108	56.108
Vapor pressure		
at 70 F (21.1 C)	15.20 psig	104.80 kPa, gage
at 100 F (37.8 C)	35.05 psig	241.66 kPa, gage
at 115 F (46.1 C)	48.10 psig	331.64 kPa, gage
at 130 F (54.4 C)	63.61 psig	438.58 kPa, gage
Density of the gas		
at 70 F (21.1 C) and 1 atm	0.15000 lb/ft^3	2.4028 kg/m^3
Specific gravity of the gas		
at 70 F and 1 atm (air = 1)	1.9368	1.9368
Specific volume of the gas		
at 60 F (15.6 C) and 1 atm	6.5245 ft^3/lb	0.4073 m^3/kg
Density of liquid at saturation pressure		
at 60 F (15.6 C)	37.97 lb/ft^3	608.22 kg/m^3
at 70 F (21.1 C)	37.47 lb/ft^3	600.21 kg/m^3
at 105 F (40.6 C)	35.89 lb/ft^3	574.90 kg/m^3
at 115 F (46.1 C)	35.43 lb/ft^3	567.53 kg/m^3
at 130 F (54.4 C)	34.71 lb/ft^3	556.00 kg/m^3
Boiling point at 1 atm	33.58 F	0.8777 C
Freezing point at 1 atm	-157.96 F	-105.53 C
Critical temperature	311.83 F	155.46 C
Critical pressure	595 psia	4102 kPa, abs.
Critical density	14.7 lb/ft^3	235.47 kg/m^3
Latent heat of vaporization		
at 33.58 F (0.8777 C)	174.39 Btu/lb	405.63 kJ/kg
Latent heat of fusion		
at -157.96 F (-105.53 C)	0.16483 Btu/lb	0.38339 kJ/kg
Specific heat of gas		
at 60 F (15.6 C) and 1 atm		
C_p	0.3618 Btu/(lb)(F)	1.5147 kJ/(kg)(C)
C_v	0.3264 Btu/(lb)(F)	1.3666 kJ/(kg)(C)
Ratio of specific heats	1.108	1.108
Weight of liquid at saturation		
pressure at 60 F (15.6 C)	5.076 lb/gal	608.239 kg/m^3
Heat of combustion		
gross	3163 Btu/ft^3	117850 kJ/m^3
net	2957 Btu/ft^3	110175 kJ/m^3

PHYSICAL CONSTANTS

Isobutene

	U.S. Units	Metric Units
International symbol		
Molecular weight	56.108	56.108
Vapor pressure		
at 70 F (21.1 C)	23.85 psig	164.44 kPa, gage
at 100 F (37.8 C)	48.04 psig	331.22 kPa, gage
at 115 F (46.1 C)	63.78 psig	439.75 kPa, gage
at 130 F (54.4 C)	82.33 psig	567.65 kPa, gage
Density of the gas		
at 70 F (21.1 C) and 1 atm	0.14957 lb/ft^3	2.3959 kg/m^3
Specific gravity of the gas		
at 70 F and 1 atm (air = 1)	1.997	1.997
Specific volume of the gas		
at 60 F (15.6 C) and 1 atm	6.545 ft^3/lb	0.4086 m^3/kg
Density of liquid at saturation		
pressure		
at 60 F (15.6 C)	37.37 lb/ft^3	598.61 kg/m^3
at 70 F (21.1 C)	36.90 lb/ft^3	591.08 kg/m^3
at 105 F (40.6 C)	35.25 lb/ft^3	564.65 kg/m^3
at 115 F (46.1 C)	34.76 lb/ft^3	556.80 kg/m^3
at 130 F (54.4 C)	34.01 lb/ft^3	544.79 kg/m^3
Boiling point at 1 atm	19.59 F	-6.8944 C
Freezing point at 1 atm	-220.61 F	-140.34 C
Critical temperature	292.51 F	144.73 C
Critical pressure	580.2 psia	4000.34 kPa, abs.
Critical density	14.7 lb/ft^3	235.47 kg/m^3
Latent heat of vaporization		
at boiling point	169.48 Btu/lb	394.21 kJ/kg
Latent heat of fusion		
at -220.61 F (-140.34 C)	0.10020 Btu/lb	0.23307 kJ/kg
Specific heat of gas		
at 60 F (15.6 C) and 1 atm		
C_p	0.3701 Btu/(lb)(F)	1.550 kJ/(kg)(C)
C_v	0.3347 Btu/(lb)(F)	1.401 kJ/(kg)(C)
Ratio of specific heats	1.106	1.106
Weight of liquid at saturation		
pressure at 60 F (15.6 C)	4.996 lb/gal	598.65 kg/m^3
Heat of combustion		
gross	3156 Btu/ft^3	117589.24 kJ/m^3
net	2949 Btu/ft^3	109876.64 kJ/m^3

PHYSICAL CONSTANTS

Isobutane

	U.S. Units	Metric Units
International symbol		
Molecular weight	58.124	58.124
Vapor pressure		
at 70 F (21.1 C)	30.58 psig	210.84 kPa, gage
at 100 F (37.8 C)	57.87 psig	399.00 kPa, gage
at 115 F (46.1 C)	75.38 psig	519.73 kPa, gage
at 130 F (54.4 C)	95.86 psig	660.93 kPa, gage
Density of the gas		
at 70 F (21.1 C) and 1 atm	0.15474 lb/ft^3	2.4787 kg/m^3
Specific gravity of the gas		
at 70 F and 1 atm (air = 1)	2.00636	2.00636
Specific volume of the gas		
at 60 F (15.6 C) and 1 atm	6.3355 ft^3/lb	0.3955 m^3/kg
Density of liquid at saturation pressure		
at 60 F (15.6 C)	35.05 lb/ft^3	561.45 kg/m^3
at 70 F (21.1 C)	34.82 lb/ft^3	557.76 kg/m^3
at 105 F (40.6 C)	33.26 lb/ft^3	532.77 kg/m^3
at 115 F (46.1 C)	32.82 lb/ft^3	525.73 kg/m^3
at 130 F (54.4 C)	32.01 lb/ft^3	512.75 kg/m^3
Boiling point at 1 atm	10.90 F	-11.72 C
Freezing point at 1 atm	-255.29 F	-159.61 C
Critical temperature	274.96 F	134.98 C
Critical pressure	529.1 psia	3648.02 kPa, abs.
Critical density	13.8 lb/ft^3	221.05 kg/m^3
Latent heat of vaporization		
at boiling point	157.53 Btu/lb	366.41 kJ/kg
Specific heat of gas		
at 60 F (15.6 C) and 1 atm		
C_p	0.3906 Btu/(lb)(F)	1.6354 kJ/(kg)(C)
C_v	0.3564 Btu/(lb)(F)	1.4922 kJ/(kg)(C)
Ratio of specific heats	1.096	1.096
Solubility in water, vol/vol		
at 100 F (37.8 C)	0.000052	0.000052
Weight of liquid at saturation pressure at 60 F (15.6 C)	4.686 lb/gal	561.51 kg/m^3
Heat of combustion		
gross	3352.7 Btu/ft^3	124918.08 kJ/m^3
net	3001.1 Btu/ft^3	111817.83 kJ/m^3

PHYSICAL CONSTANTS

Propane

	U.S. Units	Metric Units
International symbol	LP-gas or LPG	LP-gas or LPG
Molecular weight	44.097	44.097
Vapor pressure		
at 70 F (21.1 C)	109.73 psig	756.56 kPa, gage
at 100 F (37.8 C)	173.38 psig	1195.41 kPa, gage
at 115 F (46.1 C)	212.95 psig	1468.24 kPa, gage
at 130 F (54.4 C)	258.37 psig	1781.40 kPa, gage
Density of the gas		
at 70 F (21.1 C) and 1 atm	0.11599 lb/ft^3	1.8580 kg/m^3
Specific gravity of the gas		
at 70 F and 1 atm (air = 1)	1.5223	1.5223
Specific volume of the gas		
at 60 F (15.6 C) and 1 atm	8.4515 ft^3/lb	0.5276 m^3/kg
Density of liquid at saturation pressure		
at 60 F (15.6 C)	31.59 lb/ft^3	506.02 kg/m^3
at 70 F (21.1 C)	31.20 lb/ft^3	499.78 kg/m^3
at 105 F (40.6 C)	29.33 lb/ft^3	469.82 kg/m^3
at 115 F (46.1 C)	28.70 lb/ft^3	459.73 kg/m^3
at 130 F (54.4 C)	27.77 lb/ft^3	444.83 kg/m^3
Boiling point at 1 atm	−43.67 F	−42.04 C
Freezing point at 1 atm	−305.84 F	−187.69 C
Critical temperature	206.01 F	96.672 C
Critical pressure	616.3 psia	4249.24 kPa, abs.
Critical density	13.5 lb/ft^3	216.25 kg/m^3
Latent heat of vaporization at boiling point	183.05 Btu/lb	425.77 kJ/kg
Specific heat of gas at 60 F (15.6 C) and 1 atm		
C_p	0.3881 Btu/(lb)(F)	1.625 kJ/(kg)(C)
C_v	0.3430 Btu/(lb)(F)	1.436 kJ/(kg)(C)
Ratio of specific heats	1.131	1.131
Solubility in water, vol/vol at 64 F (17.8 C)	0.065	0.065
Weight of liquid at saturation pressure at 60 F (15.6 C)	4.223 lb/gal	506.03 kg/m^3
Heat of combustion		
gross	2517.5 Btu/ft^3	93799.41 kJ/m^3
net	2316.1 Btu/ft^3	86295.45 kJ/m^3

PHYSICAL CONSTANTS

Propylene

	U.S. Units	Metric Units
International symbol	Pry	Pry
Molecular weight	42.081	42.081
Vapor pressure		
at 70 F (21.1 C)	132.81 psig	915.69 kPa, gage
at 100 F (37.8 C)	206.92 psig	1426.66 kPa, gage
at 115 F (46.1 C)	253.03 psig	1744.58 kPa, gage
at 130 F (54.4 C)	306.05 psig	2113.24 kPa, gage
Density of the gas		
at 70 F (21.1 C) and 1 atm	0.110447 lb/ft^3	1.7692 kg/m^3
Specific gravity of the gas		
at 70 F and 1 atm (air = 1)	1.4529	1.4529
Specific volume of the gas		
at 60 F (15.6 C) and 1 atm	8.875 ft^3/lb	0.554 m^3/kg
Density of liquid at saturation		
pressure		
at 60 F (15.6 C)	37.43 lb/ft^3	599.57 kg/m^3
at 70 F (21.1 C)	32.07 lb/ft^3	513.71 kg/m^3
at 105 F (40.6 C)	29.89 lb/ft^3	478.79 kg/m^3
at 115 F (46.1 C)	29.20 lb/ft^3	467.74 kg/m^3
at 130 F (54.4 C)	28.08 lb/ft^3	449.80 kg/m^3
Boiling point at 1 atm	−53.90 F	−47.72 C
Freezing point at 1 atm	−301.45 F	−185.25 C
Critical temperature	197.2 F	91.77 C
Critical pressure	670.0 psia	4619.49 kPa, abs.
Critical density	14.5 lb/ft^3	232.27 kg/m^3
Latent heat of vaporization		
at boiling point and 1 atm	188.18 Btu/lb	437.71 kJ/kg
Latent heat of fusion		
at −301.45 F (−185.25 C)	0.0673 Btu/lb	0.1565 kJ/kg
Specific heat of gas		
at 60 F (15.6 C) and 1 atm		
C_p	0.3549 Btu/(lb)(F)	1.4859 kJ/(kg)(C)
C_v	0.3077 Btu/(lb)(F)	1.2883 kJ/(kg)(C)
Ratio of specific heats	1.153	1.153
Solubility in water, vol/vol		
at 100 F (37.8 C)	0.0009	0.0009
Weight of liquid at saturation		
pressure at 60 F (15.6 C)	4.343 lb/gal	520.41 kg/m^3
Heat of combustion		
gross	2371.7 Btu/ft^3	88367.05 kJ/m^3
net	2218.3 Btu/ft^3	82651.53 kJ/m^3

DESCRIPTION

The liquefied petroleum gases are butane, isobutane, propane, propylene (propene), butylenes (butenes) and any mixtures of these hydrocarbons, in the generally accepted definition of the National Fire Protection Association. They are flammable, colorless, noncorrosive and nontoxic. Easily liquefied under pressure at atmospheric temperature, they are shipped and stored compactly as liquids and are used in gaseous and liquid form in their very large and diverse applications as fuels. Propane, isobutane and butane are among the lightest hydrocarbons in the liquid phase in the paraffin series, occurring between ethane (natural gas) and pentane (the lightest natural gasoline fraction). Propylene, isobutene, 1-butene and 2-butene are among the lightest hydrocarbons in the monoolefin series, occurring between ethene and pentene. (The ending "-ane" indicates a member of the paraffin series, while "-ene" indicates a member of the monoolefin series.) The liquefied petroleum gases in the paraffin series are chemically stable and odorless, and the DOT and other regulating bodies require artificial odorization of propane and butane (except in technical uses where the odorant would harm further processing and the odorant warning action would not be important). Propylene and the butylenes have an unpleasant odor characteristic of petroleum refinery gas or coal gas.

All the liquefied petroleum gases are soluble to varying degrees in alcohol and ether. Propane and propylene are slightly soluble in water.

GRADES AVAILABLE

The liquefied petroleum gases are supplied in various scientific and commercial grades. Their properties differ according to the grade being used. Properties are given for the research grades of these gases in the preceding tables of "Physical Constants," as noted in the tables. (Properties appearing in the tables, like those presented for gases throughout this second part of the *Handbook*, are drawn from authoritative scientific and industry sources (see Ref. 1).)

As an example of the way properties of the various commercial grades of the liquefied petroleum gases generally differ from the properties for research grades that are shown in the "Physical Constants" tables, one producer gives average vapor pressure properties for commercial grades of propane and butane as follows:

	Commercial Propane	Commercial Butane
Vapor pressure, psig		
at 70 F	124	31
at 100 F	192	59
at 105 F	206	65
at 130 F	286	97

Moreover, these gases are often used in mixtures designed to have certain desired properties; in particular, propane and butane are frequently ordered as mixtures to meet certain boiling point and other requirements of individual applications. Suppliers furnish physical constants data for the various grades and mixtures they make available.

USES

Propane and butane—known most extensively in commercial and popular terms as LP-gas or LPG—have an extremely wide range of domestic, industrial, commercial, agricultural and internal combustion engine uses. It is estimated that the two gases, unmixed and in mixtures, have several thousand industrial applications and scores more in other fields. Their very broad application stems from their occurrence as hydrocarbons between natural gas and natural gasoline and from their corresponding properties.

The liquefied petroleum gases are used:

(1) As appliance fuel for space heating, water heating, boiler heating, cooking, baking, air conditioning and refrigeration in rural or urban areas beyond the reach of gas mains.

(2) In bulk by utilities and industries (especially industries using kilns or furnaces which must be maintained continuously at given temperatures), as standby fuel to protect against failure or interruption of natural or artificial gas supply. By utilities to bridge

peak load demands for natural or synthetic or substitute natural gas, and for gas enrichment. By utilities serving rural communities from central plants.

(3) For space heating during the erection of buildings.

(4) As fuel for the entire range of industrial heating processes, especially those where the Btu value must be accurately controlled. Industrial heating process uses include heat treating, stress relieving, annealing, enamel baking and firing ceramic kilns and furnaces. For brazing, metal cutting and soldering, and also as an atmosphere-producing gas for bright annealing.

(5) As fuel for such operations as poultry brooding, cotton and grain drying, tobacco curing, crop dehydration, weed burning and orchard heating.

(6) As fuel for vehicles such as trucks, buses, taxicabs, and fork lift trucks, and for mobile farm machinery like tractors and harvesters. As fuel for stationary engines powering well pumps, electric generators, etc. Engines especially designed for LP-gas are available, and gasoline engines may readily be converted for LP-gas operation.

(7) Isobutane and the gases in the monoolefin series are used less extensively than propane and butane. Isobutane and isobutene are used in manufacturing alkylate for increasing gasoline octane ratings. Isobutene is employed in the manufacture of synthetic rubber. Propylene is used substantially in the chemical industry for synthesis in the production of a wide range of products. The butenes are used in preparing a large number of organic compounds.

PHYSIOLOGICAL EFFECTS

The liquefied petroleum gases are nontoxic. Prolonged inhalation of high concentrations has an anesthetic effect; also, due to their ability to displace oxygen in the air, they can act as simple asphyxiants. Contact between the skin and these gases in liquid form can cause freezing of tissue and results in injury similar to a thermal burn.

SAFE HANDLING AND STORAGE

Materials Suitable for Containers and Storage

Any common, commercially available metals may be used with commercial grade liquefied petroleum gases because they are noncorrosive, though installations must be designed to withstand the pressures involved and must comply with all state and local regulations. Widely accepted recommendations on storage systems and safe usage are given in "Standard for the Storage and Handling of Liquefied Petroleum Gases," Pamphlet No. 58 of the National Fire Protection Association (which is also American Standard Z106.1) (see Ref. 2). Similar recommendations for larger storage systems appear in NFPA Pamphlet No. 59 (see Ref. 3).

Storage and Handling in Normal Use

Steel tanks and large underground chambers are used for the storage of liquefied petroleum gas. Steel tanks aboveground or below ground range up to 120,000 gal (450 m^3) or more in capacity, while below-ground caverns or pits have been found to offer safe and economical storage where a suitable geological formation exists at a site of large-volume handling, such as at pipelines or marine terminals and at national gasoline plants or refineries. Widely recognized recommendations for LP-gas installations are presented in the previously cited NFPA Pamphlets No. 58 and No. 59.

Handling Leaks and Emergencies

Never use flame to test for an LP-gas leak; instead, apply soap water solution to areas suspected of leaking. Frost around valve stems, at piping joints or at other points may indicate a liquid leak.

METHODS OF SHIPMENT; REGULATIONS

Under the appropriate regulations and tariffs, liquefied petroleum gases are authorized for shipment as follows:

By Rail: In cylinders (via freight, express and as baggage), in insulated and uninsulated single-unit tank cars, and in TMU tank cars.

By Highway: In cylinders on trucks, in portable tanks, and in cargo tanks on trucks or on semi and full trailers.

By Water: In cylinders on passenger vessels and on ferry or railroad car ferry vessels (passenger or vehicle); in cylinders and portable tanks (meeting DOT 51 specifications and not over 20,000 lb (9072 kg) gross weight, with vapor pressure at 115 F (46.1 C) not exceeding the container service pressure) on cargo vessels; in authorized tank cars on trainships only, and in authorized tank trucks on trailerships and trainships only. In cylinders on barges of U.S. Coast Guard classes A, CA and CB only. In cargo tanks aboard tankships and tank barges (to maximum filling densities by specific gravity as prescribed in Coast Guard regulations).

By Air: In cylinders on cargo aircraft only up to 300 lb (140 kg) maximum net weight per cylinder.

By Pipeline: Propane and butane are also transported by pipeline from points of production to distant bulk storage facilities.

CONTAINERS

Propane and butane are authorized for shipment as liquefied compressed gases in cylinders, portable tanks and cargo tanks, and in insulated or uninsulated single-unit tank cars. DOT regulations also provide for their shipment in TMU (ton multi-unit) tanks and tank cars, but they are not generally shipped in TMU containers. The two gases are stored in large tanks aboveground and below ground, and also in very large underground chambers such as in natural salt deposits.

In addition to such storage in liquefied form under their vapor pressures at normal atmospheric temperatures, refrigerated storage in liquefied form under atmospheric pressures is used for propane and butane. Refrigerated storage systems are closed and insulated, and in them the LP-gas vapor is circulated through pumps and compressors to serve as the systems' refrigerant. Propane and butane are stored in

pits in the earth capped by metal domes as well as in underground chambers; one of the largest storage pits of this kind, located in Utah, holds propane in the millions of gallons in refrigerated liquid form.

Isobutane and the monoolefins are authorized for shipment in single-unit tank cars and truck cargo tanks, and are usually shipped in bulk units because they are generally used in large quantities. They are also authorized for shipment in cylinders, portable tanks and TMU tanks.

The maximum filling densities authorized by the DOT for liquefied petroleum gases are prescribed according to the specific gravity of the liquid material at 60 F (15.6 C) in detailed tables that are part of the DOT regulations. Producers and suppliers who charge LP-gas containers should consult these tables in the current regulations for the maximum densities to which to fill containers with the particular grades and mixtures they are handling. The lower and upper limits of the maximum filling densities authorized in the present regulations are as follows (percent water capacity by weight):

In cylinders—from 26 percent for 0.271–0.289 specific gravity, to 57 percent for 0.627–0.634 specific gravity.

In single-unit tank cars and TMU tanks—from 45.500 percent (insulated tanks, April through October) for 0.500 specific gravity, to 61.57 percent (uninsulated tanks, November through March with no storage in transit) for 0.635 specific gravity. (In addition, filling must not exceed various specified limits of pressure and liquid content at temperatures of 105 F (40.6 C) or 130 F (54.4 C) as given in DOT regulations.)

In cargo tanks and portable tanks—from 38 percent (tanks of 1200-gal or 4542 liters capacity or less) for 0.473–0.480 specific gravity, to 60 percent (tanks of over 1200-gal capacity) for 0.627 and over specific gravity (except when using fixed-length dip tube or other fixed maximum liquid level indicators). Moreover, the tank must not be liquid full at 105 F (40.6 C) if insulated nor at 115 F (46.1 C) if uninsulated, and the gage vapor pressure at 115 F must not exceed the tank's design pressure.

Cylinders

Cylinders authorized for liquefied petroleum gas service include those which comply with the following DOT specifications: 3A, 3AA, 3B, 3F, 4A, 4B, 4BA, and 4B240ET, 4B240FLW, 4BW, 4F, 4, 9, and 41. (Cylinders meeting the following DOT specifications may be continued in use, but new construction is not authorized: 3, 4B240X, 25, 26 and 38. DOT regulations also authorize shipment in several special types of small containers.)

All of these types of cylinders must be requalified by hydrostatic retest every 5 years under present DOT regulations except as follows:

(1) no periodic retest is required for 3E cylinders;

(2) 10 years is the required retest interval for type 4 cylinders; and

(3) external visual inspection may be used in lieu of hydrostatic retest for cylinders that are used exclusively for liquefied petroleum gas which is commercially free from corroding components and that are of the following types (including cylinders of these types with higher service pressures): 3A480, 3AA480, 3A480X, 3B, 4B, 4BA, 4E, 4BW, 26-240 and 260-300.

Valve Outlet Connections. Standard connection, United States and Canada (for butane, isobutane, propane, and propylene or propene)— No. 510. No. 555 is used for propane liquid withdrawal.

Cargo Tanks

Liquefied petroleum gas is authorized for shipment in truck or truck-trailer cargo tanks that comply with DOT specifications MC-330 or MC-331. Various design pressures may be used, but the gage vapor pressure at 115 F (46.1 C) of a shipment must not exceed the tank's design pressure.

Portable Tanks

Liquefied petroleum gases may be shipped in portable tanks meeting specifications DOT-51. Design pressures are the same as for cargo tanks, above.

TMU Tanks

TMU tank car shipment, though little used, is also authorized for LP-gas with pressure not exceeding 375 psi at 130 F (2590 kPa at 54 C).

Tank Cars

Bulk shipment of liquefied petroleum gas is authorized for single-unit tank cars meeting DOT specifications of the 105A series, 105A tanks of given service pressures being specified for gases having vapor pressures of stipulated maximum values at 105 F (40.6 C) or 115 F (46.1 C); single-unit tank cars in the 112A series and of the 114A340W type are also authorized under specific conditions.

METHOD OF MANUFACTURE

Butane and propane (with other hydrocarbons in the paraffin series) are recovered from "wet" natural gas, from natural gas associated with or dissolved in crude oil, and from petroleum refinery gases. They may be separated from wet natural gas or crude oil through absorption in light "mineral seal" oil, adsorption on surfaces such as activated charcoal, or by refrigeration, followed in each case by fractionation. Propylene and other gases in the monoolefin series are recovered from petroleum gases by fractionation.

REFERENCES

1. NGPA 2145—Physical Constants for Paraffin Hydrocarbons; NGPSA Engineering Data Book, Ninth Edition, 1972; API Technical Data Book; and ASTM Data Series 4.
2. "Standard for the Storage and Handling of Liquefied Petroleum Gases" (Pamphlet No. 58), National Fire Protection Association, 470 Atlantic Ave., Boston, MA 02210.
3. "Standard for the Storage and Handling of Liquefied Petroleum Gases at Utility Gas Plants" (Pamphlet No. 59), National Fire Protection Association.

Methane

CH$_4$
Synonyms: Marsh gas, methyl hybride
DOT Classification: Flammable gas; flammable gas label

PHYSICAL CONSTANTS

	U.S. Units	Metric Units
International symbol	CH$_4$	CH$_4$
Molecular weight	16.042	16.042
Density of the gas at 60 F (15.6 C) and 1 atm	0.04235 lb/ft^3	0.6784 kg/m^3
Specific gravity of the gas at 60 F (15.6 C) and 1 atm (air = 1)	0.55491	0.55491
Specific volume of the gas at 60 F (15.6 C) and 1 atm	23.6113 ft^3/lb	1.47400 m^3/kg
Density of liquid at boiling point	26.57 lb/ft^3	425.61 kg/m^3
Boiling point at 1 atm	-258.68 F	-161.49 C
Freezing point at 1 atm	-296.7 F	-182.61 C
Critical temperature	-115.78 F	-82.100 C
Critical pressure	673.1 psia	4640.86 kPa, abs.
Critical density	10.09 lb/ft^3	161.63 kg/m^3
Triple point	-296.5 F at 1.69 psia	-182.5 C at 11.65 kPa, abs.
Latent heat of vaporization at boiling point and 1 atm	219.22 Btu/lb	509.91 kJ/kg
Latent heat of fusion at -296.5 F (-182.5 C)	0.05562 Btu/lb	0.1294 kJ/kg
Specific heat of ideal gas at 60 F (15.6 C) and 1 atm		
C_p	0.5271 Btu/(lb) (F)	2.207 kJ/(kg) (C)
C_v	0.4032 Btu/(lb) (F)	1.688 kJ/(kg) (C)
Ratio of specific heats	1.307	1.307
Weight of liquid at boiling point	3.552 lb/gal	425.6 kg/m^3

	U.S. Units	Metric Units
Gross heat of combustion of the real gas at 60 F (15.6 C) and 1 atm	1011.6 Btu/ft^3	37 691.15 kJ/m^3
Net heat of combustion of the real gas at 60 F (15.6 C) and 1 atm	910.77 Btu/ft^3	33 934.33 kJ/m^3
Air required for combustion at 60 F and 1 atm		
Per ft^3 of the real gas	9.563 ft^3 air	0.2708 m^3 air
Per lb of the real gas	17.233 lb air	7.8167 kg air
Flammable limits in air	5.0–15 percent	5.0–15 percent
Flash point	–306 F	–188 C

DESCRIPTION

Methane is a colorless, odorless, tasteless flammable gas. It is the first member of the paraffin (aliphatic or saturated) series of hydrocarbons. It is soluble in alcohol or ether, and slightly soluble in water. Methane is shipped as a nonliquefied compressed gas in cylinders at pressures up to 6000 psig at 70 F (15,615 kPa gage at 21.1 C).

Methane is a major constituent of coal gas and is present to an extent in air in coal mines.

GRADES AVAILABLE

Methane is typically available for commercial and industrial purposes in a C.P. Grade (minimum purity of 99 mole percent), a technical grade (minimum purity of 98.0 mole percent), and a commercial grade which is actually Tennessee natural gas as it is received from the pipeline (there is no guaranteed purity but methane content usually runs about 93 percent or better). A typical analysis for commercial grade methane is as follows:

Methane	93.63 mole percent
Carbon dioxide	0.70 mole percent
Nitrogen	0.47 mole percent
Ethane	3.58 mole percent
Propane	1.02 mole percent
Isobutane	0.21 mole percent
n-Butane	0.19 mole percent
Isopentane	0.06 mole pecent
n-Pentane	0.06 mole percent
Hexane	0.02 mole percent
Heptanes plus	0.06 mole percent

Tertiary-butyl mercaptan is added in trace amounts as an odorizer. This grade of gas has sulfur content of 0.002 grains/100 ft^3 (0.0457 mg/m^3) and a typical gross heating value of 1044 Btu/ft^3 (37.28 kJ/m^3).

USES

Methane in natural gas serves very widely as a fuel. In the chemical industry, it is used heavily as a raw material for making important products that include acetylene, ammonia, ethanol and methanol; its chlorination also yields carbon tetrachloride, chloroform, methyl chloride and methylene chloride. The burning of high-purity methane is used to make carbon black of special quality for electronic devices.

The use of natural gas as a motor fuel handled as a compressed gas in high-pressure cylinders or liquid dewars is gaining wider acceptance.

PHYSIOLOGICAL EFFECTS

Methane is generally considered nontoxic. Coal miners inhale concentrations of up to 9 percent methane in air without apparent ill effects; inhalation of higher concentrations eventually causes a feeling of pressure on the forehead and eyes, but the sensation ends after returning to fresh air. Methane is a simple asphyxiant.

SAFE HANDLING AND STORAGE

Materials Suitable for Handling and Storage

Methane is noncorrosive, and may be contained by any common, commercially available metals. Handling equipment must, however, be designed to withstand safely the pressures involved.

Storage and Handling in Normal Use

All the precautions necessary for the safe handling of any flammable compressed gas must be observed in working with methane. Any shipping mode or regulation applicable for methane may also apply for natural gas (see also Part I, Chapter 4, Section A).

Handling Leaks and Emergencies

See Part I, Chapter 4, Section A.

METHODS OF SHIPMENT; REGULATIONS

Under the appropriate regulations and tariffs, methane and natural gas are authorized for shipment as follows:

By Rail: In cylinders (via freight or express). As a liquid in tank cars designed for cryogenic gases, shipped under DOT exemption.

By Highway: In cylinders on trucks. As a liquid in tanks designed for cold compressed gases liquefied at low temperatures.

By Water: In cylinders aboard cargo vessels only. In cylinders on barges of U.S. Coast Guard classes A, CA and CB only. As a liquid in cargo tankers or in barge tanks designed for cold compressed gases liquefied at low temperatures.

By Air: In cylinders aboard cargo aircraft only up to 300 lbs (140 kg) maximum net weight per cylinder.

By Pipeline: As gas in high-pressure pipelines or as a liquid in lines designed for cold compressed gases liquefied at low temperatures.

CONTAINERS

Methane may be shipped in any cylinders of the types authorized for nonliquefied compressed gases. Bulk industrial users of methane usually receive it in natural gas by pipeline, and purify it if necessary for further processing. Transatlantic shipment of methane has been made in ships carrying it liquefied in insulated tanks at a temperature of some -260 F (-162 C).

Filling Limits: The maximum filling densities permitted for methane or natural gas in cylinders at 70 F (21.1 C) are the authorized service pressures of the cylinders into which it is charged.

Cylinders

Any cylinders authorized for the shipment of a nonliquefied compressed gas may be used in methane service under DOT regulations, but cylinders of the 3A and 3AA types are probably those most commonly used for methane. Authorized cylinders for methane service include those meeting DOT specifications 3A, 3AA, 3B, 3C, 3D, 3E, 4, 4A, 4BA and 4C; cylinders meeting specifications, 3, 7, 25, 26, 33 and 38 may also be continued in methane service, but new construction is not authorized.

All types of cylinders authorized for methane shipment must be requalified by periodic hydrostatic retest every 5 years under present regulations, with the following exceptions: DOT-4 cylinders, every 10 years; and no periodic retest is required for cylinders of types 3C, 3E, 4C and 7.

Valve Outlet and Inlet Connections. Standard connections, United States and Canada— up to 3000 psi (20,684 kPa), No. 350; 3000-10,000 psi (20,684-68,948 kPa) No. 677, see Part I, Chapter 7, Section A, paragraph 2.1.7; cryogenic liquid withdrawal, No. 450. Alternate standard connection, United States and Canada —3000-10,000 psi, No. 677.

METHOD OF MANUFACTURE

Methane is produced commercially from natural gas by absorption or adsorption methods of purification; supercooling and distillation are sometimes employed to secure methane of very high purity. Some California natural gas wells produce methane of high purity. It can also be obtained by cracking petroleum fractions.

Methylamines (Anhydrous)

Monomethylamine (Synonyms—Methylamine, aminomethane): CH_3NH_2
Dimethylamine: $(CH_3)_2NH$
Trimethylamine: $(CH_3)_3N$
DOT Classification: Flammable gas; flammable gas label

PHYSICAL CONSTANTS

Monomethylamine

	U.S. Units	Metric Units
International symbol	CH_3NH_2	CH_3NH_2
Molecular weight	31.058	31.058
Vapor pressure at 68 F (20 C)	14.1 psia	97.2161 kPa, abs.
Density of the gas at 68 F (20 C) and 1 atm	41.32 lb/ft³	661.88 kg/m³
Specific gravity of the gas at 59 F and 1 atm (air = 1)	1.07	1.07
Specific volume of the gas at 70 F (21.1 C) and 1 atm	12.1 ft³/lb	0.755 m³/kg
Density of liquid		
at 68 F (20 C)	41.35 lb/ft³	662.36 kg/m³
at 70 F (21.1 C)	41.4 lb/ft³	663.164 kg/m³
at 105 F (40.5 C)	40.1 lb/ft³	642.340 kg/m³
at 115 F (46.1 C)	39.8 lb/ft³	637.535 kg/m³
at 130 F (54.4 C)	39.2 lb/ft³	627.924 kg/m³
Boiling point at 1 atm	20.6 F	-6.33 C
Freezing point at 1 atm	-136.3 F	-93.5 C
Critical temperature	314.4 F	156.9 C
Critical pressure	1081.9 psia	7459.44 kPa, abs.
Critical density	13.47 lb/ft³	215.77 kg/m³
Latent heat of vaporization at boiling point	357.5 Btu/lb	831.5 kJ/kg
Latent heat of fusion at freezing point	84.95 Btu/lb	197.6 kJ/kg

	U.S. Units	Metric Units
Specific heat of gas at 77 F (25 C) and 1 atm		
C_p	0.4343 Btu/(lb) (F)	1.818 kJ/(kg) (C)
C_v	0.425 Btu/(lb) (F)	1.78 kJ/(kg) (C)
Ratio of specific heats	1.202	1.202
Solubility in water, wt/wt at 77 F (25 C), 1 atm	1.08 percent	1.08 percent
Weight of liquid at 68 F (20 C)	5.55 lb/gal	665 kg/m³
Autoignition temperature in air	806 F	430 C
Flammable limits in air, by volume	4.9-20 percent	4.9-20 percent

PHYSICAL CONSTANTS

Dimethylamine

	U.S. Units	Metric Units
International symbol	$(CH_3)_2NH$	$(CH_3)_2NH$
Molecular weight	45.08	45.08
Vapor pressure at 68 F (20 C)	-3.7 psia	-25.511 kPa, abs.
Specific gravity of the gas at 59 F and 1 atm (air = 1)	1.55	1.55
Specific volume of the gas at 70 F (21.1 C) and 1 atm	8.6 ft³/lb	0.54 m³/kg
Density of liquid		
at 70 F (21.1 C)	40.8 lb/ft³	653.553 kg/m³
at 105 F (40.5 C)	39.4 lb/ft³	631.127 kg/m³
at 115 F (46.1 C)	39.0 lb/ft³	624.720 kg/m³
at 130 F (54.4 C)	38.5 lb/ft³	616.71 kg/m³
Boiling point at 1 atm	44.4 F	6.89 C
Freezing point at 1 atm	-134 F	-92.22 C
Critical temperature	328.3 F	164.6 C
Critical pressure	760.0 psia	5240 kPa, abs.
Critical density	15.98 lb/ft³	255.97 kg/m³
Latent heat of vaporization at boiling point	252.8 Btu/lb	588.01 kJ/kg
Latent heat of fusion at freezing point	56.7 Btu/lb	131.88 kJ/kg
Specific heat of gas at 77 F (25 C) and 1 atm		
C_p	0.3819 Btu/(lb) (F)	1.599 kJ/(kg) (C)
C_v	0.3324 Btu/(lb) (F)	1.392 kJ/(kg) (C)

	U.S. Units	Metric Units
Ratio of specific heats	1.149	1.149
Solubility in water, weight percent at 140 F (60 C) and 1 atm	23.7 percent	23.7 percent
Weight of liquid at 68 F (20 C)	5.48 lb/gal	656.65 kg/m³
Autoignition temperature in air	756 F	402 C
Flammable limits in air by volume	2.8-14.4 percent	2.8-14.4 percent

PHYSICAL CONSTANTS

Trimethylamine

	U.S. Units	Metric Units
International symbol	$(CH_3)_3N$	$(CH_3)_3N$
Molecular weight	59.11	59.11
Vapor pressure at 68 F (20 C)	-1.7 psia	-11.721 kPa, abs.
Specific volume of the gas at 70 F (21.1 C) and 1 atm	6.4 ft³/lb	0.40 m³/kg
Density of liquid		
at 70 F (21.1 C)	39.4 lb/ft³	631.127 kg/m³
at 105 F (40.5 C)	38.1 lb/ft³	610.303 kg/m³
at 115 F (46.1 C)	37.7 lb/ft³	603.896 kg/m³
at 130 F (54.4 C)	37.1 lb/ft³	594.285 kg/m³
Boiling point at 1 atm	37.2 F	2.89 C
Freezing point at 1 atm	-178.8 F	-117.1 C
Critical temperature	320.2 F	160.1 C
Critical pressure	590.9 psia	4074.11 kPa, abs.
Critical density	14.55 lb/ft³	233.07 kg/m³
Latent heat of vaporization at boiling point	166.9 Btu/lb	388.2 kJ/kg
Latent heat of fusion at freezing point	47.6 Btu/lb	110.7 kJ/kg
Specific heat of gas at 77 F (25 C) and 1 atm		
C_p	0.3717 Btu/(lb) (F)	1.556 kJ/(kg) (C)
C_v	0.3139 Btu/(lb) (F)	1.314 kJ/(kg) (C)
Ratio of specific heats	1.184	1.184
Solubility in water, weight percent at 86 F (30 C) and 1 atm	47.5 percent	47.5 percent
Weight of liquid at 68 F (20 C)	5.31 lb/gal	636.28 kg/m³

	U.S. Units	Metric Units
Autoignition temperature in air	374 F	190 C
Flammable limits in air, by volume	2.0–11.6 percent	2.0–11.6 percent

DESCRIPTION

The methylamines (monomethylamine, dimethylamine and trimethylamine) are colorless, flammable and toxic gases at room temperatures and pressures in their anhydrous form. They have a distinct and disagreeable fishy odor in concentrations up to 100 ppm. In higher concentrations they have an odor like ammonia, which they resemble and from which they are derived. They are easily liquefied, and are shipped in their anhydrous form as liquefied compressed gases. They are highly soluble in water and in alcohol, ether and various other organic solvents.

Vapors of the methylamines in air can burn within certain concentration ranges. Gaseous methylamines and their solutions are alkaline materials and in sufficient concentrations can irritate and burn the skin, eyes and respiratory system.

Chemically, the methylamines are slightly stronger than ammonia as bases. They hydrate in water solutions and neutralize acids to form methylammonium salts. They do not corrode iron and steel, but do attack copper and its alloys, zinc and aluminum. The methylamines can form explosive compounds with mercury, and must never be brought into contact with mercury.

Methylamines are used and shipped both in the form of anhydrous gases and in aqueous solutions. Only the anhydrous form is treated here.

GRADES AVAILABLE

Methylamines (anhydrous) are available for commercial and industrial use in various grades having much the same component proportions from one producer to another.

USES

As sources of reactive organic nitrogen, the methylamines serve as intermediate in synthesizing pharmaceuticals, agricultural chemicals, dyes, rubber chemicals and explosives. Derivatives serve in agriculture as fungicides, insecticides and feed supplements. Derivatives have also been employed in producing antihistamines, tranquilizers and other drugs, in making dyestuffs, explosives and rocket fuel, and in curing resins. Rubber industry applications of derivatives include use as accelerators, vulcanizing agents and chain terminators in synthetic rubber production. Derivatives are also solvents for various organic plastics, resins, gums, dyes and pharmaceuticals.

PHYSIOLOGICAL EFFECTS

The methylamines are toxic, and contact with them must be avoided. They are irritating to the nose, throat and eyes in low concentrations, and require suitable gas masks and eye-protective devices for safe handling by exposed personnel. Severe exposure of the eyes may lead to loss of sight. Dermatitis results from contact of the methylamines with the skin. Inhalation of sufficiently high concentrations is followed by violent sneezing, a burning sensation of the throat with constriction of the larynx, and difficulty in breathing with congestion of the chest and inflammation of the eyes.

A TLV of 10 ppm (12 mg/m^3) for an 8-hour work day for monomethylamine and dimethylamine has been adopted by the ACGIH.

SAFE HANDLING AND STORAGE

Materials Suitable for Containers and Storage

Iron, steel, stainless steels and Monel have proven satisfactory in methylamines service.

Some plastics and elastomers also withstand their action. Copper, copper alloys (including brass and bronze), zinc (together with zinc alloys and galvanized surfaces), and aluminum are corroded by the methylamines and should not be used in direct contact with them. Mercury and the methylamines can explode on contact, and instruments containing mercury must never be used with the methylamines. Among gasket and packing materials satisfactory for use with them are compressed asbestos, polyethylene, Teflon and carbon steel or stainless steel wound asbestos.

Storage tanks for the methylamines should be made of steel and designed to comply with the Unfired Pressure Vessel Code of the ASME as well as with all state and local regulations. Safe working pressures vary with the vapor pressure-temperature relationship of the particular methylamine being stored, and with the high-temperature ranges at the plant location. Important parts of well-designed tanks include proper dual pressure-relief valves, a vapor-absorbing system, liquid level and pressure gages, liquid and vapor-transfer valves and an adequate electrical ground. Pipes, fittings, pumps, gages and other equipment should be of steel, iron or other material not subject to corrosion by the methylamines. It is best to have storage and handling installations designed with the help of engineers thoroughly familiar with the gases.

Storage and Handling in Normal Use

All the precautions necessary for the safe handling of any flammable, toxic gas must be observed with the anhydrous methylamines (see also Part I, Chapter 4, Section A).

Handling Leaks and Emergencies

Detailed information on first aid and medical treatment for persons injured by the methylamines is given in Safety Data Sheet SD-57 of the Chemical Manufacturers Association (see Ref. 1). (See also Part I, Chapter 4, Section A.)

METHODS OF SHIPMENT; REGULATIONS

Under the appropriate regulations and tariffs, the anhydrous methylamines are authorized for shipment as follows:

By Rail: In cylinders (via freight or express), and by insulated single-unit tank cars and TMU tank cars.

By Highway: In cylinders on trucks, in tank trucks, and in portable tanks on trucks.

By Water: In cylinders and portable tanks (not over 20,000 lb or 9072 kb gross weight) via cargo vessels and in railroad tank cars complying with DOT provisions on trainships only. Dimethylamine and trimethylamine are not permitted on passenger vessels, ferry vessels and railroad car ferry vessels. Monomethylamine may be shipped in cylinders on passenger vessels, ferries and railroad car ferries (including passenger or vehicle ferry vessels). The methylamines are also authorized for shipment in cylinders on barges of U.S. Coast Guard classes A and C only, and in cargo tanks aboard tankships and tank barges (with maximum filling densities by specific gravity as prescribed in Coast Guard regulations).

By Air: Aboard cargo aircraft only in appropriate cylinders up to 300 lb (140 kg) maximum net weight per cylinder.

CONTAINERS

The anhydrous methylamines are shipped as compressed liquefied gases under their own vapor pressures in cylinders, tank cars and cargo tank trucks. They are also authorized by the DOT for shipment in TMU (ton multi-unit) tank cars and portable tanks.

Filling Limits. The maximum filling densities allowed under DOT regulations for cylinders, TMU tank car tanks (DOT 106A type) and truck cargo tanks and portable tanks are (percent water capacity by weight): monomethylamine, 60 percent; dimethylamine, 59 percent; and trimethylamine, 57 percent. Corresponding maximum filling densities for single-unit tank cars (DOT 105A300-W with properly fitted loading and unloading valves) are: monomethyl-amine, 62 percent; dimethylamine, 62 percent; and trimethylamine, 59 percent.

Cylinders

Cylinders meeting the following DOT specifications are authorized for methylamines service: 3A150, 3AA150, 3B150, 4B150, 4BA225 and 3E1800. Safety relief devices are not required on cylinders charged with the methylamines.

3A, 3AA and 3B cylinders used in methylamines service must be requalified by hydrostatic retest every 5 years under present regulations. 4B, 4BA, 4BW, 4E and DOT-26-300 must be retested after expiration of the first 12-year period and each 7 years thereafter. Cylinders in compliance with 3A, 3AA, 3B, 4B, 4BA and 4BW used specifically for monomethylamine, dimethylamine and trimethylamine service free from corroding components may also be qualified by an external visual inspection described in Part I, Chapter 6, Section C of this book. Periodic hydrostatic retest is not required for 3E cylinders.

Valve Outlet Connections. Standard connection, United States and Canada—No. 705.

Cargo Tanks

Anhydrous methylamines are authorized for shipment in motor vehicle cargo tanks complying with DOT specification MC-330 and MC-331 with a minimum design pressure of 150 psig (1034 kPa gage).

Portable Tanks

Anhydrous methylamines are also authorized for shipment via motor vehicle in steel portable tanks built to DOT-51 specification and with a minimum design pressure of 150 psig (1034 kPa gage).

TMU Tanks

Authorized rail shipment of the methylamines may also be made in TMU tank cars of DOT specification 106A500X.

Tank Cars

Bulk quantities of the anhydrous methylamines are commonly shipped by rail in tank cars of DOT specification 105A300W (an insulated single-unit tank car with properly fitted loading and unloading valves).

METHOD OF MANUFACTURE

The methylamines are produced by having methyl alcohol and ammonia interact over a catalyst at high temperature.

REFERENCE

1. "Methylamines" (Safety Data Sheet SD-57), Chemical Manufacturers Association, 1825 Connecticut Ave., N.W., Washington, DC 20009.

Methyl Chloride

CH₃Cl

Synonym: Chloromethane

DOT Classification: Flammable gas; flammable gas label

PHYSICAL CONSTANTS

	U.S. Units	Metric Units
International symbol	CH₃Cl	CH₃Cl
Molecular weight	50.491	50.491
Vapor pressure		
at 70F (21.1 C)	73.4 psia	506.1 kPa, abs.
at 105 F (40.6 C)	123.7 psia	852.9 kPa, abs.
at 115 F (46.1 C)	143.7 psia	990.3 kPa, abs.
at 130 F (54.4 C)	173.7 psia	1197.6 kPa, abs.
Density of the gas		
at 70 F (21.1 C) and 1 atm	0.1330 lb/ft³	2.130 kg/m³
at -11.6 F (-24.2 C) 1 atm	0.159 lb/ft³	2.547 kg/m³
Specific gravity of the gas at 32 F		
(0 C) and 1 atm (air = 1)	1.74	1.74
Specific volume of the gas at 70 F		
(21.1 C) and 1 atm	7.5 ft³/lb	0.47 m³/kg
Density of liquid		
at boiling point	62.2 lb/ft³	996.3 kg/m³
at 70 F (21.1 C)	57.3 lb/ft³	917.9 kg/m³
at 105 F (40.6 C)	55.6 lb/ft³	890.6 kg/m³
at 115 F (46.1 C)	53.4 lb/ft³	855.4 kg/m³
at 130 F (54.4 C)	53.0 lb/ft³	849.0 kg/m³
Boiling point at 1 atm	-11.6 F	-24.2 C
Melting point at 1 atm	-143.7 F	-97.6 C
Critical temperature	289.4 F	143.0 C
Critical pressure	968.7 psia	6678.9 kPa, abs.
Critical density	23.1 lb/ft³	370.0 kg/m³
Triple point	-144 F at	-97.8 C at
	1.27 psia	8.76 kPa, abs.

	U.S. Units	Metric Units
Latent heat of vaporization at -11.6 F (-24.2 C) and 1 atm	183.42 Btu/lb	426.6 kJ/kg
Latent heat of fusion at -143.7 F (-97.6 C) and 1 atm	55.8 Btu/lb	129.8 kJ/kg
Specific heat of gas at 77 F (25 C) and 1 atm		
C_p	0.199 Btu/(lb)(F)	0.833 kJ/(kg)(C)
C_v	0.155 Btu/(lb)(F)	0.649 kJ/(kg)(C)
Ratio of specific heats	1.284	1.284
Weight of liquid at 70 F (21.1 C)	7.68 lb/gal	920.27 kg/m³
Specific gravity of liquid (water = 1)		
at −11.11 F (-23.95 C)	1.000	1.000
at 70 F (21.1 C)	0.919	0.919
Specific heat of liquid, average 5 F–86 F (-15 C-30 C)	0.376 Btu/(lb)(F)	1.574 kJ/(kg)(C)
Solubility of gas in water, vol/vol of water, at 1 atm		
at 32 F (0 C)	3.4/1	3.4/1
at 68 F (20 C)	2.2/1	2.2/1
at 86 F (30 C)	1.7/1	1.7/1
at 104 F (40 C)	1.3/1	1.3/1
Flammable limits in air, by volume	8.1–17.2 percent	8.1–17.2 percent
Autoignition temperature	1170 F	632.2 C
Flash point (open cup)	below 32 F	below 0 C

DESCRIPTION

Methyl chloride is a colorless, flammable gas with a faintly sweet, nonirritating odor at room temperatures. It is shipped as a transparent liquid under its vapor pressure of about 59 psig at 70 F (407 kPa gage at 21.1 C).

Methyl chloride burns feebly in air but forms mixtures with air that can be explosive within its flammability range.

Dry methyl chloride is very stable at normal temperatures and in contact with air, but may decompose at temperatures above 700 F (371 C) into toxic end products (hydrochloric acid, phosgene, chlorine and carbon monoxide). Methyl chloride hydrolyzes slowly in the presence of moisture with the formation of corrosive hydrochloric acid. It is slightly soluble in water and very soluble in alcohol, mineral oils, chloroform and most organic liquids.

GRADES AVAILABLE

Methyl chloride is available for commercial and industrial use in various grades having much the same component proportions from one producer to another.

USES

Methyl chloride is used as a catalyst carrier in the low-temperature polymerization of such products as the silicones and Butyl and other types of synthetic rubber; as a refrigerant gas; as a methylating agent in organic synthesis of such compounds as Grignard reagents, methyl ethers and quaternary ammonium compounds, and also as a chlorinating agent; as an extractant for greases, waxes, essential oils and resins, and as a low-temperature solvent; and as a fluid for thermometric and thermostatic equipment.

PHYSIOLOGICAL EFFECTS

Methyl chloride is toxic, and areas in which it is handled must be adequately ventilated. It is particularly dangerous in that it has no pronounced smell that can serve as a warning.

It is an anesthetic about one-fourth as potent as chloroform, and also acts as a narcotic. Inhalation of it must be avoided, for inhalation in concentrations of several hundred ppm or more leads successively to dizziness, headache, an unsteady walk, weakness, nausea and vomiting, abdominal pain, tremors, extreme nervousness, mental confusion, convulsion, unconsciousness and death. Apparent recovery from what seems a mild exposure through inhalation may be followed by serious and prolonged or even fatal aftereffects within a few days or weeks. Repeated exposures are dangerous because methyl chloride is eliminated slowly from the body, in which it is converted into hydrochloric acid and methyl alcohol (wood alcohol).

A TLV of 100 ppm (210 mg/m^3) for an 8-hour work day for methyl chloride has been adopted by the ACGIH.

Contact between the skin or the eyes and methyl chloride liquid (or vapor in a concentrated stream) must also be avoided, for such contact can result in frostbite of the tissues.

The physiological effects of methyl chloride in increasing concentrations are described in Bulletin 185 of the U.S. Public Health Service (see Ref. 1).

SAFE HANDLING AND STORAGE

Materials Suitable for Containers and Storage

Dry methyl chloride may be contained in such common metals as steel, iron, copper and bronze, but it has a corrosive action on zinc, aluminum, die castings, and magnesium alloys. Methyl chloride must not be used with aluminum, since it forms spontaneously flammable methyl aluminum compounds upon contact with that metal. No reaction occurs, however, with the drying agent, activated alumina.

Gaskets made of natural rubber and many neoprene compositions should be avoided because methyl chloride dissolves many organic materials. Pressed-fiber gaskets, including those made of asbestos, may be used with methyl chloride. Polyvinyl alcohol is unaffected by methyl chloride and its use is also recommended. Medium-soft metal gaskets may be used for applications where alternating stresses like those resulting from large temperature changes do not lead to "ironing out" and consequent leakage.

Storage and Handling in Normal Use

All the precautions necessary for the safe handling of any flammable toxic gas must be observed with methyl chloride (see also Part I, Chapter 4, Section A).

Handling Leaks and Emergencies

In detecting leaks, it is advisable to transfer the methyl chloride vapor or liquid from refrigerating units into gas-tight containers before testing the units for suspected leaks. The units may then be placed under carbon dioxide, air or nitrogen pressure and tested by the application of soapy water (or glycerine, in freezing weather) to the suspected points. Soapy water and glycerine are also recommended for testing possible leaks in cylinders and other containers. An open flame or a halide torch should not be used to detect leaks. See Part I, Chapter 4, Section A.

Detailed information on first aid in the event of methyl chloride exposure is given in Safety Data Sheet SD-40 of the Chemical Manufacturers Association (see Ref. 2).

METHODS OF SHIPMENT; REGULATIONS

Under the appropriate regulations and tariffs, methyl chloride is authorized for shipment as follows:

By Rail: In cylinders (via freight, express or baggage), in TMU tank cars, and in single-unit tank cars.

By Highway: In cylinders on trucks, in TMU

tanks on trucks, and in portable tanks and cargo tanks.

By Water: In cylinders on cargo vessels, passenger vessels, and ferry and railroad car ferry vessels (passenger or vehicle). On cargo vessels only in portable tanks (complying with DOT-51 specification) not over 20,000 lb (9072 kg) gross weight. In authorized tank cars on trainships only, and in authorized tank trucks on trailerships and trainships only. In cylinders on barges of U.S. Coast Guard classes A, CA and CB only. In cargo tanks on tankships and tank barges (to maximum filling densities by specific gravity as prescribed in Coast Guard regulations).

By Air: In cylinders aboard cargo aircraft only up to 300 lb (140 kg) maximum net weight per cylinder.

CONTAINERS

Methyl chloride is authorized for shipment in cylinders, insulated single-unit tank cars, and TMU tanks and tank cars. It is also shipped in tank barges. Though authorized for shipment in portable tanks and cargo tanks on trucks, it is rarely transported in such tanks at the present time.

Filling Limits. The maximum filling densities authorized for methyl chloride under present regulations are (percent water capacity by weight):

In cylinders—84 percent.

In TMU tanks—84 percent.

In single-unit tank cars (complying with DOT 105A300 W specifications)—86 percent.

In cargo tanks and portable tanks—84 percent by weight; 88.5 percent by volume.

Cylinders

Methyl chloride is authorized for shipment in cylinders meeting DOT specifications as follows: 3A225, 3AA225, 3B225, 4A225, 4B225, 4BA225, 4, 3E1800 and 4B240ET (cylinders which comply with DOT specifications 3, 25, 26-300 and 38 may also be continued in methyl chloride service, but new construction is not authorized).

Cylinders authorized for methyl chloride service must be requalified by hydrostatic retest every 5 years under present regulations, with the following exceptions: DOT-4 cylinders, hydrostatic retest every 10 years; no retest is required for 3E cylinders; and 4B, 4BA and 26-300 cylinders, retest every 12 years (if they are used exclusively for methyl chloride that is free from corroding components and if they are protected externally by suitable corrosion-resistant coatings such as galvanizing, painting, etc.; as an alternative, these cylinders may also be requalified by being retested every 7 years to an internal hydrostatic pressure at least two times the marked service pressure and without determination of expansions).

Valve Outlet Connections. Standard connection, United States and Canada—No. 510.

Cargo Tanks

Cargo tanks that comply with DOT specifications MC-330 or MC-331 are authorized for the shipment of methyl chloride via motor vehicle. Minimum design pressure for cargo tanks is 150 psig (1034 kPa gage).

Portable Tanks

Portable tanks meeting DOT-51 specifications and with a minimum design pressure of 150 psig (1034 kPa gage) are authorized for shipment of methyl chloride via motor vehicle.

TMU Tanks

TMU tanks complying with DOT specifications 106A500-X are authorized for methyl chloride shipment via motor vehicle and rail.

Tank Cars

Methyl chloride is authorized for shipment in insulated, single-unit tank cars which comply with DOT specifications 105A300-W (provided that interior pipes of loading are equipped with excess-flow valves of approved design).

METHOD OF MANUFACTURE

Methyl chloride is made commercially in the United States mainly by two methods: the reaction of hydrogen chloride gas or hydrochloric acid with methyl alcohol (in the presence of a catalyst to accelerate the reaction); and the chlorination of methane.

In the first process, the products are gaseous methyl chloride with unreacted hydrogen chloride and methyl alcohol and several by-products. These are removed in a series of chemical purification steps and the methyl chloride gas is compressed and dried. A small amount of air remaining in the condensate is distilled off before the liquid is charged into the shipping container.

Natural gas is the source of the methane used in the second process, in which the methane is removed by fractional distillation and reacted with chlorine. Undesired reaction products (including methylene chloride, chloroform, carbon tetrachloride, hydrochloric acid and some chlorinated hydrocarbons) are similarly removed from the methyl chloride in subsequent chemical purification steps.

REFERENCES

1. "United States Public Health Service Bulletin No. 185," U.S. Public Health Service. Available from the Superintendent of Documents, U.S. Government Printing Office, Washington, D.C.
2. "Methyl Chloride" (Safety Data Sheet SD-40), Chemical Manufacturers Association, 1825 Connecticut Ave., N.W., Washington, DC 20009.

Methyl Mercaptan

CH$_3$SH
Synonym: Methanethiol
DOT Classification: Flammable gas; flammable gas label

PHYSICAL CONSTANTS

	U.S. Units	Metric Units
International symbol	CH$_3$SH	CH$_3$SH
Molecular weight	48.107	48.107
Vapor pressure		
at 70 F (21.1 C)	11.004 psig	75.870 kPa, gage
at 105 F (40.6 C)	38 psig	262.0 kPa, gage
at 115 F (46.1 C)	47 psig	324.1 kPa, gage
at 130 F (54.4 C)	62 psig	427.5 kPa, gage
Density of the gas		
at 70 F (21.1 C) and 1 atm	0.125 lb/ft^3	2.00 kg/m^3
Specific gravity of the gas		
at 59 F (15 C) and 1 atm (air = 1)	1.66	1.66
Specific volume of the gas		
at 70 F (21.1 C) and 1 atm	8.00 ft^3/lb	0.499 m^3/kg
Density of liquid		
at 32 F (0.0 C)	56.12 lb/ft^3	899.0 kg/m^3
at 70 F (21.1 C)	54.08 lb/ft^3	866.28 kg/m^3
at 105 F (40.6 C)	52.46 lb/ft^3	840.33 kg/m^3
at 115 F (46.1 C)	51.97 lb/ft^3	832.48 kg/m^3
at 130 F (54.4 C)	51.28 lb/ft^3	821.43 kg/m^3
Boiling point at 1 atm	42.73 F	5.96 C
Melting point at 1 atm	−185.8 F	−121.0 C
Critical temperature	386.2 F	196.8 C
Critical pressure	1049.6 psia	7236.7 kPa, abs.
Critical density	20.72 lb/ft^3	331.90 kg/m^3
Latent heat of vaporization		
at boiling point	219.82 Btu/lb	511.30 kJ/kg
Latent heat of fusion		
at melting point	52.82 Btu/lb	122.86 kJ/kg

	U.S. Units	Metric Units
Specific heat of gas		
at 60 F (15.6 C) and 1 atm		
C_p	0.2458 Btu/(lb) (F)	1.029 kJ/(kg) (C)
C_v	0.2048 Btu/(lb) (F)	0.857 kJ/(kg) (C)
Ratio of specific heats	1.200	1.200
Solubility in water, weight percent		
at 59 F (15 C)	2.4 percent	2.4 percent
at 77 F (25 C)	1.3 percent	1.3 percent
Weight of liquid		
at 70 F (21.1 C)	7.230 lb/gal	866.34 kg/m^3
Flammable limits in air,		
by volume	3.9–21.8 percent	3.9–21.8 percent
Flash point (open cup)	below 0 F	below −17.8 C
Coefficient of expansion		
of liquid	0.00088 per °F	0.0016 per °C

DESCRIPTION

Methyl mercaptan is a colorless, flammable gas with an extremely strong, nauseating odor. It is toxic, but its unpleasant odor provides ample warning of its presence. The odor also makes it necessary to evacuate premises in which leaks of any quantity have occurred. Methyl mercaptan is easily liquefied under pressure at room temperatures and is water white as a liquid.

Chemically, methyl mercaptan resembles methyl alcohol in many respects and provides a means of introducing the methylthio linkage into many compounds.

GRADES AVAILABLE

Methyl mercaptan is available for commercial and industrial use in various grades having much the same component proportions from one producer to another.

USES

One of the most important uses of methyl mercaptan is in the manufacture of methionine, an important amino acid used as an animal feed supplement, especially for poultry. It is also used in the synthesis of insecticides.

PHYSIOLOGICAL EFFECTS

Methyl mercaptan, with the other lower alkyl mercaptans, has a fairly high degree of toxicity. It acts primarily on the central nervous system after inhalation, at first resulting in great stimulation and then depression. Little research has been done in mercaptan toxicity with humans because its strong odor has been thought to provide sufficient protection for personnel. However, inhalation of increasing methyl mercaptan concentrations by laboratory animals has led to increased respiration and restlessness, convulsions, muscular paralysis and finally death by respiratory paralysis. Fish show special sensitivity to methyl mercaptan, and have been fatally poisoned by as little as 0.5–3 ppm in water.

Liquid methyl mercaptan may severely irritate the skin.

A TLV of 0.5 ppm (1 mg/m^3) for an 8-hour work day for methyl mercapatan has been adopted by the ACGIH.

SAFE HANDLING AND STORAGE

Materials Suitable for Containers and Storage

Stainless steel is a satisfactory material for use with methyl mercaptan, but can be costly for entire installations. Iron and steel with stainless steel trim are successfully employed in methyl mercaptan service if proper preparatory and maintenance steps are taken. However, stainless steel must be used for components with which intermittent exposure to air or moisture cannot be avoided and in which slight corrosion could interfere with proper functioning. Such components include unloading risers, flame arrestor cores, gages and instruments. Aluminum has been considered one of the least desirable materials for use with methyl mercaptan, for caution must be exercised in selecting a proper type of aluminum to handle it under pressure. Copper and copper-bearing alloys should not be used for fixed equipment under constant exposure to methyl mercaptan.

Before continuous exposure to liquid methyl mercaptan, all iron and steel equipment should be passivated with methyl mercaptan vapor or hydrogen sulfide to coat the surfaces with ferrous sulfide. Treated surfaces must subsequently be kept free from moisture and under inert atmosphere to prevent the protective sulfide film from deteriorating. Ferrous sulfide oxidizes with a red glow in the presence of air and ignites highly flammable methyl mercaptan vapor. When cleaning or disassembling treated equipment, large quantities of water should be used to reduce the hazards of fire or explosion.

Asbestos and "Teflon" are among various materials recommended by methyl mercaptan suppliers as satisfactory for gaskets and packing.

Storage and Handling in Normal Use

Welded low-carbon steel storage tanks complying with the unfired pressure vessel Code of the ASME have proven satisfactory for use with methyl mercaptan. A recommended design pressure is 85 psig at 150 F (586 kPa gage at 65.6 C). Explosion-proof, stainless steel pumps and carbon steel or stainless steel valves and fittings are among recommended equipment. Storage and handling facilities should be designed with the help of professional personnel thoroughly familiar with methyl mercaptan.

Prolonged inhalation of methyl mercaptan

vapors must be avoided and, although its odor will become extremely disagreeable before lethal concentrations are reached, the nose is temporarily desensitized to mercaptan after initial exposure. Air-line or oxygen-type gas masks must always be on hand in operating areas.

Low flash and fire points and high volatility of methyl mercaptan gas poses another principal hazard in its handling. Personnel must avoid spills and leaks.

Methyl mercaptan severly irritates the skin and the eyes, and safety showers and eye-washing facilities must be readily available for possible emergencies. Flush the skin or eyes thoroughly with water should contact occur, and get medical attention promptly if irritation continues after washing.

Other precautions that must be observed in handling methyl mercaptan include the following:

(1) Store cylinders away from open flames and away from locations in which the gas could contact a spark.

(2) Require the wearing of rubber gloves, goggles and nonsparking shoes by personnel handling methyl mercaptan.

(3) Provide explosion-proof equipment and make sure that operating areas are well ventilated so that vapors cannot accumulate and ignite.

(4) Provide at least two men for work requiring the opening of any equipment, piping or vessels used in methyl mercaptan service so that one can summon aid if the other is overcome by fumes.

Further information about the toxic effects of mercaptan and their treatment is given in Medical Bulletin 5, 78 (1941) of the Standard Oil Company of New Jersey (see Ref. 1). (See also Part I, Chapter 4, Section A.)

Handling Leaks and Emergencies

To spills, immediately apply liquid household bleach, calcium or sodium hypochlorite in 5 percent aqueous solution. Do not use powdered bleach, which could cause fire or explosion. (See also Part I, Chapter 4, Section A).

METHODS OF SHIPMENT; REGULATIONS

Under the appropriate regulations and tariffs, methyl mercaptan is authorized for shipment as follows:

By Rail: In cylinders (via freight or express), and by single-unit tank cars and TMU tank cars.

By Highway: In cylinders on trucks, and in tank trucks, portable tanks and TMU tanks.

By Water: In cylinders and portable tanks (not over 20,000 lb or 9070 kg gross weight) via cargo vessels only; in authorized tank cars aboard steamships only; and in authorized tank trucks on trailerships and trainships only. In cylinders on barges on U.S. Coast Guard classes A and C only. In cargo tanks aboard tankships and tank barges (to maximum filling densities by specific gravity as prescribed in Coast Guard regulations).

By Air: Aboard cargo aircraft only in appropriate cylinders up to 300 lb (140 kg) maximum net weight per cylinder.

CONTAINERS

Methyl mercaptan is shipped as a compressed liquefied gas under its own vapor pressure in cylinders, single-unit tank cars and tank trucks. It is also authorized for shipment in TMU (ton multi-units) tank cars, and in TMU tanks and portable tanks on trucks.

Filling Limits. The maximum filling densities permitted for methyl mercaptan under DOT regulations are as follows (percent of water capacity by weight):

For cylinders, TMU tanks, cargo tanks and portable tanks—80 percent. (Cargo tanks and portable tanks are also authorized for a maximum filling density of 90 percent by volume.)

For single-unit tank cars—82 percent.

Cylinders

Cylinders that meet the following DOT specifications are authorized for methyl mercaptan service: 3A240, 3AA240, 3B240, 4B240, 4BA240, 4B240ET, 4BW240 and 3E1800. Cylinders of the same types with higher service pressures are also authorized. Safety relief devices are not required on cylinders charged with methyl mercaptan.

All cylinders authorized for methyl mercaptan service must be requalified by hydrostatic test every 5 years under present regulations, except that no periodic retest is required for 3E cylinders.

Valve Outlet Connections. Standard connection, United States and Canada— No. 330.

Cargo Tanks

Methyl mercaptan is authorized for shipment in motor vehicle cargo tanks conforming with DOT specifications MC-330 or MC-331, and in steel portable tanks meeting specifications DOT-51.

TMU Tanks

In addition, TMU tanks complying with DOT-106A500-X specifications and not equipped with safety relief devices of any kind are authorized for transportation by motor vehicle.

Tank Cars

Methyl mercaptan is authorized for shipment in tank cars meeting DOT specifications 105A-300-W (an insulated, single-unit tank car with properly fitted valves). One supplier ships in 6000-gal (22.7-m^3) cars of this type, and recommends unloading by pressurizing with dry nitrogen or natural gas with a maximum withdrawal rate of 30 gpm (114 1/min) that results in complete unloading withing 3 to 4 hours.

TMU Tank Cars

In addition rail shipment of methyl mercaptan is authorized for TMU tank cars complying with DOT specifications 106A500-X; these TMU tanks must not be equipped with safety relief devices of any description.

METHOD OF MANUFACTURE

Methyl mercaptan is produced primarily by chemical synthesis of methyl alcohol and hydrogen sulfide.

REFERENCE

1. V. Cristensen, "A Case of Poisoning with Mercaptans," Standard Oil Company (N.J.), Medical Bulletin 5, 78 (1941).

Nitrogen

N$_2$ DOT Classification: Nonflammable gas; nonflammable gas label

PHYSICAL CONSTANTS

	U.S. Units	Metric Units
International symbol	N$_2$	N$_2$
Molecular weight	28.013	28.013
Density of the gas at 70 F (21.1 C) and 1 atm	0.07245 lb/ft^3	1.1605 kg/m^3
Specific gravity of the gas at 70 F and 1 atm (air = 1)	0.967	0.967
Specific volume of the gas at 70 F (21.1 C) and 1 atm	13.80 ft^3/lb	0.862 m^3/kg
Density of liquid at boiling point and 1 atm	50.49 lb/ft^3	808.8 kg/m^3
Boiling point at 1 atm	-320.4 F	-195.8 C
Melting point at 1 atm	-345.7 F	-209.8 C
Critical temperature	-232.5 F	-147.0 C
Critical pressure	493 psia	3399 kPa, abs.
Critical density	19.60 lb/ft^3	314 kg/m^3
Triple point	-346.0 F at 1.82 psia	-210.0 C at 12.5 kPa, abs.
Latent heat of vaporization at boiling point	85.6 Btu/lb	199.1 kJ/kg
Latent heat of fusion at triple point	11.0 Btu/lb	25.6 kJ/kg
Specific heat of gas at 70 F (21.1 C) and 1 atm		
C_p	0.248 Btu/(lb) (F)	1.04 kJ/(kg) (C)
C_v	0.177 Btu/(lb) (F)	0.742 kJ/(kg) (C)
Ratio of specific heats	1.40	1.40
Solubility in water, vol/vol at 32 F (0 C)	0.023	0.023
Weight of liquid at boiling point	6.745 lb/gal	808.2 kg/m^3

	U.S. Units	Metric Units
Density of gas at boiling point	0.2878 lb/ft^3	4.610 kg/m^3
Liquid/gas ratio (liquid at boiling point, gas at 70 F and 1 atm), vol/vol	1/696.5	1/696.5

DESCRIPTION

Nitrogen makes up the major portion of the atmosphere (78.03 percent by volume, 75.5 percent by weight). It is a colorless, odorless, tasteless, nontoxic and almost totally inert gas, and is colorless as a liquid. Nitrogen is nonflammable, will not support combustion and is not life supporting. It combines with some of the more active metals, such as lithium and magnesium, to form nitrides, and at high temperatures it will also combine with hydrogen, oxygen and other elements. It is employed as an inert protection against atmospheric contamination in many nonwelding applications. Nitrogen is only slightly soluble in water and most other liquids, and is a poor conductor of heat and electricity. As a liquid at cryogenic temperatures it is nonmagnetic. It is shipped as a non-liquefied gas at pressures of 2000 psig (13,790 kPa gage) or above, and also as a cryogenic fluid at pressures below 200 psig (1379 kPa gage).

GRADES AVAILABLE

Table 1 presents the component maxima in ppm (v/v), unless shown otherwise, for the grades of gaseous and liquid nitrogen. (Type II references in the table denote liquid nitrogen; notations in the table's Type II entries to "see" numbered sections pertain to sections in Ref. 1, in which nitrogen commodity specifications are more fully explained.)

USES

Nitrogen has many commercial and technical applications. As a gas it is used in: agitation of color film solution in photographic processing; blanketing of oxygen-sensitive liquids, and of volatile liquid chemicals; the production of semiconductor electronic components, as a blanketing atmosphere; the blowing of foam-type plastics; the deaeration of oxygen-sensitive liquids, the degassing of nonferrous metals; food processing and packing; inhibition of aerobic bacteria growth; magnesium reduction of aluminum scrap; and the propulsion of liquids through pipelines.

Gaseous nitrogen is also used in: pressurizing aircraft tires and emergency bottles to operate landing gear; purging, in the grazing of copper tubing for air conditioning and refrigeration systems; the purging and filling of electronic devices; the purging, filling and testing of high-voltage compression cables; the purging and testing of pipelines and related instruments; and the treatment of alkyd resins in the paint industry.

Liquid nitrogen also has a great many uses, among them: the cold-trapping of materials such as carbon dioxide from gas streams (and it is commonly employed in this way in systems which produce high vacuums); as a coolant for electronic equipment, for pulverizing plastics, and for simulating the conditions of outer space; for creating a very high-pressure gaseous nitrogen (15,000 psig 103,000 kPa gage) through liquid nitrogen pumping; in food and chemical pulverization; for the freezing of expensive and highly perishable foods, such as shrimp; for the freezing of liquids in pipelines for emergency repairs; for low-temperature stabilization and hardening of metals; for low-temperature research; for low-temperature stress relieving of aluminum alloys; for the preservation of whole blood, livestock sperm, and other biologicals; for refrigerating foods in long-distance hauling as well as local delivery; for referigeration shielding of liquid hydrogen, helium and neon; for

TABLE 1. NITROGEN GRADES AVAILABLE.

Limiting Characteristics	A	B	C	D	E	F	G	H	J	K	L	M	N	P
Nitrogen Min., % (v/v) (Note)	97	99.0	99.5	99.5	99.5	99.9	99.95	99.99	99.99	99.995	99.998	99.999	99.9985	99.999
Water	None con-densed			26.3	26.3	32.0	26.3	11.4	3.5	16.2	3.5	1.5	1.5	1.5
Dewpoint, °F				−63	−63	−60	−63	−75	−90	−70	−90	−100	−100	−100
Hydrocarbons, condensed									0.1 (wt/wt) see 8.3					
Total Hydrocarbons, (as methane)					58.3			5	3					
Oxygen			0.5%	0.5%	0.5%	0.1%	500	50	50	20	10	5	1	1
Hydrogen													1	
Argon, Neon, Helium													5	
Carbon Monoxide													1	
U.S.P.		Yes												
Permanent* Particulates														

Note. Unless shown otherwise % N$_2$ includes trace quantities of neon and helium and small amounts of argon.

*See CGA Pamphlet G-10.1 "Commodity Specification for Nitrogen" for permanent particulate removal filtering requirements. (Ref. 1) the removal of skin blemishes in dermatology; and for shrink fitting of metal parts.

Liquid nitrogen also has a number of classified applications in the missile and space programs of the United States, in which it is used in large quantities.

PHYSIOLOGICAL EFFECTS

Nitrogen is nontoxic and largely inert. It can act as a simple asphyxiant by displacing needed oxygen in the air. Inhalation of it in excessive concentrations can result in unconsciousness without any warning symptoms, such as dizziness.

Gaseous nitrogen must be handled with all the precautions necessary for safety with any nonflammable, nontoxic compressed gas.

All precautions necessary for the safe handling of any gas liquefied at very low temperatures must be observed with liquid nitrogen. Severe burn-like injuries result from contact between the tissues and liquid nitrogen.

SAFE HANDLING AND STORAGE

Materials Suitable for Containers and Storage

Gaseous nitrogen is noncorrosive and inert, and may consequently be contained in systems constructed of any common metals and designed to withstand safely the pressures involved. At the temperature of liquid nitrogen, ordinary carbon steels and most alloy steels lose their ductility, and are considered unsafe for liquid nitrogen service. Satisfactory materials for use with liquid nitrogen include 18-8 stainless steel and other austenitic nickel-chromium alloys, copper, "Monel," brass and aluminum.

Storage and Handling in Normal Use

Gaseous nitrogen is often stored in high-pressure tube trailers as the storage supply. Nitrogen is also often stored in compact liquid form at the consumer's site. Liquid storage systems, which frequently include vaporizing equipment for conversion to gas, range in capacity from 25,000 (700 m³) to more than one million ft³

(283,168 m³) NTP. Liquid storage systems should be designed and installed only under the direction of engineers thoroughly familiar with liquid nitrogen equipment, and in full compliance with all state and local requirements.

METHODS OF SHIPMENT; REGULATIONS

Under the appropriate regulations and tariffs, nitrogen is authorized for shipment as follows (nitrogen gas, except where liquid nitrogen is indicated):

By Rail: In cylinders and in tank cars. Liquid nitrogen, in insulated cylinders, and in insulated tank cars not subject to DOT regulations at pressures less than 25 psig (170 kPa gage).

By Highway: In cylinders on trucks, and in tube trailers. Liquid nitrogen, in insulated cylinders and in tank trucks.

By Water: In cylinders on cargo and passenger vessels, and on ferry and railroad car ferry vessels (passenger or vehicle). In authorized tank cars on cargo vessels only. In cylinders on barges of U.S. Coast Guard classes A, BA, BB, CA and CB. Liquid nitrogen, in pressurized cylinders on cargo and passenger vessels and ferry and railroad car ferry vessels (passenger or vehicle).

By Air: In cylinders aboard passenger aircraft up to 150 lb (70 kg), and aboard cargo aircraft up to 300 lb (140 kg), maximum net weight per cylinder. Nonpressurized liquid nitrogen, aboard passenger and cargo aircraft up to about 12 gal (50 liters) maximum net contents per container; low-pressure liquid nitrogen, or pressurized liquid nitrogen, aboard cargo aircraft only up to 300 lb (140 kg) maximum net weight per container.

CONTAINERS

Nitrogen gas is authorized for shipment in cylinders, tank cars and tube trailers. Liquid nitrogen is shipped as a cryogenic fluid in insulated cylinders, insulated tank trucks and insulated tank cars.

Filling Limits. The maximum filling limits authorized for gaseous nitrogen are as follows:

In cylinders and tube trailers—the authorized service pressures marked on the cylinders or tube assemblies at 70 F (21.1 C). In the case of cylinders of specifications 3A, 3AA, 3AX and 3AAX that meet special requirements, up to 10 percent in excess of their marked service pressures.

In tank cars—in uninsulated cars of the DOT-107A type, not more than seven-tenths of the marked test pressure at 130 F (54.4 C).

The maximum filling limits authorized for liquid nitrogen are:

In cylinders that meet DOT-4L specifications—maximum filling density of 68 percent (percent of water capacity of weight) and limitation of cylinder pressure to one and one-fourth times the marked service pressure by a pressure-controlling valve (or, for 4L cylinders insulated by a vacuum, at least 15 psi (100 kPa) lower than one and one-fourth times the marked service pressure).

Liquid nitrogen shipped at pressures below 25 psig (170 kPa gage) in insulated tank trucks and tank cars, or in other insulated containers, is not subject to DOT regulations.

Cylinders

Cylinders which comply with DOT, specifications 3A and 3AA are the types usually used to ship gaseous nitrogen, but it is authorized for shipment in any cylinders approved for non-liquefied compressed gas. (These include cylinders meeting DOT specifications 3A, 3AA, 3AX, 3AAX, 3B, 3C, 3D, 3E, 4, 4A, 4B, 4BA, 4BW and 4C; in addition, continued use of cylinders complying with DOT specifications 3, 25, 26, 33, and 38 is authorized, but new construction is not authorized.)

Liquid nitrogen is authorized for shipment in cylinders which meet DOT specifications 4L.

All cylinders authorized for gaseous nitrogen service must be requalified by hydrostatic retest every 5 or 10 years under present regulations, with the following exceptions: DOT-4 cylinders, every 10 years; and no periodic retest is required for cylinders of types 3C, 3E, 4C and 7.

Also, for cylinders of the 4L type authorized

for liquid nitrogen service, no periodic retest is required for requalification.

Valve Outlet and Inlet Connections. For gas and liquid cylinders: Standard connections, United States and Canada—up to 3000 psi (20,684 kPa), No. 580; 3000–10,000 psi (20,684–68,948 kPa), No. 677, see Part I, Chapter 7, Section A, paragraph 2.1.7; cryogenic liquid withdrawal, No. 295. Alternate standard connections, United States and Canada—up to 3000 psi, Nos. 590, 555; 3000–10,000 psi, No. 677; cryogenic liquid withdrawal, No. 440. Standard yoke connection, United States and Canada (for medical use of the gas)—up to 3000 psi, No. 960.

Tank Cars

Gaseous nitrogen is authorized for rail shipment in tak cars that comply with DOT specifications 107A. DOT regulations require that the pressure to which the containers are charged must not exceed seven-tenths of the marked test pressure at 130 F (54.4 C).

Liquid nitrogen is also shipped in special insulated tank cars at pressures less than 25 psig (170 kPa gage).

Tube Trailers

Gaseous nitrogen is shipped in tube trailers with capacities ranging to more than 40,000 ft³ (10,000 m³). These trailers are built to comply with DOT cylinder specifications 3A, 3AA, 3AX or 3AAX. The trailers commonly serve as the storage supply for the user, with the supplier replacing trailers as they are emptied.

Tank Trailers

Liquid nitrogen is shipped in bulk at pressures below 25 psig (170 kPa gage) in special insulated tank trailers, with capacities in excess of 400,000 ft³ (11,000 m³).

Small Portable Containers

Liquid nitrogen is shipped and stored in small portable containers which hold quantities ranging from 1-25 gal (0.004–0.09 m³) or more. These containers encased in shells are heavily insulated; they maintain the liquid at atmospheric pressure, and are consequently not subject to DOT regulations.

METHOD OF MANUFACTURE

Nitrogen is produced commercially at air separation plants by liquefaction of atmospheric air and removal of the nitrogen from it by fractionation.

REFERENCE

1. "Commodity Specification for Nitrogen" (Pamphlet G-10.1), Compressed Gas Association, Inc.

ADDITIONAL REFERENCE

"Saturated Liquid Densities of Oxygen, Nitrogen, Argon and Para-Hydrogen" (NBS Technical Note 361), National Bureau of Standards.

Nitrous Oxide

N$_2$O
Synonyms: Nitrogen monoxide, dinitrogen monoxide, laughing gas
DOT Classification: Nonflammable gas, nonflammable gas label

PHYSICAL CONSTANTS

	U.S. Units	Metric Units
International symbol	N$_2$O	N$_2$O
Molecular weight	44.013	44.013
Vapor pressure		
at -4 F (-20 C)	262 psia	1806.42 kPa, abs.
at 32 F (0 C)	460 psia	3171.59 kPa, abs.
at 68 F (20 C)	736 psia	5074.54 kPa, abs.
at 98 F (36.7 C)	1069 psia	7370.50 kPa, abs.
Density of the gas		
at 32 F (0 C) and 1 atm	0.1230 lb/ft^3	1.9703 kg/m^3
at 68 F (20 C) and 1 atm	0.1146 lb/ft^3	1.8357 kg/m^3
Density of saturated vapor		
at boiling point	0.194 lb/ft^3	3.108 kg/m^3
at -4 F (-20 C) and 262 psia		
(1806 kPa, abs.)	2.997 lb/ft^3	48.007 kg/m^3
at 68 F (20 C) and 736 psia		
(5075 kPa, abs.)	10.051 lb/ft^3	161.002 kg/m^3
Specific gravity of the gas at 32 F		
and 1 atm (air = 1)	1.529	1.529
Specific gravity of liquid (water = 1)		
at 68 F (20 C) and 736 psia		
(5075 kPa, abs.)	0.785	0.785
Specific volume of the gas		
at 32 F (0 C) and 1 atm	8.130 ft^3/lb	0.5075 m^3/kg
at 68 F (20 C) and 1 atm	8.726 ft^3/lb	0.5447 m^3/kg
Density of Liquid		
at 70 F (21.1 C)	48.3 lb/ft^3	773.69 kg/m^3
at boiling point and 1 atm	76.6 lb/ft^3	1227.01 kg/m^3
Boiling point at 1 atm	-127.24 F	-88.47 C

	U.S. Units	Metric Units
Melting point at 1 atm	-131.5 F	-90.83 C
Critical temperature	97.7 F	36.5 C
Critical pressure	1054 psia	7267.07 kPa, abs.
Critical density	28.15 lb/ft^3	450.92 kg/m^3
Triple point	-131.5 F at	-90.83 C at
	12.74 psia	87.84 kPa, abs.
Latent heat of vaporization		
at boiling point	161.8 Btu/lb	376.3 kJ/kg
at 32 F (0 C)	107.5 Btu/lb	250.0 kJ/kg
at 68 F (20 C)	78.7 Btu/lb	183.1 kJ/kg
Latent heat of fusion at triple point	63.9 Btu/lb	148.6 kJ/kg
Specific heat of gas at 59 F (15 C)		
and 1 atm		
C_p	0.2004 Btu/(lb)(F)	0.8390 kJ/(kg)(C)
C_v	0.1538 Btu/(lb)(F)	0.6439 kJ/(kg)(C)
Ratio of specific heats	1.303	1.303
Solubility in water, vol/vol of		
water, at 1 atm		
at 32 F (0 C)	1.3/1	1.3/1
at 77 F (25 C)	0.66/1	0.66/1
Solubility in alcohol, vol/vol of		
alcohol, at 68 F (20 C) and 1 atm	3.0/1	3.0/1
Weight of liquid		
at boiling point	10.23 lb/gal	1225.82 kg/m^3
at -4 F (-20 C) and 262 psia		
(1806 kPa, abs.)	8.35 lb/gal	1000.55 kg/m^3
at 68 F (20 C) and 721 psig		
(4971 kPa gage)	6.54 lb/gal	783.66 kg/m^3
Viscosity of gas		
at 32 F (0 C)	0.0135 cP	0.135 kPa-sec
at 80 F (26.7 C)	0.0149 cP	0.149 kPa-sec
Thermal conductivity of gas		
at 32 F	0.0083 (Btu)(ft)/	
	(ft^2)(hr)(F)	
at 0 C	21.5 W/m-kelvins	

DESCRIPTION

Nitrous oxide at normal temperatures and pressures is a colorless, practically odorless and tasteless, nontoxic gas. It is shipped in liquefied form at its vapor pressure which at 70 F is about 745 psig (at 21.1 C is about 5100 kPa gage). Nitrous oxide is nonflammable, but, being a mild oxidizing agent, will support combustion of flammable materials in a manner similar to oxygen but to a lesser extent than oxygen. Under ordinary conditions, nitrous oxide is stable and generally inert. Decomposition of the pure gas in the absence of catalysts is negligible at temperatures below 1200 F (650 C). Compared to air, nitrous oxide is relatively soluble in water, alcohol, and oils and in many food products. Unlike some higher oxides of nitrogen, nitrous oxide dissolves in water without change in its acidity.

GRADES AVAILABLE

Table 1 presents the component maxima in ppm (v/v), unless shown otherwise, for types and grades of nitrous oxide (see Ref. 5). Type I is vapor and Type II is liquid.

USES

The largest use of nitrous oxide probably still is a long established one, as an inhalant type of anesthetic or analgesic gas. Extensive use of

TABLE 1. NITROUS OXIDE GRADES AVAILABLE.

Limiting Characteristics	Maxima for Types I and II
Assay—Nitrous oxide (N_2O) min. % v/v	99.0 (in liquid)
Carbon monoxide (CO)	10 (in vapor)
Nitric oxide (NO)	1 (in vapor)
Nitrogen dioxide (NO_2)	1 (in liquid)
Halogens (as chlorine) (Cl_2)	1 (in vapor)
Carbon dioxide (CO_2)	300 (in vapor)
Ammonia (NH_3)	25 (in vapor)
Water	198 (in vapor)
Dew point, °F	−33
Meets USP	yes

nitrous oxide has more recently developed in the field of pressure packaging, in which it serves as a propellant for various aerosol products, particularly with foods such as whipped cream. Other applications include its employment as a leak-detecting agent, as an oxidizing agent in blowtorches, as both a refrigerant gas and a refrigerant liquid for immersion freezing of food products, and as a chemical reagent in the manufacture of various compounds (both organic, as with detonants, and inorganic, as in obtaining nitrites from alkali metals). Nitrous oxide has also served as an ingredient in rocket fuel formulations, and its use as part of the working fluid in hypersonic wind tunnels has recently been investigated.

PHYSIOLOGICAL EFFECTS

Nitrous oxide is nontoxic and nonirritating as well as being chemically stable. When inhaled in high concentrations for a few seconds, it affects the central nervous system and may induce symptoms closely resembling alcoholic intoxication. Its colloquial name, "laughing gas," stems from the fact that some persons exhibit hilarity while in this condition.

For use as a general anesthetic in medicine, high concentrations of nitrous oxide are mixed with oxygen. Continued inhalation without an ample supply of oxygen results in simple asphyxia.

In view of its nontoxic nature, nitrous oxide has been used for a number of years for the food type of aerosols and in treating and preserving foodstuffs. It has been defined by the U.S. Food and Drug Administration as a food additive substance generally recognized as safe when used as a propellant for dairy and vegetable-fat toppings in pressurized containers.

Liquid nitrous oxide evaporates so rapidly that prolonged contact of the liquid with the skin will result in freezing or frostbite.

SAFE HANDLING AND STORAGE

Materials Suitable for Containers and Storage

Nitrous oxide is noncorrosive and may therefore be used with any of the common, commercially available metals. Because of its oxidizing action, however, care must be taken to insure freedom from oil, grease and other readily combustible materials in all equipment being prepared to handle nitrous oxide, particularly at high pressures.

Piping for nitrous oxide systems should be of steel, stainless steel, wrought iron, or brass or copper pipe; or of seamless tubing made of copper, brass or stainless steel.

Storage and Handling in Normal Use

All the precautions necessary for the safe handling of any compressed gas, and of any gas used medicinally, must be observed with nitrous oxide. In addition to these and to all applicable state and local regulations, the special rules below must be followed in handling nitrous oxide.

Never permit oil, grease or any other readily combustible substance to come in contact with cylinders or other equipment containing nitrous oxide. Oil and nitrous oxide may combine with explosive violence.

Remove any paper wrappings so that the cylinder label is clearly visible before placing cylinders in service.

Store nitrous oxide cylinders in an assigned, little-frequented location, making sure not to store them in the same room with cylinders containing reserve stocks of flammable gases. Never store medical gas cylinders of nitrous oxide in the hospital operating room.

Industrial and medical consumers of nitrous oxide store the gas either in high-pressure systems (which often employ manifolded cylinders as the source of supply) or in bulk low-pressure storage containers in liquid form at reduced temperatures. Common operating conditions for low-pressure systems are in the ranges of 250-400 psig and 0-25 F (1665-2800 kPa gage and $-18 - -3.9$ C). In addition to meeting all state and local requirements, high-pressure storage containers should comply with either the ASME Code ("Boiler and Pressure Vessel Code, Section VIII, Unfired Pressure Vessels" (see Ref. 1)) or DOT specifications and regulations, and low-pressure storage containers should

comply with the ASME Code. Special construction is necessary for operating temperatures below -20 F (-28.9 C).

Detailed recommendations for nitrous oxide storage installations are given in Pamphlet G-8.1 of Compressed Gas Association, Inc. (see Ref. 2). Installations should be designed and made by persons thoroughly familiar with nitrous oxide systems. Personnel who operate the installations must be adequately trained, and legible instructions should be maintained at operating locations.

Take care to avoid exhausting a nitrous oxide cylinder completely when using it with ether in anesthesia in order to prevent the possibility of having ether drawn back into the cylinder. Always protect nitrous oxide cylinders against feedback of any other gases or foreign material by suitable traps or check valves in lines to which the cylinders are connected.

Do not transfer nitrous oxide from one cylinder to another. Instead, always return the cylinders to charging plants for refilling under recognized safe practices.

Persons responsible for the use of nitrous oxide in hospitals or other facilities should see Part I, Chapter 4, Section C. Recommendations pertaining to nitrous oxide, particularly in hospitals, are also presented in two publications of the National Fire Protection Association, NFPA No. 56F and NFPA No. 56A (see Refs. 3 and 4).

Handling Leaks and Emergencies

See Part I, Chapter 4, Sections A and C.

METHODS OF SHIPMENT; REGULATIONS

Under the appropriate regulations and tariffs, nitrous oxide is authorized for shipment as follows:

By Rail: In cylinders (via freight, express and baggage), and in portable tanks.

By Highway: In cylinders, cargo tanks and portable tanks.

By Water: In cylinders aboard passenger vessels, cargo vessels, and ferry and railroad car ferry vessels (either passenger or vehicle for both types of ferries). Also in authorized tank motor vehicles aboard cargo vessels and ferry and railroad car ferry vessels (passenger or vehicle for both ferry types); and aboard cargo vessels and railroad car ferry vessels (passenger or ferry) in portable tanks meeting DOT-51 specifications (if shipped as stowage, the tanks must be not over 20,000 lb (9072 kg) gross weight per tank, and must not be equipped with fixed length dip tube gaging devices). In cylinders on barges of U.S. Coast Guard classes A, BA, BB, CA and CB only.

By Air: In cylinders aboard passenger aircraft up to 150 lb (70 kg), and in cylinders aboard cargo aircraft up to 300 lb (140 kg), maximum net weight per cylinder.

CONTAINERS

Cylinders, cargo tanks and portable tanks are authorized for the shipment of nitrous oxide. It is transported as a liquefied compressed gas under high pressure in cylinders, and at lower pressures and reduced temperatures in refrigerated cargo tanks and insulated portable tanks.

Filling Limits. The maximum filling densities authorized for nitrous oxide are as follows:

In cylinders—68 percent (percent water capacity by weight); or 75 percent in cylinders made before February 1, 1917, that have less than 12 lb (5.4 kg) water capacity and are known to have passed a pressure test of not less than 3500 psi (24,000 kPa).

In cargo tanks and portable tanks—95 percent by volume, with the additional requirement that tanks be equipped with suitable pressure-controlling devices, and that the vapor pressure at 115 F (46.1 C) must not exceed the design pressure of the tank.

Cylinders

Cylinders that meet the following specifications are authorized for nitrous oxide service: 3A1800, 3AA1800, 3E1800 and 3HT200 (DOT-3 cylinders may also be continued in nitrous oxide service, but new construction

is not authorized; also, 3HT cylinders are restricted to aircraft use only, and in shipment must be boxed in strong outside containers). The manifolding of cylinders transporting nitrous oxide is permitted under DOT regulations if each cylinder is individually equipped with an approved safety relief device, and if all cylinders are supported and held together as a unit by structurally adequate means.

Cylinders authorized for nitrous oxide service must be requalified by hydrostatic retest every 10 years under present regulations, except that periodic retest is required every 3 years for 3HT cylinders, and no periodic retest is required for cylinders of the 3E type. (3HT cylinders must also be withdrawn from service after a service life of 12 years of 4380 pressurizations.)

Valve Outlet and Inlet Connections. Standard connection, United States and Canada—No. 326. Standard yoke connection, United States and Canada (for medical use of the gas)—No. 910.

Medical Gas Cylinders

See Part I, Chapter 4, Section C for an identification and description of standard styles of medical gas cylinders used for nitrous oxide, together with an explanation of their safe handling.

Cargo Tanks and Portable Tanks

Nitrous oxide is authorized for shipment by the DOT in portable tanks that comply with specification DOT-51, and in motor vehicle cargo tanks meeting specifications MC-330 or MC-331. (Cargo tanks meeting DOT specifications MC-320 may be continued in service if qualified by periodic retest as required, but new construction is not authorized.) The minimum design pressure stipulated for portable or cargo tanks is 200 psig (1379 kPa gage) except that it may be reduced to 100 pisg or 689 kPa gage (or the controlled pressure, whichever is greater), if the tanks are also designed to comply with the requirements for low-temperature operation of the ASME Boiler and Pressure Vessel Code,

Section VIII, Unfired Pressure Vessels (see Ref. 1). The maximum service pressure authorized for portable or cargo tanks under current DOT regulations is 500 psig (3447 kPa gage).

Portable tanks for nitrous oxide service must be lagged with a noncombustible insulating material in compliance with DOT regulations (thick enough so that conductance does not exceed 0.08 Btu/ft^2/hr/$^\circ$F temperature differential determined at 60 F). Requirements for safety relief devices, excess flow valves, piping, valves, fittings and accessories must also be met by nitrous oxide portable tanks. One or more pressure-controlling devices may be installed to allow controlled escape of gas above a maximum operating temperature (such escape also exerts a self-refrigerating effect). Coils for refrigerating or heating or both may also be used, as may liquid level gaging devices (and additional gaging devices), but the coils and gaging devices must meet specified requirements if they are used. Portable tanks must be requalified by hydrostatic retest every 5 years under current regulations, and must comply with various other requirements concerning such matters as their certification, registration and repair. As with portable tanks for any compressed gas, those used for nitrous oxide must be of more than 1000 lb (453 kg) water capacity, and must be equipped with skids, mountings or other accessories to provide for moving the tanks with handling equipment.

For refrigerated cargo tanks in nitrous oxide service, similar regulations apply to insulation, safety relief and pressure-control devices, refrigerating and heating coils, excess flow valves, piping, valves, fittings, accessories and certification and registration. For a cargo tank, it is further required that all inlet and outlet valves (except safety relief devices) must be marked to show whether they end in liquid or gas when the tank is at maximum filling density; it is also required that the tank must be fitted with a pressure gage having a shutoff valve between it and the tank (such a gage need be used only in the filling operation). The refrigerating unit may be mounted on the motor vehicle if desired. A manufacturer's data report on each cargo tank (as well as on each portable tank) must be kept in the files of the motor carrier operating the tank while the carrier has the tank in service. (Special inspection requirements for specified kinds of cargo tanks are also set forth in Section 77.824 of "Bureau of Explosives' Tariff No. BOE-6000, Publishing Hazardous Materials Regulations of the Department of Transportation.")

METHOD OF MANUFACTURE

The only practical commercial method yet developed for manufacturing nitrous oxide is by the thermal decomposition of ammonium nitrate, which yields nitrous oxide and water in the primary reaction. However, a number of side reactions also occur, depending upon the temperature of decomposition, and these are catalyzed by the presence of certain contaminants and metals or metallic compounds. The impurities formed, mostly the higher oxides of nitrogen, are highly toxic. Accordingly, after the water is condensed out, the gas is passed through a series of scrubbing towers to remove the impurities so that the final product contains only a small amount of nitrogen. In addition, the gas is usually passed through a bed of desiccant after compression in order to dry it.

REFERENCE

1. "ASME Boiler and Pressure Vessel Code, Section VIII, Unfired Pressure Vessels," American Society of Mechanical Engineers, 345 E. 47th St., New York, NY 10017.
2. "Standard for the Installation of Nitrous Oxide Systems at Consumer Sites" (Pamphlet G-8.1), Compressed Gas Association, Inc.
3. "Standard for Nonflammable Medical Gas Systems" (NFPA No. 56F), National Fire Protection Association, 470 Atlantic Ave., Boston, MA 02210.
4. "Standard for the Use of Inhalation Anesthetics" (NFPA No. 56A), National Fire Protection Association.
5. "Commodity Specification for Nitrous Oxide" (Pamphlet G-8.2), Compressed Gas Association, Inc.

Oxygen

O$_2$ DOT Classification: Nonflammable gas; oxidizer gas label

PHYSICAL CONSTANTS

	U.S. Units	Metric Units
International symbol	O$_2$	O$_2$
Molecular weight	31.999	31.999
Density of the gas at		
70 F (21.1 C) and 1 atm	0.08279 lb/ft^3	1.326 kg/m^3
Specific gravity of the gas at		
70 F and 1 atm (air = 1)	1.1049	1.1049
Specific volume of the gas		
at 70 F (21.1 C) and 1 atm	12.08 ft^3/lb	0.7541 m^3/kg
Boiling point at 1 atm	−297.3 F	−183.0 C
Freezing point at 1 atm	−361.1 F	−218.4 C
Critical temperature	−181.4 F	−118.6 C
Critical pressure	731.4 psia	5043 kPa, abs.
Critical density	27.23 lb/ft^3	436.18 kg/m^3
Triple point	−361.8 F at	−218.8 C at
	0.0220 psia	0.1517 kPa, abs.
Latent heat of vaporization		
at boiling point	91.7 Btu/lb	213 kJ/kg
Latent heat of fusion at −361.1 F		
(−218.4 C) melting point	5.98 Btu/lb	13.9 kJ/kg
Specific heat of gas		
at 70 F (21.1 C) and 1 atm		
C_p	0.219 Btu/(lb)(F)	0.917 kJ/(kg)(C)
C_v	0.156 Btu/(lb)(F)	0.653 kJ/(kg)(C)
Ratio of specific heats	1.400	1.400
Solubility in water, vol/vol		
at 32 F (0 C)	0.0489	0.0489
Weight of liquid		
at boiling point	9.52 lb/gal	1141 kg/m^3
Density of gas		
at boiling point	0.2959 lb/ft^3	4.740 kg/m^3

	U.S. Units	Metric Units
Density of liquid at boiling point	71.27 lb/ft^3	1142 kg/m^3
Liquid/gas ratio (liquid at boiling point, gas at 70 F and 1 atm), vol/vol	1/860.6	1/860.6

DESCRIPTION

Oxygen, the colorless, odorless, tasteless elemental gas that supports life and makes combustion possible, constitutes about a fifth of the atmosphere (20.99 percent by volume; by weight, almost a fourth—23.2 percent). It is a transparent, pale blue liquid slightly heavier than water at temperatures ranging below some −300 F (−184 C). All elements but the inert gases combine directly with oxygen, usually to form oxides. However, oxidation of different elements occurs over a wide range of temperatures, with phosphorus and magnesium igniting spontaneously in air at ambient temperatures and the noble metals oxidizing only at very high temperatures.

All materials that are flammable in air burn much more vigorously in oxygen. Some combustibles, such as oil and grease, burn with nearly explosive violence in oxygen if ignited. Pure oxygen itself is nonflammable.

Oxygen is shipped as a nonliquefied gas at pressures of 2000 psig (13,790 kPa gage) or above, and also as a cryogenic gas at pressures below 200 psig (1379 kPa gage).

GRADES AVAILABLE

Table 1 presents the component maxima in ppm (v/v), unless shown otherwise, for the types and grades of oxygen. Gaseous oxygen is denoted as Type I and liquefied oxygen as Type II. A blank indicates no maximum limiting characteristic (see Ref. 1).

USES

Oxygen's major uses stem from its life-sustaining and combustion-supporting properties.

It is used extensively in medicine for therapeutic purposes, for resuscitation in asphyxia, and with other gases in anesthesia. It is also used in high-altitude flying, deep-sea diving, and as both an inhalant and a power source in the United States space program. Industrial applications include its very wide utilization with acetylene, hydrogen and other fuel gases for such purposes as metal cutting, welding, hardening, scarfing, cleaning and dehydrating. Oxygen helps increase the capacity of steel and iron furnaces on a growing scale in the steel industry. One of its major uses is in the production of synthesis gas (a hydrogen-carbon monoxide mixture) from coal, natural gas or liquid fuels; synthesis gas is in turn used to make gasoline, methanol and ammonia. Oxygen is similarly employed in manufacturing some acetylene through partial oxidation of the hydrocarbons in methane. It is also used in the production of nitric acid, ethylene and other compounds in the chemical industry.

PHYSIOLOGICAL EFFECTS

The inhalation of gaseous oxygen has a tonic effect on the human system rather than any toxic effect, and its tonic properties have led to may therapeutic applications of oxygen. Inhalation of high concentrations of oxygen at atmospheric pressure for a few hours has produced no observable harmful effects. Medical gas labels now carry a warning against high concentrations of oxygen for more than 5 hours without interruption. Exposures to oxygen at higher pressures for prolonged periods have been found to affect neuromuscular coordination and attentive powers.

TABLE 1. OXYGEN GRADES AVAILABLE.

Limiting Characteristic	Type I (Gaseous) GRADES						Type II (Liquefied) GRADES			
	A	B	C	D	E	F	A	B	C	D
Oxygen Min. % (v/v)	99.0	99.5	99.5	99.5	99.6	99.995	99.0	99.5	99.5	99.5
Inerts				Balance		Balance				Balance
Odor	None			None			None			None
Water		None Condensed	50	6.6	8	1.0		6.6	26.3	6.6
Dewpoint F			-54.5	-82	-80	-105		-82	-63.5	-82
Total Hydrocarbons at Methane					50	1.0			67.7	
Methane				50						25
Ethane & Other Hydrocarbons as Ethane				6						3
Ethylene				0.4						0.2
Acetylene				0.1		0.05			0.62	0.05
Carbon Dioxide				10		1.0				5
Carbon Monoxide						1.0				
Nitrous Oxide				4		0.1				2
Halogenated Refrigerants				2						1
Solvents				0.2						0.1
Other by Infrared				0.2						0.1
USP	Yes						Yes			
Permanent Particulates									1.0 mg/1 1 mm	1.0 mg/1 1mm
Fibers (40μ diameter)										6 mm

SAFE HANDLING AND STORAGE

Materials Suitable for Containers and Storage

Gaseous oxygen is noncorrosive and may consequently be contained in systems constructed of any common metals and designed to withstand safely the pressures involved. At the temperature of liquid oxygen, ordinary carbon steels and most alloy steels lose their ductility and are considered unsatisfactory for liquid oxygen service. Satisfactory materials for use with liquid oxygen include 18.8 stainless steel and other austenitic nickel-chromium alloys, copper, "Monel," brass and aluminum. Care must be taken to remove all oil, grease and other combustible material from piping systems and containers before putting them into oxygen service. Cleaning methods employed by manufacturers of oxygen equipment are described in "Cleaning Equipment for Oxygen Service" (see Ref. 2).

Storage and Handling in Normal Use

All combustible materials—especially oil and greases—must be kept from contact with high-oxygen concentrations, and all possible sources of ignition must be safely enclosed or kept completely away from either gaseous or liquid oxygen.

All the precautions necessary for the safe handling of any compressed gas must be observed with gaseous oxygen. In addition, liquid oxygen must also be handled with all the precautions required for safety with any cryogenic fluid. Contact between the skin and liquid oxygen, or uninsulated piping or vessels containing it, can cause severe burn-like injuries (see also Part I, Chapter 4, Sections A and C).

High-pressure cylinders, tube trailers and tank cars often serve as the storage supply for gaseous oxygen. Oxygen is also frequently stored in compact liquid form at the user's site. Standards to insure safety with oxygen systems at user sites are set forth in pamphlets of the National Fire Protection Association (see Refs. 3, 4, and 5).

Handling Leaks and Emergencies

See Part I, Chapter 4, Sections A and C.

METHODS OF SHIPMENT; REGULATIONS

Under the appropriate regulations and tariffs, oxygen is authorized for shipment as follows (gaseous oxygen, except where liquid oxygen is noted):

By Rail: In cylinders and tank cars, for both gaseous and liquid oxygen (special insulated cylinders and tank cars for liquid oxygen are not subject to DOT regulations at pressures below 25 psig or 170 kPa gage).

By Highway: In cylinders on trucks, and in tube trailers. Liquid oxygen in insulated cylinders, and in tank trucks or trailers not subject to DOT regulations are pressures under 25 psig (170 kPa gage).

By Water: In cylinders on cargo and passenger vessels, and on ferry and railroad car ferry vessels (passenger or vehicle). In authorized tank cars on cargo vessels only. In cylinders on barges of U.S Coast Guard classes A, AB, BB, CA and CB. Pressurized liquid oxygen, in cylinders on cargo and passenger vessels, and on ferry and railroad car ferry vessels (passenger or vehicle).

By Air: In cylinders aboard passenger aircraft up to 150 lb (70 kg), and aboard cargo aircraft up to 300 lb (140 kg), maximum net weight per cylinder. Liquid oxygen, either pressurized or nonpressurized, is not accepted for air shipment under present regulations.

CONTAINERS

Gaseous oxygen is authorized for shipment in cylinders, tank cars and tube trailers. Liquid oxygen is shipped as a cryogenic gas in insulated cylinders, insulated tank trucks and insulated tank cars.

Filling Limits. The maximum filling limits authorized for gaseous oxygen in shipment are as follows:

In cylinders and tube trailers—the authorized service pressures marked on the cylinders or tube assemblies at 70 F (21.1 C). In the case of cylinders of specifications 3A, 3AA, 3AX and 3AAX that meet special requirements, up to 10 percent in excess of their marked service pressures.

In tank cars—not more than $\frac{7}{10}$ of the marked test pressure at 130 F (54.4 C) in uninsulated cars of the DOT 107A type.

The maximum filling limits authorized for liquid oxygen are:

In cylinders that meet DOT-4L200 specification —96 percent (percent water capacity by weight).

Liquid oxygen shipped in insulated truck cargo tanks or in other insulated containers at pressures below 25 psig (170 kPa gage) is not subject to DOT regulations.

Cylinders

Cylinders meeting DOT specifications 3A or 3AA are the types usually used to ship gaseous oxygen, but oxygen is authorized for shipment in any cylinders designated for nonliquefied compressed gases. (These include cylinders which comply with DOT specifications 3A, 3AA, 3AX, 3AAX, 3B, 3C, 3D, 3E, 4, 4A, 4B, 4BA, 4BW and 4C; in addition, cylinders meeting DOT specifications 3, 7, 25, 26, 33 and 38 may be continued in gaseous oxygen service, but new construction is not authorized.)

Liquid oxygen is authorized for shipment in cylinders which meet DOT specification 4L200.

Small cylinders of special sizes for oxygen used in medicine (alone or in mixtures with helium or carbon dioxide) are described in Part I, Chapter 4, Section C; the section also explains the applications and safe handling of oxygen used medicinally.

All cylinders authorized for gaseous oxygen service must be requalified by hydrostatic retest every 5 or 10 years under present regulations, with the following exceptions DOT-4 cylinders, every 10 years; and no periodic retest is required for cylinders of types 3C, 3E, 4C and 7.

Cylinders of the 4L200 type authorized for liquid oxygen service are also exempt from periodic retest requirements at present.

Valve Outlet and Inlet Connections. Standard connections, United States and Canada—up to 3000 psi (20,684 kPa), No. 540; cryogenic liquid withdrawal, No. 440. Alternate standard connection, United States and Canada—cryogenic liquid withdrawal, No. 295. Standard yoke connection, United States and Canada (for medical use of the gas)—No. 870.

Small Portable Containers

Liquid oxygen is shipped and stored in small portable containers which hold quantities ranging from 1 to more than 25 gal (0.004–0.09 m³). These containers are encased in steel shells and are heavily insulated; they maintain the liquid at atmospheric pressure, and are consequently not subject to DOT regulations.

Tube Trailers and Cargo Tanks on Trucks or Trailers

Gaseous oxygen is shipped in tube trailers with capacities in excess of 40,000 ft³ (1133 m³). The trailers commonly serve as the storage supply for the user, with empty trailers being replaced periodically by the supplier.

Liquid oxygen is shipped in bulk at pressures below 25 psig (170 kPa gage) in special insulated cargo tanks on trucks and truck trailers with capacities in excess of 400,000 ft³ (11,330 m³).

Tank Cars

Gaseous oxygen is authorized for rail shipment in tank cars meeting DOT specification 107A.

Liquid oxygen is also shipped in special insulated tank cars at pressures below 25 psig (170 kPa gage).

METHOD OF MANUFACTURE

Almost all commercial oxygen is produced at air separation plants by liquefaction of atmospheric air and removal of the oxygen from it by fractionation. Very small quantities are produced by the electrolysis of water.

REFERENCES

1. "Commodity Specification for Oxygen" (Pamphlet G-4.3), Compressed Gas Association, Inc.
2. "Cleaning Equipment for Oxygen Service" (Pamphlet G-4.1), Compressed Gas Association, Inc.
3. "Standard for the Design and Installation of Oxygen-Fuel Gas Systems for Welding and Cutting" (NFPA Pamphlet No. 51), National Fire Protection Association, 470 Atlantic Ave., Boston, MA 02210.
4. "Nonflammable Medical Gas Systems" (NFPA Pamphlet No. 56F), National Fire Protection Association.
5. "Standard for Bulk Oxygen Systems at Consumer Sites" (NFPA Pamphlet No. 50), National Fire Protection Association.

ADDITIONAL REFERENCE

"Standard Density Data, Atmospheric Gases and Hydrogen" (Pamphlet P-6), Compressed Gas Association, Inc.

"Industrial Practices for Gaseous Oxygen Transmission and Distribution Piping Systems" (Pamphlet G-4.4), Compressed Gas Association, Inc.

Phosgene

COCl$_2$
Synonyms: Carbonyl chloride, carbon oxychloride
DOT Classification: Poison A; poison gas label

PHYSICAL CONSTANTS

	U.S. Units	Metric Units
International symbol	COCl$_2$	COCl$_2$
Molecular weight	98.92	98.92
Vapor pressure		
at 70 F (21.1 C)	9.1 psig	63 kPa, gage
at 105 F (40.6 C)	29.9 psig	206 kPa, gage
at 115 F (46.1 C)	38.1 psig	263 kPa, gage
at 130 F (54.4 C)	52.1 psig	359 kPa, gage
Density of the gas at 70 F (21.1 C) and 1 atm	0.26 lb/ft^3	4.16 kg/m^3
Specific gravity of the gas at 68 F (20 C) and 1 atm (air = 1)	3.5*	3.5*
Specific volume of the gas at 70 F (21.1 C) and 1 atm	3.9 ft^3/lb	0.24 m^3/kg
Density of liquid		
at 68 F (20 C)	86.65 lb/ft^3	1388 kg/m^3
at 70 F (21.1 C)	86.47 lb/ft^3	1385 kg/m^3
at 105 F (40.6 C)	83.47 lb/ft^3	1337 kg/m^3
at 115 F (46.1 C)	82.66 lb/ft^3	1324 kg/m^3
at 130 F (54.4 C)	81.41 lb/ft^3	1304 kg/m^3
Boiling point at 1 atm	46.76 F	8.200 C
Freezing point at 1 atm	-198.0 F	-127.8 C
Critical temperature	359.4 F	181.9 C
Critical pressure	823 psia	5674 kPa, abs.
Critical density	32.5** lb/ft^3	521 kg/m^3

*Other source gives 3.4.
**Other source gives an equivalent value of 32.46.

	U.S. Units	Metric Units
Latent heat of vaporization at boiling point	106.2*** Btu/lb	247.0 kJ/kg
Latent heat of fusion at melting point	24.95**** Btu/lb	58.03 kJ/kg
Specific heat of gas at 77 F (25 C) and 1 atm		
C_p	0.1393 Btu/(lb) (F)	0.5832 kJ/(kg) (C)
C_v theoretical	0.1189 Btu/(lb) (F)	0.4978 kJ/(kg) (C)
Ratio of specific heats	1.171	1.171
Solubility in water, vol/vol	decomposes	decomposes
Weight of liquid at 68 F (20 C)	11.57 lb/gal	1386 kg/m^3
Specific gravity of liquid at 68 F (20 C) (water = 1)	1.388	1.388
Heat capacity of liquid at boiling point	0.244 Btu/(lb) (F)	1.02 kJ/(kg) (C)
Coefficient of expansion at 32 F (0 C)	0.002207 per °F	0.003973 per °C

***Other source gives an equivalent value of 106.12.
****Other source gives 24.838.

DESCRIPTION

Phosgene is a nonflammable colorless gas more than three times as heavy as air, and is designated a poison of the class A or "extremely dangerous" group by the DOT. Phosgene under pressure is a colorless to light yellow liquid. It has its own characteristic odor which is often stifling or suffocating and strong, but sometimes not unpleasant, depending on the concentration; it has been said to resemble sour green corn or moldy hay in odor when greatly diluted with air. Phosgene vapors strongly irritate the eyes.

Completely dry, pure phosgene is stable at ordinary temperatures. It dissociates into its component parts, carbon monoxide and chlorine, at elevated temperatures, to an extent ranging from 0.45 percent dissociation at 214 F (101.1 C) to 100 percent at 1472 F (800 C).

Phosgene is slightly soluble in water and is slowly hydrolyzed by water to form corrosive hydrochloric acid and carbon dioxide.

For other solvents, phosgene dissolves as follows (parts per 100 parts of solvent by weight, at 1 atm and 68 F or 20 C, or as indicated): carbon tetrachloride, 28; chloroform, 59; glacial acetic acid, 62; Russian mineral oil, 35; chlorinated paraffin, 81; ethyl acetate, 98; benzene, 99; toluene (at 63 F or 17.2 C), 244; xylene (at 54 F or 12.2 C), 457; and chlorobenzene (at 54 F or 12.2 C), 422.

GRADES AVAILABLE

Phosgene is available for commercial and industrial use in various grades having much the same component proportions from one producer to another.

USES

Phosgene is used mainly as an intermediate in the manufacture of many types of compounds (including: barbiturates; chloroformates and thiochloroformates; carbamoyl chlorides, acid chlorides and acid anhydrides; carbamates; carbonates and pyrocarbonates; urethanes; ureas, azo-urea dyes, triphenylmethane dyes and substituted benzophenones; isocyanates and isothiocyanates; carbazates and carbohydrazides; malonates; carbodiimides; and oxazolidinediones). It is also used in bleaching sand for glass manufacture, and as a chlorinating agent.

PHYSIOLOGICAL EFFECTS

Phosgene is a lung irritant and also attacks other parts of the respiratory system. Low concentrations in air cause watering of the eyes and coughing which may result in a thin, frothy expectoration. High concentrations cause greater distress. Phosgene is more than ten times as toxic as chlorine. Exposure to concentrations of 3 to 5 ppm can cause an irritation of the eyes and throat, with coughing; 25 ppm represents a dangerous exposure if prolonged for 30 to 60 minutes; 50 ppm proves rapidly fatal even after short exposure; and about 120 ppm (0.5 mg per liter) is lethal within 10 minutes.

A TLV of 0.1 ppm (0.4 mg/m^3) for an 8-hour work day for phosgene has been adopted by the ACGIH.

One serious difficulty with the treatment of persons exposed to phosgene is that symptoms may not appear until hours after the exposure. The delayed action of phosgene can be particularly injurious if the victim performs violent exercise after having been exposed.

All persons who have been gassed with phosgene must be examined by a physician as soon as possible, because serious symptoms may develop at a later stage.

SAFE HANDLING AND STORAGE

Materials Suitable for Containers and Storage

Anhydrous phosgene in the liquid state is compatible with a variety of common metals, including aluminum (of 99.5 percent purity), copper, pure iron or cast iron, steel (including cast steel and chrome-nickel steels), lead (up to 250 F or 121.1 C), nickel and silver; it is also compatible with platinum and platinum alloys in instruments. Nonmetallic materials with which liquid anhydrous phosgene is also compatible include acid-resistant linings (ceramic plates and carbon blocks), enamel on cast iron or glass-lined steel, Jena special glass (as well as "Pyrex" or "Kimax"), porcelain, quartzware, granite or basalt natural stone, stoneware and "Teflon."

In the presence of moisture, phosgene is not compatible with copper, steel, nor pure or cast iron. Detailed data on the corrosion resistance of various materials to phosgene under a range of conditions are given in a 1960 survey of the Shell Development Company (see Ref. 1).

Steel piping with seamless fittings is recommended for handling phosgene as a general rule, and pipe no smaller than $\frac{3}{4}$-in. (1.9 cm) nominal size should be used to insure rigidity and minimize possible leaks. For pipe size up to 4-in. (10.2 cm), schedule 80 seamless (or alloy steel to ASTM A333 GR3) piping is recommended; 6-in. diameter (15.2 cm) schedule 40 seamless may be used as a larger pipe size. Screwed or flanged joints should be kept to the minimum, and cast iron or malleable iron fittings and valves should not be used; nonarmored porcelain valves must not be used, regardless of the pressure with either liquid or gas phosgene. Only O.S.Y. or rising stem valves are recommended, to reduce the possibility of accidents; nonindicating valves should not be used. "Monel" is the material generally used in manually operated valves for disk, seat and stem.

A pipe joint compound composed of litharge and glycerin is recommended, as are bonded asbestos fiber chemical lead 2 to 4 percent antimony, or "Teflon" envelope for flat gaskets, depending on the temperature. Detailed recommendations on these and other materials for various purposes in phosgene service may be obtained from phosgene manufacturers.

Storage and Handling in Normal Use

Since phosgene is a highly toxic gas, appropriate precautions must be taken in its storage and handling as with any poison gas. It is imperative to prevent moisture from entering any closed phosgene container because of the formation of hydrochloric acid and carbon dioxide which could build up sufficient pressure and rupture the container (see also Part I, Chapter 4, Section A).

Handling Leaks and Emergencies

Field neutralization of phosgene, in emergencies or after possible gas warfare use, is accomplished with alkali or alkali solutions. The reaction of phosgene with an ammonia solution (which forms urea) is particularly effective (see also Part I, Chapter 4, Section A).

Leak Detection. Suspected leaks should be investigated only by personnel who are wearing gas masks of an approved type (these are gas masks approved for protection against phosgene by the U.S. Bureau of Mines; the Bureau recommends either an acid-gas-and-organic-vapor mask or a universal mask for concentrations of less than 2 percent, and masks with a self-contained breathing apparatus for larger concentrations or sites otherwise deficient in oxygen). Phosgene users should consult U.S. Bureau of Mines publications for recommendations of necessary masks (see Ref. 2).

Personnel wearing masks can easily detect phosgene leaks with ammonia vapor devices, as phosgene produces white fumes in the presence of ammonia. In case of leakage around the valve stem, an operator should tighten down on the valve packing nut only with the special wrench supplied with the cylinder for this purpose.

First Aid for Phosgene Exposure. First summon a physician to examine any person exposed to phosgene. Then take the following steps:

(1) Remove contaminated clothing if impregnated with phosgene.

(2) Do not permit the patient to walk.

(3) Keep the patient at rest. An occasional change of position from lying to sitting may be beneficial. Keep the patient calm and have him try to suppress desires to cough.

(4) Have the patient in a flat position. The head may be elevated. Warm patient with blankets.

(5) Cover eyes with a wet towel. Silicone oil is irritating but not harmful to the eyes.

(6) Attach bottle of silicone spray to atom-izer tube on suction machine. Check to see if it is working properly.

(7) Spray silicone mist with each inhalation for 10 minutes. Wait 15 minutes and repeat as above. Do this for four consecutive times giving 40 minutes total of silicone spray.

(8) If necessary administer pure oxygen through a mask to the patient.

(9) If and only if breathing has ceased, apply artificial respiration by the Schaefer prone-pressure method (which will help to empty the lungs of fluid), and at the same time have an assistant administer as in (8) above.

Customary Steps in Medical Treatment. Physicians usually observe the following practices in the treatment of rarely occurring cases of phosgene gassing:

(1) Absolute rest is essential, and the patient should be transported to the hospital in an ambulance.

(2) The patient should be watched very closely for 24 hours, because serious lung edema can develop suddenly in patients who are apparently normal.

(3) Continuous administration of pure oxygen by means of a mask is of the utmost importance, and may be needed for several days.

(4) Cardiac weakness is often apparent and coramine (1 cc), or camphor in oil if coramine is not available, should be given every 4 hours.

(5) Venesection should be performed only if there is definite evidence of embarrassment of the right heart, and not for cyanosis alone.

(6) If the pulse is rapid and feeble, the heart should be fully digitalized.

METHODS OF SHIPMENT; REGULATIONS

Under the appropriate regulations and tariffs, phosgene is authorized for shipment as follows:

By Rail: In cylinders (via freight only), and in TMU tank cars.

By Highway: In cylinders on trucks, and in TMU tanks on trucks and on full or semitrailers (provided that tanks are securely chocked or clamped to prevent shifting, and that adequate facilities are available for handling tanks where transfer in transit is necessary).

By Water: In cylinders and TMU tank cars authorized by the DOT on cargo vessels only. In authorized cylinders on barges of U.S. Coast Guard classes A, BA, BB, CA and CB.

By Air: Not accepted for shipment.

CONTAINERS

Phosgene is authorized for shipment by the DOT in cylinders and in TMU tanks, which usually have net capacities of 150 lb (or less) and 2000 lb (68 kg or less and 907 kg), respectively.

Filling Limits. The maximum filling densities permitted for phosgene are as follows:

In cylinders—125 percent (percent of water capacity by weight) plus the requirement that the cylinder must not contain more than 150 lb (68 kg) of phosgene.

In TMU tanks—not liquid full at 130 F (54.5 C).

Cylinders

Cylinders that meet DOT specifications 3D and that are of not over 125-lb (56.7 kg) water capacity (nominal) are authorized for the shipment of phosgene. DOT-3A, 3AA or 3E cylinders rated for 1800 psia (12.41 MPa) minimum pressure may also be used in phosgene service. (Cylinders meeting specifications DOT-33 may also be continued in phosgene service, but new construction is not authorized.) Unless 3D or 33 cylinders have valve-protection extension rings, the cylinders must be packed in wooden boxes as prescribed; if gaskets are used between the caps and cylinder necks, they must be renewed for each shipment even though they may appear to be in good condition. 3E1800 cylinders must be packaged in wooden or metal boxes.

Each filled cylinder must show absolutely no leakage in an immersion test made before shipment. Cylinders must be tested without their valve-protection caps. For the test, the cylinder and valve must be kept submerged in a bath of water heated to approximately 150 F (65.6 C) for at least 30 minutes, and frequent examinations must be made during that time to note any escape of gas. The cylinder valve must not be loosened after the test and before shipment.

Whether boxed or protected with caps or cap-protection rings, the unit must be able to pass a drop test (6 ft or 1.8 m drop onto a concrete floor impacting at the weakest point).

Valves must be either a nonperforated diaphragm type (with outlets sealed with a solid metal cap or plug) or a packed type provided the assembly is made gas tight by means of a seal cap to the valve body or to the cylinder.

All the precautions necessary for the safe handling, shipping and storage of any compressed gas cylinders must of course be observed with phosgene cylinders.

The 3D, 3A1800, 3AA1800 and 33 types of cylinders authorized for phosgene service must be requalified by hydrostatic retest every 5 years under present regulations. For 3E1800 cylinders no hydrostatic retest is required.

Warming Cylinders to Help Remove Contents. Phosgene cylinders may be heated by warm air or warm water to facilitate removal of their contents. Never use steam, boiling water or direct flame for this purpose. Never under any circumstances allow the outside of a cylinder to reach temperatures above 125 F (51.7 C).

Valve Outlet Connections. Standard connections, United States and Canada—No. 750.

TMU Tanks

TMU tanks that meet DOT specifications 106A500-X are authorized for phosgene service. (TMU tanks meeting specifications 106A500 may also be continued in phosgene service, but new construction is not authorized.) These tanks may be shipped either by rail or by motor vehicle. The tanks must be equipped with gas-tight valve-protection caps approved by the Bureau of Explosives, and they must not be equipped with safety relief devices of any type. TMU tanks equipped in this way are authorized only for phosgene.

METHOD OF MANUFACTURE

Phosgene is produced commercially by passing chlorine and carbon monoxide (in excess) over activated carbon as a catalyst under carefully controlled conditions.

REFERENCES

1. "Corrosion Data Survey" (P-3), 1960 edition, Shell Development Company.
2. Circular No. 7885 (pp. 5, 6, 8, 9); Supplemental List lC7885, pp. 2, 3, U.S. Bureau of Mines.

Rare Gases of the
Atmosphere

Argon (see separate section)
Helium (see separate section)
Krypton: Kr
Neon: Ne
Xenon: Xe
DOT Classification (neon and krypton only): Nonflammable gas; nonflammable
  gas label

PHYSICAL CONSTANTS

Krypton

	U.S. Units	Metric Units
International symbol	Kr	Kr
Molecular weight	83.80	83.80
Density of the gas at 70 F (21.1 C) and 1 atm	0.2172 lb/ft³	3.479 kg/m³
Specific gravity of the gas at 70 F and 1 atm (air = 1)	2.899	2.899
Specific volume of the gas at 70 F (21.1 C) and 1 atm	4.604 ft³/lb	0.287 m³/kg
Boiling point at 1 atm	−244.0 F	−153.4 C
Melting point at 1 atm	−251 F	−157 C
Critical temperature	−82.8 F	−63.8 C
Critical pressure	798.0 psia	5502 kPa, abs.
Critical density	56.7 lb/ft³	908 kg/m³
Triple point	−251.3 F at 10.6 psia	−157.4 C at 73.2 kPa, abs.
Latent heat of vaporization at boiling point	46.2 Btu/lb	107.5 kJ/kg
Latent heat of fusion at triple point	8.41 Btu/lb	19.57 kJ/kg

	U.S. Units	Metric Units
Specific heat of gas at 70 F (21.1 C) and 1 atm		
C_p	0.060 Btu/(lb) (F)	0.251 kJ/(kg) (C)
C_v	0.035 Btu/(lb) (F)	0.146 kJ/(kg) (C)
Ratio of specific heats	1.69	1.69
Solubility in water, vol/vol at 68 F (20 C)	0.0594	0.0594
Weight of liquid at normal boiling point	20.15 lb/gal	2415 kg/m^3
Density, liquid, at boiling point	150.6 lb/ft^3	2412.38 kg/m^3
Liquid/gas ratio (liquid at boiling point, gas at 70 F and 1 atm), vol/vol	1/693.4	1/693.4

PHYSICAL CONSTANTS

Neon

	U.S. Units	Metric Units
International symbol	Ne	Ne
Molecular weight	20.183	20.183
Density of the gas at 70 F (21.1 C) and 1 atm	0.05215 lb/ft^3	0.83536 kg/m^3
Specific gravity of the gas at 70 F and 1 atm (air = 1)	0.696	0.696
Specific volume of the gas at 70 F (21.1 C) and 1 atm	19.18 ft^3/lb	1.197 m^3/kg
Boiling point at 1 atm	−410.9 F	−246.0 C
Melting point at 1 atm	−415.6 F	−248.7 C
Critical temperature	−379.8 F	−228.8 C
Critical pressure	384.9 psia	2654 kPa, abs.
Critical density	30.15 lb/ft^3	483 kg/m^3
Triple point	−415.4 F at 6.29 psia	−248.6 C at 43.4 kPa, abs.
Latent heat of vaporization at boiling point	37.08 Btu/lb	86.3 kJ/kg
Latent heat of fusion at triple point	7.14 Btu/lb	16.6 kJ/kg
Specific heat of gas at 70 F (21.1 C) and 1 atm		
C_p	0.25 Btu/(lb) (F)	1.05 kJ/(kg) (C)
C_v	0.152 Btu/(lb) (F)	0.636 kJ/(kg) (C)
Ratio of specific heats	1.64	1.64
Solubility in water, vol/vol at 68 F (20 C)	0.0105	0.0105

	U.S. Units	Metric Units
Weight of liquid at normal boiling point	10.07 lb/gal	1207 kg/m³
Density, saturated vapor at 1 atm	0.5862 lb/ft³	9.390 kg/m³
Density, gas, at boiling point	0.6068 lb/ft³	9.7200 kg/m³
Density, liquid, at boiling point	75.35 lb/ft³	1207 kg/m³
Liquid/gas ratio (liquid at boiling point, gas at 70 F and 1 atm), vol/vol	1/1445	1/1445

PHYSICAL CONSTANTS

Xenon

	U.S. Units	Metric Units
International symbol	Xe	Xe
Molecular weight	131.3	131.3
Density of the gas at 70 F (21.1 C) and 1 atm	0.3416 lb/ft³	5.472 kg/m³
Specific gravity of the gas at 70 F and 1 atm (air = 1)	4.560	4.560
Specific volume of the gas at 70 F (21.1 C) and 1 atm	2.927 ft³/lb	0.183 m³/kg
Boiling point at 1 atm	−162.6 F	−108.1 C
Melting point at 1 atm	−168 F	−111 C
Critical temperature	61.9 F	16.6 C
Critical pressure	847.0 psia	5840 kPa, abs.
Critical density	68.67 lb/ft³	1100 kg/m³
Triple point	−169.2 F at 11.84 psia	−111.8 C at 81.6 kPa, abs.
Latent heat of vaporization at boiling point	41.4 Btu/lb	96.3 kJ/kg
Latent heat of fusion at triple point	7.57 Btu/lb	17.6 kJ/kg
Specific heat of gas at 70 F (21.1 C) and 1 atm		
C_p	0.038 Btu/(lb) (F)	0.159 kJ/(kg) (C)
C_v	0.023 Btu/(lb) (F)	0.096 kJ/(kg) (C)
Ratio of specific heats	1.667	1.667
Solubility in water, vol/vol at 68 F (20 C)	0.108	0.108
Weight of liquid at −162.6 F (−108.11 C) normal boiling point	25.51 lb/gal	3057 kg/m³
Density, liquid, at boiling point	190.8 lb/ft³	3056.3221 kg/m³

	U.S. Units	Metric Units
Liquid/gas ratio (liquid at boiling point, gas at 70 F and 1 atm), vol/vol	1/558.5	1/558.5

DESCRIPTION

Krypton, neon and xenon are rare atmospheric gases. Each is odorless, colorless, tasteless, nontoxic, monatomic and chemically inert. All three together constitute less than 0.002 percent of the atmosphere, with approximate concentrations in the atmosphere of 18 ppm for neon, 1.1 ppm for krypton, and 0.09 ppm for xenon. Few users of the three gases need them in bulk quantities and the three are shipped most often in single cylinders and liter flasks.

Radon, a rare gas, is not treated in the *Handbook* because it has little or no practical application at present. It is radioactive, and is the heaviest gas known (density at 70 F and 1 atm, 0.61 lb/ft^3; at 21.1 C and 1 atm, 9.8 kg/m^3).

Among the rare gases, neon, krypton and xenon in particular ionize at lower voltages than other gases, and the brilliant, distinctive light they emit while conducting electricity in the ionized state amounts for one of their primary uses. Their characteristic colors as ionized conductors are: neon, red; krypton, yellow-green; and xenon, blue to green (and similarly, argon, red or blue; and helium, yellow).

GRADES AVAILABLE

For commercial and industrial purposes, neon is available in "purified" grades that typically have a minimum purity of 99.7 mole percent.

USES

Neon, krypton and xenon are used principally to fill lamp bulbs and tubes. The electronics industry uses them singly or in mixtures in many types of gas-filled electron tubes (among them, voltage regulator tubes, starter tubes, phototubes, counter tubes, T. R. tubes, xenon thryatron tubes, half-wave xenon rectifier tubes and Geiger–Muller tubes). Large quantities of neon (as well as of atmospheric helium and specially purified argon) are employed as fill gases in illuminated signs. Small quantities of krypton and xenon are used for special effects. In the lamp industry the three gases serve as fill gas in specialty lamps, neon glow lamps, 100-watt fluorescent lamps, ultraviolet sterilizing lamps and very high output lamps. The three gases have additional applications in the atomic energy field as fill gas for ionization chambers, bubble chambers, gaseous scintillation counters and other detection and measurement devices.

PHYSIOLOGICAL EFFECTS

Neon, krypton and xenon are nontoxic. As gases, they can act as simple asphyxiants by displacing air, and they cannot be detected by odor or color. Liquefied neon is used in substantial quantities; the liquid and the cold gas evolving from it can produce severe burns upon contact with the skin and other tissues.

SAFE HANDLING AND STORAGE

Materials Suitable for Containers and Storage

These three inert gases may be used with containers and equipment made of any common metals. Installations must be designed to meet all requirements for the pressures involved.

Storage and Handling in Normal Use

See Part I, Chapter 4, Section A.

Handling Leaks and Emergencies

See Part I, Chapter 4, Section A.

METHODS OF SHIPMENT; REGULATIONS

Under the appropriate regulations and tariffs, neon, krypton and xenon are authorized for shipment as follows:

By Rail: In cylinders (freight or express).

By Highway: In cylinders on trucks.

By Water: In cylinders via cargo and passenger vessels, and in ferry and railroad car ferry vessels (either passenger or vehicle). In cylinders on barges of U.S. Coast Guard classes A, BA, BB and C.

By Air: For gaseous neon and krypton, aboard passenger aircraft in appropriate cylinders up to 150 lb (70 kg) maximum net weight per cylinder, and aboard cargo aircraft in appropriate cylinders up to 300 lb (140 kg) maximum net weight per cylinder. Liquefied neon, either pressurized or low pressure, is authorized for shipment aboard cargo aircraft only in containers meeting specific requirements up to 300 lb (140 kg) maximum net weight per container. Xenon is shipped by air as a "Nonflammable compressed gas, n.o.s. (not otherwise specified)."

CONTAINERS

Neon, krypton and xenon are shipped most often in individual cylinders or in liter quantities in glass flasks, since they are seldom used in bulk quantities. Neon is also shipped as a liquefied, cryogenic gas in special insulated containers under DOT exemption.

Filling Limits. The maximum authorized filling limit for neon, krypton and xenon in approved types of cylinders is the marked service pressure of the cylinder at 70 F (21.1 C) (or, in the case of only 3A and 3AA cylinders meeting additional specified requirements, 10 percent in excess of the marked service pressure). Maximum filling limits for the specific commercial grade of xenon being used must be determined experimentally, because commercial grades of xenon deviate markedly in physical properties from the ideal gas at elevated pressures.

Cylinders

Neon, krypton and xenon are authorized for shipment in cylinders of any type approved by the DOT for nonliquefied compressed gases. (These are cylinders meeting DOT specifica-

tions 3A, 3AA, 3B, 3C, 3D, 3E, 4, 4A, 4B, 4BA, 4BW, 4C and 39; also, cylinders meeting DOT specifications 3, 7, 25, 26, 33 and 38 may be continued in service with these gases, but new construction is not authorized).

Krypton and xenon are shipped in cylinders by rail, highway or water as "Nonflammable compressed gas, n.o.s. (not otherwise specified.)"

Under present regulations, cylinders of all types authorized for service with neon, krypton and xenon must be requalified by hydrostatic test every 5 years with the following exceptions: DOT-3A and 3AA used exclusively for krypton, neon and xenon may be retested every 10 years instead of every 5 years; DOT-4 may be retested every 10 years; and DOT-3C, 3E, 4C and 7 require no periodic retest.

Valve Outlet and Inlet Connections. Standard connections, United States and Canada— up to 3000 psi (20,684 kPa), No. 580; 3000– 10,000 psi (20,684–68,948 kPa) No. 677, see Part I, Chapter 7, Section A, paragraph 2.1.7; cryogenic liquid withdrawal (neon only), No. 792. Alternate standard connection, United States and Canada—3000–10,000 psi, No. 677.

METHOD OF MANUFACTURE

Neon, krypton and xenon are produced commercially at air separation plants in two stages—an initial stage of partial separation by liquefaction and fractional distillation, and a final purification stage requiring complex processing.

REFERENCES

1. G. A. Cook, Ed., *Argon, Helium and the Rare Gases,* 2 vols., pp. 435–437. New York: John Wiley & Sons (Interscience), 1961.

ADDITIONAL REFERENCE

"Standard Density Data, Atmospheric Gases and Hydrogen" (Pamphlet P-6), Compressed Gas Association, Inc.

Sulfur Dioxide

SO_2
Synonym: Sulfurous acid anhydride
DOT Classification: Nonflammable gas; nonflammable gas label

PHYSICAL CONSTANTS

	U.S. Units	Metric Units
International symbol	SO_2	SO_2
Molecular weight	64.06	64.06
Vapor pressure		
at 70 F (21.1 C)	34.4 psig	237 kPa, gage
at 105 F (40.6 C)	76.6 psig	528 kPa, gage
at 115 F (46.1 C)	92.6 psig	638 kPa, gage
at 130 F (54.4 C)	121.1 psig	835 kPa, gage
Density of the gas		
at 32 F (0 C) and 1 atm	0.1827 lb/ft^3	2.927 kg/m^3
Specific gravity of the gas		
at 32 F and 1 atm (air = 1)	2.2636	2.2636
at 70 F and 1 atm (air = 1)	2.262	2.262
Specific volume of the gas		
at 70 F (21.1 C) and 1 atm	5.90 ft^3/lb	0.368 m^3/kg
Density of liquid		
at 70 F (21.1 C)	86.06 lb/ft^3	1379 kg/m^3
at 105 F (40.6 C)	82.55 lb/ft^3	1322 kg/m^3
at 115 F (46.1 C)	81.50 lb/ft^3	1306 kg/m^3
at 130 F (54.4 C)	79.94 lb/ft^3	1281 kg/m^3
Boiling point at 1 atm	14.0 F	−10.0 C
Melting point at 1 atm	−103.9 F	−75.5 C
Critical temperature	314.82 F	157.12 C
Critical pressure	1141.5 psia	7870.4 kPa, abs.
Critical density	32.7 lb/ft^3	524 kg/m^3
Triple point	−103.9 F at	−75.46 C at
	0.2429 psia	1.675 kPa, abs.
Latent heat of vaporization		
at 70 F (21.1 C)	155.5 Btu/lb	361.7 kJ/kg

	U.S. Units	Metric Units
Latent heat of fusion		
at −103.9 F (−75.5 C)	49.70 Btu/lb	115.5 kJ/kg
Specific heat of gas		
at 77 F (25 C) and 1 atm		
C_p	0.149 Btu/(lb) (F)	0.622 kJ/(kg) (C)
C_v	0.116 Btu/(lb) (F)	0.485 kJ/(kg) (C)
Ratio of specific heats	1.28	1.28
Solubility in water, NTP vol/vol		
at 77 F (25 C)	32.79	32.79
Weight of liquid		
at 70 F (21.1 C)	11.53 lb/gal	1382 kg/m^3
Specific gravity of liquid		
at 32 F (0 C)	1.434	1.434
Solubility in water, by weight		
at 32 F (0 C)	18.59 percent	18.59 percent
at 68 F (20 C)	10.14 percent	10.14 percent
at 86 F (30 C)	7.25 percent	7.25 percent
at 104 F (40 C)	5.13 percent	5.13 percent

DESCRIPTION

Sulfur dioxide is a nonflammable, colorless gas that has a characteristic, pungent odor and is highly irritating at room temperatures and atmospheric pressures. At temperatures below 14 F, or under moderate pressures, sulfur dioxide is a colorless liquid. It is more than twice as heavy as air in gaseous form and roughly one and a half times the weight of water as a liquid, and consists by weight of 50.05 percent sulfur and 49.95 percent oxygen. It is shipped as a liquefied compressed gas under its vapor pressure of some 35 psig at 70 F (240 kPa, gage at 21.1 C).

Chemically, sulfur dioxide in aqueous solution is an outstanding oxidizing or reducing agent. Dry sulfur dioxide is not corrosive to ordinary metals. However, zinc will react with sulfur dioxide containing minute quantities of moisture, and most common metals will be corroded by sulfur dioxide holding sufficient amounts of moisture. Sulfur dioxide dissolved in water will form sulfurous acid, which is unstable toward heat, and decreasing proportions of sulfur dioxide go into solution in water as temperature increases.

Table 1 gives the vapor pressure, volume, density and latent heat properties of sulfur dioxide in containers over a range of temperatures.

GRADES AVAILABLE

Sulfur dioxide is available for commercial and industrial use in various grades having much the same component proportions from one producer to another.

USES

Sulfur dioxide's useful properties as a refrigerant, fumigant, preservative, bleach, antichlor, etc. are utilized in a diverse group of industries. Small quantities of sulfur dioxide are used by the refrigerating and air-conditioning industries. The petroleum industry consumes sulfur dioxide in the Edeleanu Process for the refining of kerosene and light lubricating oils. Another principal application of sulfur dioxide is found in the manufacture of sulfite pulp for paper and artificial silk.

Other uses of sulfur dioxide include its utilization in the multiple role of preservative, bleach

TABLE 1. VAPOR PRESSURE, VOLUME, DENSITY AND LATENT HEAT OF SULFUR DIOXIDE IN CONTAINERS AT VARIOUS TEMPERATURES (SEE REF. 1).

Temperature (°F.)	Temperature (°C.)	Vapor Pressure (psia)	Vapor Pressure (psig)	Volume (cubic feet per pound) Liquid	Volume Vapor	Density (pounds per cubic foot) Liquid	Density Vapor	Latent Heat (Btu/lb.)
—40	—40.0	3.12	*23.6"	.01044	22.2	95.79	.04505	178.4
—20	—28.9	5.88	*18.0"	.01062	12.5	94.16	.08000	174.4
0	—17.8	10.26	*9.07"	.01082	7.35	92.42	.13605	170.3
10	—12.2	13.3	*2.85"	.01092	5.77	91.58	.17331	168.3
20	— 6.7	16.9	2.2	.01103	4.59	90.66	.21786	166.3
30	— 1.1	21.3	6.6	.01114	3.70	89.77	.27027	164.2
40	4.4	26.6	11.9	.01125	3.02	88.89	.33113	162.2
50	10.0	32.9	18.2	.01137	2.48	87.95	.40323	160.0
60	15.6	40.3	25.6	.01149	2.05	87.03	.48780	157.8
70	21.1	49.1	34.4	.01162	1.70	86.06	.58824	155.5
80	26.7	59.3	44.6	.01175	1.42	85.11	.70423	153.1
90	32.2	71.0	56.3	.01189	1.20	84.10	.83333	150.7
100	37.8	84.1	69.4	.01204	1.02	83.06	.98039	148.2
110	43.3	99.1	84.4	.01219	.868	82.03	1.15207	145.7
120	48.9	116.3	101.6	.01235	.746	80.97	1.34048	143.0
130	54.4	135.8	121.1	.01251	.646	79.94	1.54799	140.0
140	60.0	157.7	143.0	.01269	.554	78.80	1.80505	137.1

*Indicates inches of mercury below atmospheric pressure.

and fumigant as it is used to preserve fruits, to bleach fruits, sugar and grains and to fumigate vermin-infested grains. Sulfur dioxide in the textile industry is employed as an antichlor and sour in bleaching and is used in the preparation of sodium hydrosulfite for dyeing and printing. As an antichlor it is also used in water treatment to remove objectionable odors remaining after purification. Liquid sulfur dioxide is used in the preparation of chlorine dioxide for pulp bleaching and other uses. Other uses include its employment in tanning leather, metal refining, fumigating ships and as a catalyst and reagent in the manufacture of various resins and plastics.

Large quantities of sulfur dioxide are used in the southwestern United States in a process which involves the addition of sulfur dioxide to irrigating water to increase crop yields in alkaline soils.

PHYSIOLOGICAL EFFECTS

Sulfur dioxide is an extremely irritating gas which is practically irrespirable to those unaccustomed to it. It readily elicits respiratory reflexes. It affects the upper respiratory tract, but with deeper breathing affects the lower system also. Four parts per million can be readily detected by odor but as the nose becomes accustomed to it, the amount necessary to produce a reflex respiratory defense response increases. In higher concentrations the severely irritating effect of the gas makes it inconceivable that a person would remain in a sulfur dioxide contaminated atmosphere unless he were trapped and unable to flee. Since the gas serves in this way as its own warning agent, cases of severe exposure to sulfur dioxide are not numerous.

Exposure to sulfur dioxide gas in low concentrations produces an irritating effect on the mucous membranes of the eyes, nose, throat and lungs due to the formation of sulfurous acid as the gas comes in contact with the moisture on these surfaces. This irritation is accompanied by a desire to cough which must be controlled to minimize injury. Exposure to higher concentrations produces a suffocating effect due to the closing of the glottis to shut out the gas. Table 2 shows the physiological response to various concentrations of sulfur dioxide. There is no evidence to indicate that exposure to sulfur dioxide in allowable concentrations produces a cumulative effect.

A TLV of 5 ppm by volume (13 mg/m³ at

TABLE 2. PHYSIOLOGICAL RESPONSE TO
VARIOUS CONCENTRATIONS
OF SULFUR DIOXIDE.

	Parts of Sulfur Dioxide per Million Parts of Air
Least detectable odor*[†]	3 to 5
Least amount causing immediate irritation to the eyes[†]	20
Least amount causing immediate irritation to the throat[†]	8 to 12
Least amount causing coughing[†]	20
Dangerous for even short exposure[‡]	400 to 500

*U.S. Dept. of Interior, Bureau of Mines, Bulletin 98, 1915.

[†]A. Fieldner and S. Katz, *Eng. and Mining J.*, 1919, 107, page 693.

[‡]Lehman, *Arch. f. Hyg.*, 18, 180.

32 F and 1 atm) for an 8-hour work day for sulfur dioxide has been adopted by the American Conference of Governmental Industrial Hygienists (ACGIH).

Exposure to escaping sulfur dioxide liquid will result in a freezing action of the skin in addition to any effects of inhalation of gas. This freezing effect is the natural result of the escape of any liquefied refrigerant in the lower boiling point range.

SAFE HANDLING AND STORAGE

Materials Suitable for Containers and Storage

Steel and other common metals except zinc have been found to give satisfactory service with dry sulfur dioxide. Among materials suitable with moisture-bearing sulfur dioxide are certain stainless steels (such as type 316) and lead.

Storage and Handling in Normal Use

All the precautions necessary for the handling of any nonflammable, highly irritating compressed gas must be taken with sulfur dioxide (see also Part I, Chapter 4, Section A).

Handling Leaks and Emergencies

Since sulfur dioxide represents a panic hazard, and in a sufficiently high concentration may cause injury or even death from suffocation, provisions should be made in advance by the sulfur dioxide user for action in an emergency. All employees handling sulfur dioxide should be impressed with the potential danger which it represents and should be trained in its safe handling. In addition they should be provided with personal protective equipment for use in an emergency and should be drilled until they are familiar with its use. This protective equipment should include a gas mask of a type approved by the Bureau of Mines for sulfur dioxide service. Care should be taken to assure that masks are kept in proper working order and that they are stored so as to be readily available in case of need. Canister-type masks are unsafe for high concentrations of sulfur dioxide and employees should be warned of possible failure of this type of mask in the event of a really serious leak. Self-contained breathing apparatus or a mask with a long air hose and outside source of air may be required under extreme conditions. Other protective equipment provided should include goggles or large lens spectacles to eliminate the possibility of liquid sulfur dioxide coming in contact with the eyes and causing possible injury.

If sulfur dioxide should be released, the irritating effect of the gas will force personnel to leave the area long before they have been exposed to dangerous concentrations. To facilitate their rapid evacuation there should be sufficient well-marked, easily accessible exits and indicators of wind direction so that evacuation routes can be upwind or crosswind. If despite all precautions a person should be trapped in a contaminated atmosphere, he should breathe as little as possible and open his eyes only when necessary. Since sulfur dioxide gas is heavier than air a trapped person should seek a high position to take advantage of lower gas concentrations at that level.

Since sulfur dioxide neither burns nor supports combustion, there is no danger of fire or explosion due to igniting gas or liquid. Should fire break out due to some other cause in an area containing sulfur dioxide, every effort should be made to remove the containers from the area to prevent overheating which would lead to melting of the fuse plugs and release of the sulfur dioxide. If they cannot be removed, firemen should be informed of their location. Whenever it is necessary to enter a storage tank or other vessel containing the sulfur dioxide or one that has previously contained liquid sulfur dioxide, a self-contained respirator, gas mask or air pack should be worn; and the person entering the tank should be connected to an attendant by a rope even though the tank has been thoroughly purged. It is further advisable that the person entering the tank first apply a protective cream to those parts of the body which tend to perspire to avoid irritation of the skin and that gas-tight goggles or face mask be worn to avoid irritation of the eyes.

Detecting Sulfur Dioxide Leaks. A sulfur dioxide leak indicates its presence by the characteristic, pungent odor of sulfur dioxide gas. The location of even the smallest leak may be readily determined by means of ammonia vapor dispensed from an aspirator or squeeze bottle in the region where a leak is suspected, or by the use of an ammonia swab, prepared by securing a small piece of cloth or sponge to a wire and soaking it in a strong solution of aqua ammonia. When the ammonia vapor or swab is passed over points of suspected leaks, dense white fumes form near the leak where the sulfur dioxide and ammonia come in contact. Leaks may also be less satisfactorily detected by applying oil or soap solution to joints and noting where bubbles of escaping gas appear.

Measures in Case of Leakage. Where quantities of sulfur dioxide are being handled, possible hazards due to ruptured lines, broken gage glasses, leaking joints, etc., must be considered. When a leak does occur, only an authorized employee should attempt to stop it; and, if there is any question as to the seriousness of the leak, a suitable gas mask should be worn (see above, Handling Leaks and Emergencies). In general, where serious leaks develop, the person responsible, even when equipped with a suitable mask, should remain in the contaminated area only long enough to close the necessary valves and to make emergency adjustments.

Leaks which might develop are ordinarily not serious and can be readily controlled. Where leaks do occur, the supply of sulfur dioxide should be shut off by closing the appropriate valve. Leaks at unions or other fittings may often be eliminated by tightening the connection. If corrosion is indicated, care must be taken to empty the lines before working on them, as a broken fitting might lead to a serious loss of sulfur dioxide before the supply valves could be shut off.

Handling of Leaking Containers. Although serious leaks in shipping containers rarely occur, careless handling will sometimes result in this condition. Cylinder and TMU tank valves are made of brass and, if struck by a heavy object, might be broken. Leaks may also occur from carelessness in heating cylinders or drums, resulting in melting of a fuse plug and discharge of the contents. Care in handling will eliminate these dangers. Occasionally, leaks may develop in the valve packing, but these can ordinarily be checked by tightening the packing nut.

In the event of a leaking container, it should, wherever possible, be moved to an open area where the hazard due to escaping sulfur dioxide will be minimized. If a container is discharging too freely to permit movement, it should, if possible, be arranged in such a position that the leak will be at the top, thus discharging gaseous sulfur dioxide and not liquid. If large quantities of gas can be withdrawn rapidly into equipment, or satisfactorily vented, the evaporation will often lower the pressure in the container to such a point that it can be moved to the open without difficulty, or possibly the leak can be repaired by persons provided with suitable masks. If the leak still prevents removal of the container to an open area, gaseous or liquid sulfur dioxide can be vented into a solution of lime, caustic soda or other alkaline material. One pound of sulfur dioxide is equivalent to about 2 lb of lime or $1\frac{1}{2}$ lb of caustic soda.

If the above information proves insufficient under unusual circumstances, call the shipper at once.

First-Aid Suggestions. Any person who has

been burned or overcome by sulfur dioxide should be placed under a physician's care immediately. Prior to the physician's arrival the following first-aid suggestions are presented, based upon what is believed to be common practice in industry. Their adoption in a particular case should of course be subject to prior endorsement by the user's medical adviser.

(a) Remove patient burned or overcome by gas to an area free from fumes, preferably a warm room (about 70 F or 21.1 C).

(b) Quickly remove sulfur dioxide saturated clothing to prevent continued inhalation of fumes.

(c) Place patient in reclining position with head and shoulders slightly elevated. Keep the patient warm, providing blankets or other covers if necessary. Keep the patient quiet and urge him to resist the desire to cough if possible. Rest is essential.

For asphyxiation. Where breathing is weak, administer mixtures of carbon dioxide and oxygen or carbon dioxide and air, containing not more than 5 percent of carbon dioxide. Give the mixture continuously until all sounds of "gurgling" have ceased or until the lungs have cleared. (Apparatus required for this treatment is on the market.)

If breathing has ceased, start artificial respiration immediately, using slow, even motions, not to exceed 14 movements per minute. If possible assist respiration with an inhalator. Continue artificial respiration until ordered to stop by a physician or until the patient is breathing normally.

For the eyes. If liquid sulfur dioxide enters the eye, wash immediately with large quantities of water (keeping the lids open) for at least 15 minutes. Seek medical attention as soon as possible.

For the skin. Immediately remove any clothing splashed with sulfur dioxide and wash the affected skin areas with large quantities of water. Cover burn areas with sterile gauze and keep patient warm. Treat for shock. Burns more serious than first degree should be dressed and treated by a physician (see also Part I, Chapter 4, Section A).

METHODS OF SHIPMENT; REGULATIONS

Under the appropriate regulations and tariffs, sulfur dioxide is authorized for shipment as follows:

By Rail: In cylinders (via freight or rail express with a maximum of 300 lb or 140 kg) in single-unit tank cars and in TMU tank cars.

By Highway: In cylinders on trucks, in cargo tanks and in portable tanks and TMU tanks.

By Water: In cylinders on cargo and passenger vessels, and on ferry and railroad car ferry vessels (passenger or vehicle). In authorized truck cargo tanks on cargo vessels only, and on ferry vessels (passenger or vehicle). In authorized single-unit and TMU tank cars, and in portable tanks (meeting DOT-51 specifications, not over 80,000 lb (36,000 kg) gross weight, and not equipped with fixed-length dip-tube gaging devices), on cargo vessels only and on railroad car ferry vessels (passenger or vehicle). In cylinders on barges of U.S. Coast Guard classes A, BA, BB, CA and CB.

By Air: In cylinders aboard cargo aircraft only up to 300 lb (140 kg) maximum net weight per cylinder.

CONTAINERS

Sulfur dioxide is authorized for shipment in cylinders, single-unit tank cars, TMU tank cars, truck cargo tanks, and portable tanks and TMU tanks on trucks.

In order to withdraw sulfur dioxide liquid or vapor from shipping or storage containers, it is often necessary to warm the contents or to employ other special methods in ways which have proven safe and practical. These withdrawal methods are described in CGA Pamphlet G-3 (see Ref. 2).

Filling Limits. The maximum filling limits for all types of containers authorized for sulfur dioxide shipment under DOT regulations is the maximum filling density of 125 percent (percent water capacity by weight). This is roughly equivalent to the filling density of 87.5 percent by volume that is authorized as an alternative for truck cargo tanks and portable tanks.

Cylinders

Sulfur dioxide is authorized for shipment in cylinders which comply with the following DOT specifications: 3A225, 3AA225, 3B225, 4A225, 4B225, 4BA225, 4B240ET, 4 and 3E1800 (cylinders that meet DOT specifications 3, 25, 26 and 38 may also be continued in sulfur dioxide service, but new construction is not authorized).

All types of cylinders authorized for sulfur dioxide service must be requalified by hydrostatic retest every 5 years under present regulations, with the following exceptions: DOT-4 cylinders, every 10 years; and no periodic retest is required for cylinders of type 3E.

Valve Outlet and Inlet Connections. Standard connection, United States and Canada—No. 668. Alternate standard connection, United States and Canada—Nos. 660, 165 and 182.

Cargo Tanks

Cargo tanks meeting specifications MC-330 or MC-331 are authorized for motor vehicle shipment of sulfur dioxide. The minimum design pressure for cargo tanks not over 1200 gal (4.5 m³) must be 150 psig (1000 kPa, gage) plus a corrosion allowance and other stipulated requirements which must be met. For tanks exceeding 1200 gal (4.5 m³) water capacity, the minimum design pressure must be 125 psig (862 kPa, gage) plus a corrosion allowance.

Portable Tanks

Portable tanks complying with specification DOT-51 are authorized for motor vehicle shipment. Portable tanks must meet the same design specifications according to water capacity as cargo tanks except an optional DOT-51 portable tank of 1000-2000 lb (450-900 kg) water capacity and with fusible plugs at each end must have a minimum design pressure of 225 psig (1550 kPa, gage) and these tanks must be filled by weight.

Single-Unit Tank Cars

Insulated single-unit tank cars complying with DOT specifications 105A200-W are authorized for rail shipment of sulfur dioxide in bulk.

TMU Tank Cars

TMU tank cars meeting DOT specification 106A500-X or 110A500-W are authorized for rail shipment or motor vehicle transport of sulfur dioxide.

Storage and Storage Tanks

Sulfur dioxide must be stored as well as handled with all the precautions necessary for safety with any compressed gas. Cylinders and other portable containers in particular should not be stored in subsurface locations, since sulfur dioxide gas, being heavier than air might accumulate in the bottom of the storeroom and not be carried away by drafts from openings located above the surface level.

TMU tank containers should be stored on their sides and blocked to prevent rolling. A convenient horizontal storage rack may be made by supporting the drums at each end on a railroad rail or I-beam. Valve-protecting plates or hoods should be in place at all times that containers are not in use.

In general, the storing of sulfur dioxide in cargo and portable tanks involves the same factors as does its storage in cylinders or multi-unit tank car tanks.

Sulfur dioxide consumers who require bulk storage equipment should secure competent engineering assistance in its design, construction and installation. Sulfur dioxide suppliers will often be able to provide valuable advice in connection with sulfur dioxide storage.

Sulfur dioxide bulk storage containers should be designed and constructed in accordance with recognized engineering practice as exemplified by Section VIII of the Boiler and Pressure Vessel Code of the American Society of Mechanical Engineers. Since it is possible that the design of the vessel, its capacity and its location may be influenced by state or municipal regulations and insurance restrictions, a thorough investigation of pertinent restrictions should be made prior to beginning construction of the vessel.

After the storage vessel has been installed, competent technical personnel should be trained in the operation and maintenance of this equipment. Personnel should be impressed with the great importance of preventing overfilling tanks with sulfur dioxide with the consequent risk of the vessel becoming liquid-full and failing due to hydrostatic pressure. The amount of liquid sulfur dioxide that may be stored safely in a vessel may be determined in two ways:

(a) By weight at 125 percent of the weight of water that the container will hold at 60 F (15.6 C).

(b) By volume in accordance with the provisions of Table 3.

Each storage container, except containers mounted on scales, should be equipped with a liquid-level gaging device to permit ready determination of the amount of liquid in the tank at any time. Liquid-level gaging devices should conform to the following recommendations:

Liquid-level indicating devices should be designed and installed so as to permit reading the liquid level within plus or minus 1 percent of the capacity of the container from full tank level down to at least 20 percent below full tank.

TABLE 3. MAXIMUM SAFE VOLUME OF LIQUID SULFUR DIOXIDE IN A STORAGE TANK AT VARIOUS TEMPERATURES.

Temperature of Liquid Sulfur Dioxide in Tank (°F)	Maximum Safe Volume Liquid Sulfur Dioxide in % of Full Volume at 125% Filling Density
30	86
40	87
50	88
60	89
70	90
80	91
90	92
100	93

Readings below this level are usually desirable for other purposes but are not necessary for avoidance of overfilling.

Gage glasses may be used as liquid-level devices if certain precautions are observed. They should be protected by solid transparent shields and guards and should be provided with excess flow valves or with weighted shutoff cocks that must be opened to take a reading and will close automatically. Each gage glass should be provided with a drain valve and should be drained after each reading to a point where it is no more than 85 percent full. This should be done to prevent bursting that could be caused by thermal expansion. Gage glasses should be located so that the glass and its content do not differ greatly in temperature from the contents of the container.

All gaging devices should be arranged so that the maximum liquid level to which the container may be filled safely is readily determinable. For this purpose the information presented in Table 3 should be duplicated on a metal plate mounted at the operating position of the gaging device.

Gaging devices that require bleeding of sulfur dioxide to the atmosphere, such as rotary tube, fixed tube and slip tube, should be designed so that the bleed valve maximum opening is not larger than a No. 54 drill size, unless provided with excess flow valve.

Gaging devices should be designed for a working pressure of at least 200 psig (1380 kPa, gage).

METHOD OF MANUFACTURE

In North America sulfur dioxide is produced by the combustion of sulfur in burners of special design, by burning pyrites, as a by-product of smelter operations and as a by-product of chemical operations. Sulfur dioxide can be purified by passing the gas into water which dissolves it and certain impurities. This liquor is then heated to drive off the sulfur dioxide, the liberated gas being dried and liquefied.

The gas thus produced is sold in two grades: the commercial grade and the refrigeration grade. The refrigeration grade is purified to contain not more than 0.005 percent of moisture, its quality being controlled by constant sampling and analysis to meet the exacting specifications of the refrigeration industry. The commercial grade contains not more than 0.010 percent of moisture but is not considered suitable for refrigerating machines.

REFERENCES

1. Rynning, D. F., and Hurd, C. O., "Thermodynamic Properties of Sulfur Dioxide," *Trans. Amer. Inst. Chem. Eng.*, **41**, No. 3, June 25, 1945.
2. "Sulfur Dioxide" (Pamphlet G-3), Compressed Gas Association, Inc.

Sulfur Hexafluoride

SF$_6$ DOT Classification: Nonflammable gas; nonflammable gas label

PHYSICAL CONSTANTS

	U.S. Units	Metric Units
International symbol	SF$_6$	SF$_6$
Molecular weight	146.05	146.05
Vapor pressure		
at 70 F (21.1 C)	325 psia	2240.80 kPa, abs.
at 105 F (40.6 C)	487 psia	3357.75 kPa, abs.
at 115 F (46.1 C)	540 psia	3723.17 kPa, abs.
at 130 F (54.4 C)	765 psia	5274.49 kPa, abs.
Density of the gas at 70 F		
(21.1 C) and 1 atm	0.400 lb/ft^3	6.41 kg/m^3
Specific gravity of the gas		
at 68 F and 1 atm (air = 1)	5.106	5.106
Specific gravity of the liquid		
at 130 F (54.4 C) (water = 1)	0.728	0.728
Specific volume of the gas		
at 70 F (21.1 C) and 1 atm	2.5 ft^3/lb	0.16 m^3/kg
Density of liquid under its own		
approximate vapor pressure		
at 70 F (21.1 C)	73.9 lb/ft^3	1183.76 kg/m^3
at 105 F (40.6 C)	84.9 lb/ft^3	1359.97 kg/m^3
Sublimation point at 1 atm	−82.8 F	−63.8 C
Melting point at 325 psia		
(2241 kPa, abs.)	−59.4 F	−50.8 C
Critical temperature	114.2 F	45.67 C
Critical pressure	546.59 psia	3768.61 kPa, abs.
Critical density	45.5 lb/ft^3	728.84 kg/m^3
Latent heat of vaporization		
at 32 F (0 C) and 32.5 psia		
(224 kPa, abs.)	37.3 Btu/lb	86.8 kJ/kg

	U.S. Units	Metric Units
Latent heat of fusion at −59.4 F (−50.8 C) and 32.5 psia (224 kPa, abs.)	14.8 Btu/lb	34.4 kJ/kg
Latent heat of sublimation at −82.8 F (−63.8 C) and 1 atm	69.6 Btu/lb	161.9 kJ/kg
Specific heat of gas at 70 F (21.1 C) and 1 atm C_p	0.16 Btu/(lb) (F)	0.67 kJ/(kg) (C)
Solubility in water, vol/vol of water at 77 F (25 C)	0.001/1	0.001/1
Weight of liquid at 68 F (20 C) and 315 psia (2172 kPa, abs.)	11.4 lb/gal	1366.0 kg/m^3

DESCRIPTION

Sulfur hexafluoride is a colorless, odorless, nontoxic, nonflammable gas that has high dielectric strength and serves widely as an insulating gas in electrical equipment. At atmospheric pressures it sublimes directly from the solid to the gas phase and does not have a stable liquid phase unless under a pressure of more than 32 psia (221 kPa, abs.). It is shipped as a liquefied compressed gas at its vapor pressure of about 310 psig at 70 F (2137 kPa, gage at 21.1 C). One of the most chemically inert gases known, it is completely stable in the presence of most materials to temperatures of about 400 F (204 C), and has shown no breakdown or reaction in quartz at 900 F (482 C). Sulfur hexafluoride is slightly soluble in water and oil. No change in pH occurs when distilled water is saturated with sulfur hexafluoride.

GRADES AVAILABLE

Sulfur hexafluoride is available for commercial and industrial use in various grades having much the same component proportions from one producer to another.

USES

Sulfur hexafluoride is employed extensively as a gaseous dielectric in various kinds of electrical power equipment, such as switch gears, transformers and circuit breakers. It has also been used as a dielectric at microwave frequencies, and as an insulating medium for the power supplies of high-voltage machines.

PHYSIOLOGICAL EFFECTS

Sulfur hexafluoride is completely nontoxic, and in fact has been used medically with humans in cases involving pneumoperitoneum, or the introduction of gas into the abdominal cavity. It can act as a simple asphyxiant by displacing the amount of oxygen in the air necessary to support life.

A TLV of 1000 ppm (6000 mg/m^3) for an 8-hour work day for sulfur hexafluoride has been adopted by the ACGIH.

SAFE HANDLING AND STORAGE

Materials Suitable for Containers and Storage

Sulfur hexafluoride is noncorrosive to all metals. It may be partially decomposed if subjected to an electrical discharge. Some of the breakdown products are corrosive, this corrosion being enhanced by the presence of moisture or high temperatures. Sulfur hexafluoride decomposes very slightly in the presence of certain metals at temperatures in excess of 400 F (204 C), and this effect is most pronounced with silicon and carbon steels. Such breakdown, presumably catalyzed by the metals, is of the order of only several tenths of 1 percent over one year. Among metals with which decomposition at elevated temperatures does not occur are aluminum, copper, brass and silver.

Most common gasket materials, including asbestos, neoprene and natural rubber, are suitable for sulfur hexafluoride service.

Storage and Handling in Normal Use

Lower fluorides of sulfur, some of which are toxic, may be produced if sulfur hexafluoride is subjected to electrical discharge, and inhalation of the gas after electrical discharge must be guarded against (see Part I, Chapter 4, Section A).

Handling Leaks and Emergencies

Standard fluorocarbon detecting devices can be employed to find sulfur hexafluoride leaks, and can identify apertures leaking quantities as small as a half-ounce (14 g) per year (see Part I, Chapter 4, Section A).

METHODS OF SHIPMENT; REGULATIONS

Under the appropriate regulations and tariffs, sulfur hexafluoride is authorized for shipment as follows:

By Rail: In cylinders.

By Highway: In cylinders on trucks, or in tube trailers.

By Water: In cylinders on cargo vessels, passenger vessels, ferry vessels (passenger or vehicle ferry vessels). In cylinders on barges of U.S. Coast Guard classes A, BA, BB, CA and CB.

By Air: In cylinders aboard passenger aircraft up to 150 lb (70 kg), and aboard cargo aircraft up to 300 lb (140 kg), maximum net weight per cylinder.

CONTAINERS

Sulfur hexafluoride is authorized for shipment in cylinders under DOT regulations.

Filling Limits. The maximum filling density permitted for sulfur hexafluoride in cylinders is 120 percent (percent water capacity by weight).

Cylinders

Cylinders that meet the following DOT specifications are authorized for sulfur hexafluoride service; 3A1000, 3AA1000 and 3E1800 (DOT-3 cylinders may also be continued in use, but new construction is not authorized).

Cylinders authorized for sulfur hexafluoride service must be requalified by hydrostatic retest every 10 years under present regulations with the exception of 3E cylinders, for which no periodic retest is required.

Valve Outlet Connections. Standard connection, United States and Canada—No. 590.

METHOD OF MANUFACTURE

Sulfur hexafluoride is made commercially by the direct fluorination of molten sulfur. Some lower fluorides formed in the process are scrubbed out with various caustic solutions, and the commercial product is more than 99 percent pure. Common impurities include small amounts of carbon tetrafluoride, nitrogen and water vapor.

Vinyl Chloride

CH$_2$CHCl (or C$_2$H$_3$Cl)
Synonyms: Chloroethylene, chloroethene
DOT Classification: Flammable gas; flammable gas label

PHYSICAL CONSTANTS

	U.S. Units	Metric Units
International symbol	C$_2$H$_3$Cl	C$_2$H$_3$Cl
Molecular weight	62.50	62.50
Vapor pressure		
at 70 F (21.1 C)	35.3 psig	243.38 kPa, gage
at 105 F (40.6 C)	75.3 psig	519.18 kPa, gage
at 115 F (46.1 C)	90.3 psig	622.60 kPa, gage
at 130 F (54.4 C)	114.3 psig	788.07 kPa, gage
Density of the gas at 70 F (21.1 C) and 1 atm	0.160 lb/ft^3	2.56 kg/m^3
Specific gravity of the gas at 59 F (15 C) and 1 atm (air = 1)	2.15	2.15
Specific volume of the gas at 70 F (21.1 C) and 1 atm	6.25 ft^3/lb	0.390 m^3/kg
Density of liquid		
at 70 F (21.1 C)	56.71 lb/ft^3	908.41 kg/m^3
at 105 F (40.6 C)	54.38 lb/ft^3	871.08 kg/m^3
at 115 F (46.1 C)	53.69 lb/ft^3	860.03 kg/m^3
at 130 F (54.4 C)	52.61 lb/ft^3	842.73 kg/m^3
Boiling point at 1 atm	7.93 F	−13.4 C
Melting point at 1 atm	−245 F	−153.9 C
Critical temperature	317.1 F	158.4 C
Critical pressure	774.7 psia	5341.37 kPa, abs.
Critical density	23.1 lb/ft^3	370.03 kg/m^3
Triple point (estimated)	−240.7 F at 0.00018 psia	−151.5 C at 0.00124 kPa, abs.
Latent heat of vaporization at boiling point	143.7 Btu/lb	334.25 kJ/kg

	Metric Units	U.S. Units
Latent heat of fusion at melting point	32.65 Btu/lb	75.94 kJ/kg
Specific heat of gas at 77 F (25 C) and 1 atm C_p	0.205 Btu/(lb) (F)	0.858 kJ/(kg) (C)
Solubility in water, wt/wt of water, at 77 F (25 C)	0.0011/1	0.0011/1
Weight of liquid at 70 F (21.1 C)	7.58 lb/gal	908.28 kg/m^3
Specific gravity of liquid at 68 F (20 C) (water = 1)	0.9121	0.9121
Flammable limits in air, by volume	4.0–22.0 percent	4.0–22.0 percent
Autoignition temperature	881.6 F	472.0 C
Flash point (open cup)	–108 F	–77.8 C

DESCRIPTION

Vinyl chloride is a colorless, flammable gas with a sweet ethereal odor. It is shipped as a liquefied compressed gas, and the colorless or water-white liquid is so highly volatile that prolonged contact of it with the skin results in freezing or frostbite through evaporation. In addition, vinyl chloride irritates the skin and the eyes on contact, while inhalation of concentrations of more than 500 ppm produces mild anesthesia.

Dry vinyl chloride does not corrode metals at normal temperatures and pressures, but vinyl chloride accelerates the corrosion of iron and steel in the presence of moisture at elevated temperatures.

The most important of the vinyl monomers, vinyl chloride polymerizes readily when exposed to air, sunlight, heat or oxygen, although it is otherwise quite stable chemically.

GRADES AVAILABLE

Vinyl chloride is available for commercial and industrial use in various grades having much the same component proportions from one producer to another.

USES

Vinyl chloride is used heavily as a monomer raw material for the polymerization of vinyl resins in making polyvinyl chloride and copolymers. It is also used in organic synthesis.

PHYSIOLOGICAL EFFECTS

Vinyl chloride acts as a general anesthetic in concentrations of well over 500 ppm. This was considered an acceptable limit until early in 1974 when it was suspected of having a carcinogenic effect on a small percentage of workers in polyvinyl chloride plants. As of April 1, 1975 a maximum allowable concentration of 1 ppm was established for an 8-hour work day as a temporary standard by the OSHA. Otherwise the "Carrier-Suspect Agent" label must be used.

Vinyl chloride can irritate or damage the eyes on contact. Liquid vinyl chloride also irritates the skin and can freeze the skin on prolonged contact.

SAFE HANDLING AND STORAGE

Materials Suitable for Containers and Storage

Steel is recommended for use with vinyl chloride, though vinyl chloride and moisture speed its corrosion at elevated temperatures. Stainless steel is also an acceptable material to use with vinyl chloride. Copper and copper alloys must not be employed in contact with vinyl chloride. Acetylene may be present as an impurity in vinyl chloride and can form an explosive acetylide when exposed to copper.

Storage and Handling in Normal Use

Precautions required for the safe handling of any flammable gas must be observed with vinyl chloride (see also Part I, Chapter, 4, Section A).

Steel is recommended for all piping, storage tanks and equipment used with vinyl chloride. Valves in particular must not contain copper or copper alloys. Adequate electrical grounding and ditching or diking in tank areas to control the liquid in the event of vessel rupture are among recommended precautions. Ditching is preferable because the material should be retained at a location not directly beneath or surrounding the storage tanks. Installations must be designed to comply with requirements for unfired pressure vessels and all state and local regulations.

Some authorities believe that the odor of vinyl chloride does not provide adequate warning of its presence in concentrations sufficient to produce dizziness and unconsciousness, and urge special caution against leaks and poor ventilation.

Personnel must wear chemical safety goggles when there is any chance of having the liquid or saturated vapor come in contact with the eyes.

Handling Leaks and Emergencies

If a leak does occur workers should use full masks with self-contained breathing apparatus. If vinyl chloride comes in contact with the skin, the skin must be washed with large amounts of running water. The eyes must similarly be washed for at least 15 minutes if vinyl chloride gets into them, and, except for cases of only very minor irritation, should be treated immediately by an eye specialist (see also Part I, Chapter 4, Section A).

METHODS OF SHIPMENT; REGULATIONS

Under the appropriate regulations and tariffs, vinyl chloride is authorized for shipment as follows:

By Rail: In cylinders (freight or express), and in single-unit tank cars and TMU tank cars.

By Highway: In cylinders on trucks, and in tank trucks and TMU tanks on trucks.

By Water: In cylinders via cargo vessels only, in authorized tank cars aboard trainships only, and in authorized containers on trucks aboard trailerships and trainships only. In cylinders on barges of U.S. Coast Guard classes A and C only. In cargo tanks aboard ships and barges (with special warning signs and manning standards as required) and aboard tankships (to maximum filling densities by specific gravity as specified in Coast Guard regulations).

CONTAINERS

Vinyl chloride is shipped as a liquefied compressed gas in cylinders, single-unit tank cars and TMU (ton multi-unit) tank cars, and in tank trucks and TMU tanks on trucks. For all these types of containers, it is required that all parts of valves and safety relief devices in contact with the contents of the container must be of a metal or other material (suitably treated if necessary) which will not cause formation of any acetylides.

Filling Limits. The maximum filling densities authorized by the DOT for vinyl chloride are as follows (percent water capacity by weight):

In cylinders—84 percent.

In single-unit tank cars—of the type DOT-105A200-W—87 percent; of the types DOT-112A340-W and 114A340-W—86 percent.

In TMU tanks of the 106A500-X type and in MC-330 or MC-331 cargo tanks—84 percent.

Cylinders

Cylinders that meet the following DOT specifications are authorized for vinyl chloride service: 3A150, 3AA150, 3E1800, 4B150 (without brazed seams) and 4BA225 (without brazed seams). (Cylinders meeting specifications DOT-25 may be continued in service, but new construction is not authorized.)

Under present regulations, cylinders of all types authorized for vinyl chloride service must be requalified by hydrostatic test every 5 years with the exception of type 3E (for which periodic hydrostatic retest is not required).

Valve Outlet and Inlet Connections. Standard connection, United States and Canada—No. 510. Alternate standard connection, United States and Canada—No. 290.

Cargo Tanks

Shipment of vinyl chloride is authorized by the DOT in cargo tanks on trucks meeting DOT specifications MC-330 or MC-331 and having a minimum design pressure of 150 psig (1030 kPa, gage).

TMU Tanks

Shipment of vinyl chloride is authorized in TMU tanks of DOT specifications 106A500-X on trucks or by rail.

Tank Cars

Vinyl chloride is authorized for shipment in single-unit tank cars meeting DOT specifications 105A200-W, 112A340-W or 114A340-W (provided that they have properly fitted loading and unloading valves; also, for cars of the 105A200-W type built before January 1, 1975, openings in tank heads to facilitate application of nickel linings are authorized).

METHOD OF MANUFACTURE

Most vinyl chloride monomer today is made via a three-step process employing ethylene oxyhydrochlorination. A small amount is made by the reaction of acetylene and hydrogen chloride, either as liquids or gases, with a copper chloride catalyst in the liquid process and a mercury catalyst in the gas process.

Vinyl Methyl Ether

CH$_2$:CHOCH$_3$ (or C$_3$H$_6$O)
Synonym: Methyl vinyl ether
DOT Classification (in inhibited form): Flammable gas; flammable gas label

PHYSICAL CONSTANTS

	U.S. Units	Metric Units
International symbol	C$_3$H$_6$O	C$_3$H$_6$O
Molecular weight	58.08	58.08
Vapor pressure at 70 F (21.1 C)	8.80 psig	60.67 kPa, gage
Density of the gas at 70 F (21.1 C) and 1 atm	0.149 lb/ft^3	2.387 kg/m^3
Specific gravity of the gas at 32 F (0 C) and 1 atm (air = 1)	1.99	1.99
Specific gravity of liquid at 68 F (20 C)(water = 1)	0.75	0.75
Specific volume of the gas at 70 F (21.1 C) and 1 atm	6.701 ft^3/lb	0.4183 m^3/kg
Density of liquid		
at 70 F (21.1 C)	46.66 lb/ft^3	747.42 kg/m^3
at 105 F (40.6 C)	45.05 lb/ft^3	721.63 kg/m^3
at 115 F (46.1 C)	44.60 lb/ft^3	714.42 kg/m^3
at 130 F (54.4 C)	43.91 lb/ft^3	703.37 kg/m^3
Boiling point at 1 atm	41.9 F	5.50 C
Freezing point at 1 atm	-187.6 F	-122.0 C
Solubility in water, wt/wt of water, at 68 F (20 C)	0.0097/1	0.0097/1
Weight of liquid at 70 F (21.1 C)	6.24 lb/gal	747.72 kg/m^3
Flammable limits in air at 77 F (25 C), by volume	2.6–39.0 percent	2.6–39.0 percent
Flash point (open cup)	-69 F	-56.1 C

DESCRIPTION

Vinyl methyl ether is a colorless, flammable gas that has a sweet, pleasant odor and is heavier than air. It will react violently with itself in polymerization unless inhibited by low temperature or pressure, or both, or by the addition of an inhibiting agent (among inhibitors or stabi-

457

lizers used are triethanolamine, dioctylamine and solid potassium hydroxide pellets). Colorless or water-white as a liquid, it is shipped as a liquefied compressed gas at about 9 psig at 70 F (62 kPa, gage at 21.1 C).

GRADES AVAILABLE

Vinyl methyl ether is available for commercial and industrial use in various grades having much the same component proportions from one producer to another.

USES

Vinyl methyl ether is used in producing copolymers that are ingredients in coatings and lacquers, as a modifier for alkyl and polystyrene resins, as a plasticizer for nitrocellulose and other plastics, and as a raw material in the manufacture of plastics and synthetic resins.

PHYSIOLOGICAL EFFECTS

Little has been established concerning the physiological effects of vinyl methyl ether. Inhalation and skin contact are believed to pose minimal hazards. Avoidance of prolonged inhalation of the gas in areas where its odor can be detected is advised, for inhalation is thought to produce mild anesthesia. Contact of vinyl methyl ether with the eye produces no permanent damage even if untreated, according to one authority. Sufficiently long contact by the highly volatile liquid vinyl methyl ether can cause freezing or frostbite of the skin.

SAFE HANDLING AND STORAGE

Materials Suitable for Containers and Storage

Steel, stainless steel, aluminum, "Hastelloys" and nickel and its alloys are among metals used with vinyl methyl ether, which is noncorrosive. Copper and copper alloys must not be used with vinyl methyl ether in which acetylene is present as an impurity (because acetylene can form explosive acetylides on exposure to cop-

per and its alloys). Vinyl methyl ether is unsafe with any materials not properly cleaned to remove acidic salts from their surface.

Mild steel is among the materials recommended for vinyl methyl ether storage tanks.

Storage and Handling in Normal Use

Precautions necessary for the safe handling of any flammable gas must be observed with vinyl methyl ether (see Part I, Chapter 4, Section A).

Vinyl methyl ether must also be stored under an inert atmosphere. It can form peroxides in the presence of air or oxygen; samples containing peroxides must not be distilled to dryness.

Storage tanks must be designed for the vapor pressure developed at the highest storage temperature expected. One satisfactory design might consist of a mild steel tank built for a 50–100 psig (345–690 kPa, gage) working pressure and fitted with adequate relief valves and cooling coils to prevent excessive loss through the relief valves during summers. All acidic materials and water must be removed from equipment handling uninhibited vinyl methyl ether, particularly during distillation, to prevent accidental polymerization. Vinyl methyl ether can be handled at atmospheric pressure as a gas or, if kept cooled to 32–36 F (0–2.2 C), as a liquid.

Handling Leaks and Emergencies

Should vinyl methyl ether ignite, recommended fire-extinguishing agents include dry chemicals, carbon dioxide and water spray. Should fire occur in or near a processing or storage area, do not put water into a closed system (see Part I, Chapter 4, Section A).

METHODS OF SHIPMENT; REGULATIONS

Under the appropriate regulations and tariffs, inhibited vinyl methyl ether is authorized for shipment as follows:

By Rail: In cylinders (freight, or express in one outside container to a maximum of 20 lb or 9 kg), and in single-unit tank cars and TMU tank cars.

By Highway: In cylinders on trucks, and in TMU tanks on trucks.

By Water: In cylinders via cargo vessels only, and in authorized tank cars via trainships only. In cylinders on barges of U.S. Coast Guard classes A and C only.

By Air: Aboard cargo aircraft only in appropriate cylinders up to 22 lb (10 kg) maximum net weight per cylinder.

CONTAINERS

Inhibited vinyl methyl ether is shipped as a compressed liquefied gas in cylinders, in single-unit tank cars, in TMU (ton multi-unit) tank cars, and in TMU tanks on trucks. For all these types of containers, it is required that all parts of valves and safety relief devices in contact with the contents of the container must be of a metal or other material (suitably treated if necessary) which will not cause formation of any acetylides.

Vinyl methyl ether without an inhibitor is also shipped in cylinders, TMU tanks and single-unit tank cars by exemption of the DOT.

Filling Limits. The maximum filling density authorized for inhibited vinyl methyl ether in cylinders, single-unit tank cars and TMU tanks is 68 percent (percent water capacity by weight).

Cylinders

Cylinders that meet the following DOT specifications are authorized for vinyl methyl ether service: 3A150, 3AA150, 3B150, 3E1800, 4B150 (without brazed seams) and 4BA225 (without brazed seams). (Cylinders meeting specification DOT-25 may be continued in service, but new construction is not authorized.)

Under present regulations, cylinders of all types authorized for vinyl methyl ether service must be requalified by hydrostatic test every 5 years with the exception of type 3E (for which periodic hydrostatic retest is not required).

Valve Outlet Connections. Standard connection, United States and Canada—No. 510.

TMU Tanks

Inhibited vinyl methyl ether is authorized for shipment on trucks or by rail in TMU tanks of DOT specifications 106A500-X.

Single-Unit Tank Cars

Inhibited vinyl methyl ether is authorized for shipment in single-unit tank cars meeting DOT specifications 105A100-W (provided that they have properly fitted loading and unloading valves). Single-unit tank cars meeting DOT specifications 105A100 may also be continued in service with inhibited vinyl methyl ether, but new construction is not authorized.

METHOD OF MANUFACTURE

Vinyl methyl ether is produced by the combination of acetylene and methyl alcohol in the presence of a catalyst. It is also made commercially by combining acetaldehyde with methanol and cracking the acetal formed to give the ether.

Gas Mixtures

DESCRIPTION

In general terms, the properties of gas mixtures are directly related to the individual products that make up the mixture, taking into consideration the concentrations of each product within the mixture. Gas mixtures fall into the categories of inerts, oxidants, radioactive gases, flammables, pyrophorics, corrosives and/or poisons. Likewise, the pressures at which these mixtures can be made without condensation occurring and the values assignable to these mixtures are a function of the physical and chemical properties of the gases making up the mixtures.

Gas mixtures exist in the form of gas-gas mixtures, gas-liquid mixtures and liquid-liquid mixtures.

Gas-gas mixtures are mixtures such as nitrogen/hydrogen in which the two gases exist in the gas phase at room temperature and nominal cylinder pressures. It is important to note that in this type of gas mixture, once the gases are mixed, they remain homogeneous unless a given reaction or liquefaction of one of the components occurs.

The second category of mixtures, gas-liquid mixtures, exemplified by nitrogen over a liquid hydrocarbon, exists in variable concentrations depending upon the temperature and pressures of the constituents as well as total pressure. These mixtures within a cylinder exist in two forms: the "gas dissolved in the liquid" and the gas-gas resultant mixture over the liquid.

The third category of mixtures is liquid-liquid mixtures. An example is ethylene oxide in flurocarbon-12. In this mixture the two components exist in the liquid phase within the cylinder, and like the gas-liquid mixture, have a vapor phase above the liquid. It is important to note that in this type of mixture, the vapor phase above the liquid is quite different in composition from the liquid. Removal of the vapor phase will cause fractionation of the liquid mixture and a change in the concentration of the components of that mixture. In this type of mixture, full-length eductor tubes are normally provided so that the liquid mixture can be withdrawn and a relatively uniform mixture is attainable throughout the withdrawal of the contents.

GRADES AVAILABLE

Specifications for gas mixtures depend upon the particular method of manufacture and customer requirements. These specifications are normally given in terms of the blending accuracy and/or analytical tolerances which are determined as applicable to each mixture.

The specifications are generally expressed in terms of volume/volume percent for gases and weight/weight percent for liquids.

The specifications range from very general (for certain industrial applications) to very stringent (in certain critical applications). The exact composition of each mixture specification may include only the mixture composition or be as complete as to describe the various impurities possible as well as the tolerance levels of each. Again, these parameters are an individual requirement and are governed by consumer need and production capabilities.

Gas mixtures produced for critical applications are normally analyzed to insure compliance with the specifications. Specifications applicable to such situations are a function of analytical capabilities.

Specifications for gas mixtures are likewise a function of the physical state of the gas (vapor, liquid, solid) in the supply container. Single-phase mixtures pose few problems in being adequately described. However, multiphase mixtures such as gas/liquid mixtures are difficult to describe due to changing compositions which are possible with varying conditions such as pressure and temperature. Such gas mixtures frequently vary in composition during the withdrawal of the contents and specifications applicable to such mixtures must address this problem.

USES

Mixtures of gases and/or liquid in cylinders provide convenient packages for a wide variety of uses, some of which are briefly noted below.

Mixtures of gases are used to reduce the toxicity or flammability of a highly active material and, similarly, to modify the physical properties of a pure material by mixing it with other materials.

Typical applications include those shown below. These examples are common but no attempt to compile a complete list has been made.

Analytical Instruments. Pure gases and mixtures are used for the operation and/or calibration of analytical instruments. Zero gases (which are usually pure materials but are certified to contain negligible or known concentrations of a component of interest) are used for flame ionization detector instruments that require gases of low hydrocarbon content to achieve maximum sensitivity. "Zero air" may be used both to zero the instrument and to provide an oxidant of low hydrocarbon content for the operation of the analyzer.

Span gases, which usually contain a minor component of interest at a known concentration level, permit the analytical device to be calibrated at a value corresponding to that concentration.

Fuel gas is used in instruments such as those with flame ionization detectors which require the supply of fuel plus certain physical properties of the diluent. A mixture of 40 percent hydrogen in helium is common.

Sterilizing. Dilutions of ethylene oxide such as 12 percent with fluorocarbon-12, or 10 to 20 percent ethylene oxide with carbon dioxide, are commonly used in sterilizing. Another sterilant is propylene oxide. Such mixtures are regulated by the U.S. Environmental Protection Agency.

Medical Applications. A low concentration of carbon monoxide in air is used for lung diffusion tests. Various percentage mixtures of carbon dioxide in oxygen are used for blood gas analyzer calibration. Cyclopropane, nitrous oxide and certain other chemically active gases are used separately and as mixtures for anesthesia.

Nuclear Counter Gases. A mixture of 0.95 percent isobutane in helium and 1.3 percent butane in helium are used as Geiger gas. Mixtures of 4 percent isobutane in helium and 10 percent methane in argon are used in proportional counter gases.

Electronic Component Manufacture. Gases which include low levels of dopants such as arsine, phosphine, diborane and others in hydrogen, helium, argon or nitrogen are used for making silicon semiconductors. Highly purified ammonia and mixtures are used for nitriding. Silane, either pure or in mixtures is used for silicon deposition. Gas mixtures similar to the ones above but with higher percentage levels of arsine and phosphine are common in manufacture of gallium semiconductors.

Welding. Mixtures of oxygen with argon, carbon dioxide and helium-nitrogen are used for welding.

Chilling. Refrigerants such as methyl chloride and mixtures of flurocarbons are used for chilling.

Fumigating. Methyl bromide and ethylene oxide are used in fumigating.

Chemical Testing. Gas mixtures are used as atmospheres by the American Society for Testing and Materials.

Metallurgical Applications. Hydrogen-nitrogen is used as annealing gas.

Leak Detection. A low percentage of fluorocarbons such as fluorocarbon-12 or helium in air or nitrogen is used for leak detection.

Catalyst Addition. Nitrogen oxides and boron trifluoride in nitrogen are used for catalyst addition.

Illumination. Mixtures of rare gases with helium, nitrogen or argon are used in lamps, signs, electronic tubes and other devices.

Radioactive Gas Mixtures. Krypton-85, tritium and carbon-14 are common in traces added to various gases.

Spark Chamber Gases. Generally, a mixture of 70 to 95 percent neon in helium is used in spark chambers.

Pollution Control. Small amounts of nitric oxide, nitrogen dioxide, sulfur dioxide, carbon monoxide, vinyl chloride, various hydrocarbons and others are added to air or other gases to calibrate analytical instruments for emission measurements.

Laser Mixtures. Mixtures of rare gases with nitrogen and carbon dioxide with helium are used for lasers.

X-Ray Fluorescence Spectroscopy. Mixtures containing small amounts of butane in a mixture of helium and neon are used in X-ray fluorescence spectroscopy.

Electron Capture. Mixture of 5 percent methane in argon is used in electron capture.

Special Carrier Gases. These include mixtures of 8.5 percent hydrogen in helium.

PHYSIOLOGICAL EFFECTS

In handling gas mixtures the physiological properties of each component must be considered. The user should become familiar with the properties of the individual components. Some components may be simple asphyxiants, others, in very low concentrations, may cause death. See Refs. 1, 2 and 3 for additional works on the physiological effects of various gases.

Once the properties of the individual components are known the hazards of the mixture may be considered by use of the methods described in Refs. 4 and 5.

SAFE HANDLING AND STORAGE

Materials Suitable for Containers and Storage

Gas mixtures possess many of the chemical and physical properties of the constituent gases. Therefore, when selecting a valve, regulator or other piping component, the material of construction should be selected for chemical compatability with all of the constituent gases. Particular attention should be paid to both the chemical and physical compatability of seal materials, particularly "O ring" materials. Often, a seal material is chemically compatible with a gas but might swell excessively and prevent the component from operating properly.

The selection of the proper material or materials of construction is complicated by the presence of several components each having its own requirement. Since this is a difficult process, the selection of proper materials of construction should be requested from the supplier of the gases.

Storage and Handling in Normal Use

Special care should be taken with gas mixtures that are flammable and/or poisonous. All the precautions necessary for safe handling of a flammable or nonflammable compressed gas, depending on the gases involved, should be taken for gas mixtures.

Serious accidents may result from the misuse, abuse or mishandling of compressed gas mixture cylinders. Workers assigned to the handling of cylinders under pressure should be carefully trained and should work only under competent supervision. Knowledge of the following rules will help control hazards in the handling of compressed gas cylinders.

Cylinders should be stored on a level floor. One common type of storage house consists of a shed roof with sidewalls extending approximately halfway down from the roof and a dividing wall between one kind of gas and another.

Cylinder storage should be planned so that cylinders will be used in the order in which they are received from the supplier. Group together cylinders which have the same contents. Smoking should be prohibited. Wiring should be conduit. Electric lights should be in a fixed position and enclosed in glass or other transparent material to prevent gas from contacting lighted sockets or lamps and should be equipped with guards to prevent breakage. Electric switches should be located outside the room. A direct flame or electric arc should never be permitted to contact any part of a compressed gas cylinder.

Because of their shape, smooth surface and weight, cylinders are difficult to carry by hand. When cylinders must be moved without the aid of a cart or other mechanical means, use some type of a carrying device. Cylinders must never be dragged.

Load cylinders to be transported to allow as little movement as possible. Secure them to prevent violent contact or upsetting.

Always consider cylinders as full and handle them with corresponding care. Accidents have resulted when containers under partial pressure were thought to be empty.

Fuel gas cylinders in which leaks occur should be taken out of use immediately and the following steps should be taken. Close the valve and take the cylinder outdoors well away from any source of ignition. Properly tag the cylinder plainly and notify the supplier. If the leak occurs at a fuse plug or other safety device, take the cylinder outdoors well away from any source of ignition, open the cylinder valve slightly and permit the fuel gas to escape slowly. Post warnings against approaching the cylinder with lighted cigarettes or other sources of ignition.

Use soapy water or suitable leak detector to detect gas leaks.

Users of toxic gases should keep masks available for persons authorized to act in emergencies. Persons overcome by toxic gases should be removed from the affected area, kept warm and given first-aid treatment pending the arrival of a physician (see also Part I, Chapter 4, Section A).

Handling Leaks and Emergencies

See Refs. 6 and 7 for methods for first aid.

The first aid and medical treatments described in the reference section are based upon what is believed to be current practice, but they should be reviewed and amplified by a competent medical advisor before adoption.

Since most medical doctors are not familiar with physiological effects of gases and related treatment, it is very important that the user have as much information as possible, both for his own use and for presentation to a doctor, should the need arise (see also Part I, Chapter 4, Section A).

METHODS OF SHIPMENT; REGULATIONS

Shipment of compressed gas mixtures is regulated by the Department of Transportation in Title 49, Code of Federal Regulations (hereafter referred to as Title 49). Air, water and export movements are additionally regulated by other instrumentalities. Certain gases are also subject to Food and Drug Administration and Environmental Protection Agency rules. The regulations are applicable to the mixture components individually with additional consideration given to the following:

Gases Capable of Combining Chemically. A cylinder charged with compressed gas must not contain gases or materials that are capable of combining chemically with each other or with the cylinder material so as to endanger its serviceability (Title 49, 173.301).

Corrosive Liquids. A cylinder that previously contained a commodity classified as a corrosive liquid must not be used for the transportation of any compressed gas unless the requirements listed in the tariff are complied with before the subsequent initial filling with the compressed gas (Title 49, 173.34E16).

Ownership of Container. A container charged with a compressed gas must not be shipped unless it was charged by or with the consent of the owner of the container (Title 49, 173.301).

Detailed Requirements. A mixture of a compressed gas and any other material must be shipped as a compressed gas if the mixture is composed of any material or mixture having in the container an absolute pressure exceeding 40 psi at 70 F (280 kPa at 21.1 C) or, regardless of the pressure at 70 F, having an absolute pressure exceeding 104 psi at 130 F (717 kPa at 54 C); or any liquid flammable material having a vapor pressure exceeding 40 psia at 100 F (280 kPa, abs. at 38 C) as determined by ASTM Test D-323 (Title 49, 173.305 and 173.300).

Nonpoisonous and Nonflammable Mixtures. Mixtures containing a compressed gas or gases including insecticides, whose mixtures are nonpoisonous and nonflammable under this part must be shipped in cylinders prescribed in the tariff for use with the pure gases or in accordance with specification 2P (Title 49, 173.305C and 173.304A).

Poisonous Mixtures. Mixtures containing any poisonous article Class A (extremely dangerous poisons) or Class C (tear gases or irritating substances) in such proportions that the mixtures would be classified as a poison must be shipped in containers as authorized for such poisonous articles. Extremely dangerous poisons, Class A, are poisonous gases or liquids of such nature that a very small amount of the gas, or vapor of the liquid, mixed with air is dangerous to life. This class includes the following:

(a) bromacetone;
(b) cyanogen;
(c) cyanogen chloride containing less than 0.9 percent water;
(d) diphosgene;
(e) ethyldichlorarsine;
(f) hydrocyanic acid;
(g) lewisite;
(h) methyldichlorasine;
(i) mustard gas;
(j) nitrogen peroxide (tetroxide);
(k) phenylcarbylamine chloride;
(l) phosgene (diphosgene);
(m) nitrogen tetroxide–nitric oxide mixtures containing up to 33.2 percent weight nitric oxide.

Poisonous gases or liquids Class A must not be offered for transportation by rail express (Title 49, 173.305D, 173.326A and 173.381A).

CONTAINERS

A variety of containers are described in DOT regulations that may be used to receive gases or mixtures of gases. These include 3A, 3AX, 3AA, 3AAX, 3B, 3BN, 3C, 3D, 3E, 3A480X, 3T, 4, 4A, 4B, 4BA, 4C, 4B240-FLW, 4B240-ET, 4B240-X, 4AA480, 4BW and 39. Other containers are specified but because of their limited use, they are not covered here. Containers having the same specifications are manufactured in a variety of sizes. For instance, cylinders are available with a specification 3AA2015 rating in sizes that will allow capacities of 6-200 ft^3 (0.2-6 m^3).

It is essential that anyone preparing gas mixtures be aware of the types of containers that can or must be used as well as the limitations to the containers and the gases they will receive. Attention to DOT regulations is an absolute requirement (Title 49, Section 173.301).

The filling limits for gas mixtures vary with the different gases. Title 49 describes the filling limits for the different gases. There are certain conditions governing these filling limits, which again make it mandatory that the regulations be adhered to. Class A poisons have a number of restrictions.

In making liquid mixtures, the permissible filling limits are limited to an amount of the combined liquids that will not completely fill the container at 130 F (54.4 C).

Cylinders

Various cylinders are authorized for gas mixtures depending on the gases involved (see Title 49).

Cylinder requalification depends upon the type of cylinder and compressed gases being used (see Title 49).

Valve Outlet Connections. The procedure in selecting a valve for a given mixture has been incorporated in CGA Pamphlet V-1 (see Ref. 8). When in doubt about the proper valve outlet selection of a gas mixture cylinder, consult the supplier of the gas.

The selection of a valve that has the proper outlet connection has been made easier by the efforts of the Compressed Gas Association. This method does not specify the materials of construction for the valve, it does not specify what mixtures can be safely made nor does it specify the requirements regarding safety assemblies on the valve. It only intends to allow proper selection of the valve outlet that will safely accommodate the mixture.

Tube Trailers

Nonliquefied gas mixtures are authorized for motor vehicle transport in tube trailers meeting DOT specifications 3A, 3AX, 3AA, 3AAX or 3T depending upon the gas mixture.

METHOD OF MANUFACTURE

Gas mixtures are prepared by: direct preparation in the shipping container; mixing in bulk preparation vessels followed by compressing into the shipping containers; dynamically blending and mixing immediately prior to filling the shipping container or prior to transport by pipeline direct to the end user. The first two methods represent a batch preparation, while the latter method is a continuous process, usually reserved for special applications, and will not be discussed here.

A batch preparation of a gas mixture is typically made by monitoring one or more of the following as each component is added: pressure gain, weight gain or volume added. In each case, calculations must be made for the components involved, which depend upon their physical properties. The final accuracy achieved varies greatly depending upon the sophistication of measurement and the equipment used. The final concentration is expressed as either mole percent or volume percent. Weight percent is usually used for liquid mixtures. Volume percent (at atmospheric pressure) is essentially identical to mole percent for most gas mixtures.

Gas mixtures prepared by the batch process do not readily mix of their own accord to a homogeneous composition throughout the container. This is particularly true of vertical standing cylinders of high length-to-diameter ratio filled to high pressure. Such cylinders can exhibit stratified composition for many days. Mixing can be accelerated by placing the longest axis of the cylinder horizontal, which gives a greater area for diffusion between layers of different density. Also, nonuniform heating or cooling the container so as to induce convection currents within, rolling the container while horizontal along its longitudinal axis, or other means of causing agitation in the cylinder either during or after filling, may be used to speed mixing.

After mixing, gas mixtures destined for critical applications should be analyzed to determine the actual composition of each component achieved in the preparation process. The techniques used in gas analysis are many and varied. This subject is adequately treated in many textbooks and will not be covered here.

Once a gas mixture has been mixed to a homogeneous state, normal molecular motion is sufficient to maintain a composition indefinitely. However, other factors can cause composition changes. Mixtures containing one or more condensable components, when exposed to temperatures below their condensation point, will liquefy within the container and remixing may be required after the container is warmed. Likewise, mixtures containing components which may react together or with the vessel will have altered compositions.

A very great hazard in the preparation of gas mixtures is the indiscriminate mixing of gases which may react or decompose explosively. In theory, mixtures of flammable and oxidizing gases (i.e., methane and oxygen) could be prepared at concentrations up to the lower explosive limit. Such a practice would be extremely dangerous. First, the lower explosion limit for most binary mixtures is known only at atmospheric pressure. The actual value at cylinder pressure may be considerably lower than that at atmospheric pressure. Secondly, a concentration level may be entirely safe in the final mixture but the lower explosive limits may be greatly exceeded during the preparation of the mixture. The mixing of potentially reactive gases should be attempted only by those with knowledge and experience in that field.

In addition to hazards of indiscriminate mixing, the prepration of mixtures requires knowledge of the hazards associated with each individual component of the mixture. Appropriate precautions must be taken for each hazard as required in the handling of each pure component.

REFERENCES

1. Sax, N. Irving, "Dangerous Properties of Industrial Materials," Van Nostrand Reinhold, New York, 4th ed., 1975.
2. Glenson, Gossolin and Hodges, "Clinical Toxicology of Commercial Products," Williams & Wilkins Co., Baltimore, MD.
3. "Threshold Limit Values of Airborne Contaminants," American Conference of Governmental Industrial Hygienists, 1014 Broadway, Cincinnati, OH 45202.
4. Part 1910—Subpart G, "Occupational Safety and Health, Vol. 1, General Industry Standards and Interpretations," U.S. Department of Labor—OSHA, OSHA 2077.
5. Table 2—Appendix B, "Accidental Prevention Manual for Industrial Operations," 6th ed., National Safety Council, Chicago, IL.
6. "Effects of Exposure to Toxic Gases, First Aid and Medical Treatment," Matheson Gas Products, East Rutherford, NJ.
7. "Chemical Safety Data Sheet (Specific Gas)," Chemical Manufacturers Association, 1825 Connecticut Avenue, N.W., Washington, D.C.
8. "American National-Canadian Standard Compressed Gas Cylinder Valve Outlet and Inlet Connections" (Pamphlet V-1), Compressed Gas Association, Inc.

Appendices

The following appendices to the *Handbook* present supplementary information important to users, distributors, producers and others concerned with the compressed and liquified gases.

Appendix A summarizes, in three tables, certain significant features of compressed gas regulation (by states) of the United States. Appendix B consists of a complete list of the present service publications of Compressed Gas Association, Inc. The CGA regularly adds to and revises these publications, and current information about them may be obtained on request to the Association.*

The summary information given in Appendix A on regulations of the states is necessarily general and subject to change subsequent to the time of its collection. As a result, in its application to any specific situation, it should be taken as indicative, rather than conclusive. The CGA secured the data from sources believed to be fully authoritative, but the Association and the publisher can assume no responsibility in connection with use of the data. However, the summary outlined in the tables represents the

*500 Fifth Avenue, New York, NY 10110.

most comprehensive collection of information on state regulation that has yet been made available to the public. The major purposes that the three tables of the appendix are designed to serve are as follows:

Table 1, on the use by states of compressed gas standards developed by the National Fire Protection Association, is intended to aid persons responsible for the location, design and installation of systems, including piping, containers, instruments and equipment to handle compressed gases.

Table 2, on state laws pertaining to the Boiler and Pressure Vessel Code, Section VIII, of the American Society of Mechanical Engineers (the Unfired Pressure Vessel Code), should prove helpful primarily to persons responsible for compressed gas storage installations (because compressed gas storage tanks are commonly built to the requirements of this ASME Code).

Table 3, on state adoptions of the Department of Transportation (DOT) hazardous materials regulations is conceived as a general guide primarily for persons responsible for the transportation of compressed gases within state borders.

467

Summary of Selected State Regulations and Codes Concerning Compressed Gases

TABLE 1. STATE AND PROVINCE ADOPTION OF THE NATIONAL FIRE PROTECTION ASSOCIATION CODES.*

Note. The titles of the NFPA Standards identified by number in the table are:

50 Bulk Oxygen Systems
50A Gaseous Hydrogen Systems
50B Liquefied Hydrogen Systems
51 Oxygen-Fuel Gas Systems for Welding and Cutting
51A Acetylene Cylinder Charging Plants
51B Cutting and Welding Processes
54 National Fuel Gas Code
56A Inhalation Anesthetics
56B Respiratory Therapy
56C Laboratories in Health-Related Institutions
56D Hyperbaric Facilities
56E Hypobaric Facilities
56F Nonflammable Medical Gas Systems
57 Fumigation
58 Storage and Handling of Liquefied Petroleum Gases
59 Liquefied Petroleum Gases at Utility Gas Plants
59A Storage and Handling of Liquefied Natural Gas

NFPA Code	50	50A	50B	51	51A	51B	54	56A	56B	56C	56D	56E	56F	57	58	59	59A
Alabama	Yes '74	Yes '73	Yes '73	Yes '74	Yes '74	Yes '71	Yes '74	Yes '73	Yes '73	Yes '73	Yes '70	Yes '72	Yes '74	Yes '73	Yes '74	Yes '74	Yes '72
Alaska	Yes '74	Yes '73	Yes '73	Yes '74	Yes '74	Yes '71	Yes '74	Yes '73	Yes '76	Yes '73	Yes '76	Yes '72	Yes '74	Yes '73	Yes '76	Yes '76	Yes '75
Arizona	Inf.	Yes '73	Yes '73	Yes '74	Inf.	Inf.	Yes '74	Yes '73	Inf.	Inf.	Inf.	Inf.	Inf.	Inf.	Yes '74	Yes '74	Inf.
Arkansas	Inf.	Inf.	Inf.	Yes '74	Inf.	Inf.	Inf.	Yes '73	Inf.	Inf.	Inf.	Inf.	Inf.	Yes '73	Inf.	Inf.	Inf.
California	Yes '73	Inf.	Inf.	Inf.	Inf.	Inf.	Inf.	Yes '71	Yes '68	Yes '70	Yes '70	Inf.	Yes '70	No	Inf.	Inf.	Inf.
Colorado	—	—	—	—	—	—	—	—	—	—	—	—	—	—	—	—	—
Connecticut	Inf.	Inf.	Inf.	Inf.	Inf.	Inf.	Yes '69	Yes '69	Inf.	Yes '70	Yes '70	Inf.	Yes '70	Inf.	Yes '69	Yes '76	Yes '75
Delaware	Yes '73	Yes '73	Yes '73	Yes '73	Yes '73	Yes '71	Yes '69	Yes '73	Yes '73	Yes '73	Yes '70	Yes '72	Yes '73	Yes '73	Yes '72	Yes '68	Yes '72
Dist. of Columbia	Inf.	Inf.	Inf.	Inf.	Inf.	Inf.	Inf.	Inf.	Inf.	Inf.	Inf.	Inf.	Inf.	Inf.	Inf.	Inf.	Inf.
Florida	Yes '74	Yes '73	Yes '73	Yes '74	Yes '74	Yes '71	Yes '74	Yes '73	Yes '73	Yes '73	Yes '70	Yes '72	Yes '74	Yes '73	Yes '74	Yes '74	Yes '75
Georgia	Yes '65	Yes '63	No	Yes '74	Yes '64	Yes '62	Yes '64	Yes '65	No	No	No	No	No	Yes '39	Yes '67	Yes '63	Yes '67
Hawaii	—	—	—	—	—	—	—	—	—	—	—	—	—	—	—	—	—
Idaho	Yes '74	Yes '73	Yes '73	Yes '74	Yes '74	Yes '71	Yes '74	Yes '73	Yes '73	Yes '73	Yes '70	Yes '72	Yes '74	Yes '73	Yes '74	Yes '74	Yes '72
Illinois	Yes '74	Inf.	Inf.	Yes '69	Inf.	Yes '77	Yes '69	Yes '73	Inf.	Yes '73	Yes '76	Yes '77	Yes '77	Inf.	No	Yes '68	Inf.
Indiana	Inf.	Inf.	Inf.	Inf.	Inf.	Inf.	Yes '74	Inf.	Inf.	Inf.	Inf.	Inf.	Inf.	Inf.	Yes '76	Inf.	Yes '75
Iowa	Inf.	Inf.	Inf.	Inf.	Inf.	Inf.	Inf.	Inf.	Inf.	Inf.	Inf.	Inf.	Inf.	Inf.	Yes '76	Yes '76	Yes '75
Kansas	—	—	—	—	—	—	—	—	—	—	—	—	—	—	—	—	—
Kentucky	Yes '73	Yes '73	Yes '73	Yes '73	Yes '73	Yes '71	Yes '74	Yes '73	Yes '73	Yes '73	Yes '70	Yes '72	Yes '73	Yes '73	Yes '74	Yes '68	Yes '72
Louisiana	Yes '74	Yes '73	Yes '73	Yes '74	Yes '74	Yes '71	Yes '74	Yes '73	Yes '73	Yes '73	Yes '70	Yes '72	Yes '74	Yes '73	Yes '74	Yes '74	Yes '72
Maine	Yes '74	Yes '73	Yes '73	Yes '74	Inf.	Yes '71	Yes '74	Yes '73	Inf.	Inf.	Yes '70	Inf.	Inf.	Inf.	Yes '74	Yes '74	Inf.

TABLE 1 (Continued)

NFPA Code	50	50A	50B	51	51A	51B	54	56A	56B	56C	56D	56E	56F	57	58	59	59A
Maryland	—	—	—	—	—	—	Yes '74	—	—	—	—	—	—	—	Yes '74	Yes '74	Yes
Massachusetts	—	—	—	—	—	—	—	—	—	—	—	—	—	—	Inf.	—	—
Michigan	Yes '74	Inf.	Inf.	Yes	Inf.	Yes '71	Yes '74	Yes '73	Inf.	Inf.	Inf.	Inf.	Inf.	Inf.	Yes '74	Yes '74	Inf.
Minnesota	Yes '73	Yes '73	Yes '73	Yes '73	Yes '73	Yes '71	Yes '69	Yes '73	Yes '73	Yes '73	Yes '70	Yes '72	Yes '73	Yes '73	Yes '72	Yes '68	Yes '72
Mississippi	—	—	—	—	—	—	—	—	—	—	—	—	—	—	—	—	—
Missouri	Yes '74	Yes '73	Yes '73	Yes '74	Yes '74	Yes '71	Yes '74	Yes '73	Yes '73	Yes '73	Yes '70	Yes '72	Yes '74	Yes '73	Yes '74	Yes '74	Yes
Montana	Inf.	No	No	Inf.	Inf.	Inf.	Inf.	Inf.	Inf.	Inf.	No	No	Inf.	Inf.	Yes	Inf.	No
Nebraska	Yes '74	Yes '74	—	Yes '69	—	—	Yes '69	Yes '73	Yes '73	Yes '73	Yes '70	—	Yes '74	—	Yes '69	Yes '68	—
Nevada	Inf.	Inf.	Inf.	Inf.	Inf.	No	Yes '74	Yes '70	No	Yes '70	Yes '70	No	Yes '70	No	Yes '74	No	No
New Hampshire	—	—	—	—	—	—	Yes '74	—	—	—	—	—	—	—	Yes '74	—	—
New Jersey	Inf.	Inf.	Inf.	Inf.	Inf.	Inf.	—	Inf.	Inf.	Inf.	Inf.	Inf.	Inf.	Inf.	Inf.	Inf.	Inf.
New Mexico	Yes '74	Yes '73	Yes '73	Yes '74	Yes '74	Yes '71	Yes '74	Yes '73	—	—	—	—	Yes '74	Yes '73	Yes '74	Yes '74	Yes '72
New York	—	—	—	—	—	—	—	—	—	—	—	—	—	—	—	—	—
N. Carolina	Inf.	Inf.	Inf.	Inf.	Inf.	Inf.	Inf.	Inf.	Inf.	Inf.	Inf.	Inf.	Inf.	Inf.	Inf.	Inf.	Inf.
N. Dakota	Yes '74	Yes '73	Yes '73	Yes '74	Yes '74	Yes '71	Yes '74	Yes '73	Yes '73	Yes '73	Yes '70	Yes '72	Yes '74	Yes '73	Yes '74	Yes '74	Yes '72
Ohio	Inf.	Inf.	Inf.	Inf.	Inf.	Inf.	Yes '74	Inf.	Inf.	Inf.	Inf.	No	Inf.	Inf.	Yes '58	Inf.	Yes '72
Oklahoma	—	—	—	—	—	—	—	—	—	—	—	—	Inf.	—	—	—	Inf.
Oregon	—	—	—	—	—	—	—	—	—	—	—	—	—	—	—	—	Inf.
Pennsylvania	—	—	—	—	—	—	—	—	—	—	—	—	—	—	—	—	—
Puerto Rico	—	Yes '69	Yes '68	Yes '69	Yes '69	Yes '62	Yes '69	Yes '73	Yes '73	Yes '73	Yes '70	Yes '72	Yes '74	Yes '73	Yes '69	Yes '74	Yes '72
Rhode Island	Yes '74	Yes '73	Yes '73	Yes '74	Yes '74	Yes '71	Yes '74	Yes '73	Yes '73	Yes '73	Yes '70	Yes '72	Yes '74	Yes '73	Yes '74	Yes '74	Yes '72
S. Carolina	Inf.	Yes '73	Inf.	Inf.	Inf.	Inf.	Yes '74	Inf.	Inf.	Inf.	Inf.	Inf.	Inf.	No	Yes '74	Yes '74	Inf.
S. Dakota	Yes '74	Yes '73	Yes '73	Yes '74	Yes '74	Yes '71	Yes '74	Yes '73	Yes '73	Yes '73	Yes '70	Yes '72	Yes '74	No	Yes '74	Yes '74	Yes '72
Tennessee	Yes '74	Yes '73	Yes '73	Yes '74	Yes '74	Yes '71	Yes '74	Yes '73	Yes '73	Yes '73	Yes '70	Yes '72	Yes '74	Yes '73	Yes '74	Yes '74	Yes '72
Texas	No	No	No	No	No	No	No	No	No	No	No	No	No	No	No	No	No
Utah	Yes '74	Yes '73	Yes '73	Yes '77	Yes '74	Yes '71	Yes '69	Yes '73	Yes '73	Yes '73	Yes '70	Yes '72	Yes '74	Yes '73	Yes '72	Yes '68	Yes '72
Vermont	Yes '73	Yes '73	Yes '73	Yes '74	Yes '74	Yes '71	Yes '74	Yes '73	Yes '73	Yes '73	Yes '70	Yes '72	Yes '73	Yes '73	Yes '74	Yes '74	Yes '72
Virginia	Inf.	Inf.	Inf.	Inf.	Inf.	Inf.	Yes '69	Inf.	Inf.	Inf.	Inf.	Inf.	Inf.	Inf.	Yes '69	Yes '74	Inf.
Washington	Inf.	Yes '69	Yes '68	Yes '69	No	No	Yes '74	Yes '70	Inf.	Yes '70	Inf.	Yes '72	Yes '70	No	Yes '69	Yes '68	Inf.
W. Virginia	Yes '74	Yes '73	Yes '73	Yes '74	Yes '74	Yes '71	No	Yes '73	Yes '76	Yes '73	Yes '76	Yes '74	Yes '74	No	Yes '76	Yes '76	Yes '75
Wisconsin	Yes '73	No	No	Yes '73	No	Yes '71	No	No	No	No	No	No	No	Yes '73	Yes '72	No	No
Wyoming	Yes '74	Yes '73	Yes '73	Yes '74	Yes '74	Yes '71	Yes '74	Yes '73	Yes '73	Yes '73	Yes '70	Yes '72	Yes '74	Yes '73	Yes '74	Yes '74	Yes '72
Alberta	Inf.	Inf.	Inf.	Inf.	Inf.	Inf.	Inf.	Inf.	Inf.	Yes '73	Inf.	Inf.	Yes '74	Inf.	Inf.	Inf.	Inf.
British Columbia	Inf.	Inf.	Inf.	Inf.	Inf.	Inf.	Inf.	Inf.	Inf.	Inf.	Inf.	Inf.	Inf.	Inf.	Inf.	Inf.	Inf.
Manitoba	—	—	—	—	—	—	—	—	—	—	—	—	—	—	—	—	—
New Brunswick	Yes '74	Yes '73	Yes '73	Inf.	Inf.	Inf.	No	Yes '73	Inf.	Yes '73	Yes '70	Yes '72	Yes '74	Inf.	Yes '74	Yes '74	No
Newfoundland (incl. Labrador)	—	—	—	—	—	—	—	—	—	—	—	—	—	—	—	—	—
Northwest Terr.	Inf.	Inf.	Inf.	Inf.	Inf.	Inf.	Inf.	Inf.	Inf.	Inf.	Inf.	Inf.	Inf.	Inf.	Inf.	Inf.	Inf.

TABLE 1 (*Continued*)

NFPA Code	50	50A	50B	51	51A	51B	54	56A	56B	56C	56D	56E	56F	57	58	59	59A
Nova Scotia	Inf.	Inf.	Yes '73	Yes '74	Yes '74	Yes '71	Yes '74	Yes '73	Yes '73	Yes '73	Yes '70	Yes '72	Yes '74	Yes '73	No	Yes '74	Inf.
Ontario	—	—	—	—	—	—	—	—	—	—	—	—	—	—	—	—	—
Prince Edw. Is.	Yes '74	Yes '73	Yes '73	Yes '74	Yes '74	Yes '71	Yes '74	Yes '73	Yes '73	Yes '73	Yes '70	Yes '72	Yes '74	Yes '73	Yes '74	Yes '74	Yes '72
Quebec	No	No	No	No	No	No	No	No	No	No	No	No	No	No	No	No	No
Saskatchewan	Inf.	Inf.	Inf.	Inf.	Inf.	Inf.	Inf.	Inf.	Inf.	Inf.	Inf.	Inf.	Inf.	Inf.	Inf.	Inf.	Inf.
Yukon Terr.	Yes '74	Yes '73	Yes '73	Yes '74	Yes '74	Yes '71	Yes '74	Yes '73	Yes '73	Yes '73	Yes '70	Yes '72	Yes '74	Yes '73	Yes '74	Yes '74	Yes '72

*Explanatory notes to Table 1.

Note 1. "Yes" entered in the table means that the standard is either adopted by reference or reproduced in a law, regulation or similar vehicle. In a number of instances, standards are included by virtue of adoption of the National Fire Codes of the NFPA, of which they are a part. The date after "Yes" indicates the edition of the standard used by the state.

Note 2. "Inf." in the table means that the standard is used informally under the broad jurisdictional powers of the regulating official or agency concerned.

Note 3. A dash in the table means that no use of the standard was reported when the information given in the table was requested. Such a response may reflect a misunderstanding of the question, or reservation on one or another specific point in a standard rather than on the entire standard.

TABLE 2. STATES (AND CITIES AND COUNTIES WITHIN STATES) AND PROVINCES THAT HAVE ADOPTED SECTION VIII OF THE ASME BOILER AND PRESSURE VESSEL CODE, OR OTHER CODES FOR UNFIRED PRESSURE VESSELS, WITH NAME AND ADDRESS OF ENFORCEMENT OFFICIALS.

Explanatory Note. Section VIII of the Boiler and Pressure Vessel Code of the ASME (American Society of Mechanical Engineers) concerns unfired pressure vessels, and the standards of the Compressed Gas Association, Inc., recommend its use for compressed gas storage containers. The information below is reprinted with permission from the "1978 Data Sheet" of the Uniform Boiler and Pressure Vessel Laws Society (57 Pratt St., Hartford, CT 06103), a nonprofit, nonpartisan technical body supported by voluntary contributions.

Key to Symbols.
VIII(1) –Pressure vessels.
VIII(2) –Pressure vessels, alternate rules.
  A–Law requires ASME construction.
  O–Have own construction code.
  N–Law does not cover.
  *–Operator's license required.
  **–Limited to specific vessels.
  ***–Pending rules and regulations.

Every effort has been made to check the reliability of the information in this table but the Society can accept no responsibility for its accuracy.

	VIII (1)	(2)	Enforcement Official & His Address
Alabama	N	N	James Stephenson, Administrator, OSHA, 600 Administrative Bldg, 64 No. Union St., Montgomery, Al. 36130
Alaska	A	A	R. D. Cather, Chief, Pressure Vessel Section, Suite 100, 650 W. International Airport Rd., Anchorage, Alaska 99502
Arizona	A	A	Bernard Geary, Chief Boiler Inspector, Division of OSHA, P.O. Box 19070, Phoenix, Az. 85007
Arkansas	A*	A	J. T. Crosby, Chief Boiler Inspector, Boiler Inspection Dept., Box 1797, Little Rock, Ar. 72203
California	A	A	P. J. Carosella, Principal Safety Engineer, Pressure Vessel Section, P.O. Box 603, San Francisco, Ca. 94101
Colorado	A	A	Bill E. Cimino, Chief Boiler Inspector, Safety Section, 1313 Sherman St., Denver, Co. 80203
Connecticut	N	N	Leo F. Alix, Dpty. Comm. of Factory Inspection, Labor Dept., 200 Folly Brook Blvd., Wethersfield, Ct. 06115
Delaware	A	A	H. S. Mauk, Director, Division of Boiler Safety, 820 N. French St., Wilmington, De. 19801
Dist. of Columbia	A*	A*	Harry R. Williams, Chief, Boiler Section, 614 H St. N.W., Rm. LL9, Washington, D.C. 20001
Florida	N	N	L. E. Price, Adm., Safety Standards Section, 1321 Executive Center Dr., E. Tallahassee, Fl. 32301
Georgia	A	N	Esters M. Shiver, Director of Inspection Division, 287 State Labor Building, Atlanta, Ga. 30334
Guam	A	N	Vinay K. Sood, P. E., P.O. Box 2950, Agana, Guam 96910
Hawaii	A	A	Richard E. Peterson, Mgr., Technical Inspection & Compliance Br., 667 Ala Moana Blvd., Rm. 910, Honolulu, Hi. 96813
Idaho	A	A	Gerald Gedes, Chairman, Industrial Commission, 317 Main St., Boise, Id. 83702
Illinois	A	A	Duane R. Gallup, Chief Boiler Inspector, 302 Armory Building, Springfield, Il. 62706

State			Chief Inspector / Address
Indiana	A	A	Robert R. Johnson, Chief Inspector, 909 State Office Bldg., 100 No. Senate Ave., Indianapolis, In. 46204
Iowa	A	A	W. W. Larsen, Boiler Supervisor, Bureau of Labor, Boiler Inspection Division, Des Moines, Ia. 50319
Kansas	N	N	E. H. Cannon, Jr., Chief, Boiler Inspection Section, 610 West Tenth, 2nd floor, Topeka, Ks. 66612
Kentucky	A**	N	C. A. Brown, Chief Blr. Inspector, Bureau of Blr. Inspection, The 127 Bldg., Hwy. 127 South, Frankfort, Ky. 40601
Louisiana	A	A	Wayne Morvant, Chief Inspector, Div. of Boiler Inspection, Dept. of Labor, Baton Rouge, La. 70804
Maine	A	A	J. W. Emerson, Chief Inspector, Div. of Blr. Insp., State Office Bldg. Room 614, Augusta, Maine 04333
Maryland	A	A	John T. Grail, Chief Inspector, Div. of Labor & Industry, 203 E. Baltimore St., Baltimore, Md. 21202
Massachusetts	A**	N	John K. Olsen, Chairman of Bd. of Boiler Rules, Div. of Inspection, McCormick Bldg., 13th Fl., Boston, Ma. 02108
Michigan	A**	N	Stanley Mierzwa, Chief, Boiler Division, Dept. of Labor, 7150 Harris Dr., Lansing, MI. 48926
Minnesota	A*	A	Henry Baron, Chief Inspector, Dept. of Labor & Ind., 444 Lafayette Rd., Rm. 567, St. Paul, Mn. 55101
Mississippi	A	A	L. A. Grant, Chief Inspector, Dept. of B & PV Inspection, 880 Lakeland Dr., Jackson, Ms. 39216
Missouri	N	N	G. B. Graven, Director, Div. of Industrial Inspection, P.O. Box 449, Jefferson City, Mo. 65101
Montana	N	N	M. B. Salazar, Chief, Bureau of Safety and Health, 815 Front St., Helena, Mt. 59601
Nebraska	A**	A**	John P. Mickels, Chief Boiler Inspector, Department of Labor, Lincoln, Nebraska 68509
Nevada	A	A	Burd O. Rohde, Dept. of Occ. Safety and Health, P.O. Box 2776, 1710 W. Charleston Blvd., Las Vegas, Nv. 89102
New Hampshire	A	N	Robert M. Duvall, Commissioner, Department of Labor, 1 Pillsbury Street, Concord, N.H. 03301
New Jersey	A	A	John L. Sullivan, Asst. Director, Labor & Industry Building, P.O. Box 1503, Trenton, N.J. 08625
New Mexico	N	N	Fred Gerber, Chief Adm., Mechanical Bd. of Constr. Industries Commission, P.O. Box 5155, Santa Fe, N.M. 87501
New York	N	N	Edwin M. Hicks, Chief Boiler Inspector, Dept. of Labor, State Campus, Albany, N.Y. 12201
North Carolina	A	A	B. L. Whitley, Director, Boiler & Pressure Vessel Div., P.O. Box 27407, Raleigh, N.C. 27611
North Dakota	A	N	Harold Gragg, Chief Boiler Inspector, Dept. of Blr. Inspection, Russell Bldg., Hwy. 83 North, Bismarck, N.D. 58505
Ohio	A	O	Richard E. Jagger, Chief, Div. of Boiler Inspection, 2323 W. 5th Ave., P.O. Box 825, Columbus, Oh. 43216
Oklahoma	N	N	Tom Monroe, Chief, Bureau of Boiler Inspection, Dept. of Labor, State Capitol, Rm. 118, Oklahoma City, Ok. 73105
Oregon	A	A	Charles H. Walters, Chief Inspection, Rm. 408-State Office Bldg., 1400 S.W. 5th Ave, Portland, Or. 97201
Panama Canal Zone	A	A	A. L. Gallin, Chief, Industrial Division, P.O. Box 5046, Cristobal, Canal Zone
Pennsylvania	A*	A	Chief, Boiler Division, Labor & Industry Building, Harrisburg, Pa. 17120
Puerto Rico	A	N	Russell R. Perez, Chief Boiler Engineer, Dept. of Labor, 414 Barbosa Ave., San Juan, P.R. 00917
Rhode Island	N	N	Raymond DeStefanis, Administrator, Div. of Occ. Safety, 220 Elmwood Av., Providence, R.I. 02907
South Carolina	N	N	Robert C. Parks, Director, Dept. of OSHA, 3600 Forest Dr., P.O. Box 11329, Columbia, S.C. 29211
South Dakota	N	N	Douglas Hague, Chief Blr. Inspector, Div. of Blr. Inspection, Dept. of Public Safety, Pierre, S.D. 57501
Tennessee	A	A	C. W. Allison, Chief Inspector, 501 Union Building-2nd Floor, Nashville, Tn. 37219
Texas	N	N	T. M. Wedemeier, Chief Inspector, Boiler Inspection Div., Box 12157, Capitol Station, Austin, Tx. 78711
Utah	A	A	Raymond K. Blosch, Chief Inspector, Industrial Comm., Safety Div., 350 E. 500 South, Salt Lake City, Ut. 84111

473

TABLE 2 (Continued)

	VIII (1)	(2)	Enforcement Official & His Address
Vermont	A	A	Albert A. Fraser, Deputy Commissioner, Dept. of Labor & Industry, State Office Bldg., Montpelier, Vt. 05602
Virginia	A	A	Vernon E. Doss, Chief Inspector, Boiler & Pressure Vessel Safety Div., P.O. Box 12064, Richmond, Va. 23241
Washington	A	A	Martin M. Forseth, Chief, Boiler and Pressure Vessel Section, 300 W. Harrison-Rm. 506, Seattle, Wa. 98119
West Virginia	N	N	Dale Morgan, Chief Boiler Inspector, Department of Labor, Charleston, W. Va. 25305
Wisconsin	A	A	John J. Duffy, Chief Inspector, Div. of Ind. Safety & Bldgs., 201 E. Washington Ave., Madison, Wi. 53702
Wyoming**	A	A	William W. Wilkins, Administrator, Department of OSHA, 200 East 8th Avenue, Cheyenne, Wy. 82001

Provinces in Canada

	VIII (1)	(2)	Enforcement Official & His Address
Alberta	A	A	A. J. R. Rees, Chf. Inspector, Dept. of Labour, 3rd Fl., Princeton Place, 10339 124th St., Edmonton, Alberta T5N 3W1
British Columbia	A	A	Brian Cole, Chief Inspector, Boiler & Pressure Vessel Branch, 501 W. 12th Ave., Vancouver, B.C. V5Z 1M6
Manitoba	A*	A	L. A. O'Morrow, Director, Mechanical & Engineering Div., Rm. 500-Norquay Bldg., Winnipeg, Manitoba R3C 0P8
New Brunswick	A*	A	D. E. Ross, Chief Inspector, Dept. of Labour, P.O. Box 6000, Fredericton, New Brunswick E3B 5H1
Newfoundland & Labrador	A*	A	A. W. Diamond, Director, Dept. of Manpower & Ind. Relations, Confederation Bldg., St. Johns, Newfoundland
Northwest Terr.	A*	N	I. H. Grabke, Chf. Inspector, Boiler & Pressure Vessel Br., Box 1320, Yellowknife, Northwest Terr. X0E 1H0
Nova Scotia	A*	A	Robert A. Yeo, Chief Inspector, Department of Labour, P.O. Box 697, Halifax, Nova Scotia B3J 2T8
Ontario	A*	A	H. J. Wright, Director, Pressure Vessels Branch, 400 University Ave.-10th Fl, Toronto, Ontario M7A 2J9
Prince Edward Is.	A	N	W. A. Miller West, Chf. Blr. & P. V. Inspector, Dept. of Labour, Box 2000, Charlottetown, P.E.I. C1A 7N8
Quebec	A*	A	R. Sauve, Asst. Director, Dept. of Labour & Manpower, 255 Cremazie East, Montreal, P.Q.
Saskatchewan	A*	N	R. V. Curry, Chief Inspector, Boiler & Pressure Vessel Unit, 1150 Rose St., Regina, Saskatchewan S4R 1Z6
Yukon Territory	A*	N	J. M. Campbell, Chief Inspector, Box 2703, Whitehorse, Yukon Terr. Y1A 2C6

Cities and Counties

	VIII (1)	(2)	Enforcement Official & His Address
Albuquerque, N.M.	N	N	G. M. Scalia, Boiler Inspection, Div. of Bldgs. & Inspections, Box 1293, Albuquerque, N.M. 87103
Buffalo, N.Y.	A	N	Thomas A. Hearn, Jr., Chief Examiner of Engineers & Boilers, 2501 City Hall, Buffalo, N.Y. 14202
Chicago, IL.	A	A	A. J. Cmeyla, Chief Inspector, Inspection Services Div., 320 N. Clark St.-Rm. 402, Chicago, Il. 60610
Dearborn, MI.	A*	N	G. M. Powers, Chief Safety Engineer, Dept. of Public Works, 4500 Maple, Dearborn, Mi. 48126
Denver, CO.	A*	A	Ken Williams, Chief Mech. Inspector, Building Dept., 1445 Cleveland Place-Rm. 204, Denver, Co. 80802

474

Location			
Des Moines, IA.	N	N	Glen L. Bowers, Building Inspection Services, City Hall, Des Moines, Ia. 50307
Detroit, MI.	A	A	Samuel Schugar, Chief Safety Engineer, Fourth Floor-City County Building, Detroit, Mi. 48226
E. St. Louis, IL.	N	A*	Joseph Iwasyszyn, Air Pollution Officer, City Hall, East St. Louis, Il. 62201
Greensboro, N.C.	A	A	Lloyd H. Doolittle, Chief Inspector, Heating Inspection Dept., Drawer W-2 Greensboro, N.C. 27402
Kansas City, MO.	A	A*	E. S. Bybee, Chief Inspector, Public Works Dept., 3410 Troost Ave., Kansas City, Mo. 64109
Los Angeles, CA.	A	A	C. J. Dannehold, Chief, Boiler Division, Rm. 495H, City Hall, Los Angeles, Ca. 90012
Memphis, TN.	A	A*	Otis R. Kyle, Chief Safety Engineer, City Hall, Room 406, 125 North Main St., Memphis, Tn 38103
Miami, Fl.	A	A*	R. Timmersman, Mechanical Div., City Bldg. Dept., Box 330708, Coconut Grove Station, Miami, Fl. 33133
Milwaukee, WI.	A	A*	G. M. Kuetemeyer, Supervisor, Safety Engineering, Municipal Bldg., Rm. 1009, Milwaukee, Wi. 53202
New Orleans, LA.	A	A*	C. Curtis Mann, Chief Mech. Inspector, Rm. 7E04, City Hall, 1200 Perido St., New Orleans, La. 70112
New York City, N.Y.	N	N	P. J. Dillon, Chief Inspector, Boiler & Licensing Div., 120 Wall St., New York, N.Y. 10005
Oklahoma City, OK.	N	N	Charles Huddleston, Boiler Inspector, Municipal Building, Oklahoma City, OK. 73102
Omaha, NB.	A*	N	Chief of Mechanical Section, P and I, Dept. of Housing & Comm. Dev., City Hall, Omaha, Nb. 68102
Phoenix, AZ.	A	A	O. J. Freed, Chf. Mech. Inspector, Mech. Inspection Section, 251 W. Washington St., Phoenix, Az. 85003
St. Joseph, MO.	A*	A	T. Zebelean, Supt. Building Regulations, City Hall, St. Joseph, Mo. 64501
St. Louis, MO.	A*	N	L. E. Stoppelman, Supv., Mechanical Equipment, Public Safety Dept., Rm-425, City Hall, St. Louis, Mo. 63103
San Francisco, CA.	A	N	Frank M. Reid, Chief Boiler Inspector, Dept. of Public Works, 450 McAllister St.-Rm. 103, San Francisco, Ca. 94102
San Jose, CA.	A	N	R. G. Eldridge, Boiler Inspector, Bureau of Fire Prevention, 476 Park Ave., San Jose, Ca. 95110
Seattle, WA.	A	A	S. B. Voris, Chief Boiler Inspector, Dept. of Buildings, 600 4th Avenue, Rm. 503, Seattle, Wa. 98104
Spokane, WA.	A	N	R. R. Reese, Director of Building, City Hall, Spokane, Wa. 99201
Tacoma, WA.	A*	N	F. W. King, Chief Boiler Inspector, 438 County-City Building, 930 Tacoma Ave., Tacoma, Wa. 98402
Tampa, FL.	A*	A	Enrique R. Marcet, Chief Inspector, Boiler Bureau, 301 N. Florida Avenue, Tampa, Fl. 33602
Tucson, AZ.	A	A	F. H. Blackmore, Chief Mechanical Inspector, Box 5547, Tucson, Arizona 85703
Tulsa, OK.	A*	A	C. T. West, Jr., Chief Boiler Inspector, 200 Civic Center, Rm. 424, Tulsa, Ok. 74103
University City, MO	A*	N	W. G. Walters, General Inspector, City Hall, University City, Mo. 63130
White Plains, NY.	N	N	Saverio Innamorato, Commissioner of Building, 255 Main Street, White Plains, N.Y. 10601
Arlington Co., VA.	A	A	Elmer Shrout, Chief Mechanical Inspector, Court House, Arlington, VA 22201
Dade Co., FL.	A	A	John D. Davidson, Boiler Inspector, Bldg. & Zoning Dept., 909 S.E. First Avenue, Miami, Fl. 33131
Fairfax Co., VA	A*	A	Larry Stahl, Chief Mechanical Inspector, 4100 Chain Bridge Road, Fairfax, Va. 22030
Jefferson Parish	A*	A	H. Schouest, Jr., Director, Dept. of Regulatory Inspection, 3300 Metairie Road, Metairie, La, 70001
St. Louis Co., MO.	A	A	Marvin P. Feuring, Chf. Mech. Inspector, St. Louis County Government Center, 7900 Forsyth Blvd., Clayton, Mo. 63105

TABLE 3. STATE ADOPTIONS OF HAZARDOUS MATERIALS REGULATIONS OF DEPARTMENT OF TRANSPORTATION (DOT) BY PARTS AS OF MARCH 27, 1979.

Key to Symbols.
  A—Adopted in toto.
  B—Adopted in part.
  C—Has similar state rule.
  D—Has no rule.

State	171	172	173	177	178	179	State	171	172	173	177	178	179
Alabama	B	B	B	B	B	D	Montana	A	A	A	A	A	A
Alaska	D	D	D	D	D	D	Nebraska	D	C	C	C	D	D
Arizona	A	A	A	A	A	A	Nevada	A	A	A	A	A	D
Arkansas	A	A	A	A	A	A	New Hampshire	C	C	C	C	C	D
California	A	A	A	A	A	A	New Jersey	C	C	C	C	C	C
Colorado	A	A	A	A	A	D	New Mexico	A	A	A	A	A	D
Connecticut	D	D	D	D	D	D	New York	A	A	A	A	A	A
Delaware	D	D	D	C	D	D	North Carolina	A	A	A	A	A	A
Dist. of Columbia	A	A	A	A	A	D	North Dakota	D	D	D	D	D	D
Florida	B	A	A	A	A	D	Ohio	A	A	A	A	A	A
Georgia	C	C	C	C	C	D	Oklahoma	D	D	D	C	C	D
Hawaii	A	A	A	A	A	A	Oregon	A	A	A	A	A	A
Idaho	A	A	A	A	A	A	Pennsylvania	A	A	A	A	A	D
Illinois	A	A	A	A	A	B	Rhode Island	A	A	A	A	A	A
Indiana	D	D	D	D	D	D	South Carolina	A	A	A	A	A	A
Iowa	A	A	A	A	A	A	South Dakota	A	A	A	A	A	A
Louisiana	D	D	D	D	D	D	Tennessee	A	A	A	A	A	A
Kansas	A	A	A	A	A	A	Texas	B	A	A	A	A	D
Kentucky	B	B	B	B	B	B	Utah	A	A	A	A	A	D
Maine	D	D	D	D	D	D	Vermont	A	A	A	A	A	A
Maryland	B	B	B	A	B	D	Virginia	C	C	C	C	C	C
Massachusetts	A	A	A	A	A	A	Washington	A	A	A	A	A	A
Michigan	A	A	A	A	A	A	West Virginia	D	D	D	D	D	D
Minnesota	A	A	A	A	A	A	Wisconsin	A	A	A	A	A	A
Mississippi	B	B	B	B	B	D	Wyoming	A	A	A	A	A	A
Missouri	A	A	A	A	A	D							

REPORT ON STATE ADOPTION OF FEDERAL MOTOR
CARRIER REQUIREMENTS HAZARDOUS MATERIALS
AND SAFETY REGULATIONS

State Alabama

Part 171	1, 2, 3
Part 172	3, 4, 5
Part 173	2, 3, 4, 5
Part 177	1, 2, 3, 4, 5
Part 178	2, 3, 4, 5
Part 179	None

Responsible Agency

1. Alabama Public Service Commission
   P.O. Box 991
   Montgomery, AL 36130
   (Regulations of the Alabama Public
   Service Commission apply to for hire
   motor carriers only.)

2. State Fire Marshal
   435 McDonough
   Montgomery, AL 36104

3. Liquified Petroleum Gas Board
   818 S. Perry St.
   Montgomery, AL 36104

4. Alabama Department of Agriculture
   and Industries
   1445 Federal Dr.
   Montgomery, AL 36107

5. Alabama Department of Health
   501 Dexter Ave.
   Montgomery, AL 36104

State Alaska

Part 171	None
Part 172	None
Part 173	None
Part 177	None
Part 178	None
Part 179	None

Responsible Agency

1. Alaska Transportation Commission
   338 Denali 10th Floor
   Anchorage, AK 99501

State Arizona

Part 171	1
Part 172	1
Part 173	1
Part 177	1
Part 178	1
Part 179	1

Responsible Agency

1. Arizona Corporation Commission
   2222 W. Encanto Blvd.
   Phoenix, AZ 85009

State Arkansas _____

Part 171 1, 2 _____
Part 172 1, 2 _____
Part 173 1, 2 _____
Part 177 1, 2 _____
Part 178 1, 2 _____
Part 179 1, 2 _____

Responsible Agency

1. Arkansas Transportation Commission
   Justice Bldg., State Capitol
   Little Rock, AR 72201
2. Arkansas Department of Public Safety
   Arkansas State Police Division
   P.O. Box 4005
   Little Rock, AR 72204

State California _____

Part 171 1, 2, 3 _____
Part 172 1, 2, 3 _____
Part 173 1, 2, 3 _____
Part 177 1, 2, 3 _____
Part 178 1, 2, 3 _____
Part 179 1, 2, 3 _____

Responsible Agency

1. California Department of Health
   714/744 P St.
   Sacramento, CA 95814
2. California Highway Patrol
   2611 26th St.
   Sacramento, CA 95818
3. California Department of Industrial
      Relations
   Division of Industrial Safety
   455 Golden Gate Ave.
   San Francisco, CA 94101

State Colorado _____

Part 171 1 _____
Part 172 1 _____
Part 173 1 _____
Part 177 1 _____
Part 178 1 _____
Part 179 None _____

Responsible Agency

1. Colorado Public Utilities Commission
   500 State Services Bldg.
   1525 Sherman St.
   Denver, CO 80203
   (Regulations promulgated by the Colorado
      Public Utilities Commission apply to for
      hire motor carriers only.)

State Connecticut _____

Part 171 None _____
Part 172 None _____
Part 173 None _____
Part 177 None _____
Part 178 None _____
Part 179 None _____

Responsible Agency

State Delaware

		Responsible Agency
Part 171	None	1. Office of the State Fire Marshal
Part 172	None	Chestnut Grove Rd.
Part 173	None	Dover, DE 19901
Part 177	1	
Part 178	None	
Part 179	None	

State District of Columbia

		Responsible Agency
Part 171	1	1. Washington Metropolitan Area
Part 172	1	Transit Authority
Part 173	1	600 5th St. N.W.
Part 177	1	Washington, DC 20001
Part 178	1	
Part 179	None	

State Florida

		Responsible Agency
Part 171	1	1. Florida Public Service Commission
Part 172	1	West Gaines St.
Part 173	1	Tallahassee, FL 32301
Part 177	1	(Regulation in Part 171 to Part 397 apply
Part 178	1	to for hire motor carriers only.)
Part 179	None	

State Georgia

		Responsible Agency
Part 171	1	1. Fire Marshal's Office
Part 172	1	Hazardous Materials Division
Part 173	1	7 Martin Luther King, Jr. Dr.
Part 177	1	Atlanta, GA 30301
Part 178	1	
Part 179	None	

State Hawaii

		Responsible Agency
Part 171	1	1. Hawaii Public Utility Commission
Part 172	1	1010 Richards St.
Part 173	1	P.O. Box 541
Part 177	1	Honolulu, HI 96309
Part 178	1	
Part 179	1	

State Idaho

		Responsible Agency
Part 171	1	1. Idaho Department of Law Enforcement
Part 172	1	P.O. Box 34
Part 173	1	Boise, ID 83731
Part 177	1	
Part 178	1	
Part 179	1	

State Illinois

		Responsible Agency
Part 171	1, 2, 3	1. Illinois Department of Transportation
Part 172	1, 2, 3	2300 S. Dirksen Parkway
Part 173	1, 2, 3	Springfield, IL 62704
Part 177	1, 2, 3	2. Illinois Department of Public Health
Part 178	1, 2, 3	535 W. Jefferson
Part 179	1	Springfield, IL 62702
		3. Illinois State Police
		401 Armory Bldg.
		Springfield, IL 62706

State Indiana

		Responsible Agency
Part 171	None	
Part 172	None	
Part 173	None	
Part 177	None	
Part 178	None	
Part 179	None	

State Iowa

		Responsible Agency
Part 171	1	1. Iowa Department of Transportation
Part 172	1	Ames, LA 50010
Part 173	1	
Part 177	1	
Part 178	1	
Part 179	1	

State Kansas

		Responsible Agency
Part 171	1	1. Kansas Corporation Commission
Part 172	1	State Office Bldg.
Part 173	1	Topeka, KS 66612
Part 177	1	
Part 178	1	
Part 179	1	

State Kentucky

Part 171	1
Part 172	1
Part 173	1
Part 177	1
Part 178	1
Part 179	1

Responsible Agency

1. Kentucky Department of Transportation
   Bureau of Vehicle Regulation
   Division of Highway Enforcement
   2nd floor State Office Bldg.
   Frankfort, KY 40601

State Louisiana

Part 171	None
Part 172	None
Part 173	None
Part 177	None
Part 178	None
Part 179	None

Responsible Agency

State Maine

Part 171	None
Part 172	1, 2
Part 173	None
Part 177	1, 2
Part 178	None
Part 179	None

Responsible Agency

1. Maine State Police
   36 Hospital St.
   Augusta, ME 04333

2. Maine Public Utilities Commission
   State House
   Augusta, ME 04333

State Maryland

Part 171	1
Part 172	3
Part 173	3
Part 177	2
Part 178	3
Part 179	None

Responsible Agency

1. Maryland Motor Vehicle Administration
   Glen Burnie, MD 21061

2. Maryland State Police
   Pikesville, MD 21208

3. Maryland Department of Transportation
   P.O. Box 8155
   Balt.-Wash. International Airport, MD 21240

State Massachusetts

Part 171	1, 2
Part 172	1, 2
Part 173	1, 2
Part 177	1, 2
Part 178	1, 2
Part 179	1, 2

Responsible Agency

1. Massachusetts Department of Public Safety
   1010 Commonwealth Ave.
   Boston, MA 02215

2. Massachusetts Registry of Motor Vehicles
   100 Nashua St.
   Boston, MA 02114

State Michigan

Part 171 1   Responsible Agency
Part 172 1   1. Michigan Public Service Commission
Part 173 1   6545 Mercantile Way
Part 177 1   Lansing, MI 48909
Part 178 1
Part 179 1

State Minnesota

Part 171 1   Responsible Agency
Part 172 1   1. Minnesota Department of Transportation
Part 173 1   Transportation Bldg.
Part 177 1   St. Paul, MN 55155
Part 178 1
Part 179 1

State Mississippi

Part 171 1   Responsible Agency
Part 172 1, 2   1. State Board of Health (Radioactive Materials
Part 173 1, 2   Radiological Health only)
Part 177 1, 2   2423 N. State St.
Part 178 1, 2   Jackson, MS 39205
Part 179 None   2. Motor Vehicle Comptroller
                                      Liquefied Gas Division
                                      Woolfolk State Office Bldg.
                                      Jackson, MS 39201

State Missouri

Part 171 1   Responsible Agency
Part 172 1   1. Missouri Public Service Commission
Part 173 1   Jefferson City, MO 65101
Part 177 1
Part 178 1
Part 179 None

State Montana

                                   Responsible Agency
Part 171 1, 2, 3   1. Montana Public Service Commission
Part 172 1, 2, 3   1227 11th Ave.
Part 173 1, 2, 3   Helena, MT 59601
Part 177 1, 2, 3   2. Montana Highway Patrol Bureau
Part 178 1, 2, 3   1014 National Ave.
                                      Helena, MT 59601

Part 179 1, 2, 3 _____

3. Gross Vehicle Weight Division
   Montana Highway Dept.
   2701 Prospect Ave.
   Helena, MT 59601

State Nebraska _____

Part 171 None _____
Part 172 1 _____
Part 173 1 _____
Part 177 1 _____
Part 178 None _____
Part 179 None _____

Responsible Agency
1. Nebraska State Fire Marshal (Explosives
   only)
   301 Centennial Mall S.
   Lincoln, NE 68508

State Nevada _____

Part 171 1 _____
Part 172 1 _____
Part 173 1 _____
Part 177 1 _____
Part 178 1 _____
Part 179 None _____

Responsible Agency
1. Nevada Public Service Commission
   505 E. King St.
   Carson City, NV 89701

State New Hampshire _____

Part 171 1 _____
Part 172 1 _____
Part 173 1 _____
Part 177 1 _____
Part 178 1 _____
Part 179 None _____

Responsible Agency
1. New Hampshire Fire Marshal
   Division of Safety Services
   Hazen Dr.
   Concord, NH 03301

State New Jersey _____

Part 171 1 _____
Part 172 1 _____
Part 173 1 _____
Part 177 1 _____
Part 178 1 _____
Part 179 1 _____

Responsible Agency
1. New Jersey Environmental (PENDING)
   Protection Department
   John Fitch Plaza
   Trenton, NJ 08625

State New Mexico

Part 171	1, 2, 3
Part 172	1, 2, 3
Part 173	1, 2, 3
Part 177	1, 2, 3
Part 178	1, 2, 3
Part 179	3

Responsible Agency

1. Motor Transportation Division
   New Mexico Department of Transportation
   P.E.R.A Bldg.
   Santa Fe, NM 87501
2. New Mexico State Police
   Albuquerque Hwy.
   Santa Fe, NM 87501
3. LPG Division
   New Mexico Department of Commerce
   and Industry
   Bataan Mem. Bldg.
   Santa Fe, NM 87501
   (Regulations in Part 171 apply to for hire
   carriers only.)

State New York

Part 171	1, 2
Part 172	1, 2
Part 173	1, 2
Part 177	1, 2
Part 178	1, 2
Part 179	1, 2

Responsible Agency

1. New York State Department of
   Transportation
   State Campus
   Albany, NY 12232
2. New York State Police Division
   Headquarters
   State Campus
   Albany, NY 12226

State North Carolina

Part 171	1
Part 172	1
Part 173	1
Part 177	1
Part 178	1
Part 179	1

Responsible Agency

1. North Carolina Utilities Commission
   430 N. Salisbury St.
   Dobbs Bldg.
   Raleigh, NC 27602

State North Dakota

Part 171	None
Part 172	None
Part 173	None
Part 177	None
Part 178	None
Part 179	None

Responsible Agency

State Ohio

Part 171 1
Part 172 1
Part 173 1
Part 177 1
Part 178 1
Part 179 1

Responsible Agency

1. Public Utilities Commission of Ohio
   180 E. Broad St.
   Columbus, OH 43215

State Oklahoma

Part 171 None
Part 172 None
Part 173 None
Part 177 1
Part 178 1
Part 179 None

Responsible Agency

1. Oklahoma LPG Board
   State Capitol Bldg.
   Oklahoma City, OK 73105

State Oregon

Part 171 1
Part 172 1
Part 173 1
Part 177 1
Part 178 1
Part 179 1

Responsible Agency

1. Oregon Public Utilities Commission
   Labor and Industries Bldg.
   Salem, OR 97310

State Pennsylvania

Part 171 1
Part 172 1
Part 173 1
Part 177 1
Part 178 1
Part 179 None

Responsible Agency

1. Pennsylvania Hazardous Substances
   Transportation Board
   Transportation and Safety Bldg.
   Harrisburg, PA 17120

State Rhode Island

Part 171 1
part 172 1
Part 173 1
Part 177 1
Part 178 1
Part 179 1

Responsible Agency

1. Rhode Island Department of Transportation
   State Office Building
   Providence, RI 02903

State South Carolina

Part 171 1
Part 172 1
Part 173 1
Part 177 1
Part 178 1
Part 179 1

1. South Carolina Public Service Commission
   Transportation Division
   P.O. Drawer 11649
   Columbia, SC 29211

State South Dakota

Part 171 1
Part 172 1
Part 173 1
Part 177 1
Part 178 1
Part 179 1

1. South Dakota Public Utilities Commission
   Pierre, SD 57501

State Tennessee

Part 171 1
Part 172 1
Part 173 1
Part 177 1
Part 178 1
Part 179 1

1. Tennessee Public Service Commission
   Cordell Hull Bldg.
   Nashville, TN 37219

State Texas

Part 171 1
Part 172 1
Part 173 1
Part 177 1
Part 178 1
Part 179 None

1. Texas Railroad Commission
   Drawer 12967
   Capitol Station
   Austin, TX 78711

State Utah

Part 171 1
Part 172 1
Part 173 1
Part 177 1
Part 178 1
Part 179 None

1. Utah Department of Transportation
   Safety Division
   State Office Bldg.
   Salt Lake City, UT 84114

State Vermont

Part 171 1, 2
Part 172 1, 2
Part 173 1, 2
Part 177 1, 2
Part 178 1, 2
Part 179 1, 2

Responsible Agency

1. Vermont Department of Transportation
   120 State St.
   Montpelier, VT 05602
2. Vermont Department of Public Safety
   Montpelier, VT 05602
   (Regulations in Part 171 to Part 398 apply
   to motor carriers transporting hazardous
   material only.)

State Virginia

Part 171 1, 2
Part 172 1, 2
Part 173 1, 2
Part 177 1, 2
Part 178 1, 2
Part 179 1, 2

Responsible Agency

1. Virginia State Corporation Commission
   P.O. Box 1197
   Richmond, VA 23209
2. Virginia State Police
   P.O. Box 27472
   Richmond, VA 23261

State Washington

Part 171 1, 2
Part 172 1, 2
Part 173 1, 2
Part 177 1, 2
Part 178 1, 2
Part 179 1, 2

Responsible Agency

1. Washington Utilities and Transportation
   Commission
   Highway License Bldg.
   Olympia, WA 98504
2. Washington State Patrol
   General Administration Bldg.
   Olympia, WA 98504
   (The Regulations in Part 171 to Part 178
   do not apply to private motor carriers
   engaged in intrastate commerce.)

State West Virginia

Part 171 None
Part 172 None
Part 173 None
Part 177 None
Part 178 None
Part 179 None

Responsible Agency

State Wisconsin

Part 171 1
Part 172 1
Part 173 1
Part 177 1
Part 178 1
Part 179 1

Responsible Agency

1. Wisconsin Department of Transportation
   4802 Sheboygan Ave.
   Madison, WI 53702
   (Regulations in Part 171 to Part 179 apply
      to intrastate for hire motor carriers only.)

State Wyoming

Part 171 1, 2
Part 172 1, 2
Part 173 1, 2
Part 177 1, 2
Part 178 1, 2
Part 179 1, 2

Responsible Agency

1. Wyoming Public Service Commission
   Capitol Hill Bldg.
   Cheyenne, WY 82001

2. Wyoming State Highway Patrol
   N W Cheyenne
   Cheyenne, WY 82001

APPENDIX B

Service Publications of the Compressed Gas Association

PAMPHLET SERIES

Cylinders

C-1 Methods for Hydrostatic Testing of Compressed Gas Cylinders. DOT regulations require certain cylinders to be periodically retested to requalify them for continued service. Pamphlet details requirements for test equipment, method of operation and test records.

C-2 Recommendations for the Disposition of Unserviceable Compressed Gas Cylinders. Outlines procedures for the safe disposition of unserviceable cylinders including discharge of cylinder content and methods of handling when content cannot be discharged.

C-3 Standards for Welding and Brazing on Thin Walled Containers. Standards applicable to DOT welded or brazed compressed gas cylinders designed for 500 psig service pressure or less and having 1000 lb water capacity maximum; also minimum wall thickness under $\frac{3}{8}$ in. Covers procedure and operator qualification, radiographic inspection and container repair.

C-4 American National Standard Method of Marking Portable Compressed Gas Containers to Identify the Material Contained, Z48.1. Sponsored by CGA, this standard covers containers having a water capacity up to 1000 lbs. Describes standard method of marking containers.

C-5 Cylinder Service Life—Seamless, High-Pressure Cylinder Specifications DOT-3, DOT-3A, DOT-3AA. Contains detailed methods of estimating wall thickness that can be applied with accuracy and simplicity to the retesting of compressed gas cylinders to determine their suitability for continued service. Includes information regarding elastic expansion limits and method of determining and checking "k" factors. Tabulations of commercially used "k" factors and elastic expansion limits are included.

C-6 Standards for Visual Inspection of Compressed Gas Cylinders. A guide for establishing cylinder inspection procedures and standards in order to meet cylinder inspection requirements of the DOT. Covers all types of compressed gas cylinders including those which may be visually inspected in lieu of hydrostatic retest.

C-6.1 Standards for Visual Inspection of High Pressure Aluminum Compressed Gas Cylinders. This pamphlet has been prepared as a guide for the visual inspection of aluminum compressed gas cylinders (1800 psi (12410 kPa) and over). It is general in nature and will not cover all circumstances for each individual cylinder type or lading. Each inspection agency may modify these guidelines to fit their conditions of service which may be more severe than encountered in transportation.

C-7 Guide to the Preparation of Precautionary Labeling and Marking of Compressed Gas Containers. Covers use of precautionary labels to warn of principal hazards. Includes general principles and illustrative labels for several types of gases.

C-8 Standard for Requalification of DOT-3HT Cylinders. A guide to cylinder users for establishing their own cylinder inspection procedures and standards. Intended to cover visual inspection requirements and other service life limitations prescribed by DOT regulations.

C-9 Standard Color-Marking of Compressed Gas Cylinders Intended for Medical Use in the United States. Covers the uniform method of color-marking of compressed medical gas cylinders which will facilitate the recognition of cylinders intended for medical use.

C-10 Recommendations for Changes of Service for Compressed Gas Cylinders including Procedures for Inspection and Contaminent Removal. Provides guidelines for the procedures used for changing cylinders from one gas service to another.

C-11 Recommended Practices for Inspection of Compressed Gas Cylinders at Time of Manufacture. Outlines DOT inspection requirements for cylinders as interpreted and practiced by manufacturers and inspectors. Qualification of certifying inspectors is covered as well as inspection requirements for seamless, welded and brazed and nonrefillable cylinders.

C-12 Qualification Procedure for Acetylene Cylinder Design. Presents qualification tests for use of manufacturers of acetylene cylinders and required when there is a new cylinder design or a significant design change. The tests include a proof of mechanical strength of filler test, or flashback test, an impact stability test and a fire test.

C-13T Tentative Guidelines for Visual Inspection and Requalification of Acetylene Cylinders. Provides a guide for reinspection procedure for requalification of acetylene cylinders and fillers for use by charging companies and reinspection agencies.

C-14 Procedures for Fire Testing of D.O.T. Cylinder Safety Relief Device Systems. Describes test equipment and procedures which are applicable for fire testing cylinders which are less than 500 pounds internal water volume.

Regulators and Hose Line Equipment

E-1 Standard Connections for Regulators, Torches and Fitted Hose for Welding and Cutting Equipment. Describes connections for regulator outlets, torches and fitted hose for welding and cutting equipment.

E-2 Hose Line Check Valve Standards for Welding and Cutting. Covers check valves for use at pressures of 200 psig or less, designed to fit onto the oxy-fuel gas torch inlets, or onto the outlets of both the oxygen and fuel gas regulators.

E-3 Pipeline Regulator Inlet Connection Standards. This standard covers the inlet connections to be used on removable pipeline regulators used in the welding, cutting and related process industries and where the pipeline pressure does not exceed 200 psig.

E-4 Standard for Gas Regulators for Welding and Cutting. Covers design and performance requirements for regulators intended for use in welding and cutting and allied applications to reduce supply pressure from a gas storage system, pipeline, or other source to the pressure required for the application.

E-5 Torch Standard for Welding and Cutting. This standard covers design, marking, performance and stability testing of manual oxygen-fuel gas torches (and cutting attachments) intended for cutting, welding, scarfing, heating and other allied processes.

E-6 Standard For Hydraulic Type Pipeline Protective Devices. Provides information on hydraulic type (NFPA designated P_1) pipeline protective devices including materials and construction, design requirements, marking and maintenance.

Gases

G-1 Acetylene. Contains information on the properties, manufacture, transportation, storage, handling and use of acetylene.

G-1.1 Commodity Specification for Acetylene. Covers specification requirements for acetylene. Presents data on quality verification, sampling analytical procedures and containers with supplemental tables.

G-1.2 Recommendations for Chemical Acetylene Metering. Presents data on types of meters for acetylene service, meter accessories, materials of construction for meters and meter acessories; meter installation and maintenance.

G-1.3 Acetylene Transmission for Chemical Synthesis. Comprehensively presents recommended safe practices governing design of acetylene piping systems, gas holder storage and booster systems.

G-1.5 Carbide Lime—Its Value and Its Uses. Pamphlet makes known the uses of calcium hydroxide (a by-product from acetylene generation) in agriculture, building construction, industrial and chemical processes, and miscellaneous uses.

G-1.6T Tentative Recommended Practices for Mobile Acetylene Trailer Systems. Provides information on design and construction, charging, transporting and discharging, markings and fire safety practices. Recommendations on equipment at fill stations and discharge sites is also provided.

G-2 Anhydrous Ammonia. Presents information on the properties, manufacture, transportation, storage and use of anhydrous ammonia.

G-2.1 American National Standard Safety Requirements for the Storage and Handling of Anhydrous Ammonia, K61.1. Sponsored by CGA, this standard includes standards for the location, design, construction, and operation of anhydrous ammonia systems. Sections on refrigerated storage systems, systems mounted on farm vehicles and tank motor vehicles for transportation purposes are contained. The standard does not apply to ammonia manufacturing plants, refrigerating, or air-conditioning systems.

G-2.2T Tentative Standard Method for Determining Minimum of 0.2% Water in Anhydrous Ammonia. Describes a suitable method for this purpose.

G-3 Sulfur Dioxide. Information on the properties, manufacture, transportation, storage, handling and use of sulfur dioxide.

G-4 Oxygen. Information on the properties, manufacture, transportation, storage, handling and use of oxygen. Section on liquid oxygen is included.

G-4.1 Cleaning Equipment for Oxygen Service. Pamphlet describes cleaning processes, agents, methods, inspections, tests and safety. Revised edition is a complete update of the original edition.

G-4.3 Commodity Specification for Oxygen. Covers specification requirements for all types and grades of commercially available gaseous and liquid oxygen for which an end usage has been established through combined industrial experience. Presents data on quality verification, sampling, analytical procedures, and containers with supplemental tables and charts.

G-4.4 Industrial Practices for Gaseous Oxygen Transmission and Distribution Piping Systems. Presents information on the current practices used in gaseous oxygen transmission and distribution piping systems such as encountered at various chemical facilities, steel mills, petroleum product refineries, etc.

G-5 Hydrogen. Presents information on the properties, manufacture, transportation, storage, handling and use of hydrogen and includes a section on liquefied hydrogen.

G-5.3 Commodity Specification for Hydrogen. Lists specification requirements for all types and grades of commercially available gaseous and liquid hydrogen. Treats quality verification systems, sampling, analytical procedures, with supplemental specification data.

G-6 Carbon Dioxide. Presents information on the properties, manufacture, transportation, storage, handling and use of carbon dioxide.

G-6.1 Standard for Low Pressure Carbon Dioxide Systems at Consumer Sites. This pamphlet is concerned with the minimum requirements and practices for design, construction, installation, operation and maintenance of low-pressure CO_2 supply systems at consumer sites.

G-6.2 Commodity Specification for Carbon Dioxide. Describes specification requirements for gaseous, liquid and solid carbon dioxide. Presents data on quality verification, sampling, analytical procedures and containers, with supplemental tables and charts.

G-6.3 Carbon Dioxide Cylinder Filling and Handling Procedures For Beverage Plants, NSDA TD01. This pamphlet, co-sponsored with NSDA presents detailed procedures for inspection, filling, care and proper handling, recommended shipping and storage practices, DOT marking and labeling requirements for carbon dioxide cylinders as used in the beverage industry.

G-7 Compressed Air for Human Respiration. To be used in conjunction with G-7.1 and covers special recommendations when air is intended for human use.

G-7.1 Commodity Specification for Air. Z86.1. Covers specification requirements for all types and grades of gaseous and liquid air known to be commercially available and for which an end usage has been established through combined industrial experience. Data concerning quality verification systems, sampling, analytical procedures and containers. Supplemental specification tables and charts are included.

G-8.1 Standard for the Installation of Nitrous Oxide Systems at Consumer Sites. Covers the general design principles recommended for the installation of nitrous oxide central supply systems—either cylinders connected to a common manifold, or systems fed by bulk liquid containers—at medical or industrial consumer sites, with special reference to protection of the system, patients, personnel and property from fire from sources apart from the system itself.

G-8.2 Commodity Specification for Nitrous Oxide. Describes the specification requirements for nitrous oxide manufactured by the thermal decomposition of nitrous oxide grade ammonium nitrate and filled into standard pressurized containers. Data regarding quality verification systems, sampling, analytical procedures and containers are included.

G-9.1 Commodity Specification for Helium. Covers specification requirements for helium known to be commercially available and for which an end usage has been established through combined industrial experience. Data regarding verification systems, sampling, analytical procedures, supplemental specifications, tables and charts are included.

G-10.1 Commodity Specification for Nitrogen. Designates specification requirements for all types and grades of commercially available gaseous and liquid nitrogen for which an end usage has been established through combined industrial experience. Material includes quality verification systems, analytical procedures, containers and supplemental data.

G-11.1 Commodity Specification for Argon. Lists specification requirements for all types and grades of commercially available gaseous and liquefied argon. Treats quality verification systems, sampling, analytical procedures and supplemental specification data.

Protection and Safe Handling

P-1 Safe Handling of Compressed Gases in Containers. Primarily for the guidance of users of compressed gases in cylinders, although some general precautions are included for tank car handling. Presents basic rules for safe handling and regulations applying to compressed gases.

P-2 Characteristics and Safe Handling of Medical Gases. Intended for the guidance of personnel engaged in and responsible for the handling and use of compressed medical gases.

Contains information relating to the properties of medical gases as well as information on containers and regulations governing these commodities. Recommended safe practices are included.

P-2.1 Standard for Medical-Surgical Vacuum Systems in Hospitals. Applies to piped vacuum systems in hospitals but not to systems used for vacuum cleaning or as a vacuum condensate return. Covers requirements for central vacuum supply systems with control equipment, piping system to points in the hospital where suction may be required, and suitable outlet valves at point of use.

P-2.4 Guide for Manufacture, Control and Distribution of Medical Gases in Canada. Provides guidance applicable to the manufacture, control and distribution of medical gases by manufacturers in Canada.

P-4 Safe Handling of Cylinders by Emergency Rescue Squads. A guide to emergency rescue squads, such as fire and police department rescue units, volunteer ambulance units, and industrial plant first-aid departments which may be called upon to handle medical gas cylinders. Contains recommended safe practices for handling medical gases and information concerning pertinent regulations.

P-5 Suggestions for the Care of High-Pressure Air Cylinders for Underwater Breathing. A guide to users of underwater breathing air cylinders and to those responsible for the care, maintenance and recharging of such cylinders.

P-6 Standard Density Data Atmospheric Gases and Hydrogen. Density data recommended in this pamphlet were developed by the Compressed Gas Association to provide uniform values of liquid and gas density of atmospheric gases and hydrogen for the benefit of suppliers and users of these commodities. Tables present standard density data and volumetric conversion factors.

P-7T Tentative Standard for Requalification of Cargo Tank Hose Used in the Transfer of Compressed Gases. This CGA pamphlet was published as a guide for cargo tank hose users in establishing hose requalification procedures and standards in compressed gas transfer service applications.

P-8 Safe Practices Guide for Air Separation Plants. Applies to safety in the design, construction, location, installation, and operation of cryogenic air separation plants regardless of whether product is withdrawn as gas or liquid or both.

P-9 The Inert Gases Argon, Nitrogen and Helium. Presents general information regarding the characteristics and handling of the inert gases Argon, Nitrogen and Helium. It is an addition to the series of specific compressed gases.

Pressure-Relief Devices

S-1.1 Pressure-Relief Device Standards—Cylinders for Compressed Gases. Minimum recommended requirements for safety relief devices for use on DOT cylinders having capacities of 1000 lb of water or less. Describes the various types of safety relief devices, their design, construction and application for various cylinders. Also covers relief device test and maintenance requirements.

S-1.2 Pressure-Relief Device Standards— Cargo and Portable Tanks for Compressed Gases. Minimum recommended requirements for safety relief devices for use on cargo tanks (tank trucks) and portable tanks (skid tanks) designed to DOT specifications. These requirements are recommended for application to cargo and portable tanks that do not come within DOT or BTC jurisdiction.

S-1.3 Pressure-Relief Device Standards—Compressed Gas Storage Containers. Minimum recommended requirements for safety relief devices for storage containers constructed in accordance with the ASME or API-ASME Codes.

S-4 Recommended Practice for the Manufacture of Fusible Plugs. Intended to guide the manufacturer in the production of fusible plugs to meet the requirements of the "Safety Relief Device Standards," Pamphlet S-1. Contains minimum recommended requirements for operation, application and testing of devices.

Valve Connections

M-1 Standard for 22 mm Anesthesia Breathing Circuit Connectors. Contains the dimensions and profile of male and female adult anesthesia breathing circuit connectors. Covers the marking of anesthesia breathing circuit connectors with regard to direction of gas flow.

V-1 American National, Canadian and CGA Standard Compressed Gas Cylinder Valve Outlet and Inlet Connections; ANSI B57.1; CSA B96. (Supersedes 1965 edition and Proposed Revision issued July 1975.) Contains the American, Canadian and CGA Standards for Cylinder Valve Outlet and Inlet Connections. Detailed dimensioned drawings of 56 valve outlet connections for 207 different products. The standard covers threaded connections, yoke outlets and the Pin-Index Safety System for flush outlet valves of the yoke type used for medical gases.

V-5 Diameter-Index Safety System. Describes a standard to provide noninterchangeable connections where removable exposed threaded connections are employed in conjunction with individual gas lines of medical gas administering equipment at pressures of 200 psig, or less, such as outlets from medical gas regulators and connectors for anesthesia, resuscitation and therapy apparatus. Detailed dimensioned drawings are included. In addition to the medical gases covered by the Pin-Index Safety System, this standard includes air and suction.

V-6 Standard Cryogenic Liquid Transfer Connections. Offers cryogenic liquid connections which provide mechanical protection for product integrity. Engineering specifications and cut-away drawings are provided.

V-7T Tentative Standard Method of Determining Cylinder Valve Outlet Connections for Industrial Gas Mixtures. Provides standard procedure for selection of cylinder valve outlet connections for mixtures of two or more industrial gases. This standard supplements CGA V-1 but does not cover medical gases.

Safety Bulletins

SB-1 Hazards of Refilling Compressed Refrigerant (Halogenated Hydrocarbons) Gas Cylinders. Presents recommendations for eliminating injury to persons and property when refrigerant gas is transferred from large to small cylinders in the field—a common, but potentially hazardous practice of distributors and service organizations in the refrigerating and air-conditioning fields. Stresses the absolute need for thorough training of personnel with respect to DOT regulations, filling densities, valves, safety relief devices, etc. A copy of this bulletin should be placed in the hands of every person engaged in transfer operations.

SB-2 Oxygen-Deficient Atmospheres. Depletion of the oxygen content in air by combustion of displacement with inert gas is a potential hazard to personnel throughout industry. Pamphlet deals with recommended practices for the protection of personnel working in a potentially oxygen-deficient atmosphere.

SB-3 Compressed Gas Cylinder Filling Procedures. Emphasizes the importance of complying with the Hazardous Materials Regulations of the Department of Transportation in connection with obtaining the owners consent before charging compressed gas cylinders. Includes suggested methods of indicating cylinder ownership and authorization to charge cylinders.

SB-4 Handling Acetylene Cylinders in Fire Situations. Provides guidance in how to handle acetylene cylinders in fire situations to welders and other users and to those individuals concerned with safety and fire protection.

SB-5 Hazards of Reusing Disposable Refrigerant (Halogenated Hydrocarbon) Gas Cylinders. Provides warnings as to the dangers of reusing disposable refrigerant cylinders and cites applicable DOT regulations.

SB-6 Nitrous Oxide Security and Control. This safety bulletin is an alert to manufacturers, repackagers and distributors, and those using nitrous oxide for medical and commercial-industrial purposes. It warns of the misuse and abuse of nitrous oxide so that effective steps can be taken to prevent theft or improper use.

TB-2 Guidelines for Inspection and Repair of MC-330 and MC-331 Cargo Tanks. To assist those having the responsibility for the inspection and repair of these tanks.

Insulated Tank Trucks

CGA-341 Standard Insulated Tank Truck Specification. Contains minimum recommended requirements for insulated tank trucks for the transportation of cold liquid gases.

Index